Student Solutions Manual

Jeffery A. Cole
Anoka-Ramsey Community College

Prealgebra
Fourth Edition

Margaret L. Lial
American River College

Diana L. Hestwood
Minneapolis Community and Technical College

Addison-Wesley
is an imprint of

The author and publisher of this book have used their best efforts in preparing this book. These efforts include the development, research, and testing of the theories and programs to determine their effectiveness. The author and publisher make no warranty of any kind, expressed or implied, with regard to these programs or the documentation contained in this book. The author and publisher shall not be liable in any event for incidental or consequential damages in connection with, or arising out of, the furnishing, performance, or use of these programs.

Reproduced by Pearson Addison-Wesley from electronic files supplied by the author.

Copyright © 2010 Pearson Education, Inc.
Publishing as Pearson Addison-Wesley, 75 Arlington Street, Boston, MA 02116.

All rights reserved. This manual may be reproduced for classroom use only.
Printed in the United States of America.

ISBN-13: 978-0-321-57478-7
ISBN-10: 0-321-57478-8

1 2 3 4 5 6 BB 11 10 09 08

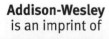

www.pearsonhighered.com

Preface

This *Student's Solutions Manual* contains solutions to selected exercises in the text *Prealgebra, Fourth Edition* by Margaret L. Lial and Diana L. Hestwood. It contains solutions to all margin exercises, the odd-numbered exercises in each section, all Relating Concepts exercises, as well as solutions to all the exercises in the review sections, the chapter tests, and the cumulative review sections.

This manual is a text supplement and should be read along *with* the text. You should read all exercise solutions in this manual because many concept explanations are given and then used in subsequent solutions. All concepts necessary to solve a particular problem are not reviewed for every exercise. If you are having difficulty with a previously covered concept, refer back to the section where it was covered for more complete help.

A significant number of today's students are involved in various outside activities, and find it difficult, if not impossible, to attend all class sessions; this manual should help meet the needs of these students. In addition, it is my hope that this manual's solutions will enhance the understanding of all readers of the material and provide insights to solving other exercises.

I appreciate feedback concerning errors, solution correctness or style, and manual style. Any comments may be sent directly to me at the address below, at jeff.cole@anokaramsey.edu, or in care of the publisher, Pearson Addison-Wesley.

I would like to thank Marv Riedesel and Mary Johnson, formerly of Inver Hills Community College, for their careful accuracy checking and valuable suggestions; and Maureen O'Connor and Courtney Slade, of Pearson Addison-Wesley, for entrusting me with this project.

Jeffery A. Cole
Anoka-Ramsey Community College
11200 Mississippi Blvd. NW
Coon Rapids, MN 55433

Table of Contents

1 Introduction to Algebra: Integers 1
    1.1   Place Value • 1
    1.2   Introduction to Signed Numbers • 2
    1.3   Adding Integers • 3
    1.4   Subtracting Integers • 8
    1.5   Problem Solving: Rounding and Estimating • 10
    1.6   Multiplying Integers • 13
    1.7   Dividing Integers • 16
           Summary Exercises on Operations with Integers • 19
    1.8   Exponents and Order of Operations • 20
           Chapter 1 Review Exercises • 24
           Chapter 1 Test • 28

2 Understanding Variables and Solving Equations 31
    2.1   Introduction to Variables • 31
    2.2   Simplifying Expressions • 34
           Summary Exercises on Variables and Expressions • 39
    2.3   Solving Equations Using Addition • 40
    2.4   Solving Equations Using Division • 45
    2.5   Solving Equations with Several Steps • 48
           Chapter 2 Review Exercises • 54
           Chapter 2 Test • 57
           Cumulative Review Exercises (Chapters 1–2) • 59

3 Solving Application Problems 63
    3.1   Problem Solving: Perimeter • 63
    3.2   Problem Solving: Area • 67
           Summary Exercises on Perimeter and Area • 73
    3.3   Solving Application Problems with One Unknown Quantity • 74
    3.4   Solving Application Problems with Two Unknown Quantities • 79
           Chapter 3 Review Exercises • 84
           Chapter 3 Test • 87
           Cumulative Review Exercises (Chapters 1–3) • 90

4 Rational Numbers: Positive and Negative Fractions 95
    4.1   Introduction to Signed Fractions • 95
    4.2   Writing Fractions in Lowest Terms • 97
    4.3   Multiplying and Dividing Signed Fractions • 100
    4.4   Adding and Subtracting Signed Fractions • 103
    4.5   Problem Solving: Mixed Numbers and Estimating • 108
           Summary Exercises on Fractions • 112
    4.6   Exponents, Order of Operations, and Complex Fractions • 113
    4.7   Problem Solving: Equations Containing Fractions • 118
    4.8   Geometry Applications: Area and Volume • 122
           Chapter 4 Review Exercises • 125
           Chapter 4 Test • 129
           Cumulative Review Exercises (Chapters 1–4) • 131

5 Rational Numbers: Positive and Negative Decimals 135

- 5.1 Reading and Writing Decimal Numbers • 135
- 5.2 Rounding Decimal Numbers • 137
- 5.3 Adding and Subtracting Signed Decimal Numbers • 140
- 5.4 Multiplying Signed Decimal Numbers • 145
- 5.5 Dividing Signed Decimal Numbers • 149
  - *Summary Exercises on Decimals* • 156
- 5.6 Fractions and Decimals • 157
- 5.7 Problem Solving with Statistics: Mean, Median, and Mode • 162
- 5.8 Geometry Applications: Pythagorean Theorem and Square Roots • 164
- 5.9 Problem Solving: Equations Containing Decimals • 166
- 5.10 Geometry Applications: Circles, Cylinders, and Surface Area • 170
  - Chapter 5 Review Exercises • 174
  - Chapter 5 Test • 183
  - Cumulative Review Exercises (Chapters 1–5) • 185

6 Ratio, Proportion, and Line/Angle/Triangle Relationships 189

- 6.1 Ratios • 189
- 6.2 Rates • 190
- 6.3 Proportions • 194
  - *Summary Exercises on Ratios, Rates, and Proportions* • 199
- 6.4 Problem Solving with Proportions • 200
- 6.5 Geometry: Lines and Angles • 205
- 6.6 Geometry Applications: Congruent and Similar Triangles • 207
  - Chapter 6 Review Exercises • 210
  - Chapter 6 Test • 217
  - Cumulative Review Exercises (Chapters 1–6) • 218

7 Percent 223

- 7.1 The Basics of Percent • 223
- 7.2 The Percent Proportion • 226
- 7.3 The Percent Equation • 230
  - *Summary Exercises on Percent* • 233
- 7.4 Problem Solving with Percent • 234
- 7.5 Consumer Applications: Sales Tax, Tips, Discounts, and Simple Interest • .. 239
  - Chapter 7 Review Exercises • 244
  - Chapter 7 Test • 248
  - Cumulative Review Exercises (Chapters 1–7) • 249

8 Measurement 253

- 8.1 Problem Solving with U.S. Customary Measurements • 253
- 8.2 The Metric System—Length • 258
- 8.3 The Metric System—Capacity and Weight (Mass) • 260
  - *Summary Exercises on U.S. Customary and Metric Units* • 263
- 8.4 Problem Solving with Metric Measurement • 263
- 8.5 Metric—U.S. Customary Conversions and Temperature • 265
  - Chapter 8 Review Exercises • 268
  - Chapter 8 Test • 272
  - Cumulative Review Exercises (Chapters 1–8) • 274

9 Graphs 279
- 9.1 Problem Solving with Tables and Pictographs • 279
- 9.2 Reading and Constructing Circle Graphs • 281
- 9.3 Bar Graphs and Line Graphs • 284
- 9.4 The Rectangular Coordinate System • 286
- 9.5 Introduction to Graphing Linear Equations • 288
- Chapter 9 Review Exercises • 292
- Chapter 9 Test • 295
- Cumulative Review Exercises (Chapters 1–9) • 296

10 Exponents and Polynomials 301
- 10.1 The Product Rule and Power Rules for Exponents • 301
- 10.2 Integer Exponents and the Quotient Rule • 302
- *Summary Exercises on Exponents* • 303
- 10.3 An Application of Exponents: Scientific Notation • 304
- 10.4 Adding and Subtracting Polynomials • 306
- 10.5 Multiplying Polynomials: An Introduction • 309
- Chapter 10 Review Exercises • 311
- Chapter 10 Test • 313
- Cumulative Review Exercises (Chapters 1–10) • 314

Whole Numbers Computation: Pretest 319

R Whole Numbers Review 321
- R.1 Adding Whole Numbers • 321
- R.2 Subtracting Whole Numbers • 323
- R.3 Multiplying Whole Numbers • 326
- R.4 Dividing Whole Numbers • 330
- R.5 Long Division • 333

Appendix: Inductive and Deductive Reasoning 339

CHAPTER 1 INTRODUCTION TO ALGEBRA: INTEGERS

1.1 Place Value

1.1 Margin Exercises

1. The whole numbers are: 502; 3; 14; 0; 60,005

2. **(a)** The 8 in 45,628,665 is in the thousands place.

   (b) The 8 in 800,503,622 is in the hundred-millions place.

   (c) The 8 in 428,000,000,000 is in the billions place.

   (d) The 8 in 2,385,071 is in the ten-thousands place.

3. **(a)** 23,605 in words: twenty-three *thousand*, six hundred five.

   (b) 400,033,007 in words: four hundred *million*, thirty-three *thousand*, seven.

   (c) 193,080,102,000,000 in words: one hundred ninety-three *trillion*, eighty *billion*, one hundred two *million*.

4. **(a)** Eighteen million, two thousand, three hundred five
   The first group name is *million*, so you need to fill *three groups* of three digits.

   $\underline{0\,1\,8}, \underline{0\,0\,2}, \underline{3\,0\,5} = 18{,}002{,}305$

   (b) Two hundred billion, fifty million, six hundred sixteen
   The first group name is *billion*, so you need to fill *four groups* of three digits.

   $\underline{2\,0\,0}, \underline{0\,5\,0}, \underline{0\,0\,0}, \underline{6\,1\,6} = 200{,}050{,}000{,}616$

   (c) Five trillion, forty-two billion, nine million
   The first group name is *trillion*, so you need to fill *five groups* of three digits.

   $\underline{0\,0\,5}, \underline{0\,4\,2}, \underline{0\,0\,9}, \underline{0\,0\,0}, \underline{0\,0\,0} = 5{,}042{,}009{,}000{,}000$

   (d) Three hundred six million, seven hundred thousand, nine hundred fifty-nine
   The first group name is *million*, so you need to fill *three groups* of three digits.

   $\underline{3\,0\,6}, \underline{7\,0\,0}, \underline{9\,5\,9} = 306{,}700{,}959$

1.1 Section Exercises

1. The whole numbers are: 15; 0; 83,001

3. The whole numbers are: 7; 362,049

5. The 2 in 61,284 is in the hundreds place.

7. The 2 in 284,100 is in the hundred-thousands place.

9. The 2 in 725,837,166 is in the ten-millions place.

11. The 2 in 253,045,701,000 is in the hundred-billions place.

13. Name the place value for each zero in

    302,016,450,098,570.

    From left to right: ten-trillions, hundred-billions, millions, hundred-thousands, and ones.

15. 8421 in words: eight thousand, four hundred twenty-one.

17. 46,205 in words: forty-six thousand, two hundred five.

19. 3,064,801 in words: three million, sixty-four thousand, eight hundred one.

21. 840,111,003 in words: eight hundred forty million, one hundred eleven thousand, three.

23. 51,006,888,321 in words: fifty-one billion, six million, eight hundred eighty-eight thousand, three hundred twenty-one.

25. 3,000,712,000,000 in words: three trillion, seven hundred twelve million.

27. Forty-six thousand, eight hundred five
    The first group name is *thousand*, so you need to fill *two groups* of three digits.

    $\underline{0\,4\,6}, \underline{8\,0\,5} = 46{,}805$

29. Five million, six hundred thousand, eighty-two
    The first group name is *million*, so you need to fill *three groups* of three digits.

    $\underline{0\,0\,5}, \underline{6\,0\,0}, \underline{0\,8\,2} = 5{,}600{,}082$

31. Two hundred seventy-one million, nine hundred thousand
    The first group name is *million*, so you need to fill *three groups* of three digits.

    $\underline{2\,7\,1}, \underline{9\,0\,0}, \underline{0\,0\,0} = 271{,}900{,}000$

33. Twelve billion, four hundred seventeen million, six hundred twenty-five thousand, three hundred ten
    The first group name is *billion*, so you need to fill *four groups* of three digits.

    $\underline{0\,1\,2}, \underline{4\,1\,7}, \underline{6\,2\,5}, \underline{3\,1\,0} = 12{,}417{,}625{,}310$

2   Chapter 1   Introduction to Algebra: Integers

35. Six hundred trillion, seventy-one million, four hundred
The first group name is *trillion,* so you need to fill *five groups* of three digits.

$\underline{6}\underline{0}\underline{0}, \underline{0}\underline{0}\underline{0}, \underline{0}\underline{7}\underline{1}, \underline{0}\underline{0}\underline{0}, \underline{4}\underline{0}\underline{0} =$
600,000,071,000,400

37. 3151 in words: three thousand, one hundred fifty-one

39. One hundred one million, two hundred eighty thousand
The first group is *millions,* so fill *three groups* of three digits.

$\underline{1}\underline{0}\underline{1}, \underline{2}\underline{8}\underline{0}, \underline{0}\underline{0}\underline{0} = 101{,}280{,}000$

41. 173,523,700 in words: one hundred seventy-three million, five hundred twenty-three thousand, seven hundred

43. Fifty-five million, eight hundred
The first group is *millions,* so you need to fill *three groups* of three digits.

$\underline{0}\underline{5}\underline{5}, \underline{0}\underline{0}\underline{0}, \underline{8}\underline{0}\underline{0} = 55{,}000{,}800$

45. 6,400,000 in words: six million, four hundred thousand every day. 2,336,000,000 in words: two billion, three hundred thirty-six million in one year.

47. Four billion, two hundred million is 4,200,000,000.

49. To make the largest possible whole number, arrange the digits from largest to smallest.

97651100 → 97,651,100

In words: ninety-seven million, six hundred fifty-one thousand, one hundred.

To make the smallest possible whole number, arrange the numbers from smallest to largest with one exception: because we must use all the digits, start with the smallest nonzero digit.

10015679 → 10,015,679

In words: ten million, fifteen thousand, six hundred seventy-nine.

50. Answers will vary.

51.

sixty-fours	thirty-twos	sixteens	eights	fours	twos	ones
1	1	1	1	1	1	1

(a) $5 = 4 + 1 =$ binary 101

(b) $10 = 8 + 2 =$ binary 1010

(c) $15 = 8 + 4 + 2 + 1 =$ binary 1111

52. (a) Answers will vary but should mention that the location or place in which a digit is written gives it a different value.

(b) $8 = 5 + 3 =$ VIII
$38 = 30 + 5 + 3 =$ XXXVIII
$275 = 200 + 50 + 20 + 5 =$ CCLXXV
$3322 = 3000 + 300 + 20 + 2 =$ MMMCCCXXII

(c) The Roman system is *not* a place value system because no matter what place it's in, M = 1000, C = 100, etc. One disadvantage is that it takes much more space to write many large numbers; another is that there is no symbol for zero.

1.2   Introduction to Signed Numbers

1.2   Margin Exercises

1. (a) "Below zero" implies a negative number.

$^-5\frac{1}{2}$ degrees

(b) "Lost 12 pounds" implies a negative number.

$^-12$ pounds

(c) "Deposit" implies a positive number.

$^+\$210.35$ or $\$210.35$

(d) "Overdrawn" implies a negative number.

$^-\$65$

(e) "Below the surface of the sea" implies a negative number.

$^-100$ feet

(f) "Won 50 points" implies a positive number.

$^+50$ points   or   50 points

2. (a) $^-2$ **(b)** 2 **(c)** 0 **(d)** $^-4$ **(e)** 4

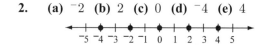

(f) $^-3\frac{1}{2}$ **(g)** $\frac{1}{2}$ **(h)** $^-1$ **(i)** 3

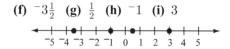

3. (a) 5 is to the *right* of 4 on the number line, so 5 is *greater than* 4. Write $5 > 4$.

(b) 0 is to the *left* of 2 on the number line, so 0 is *less than* 2. Write $0 < 2$.

(c) $^-3$ is to the *left* of $^-2$ on the number line, so $^-3$ is *less than* $^-2$. Write $^-3 < {}^-2$.

(d) $^-1$ is to the *right* of $^-4$ on the number line, so $^-1$ is *greater than* $^-4$. Write $^-1 > {}^-4$.

1.3 Adding Integers

(e) 2 is to the *right* of ⁻2 on the number line, so 2 is *greater than* ⁻2. Write 2 > ⁻2.

(f) ⁻5 is to the *left* of 1 on the number line, so ⁻5 is *less than* 1. Write ⁻5 < 1.

4. (a) |13| = 13 because the distance from 0 to 13 on the number line is 13 spaces.

(b) |⁻7| = 7 because the distance from 0 to ⁻7 on the number line is 7 spaces.

(c) |0| = 0 because the distance from 0 to 0 on the number line is 0 spaces.

(d) |⁻350| = 350 because the distance from 0 to ⁻350 on the number line is 350 spaces.

(e) |6000| = 6000 because the distance from 0 to 6000 on the number line is 6000 spaces.

1.2 Section Exercises

1. "Above sea level" implies a positive number.

$$^+29{,}035 \text{ feet or } 29{,}035 \text{ feet}$$

3. "Below zero" implies a negative number.

$$^-128.6 \text{ degrees}$$

5. "Lost a total of 18 yards" implies a negative number.

$$^-18 \text{ yards}$$

7. "Won $100" implies a positive number.

$$^+\$100 \text{ or } \$100$$

9. "Lost $6\frac{1}{2}$ pounds" implies a negative number.

$$^-6\frac{1}{2} \text{ pounds}$$

11. Graph ⁻3, 3, 0, ⁻5

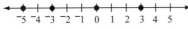

13. Graph ⁻1, 4, ⁻2, 5

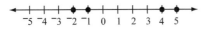

15. Graph $^-4\frac{1}{2}, \frac{1}{2}, 0, ^-8$

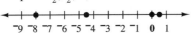

17. 10 is to the *right* of 2 on the number line, so 10 is *greater than* 2. Write 10 > 2.

19. ⁻1 is to the *left* of 0 on the number line, so ⁻1 is *less than* 0. Write ⁻1 < 0.

21. ⁻10 is to the *left* of 2 on the number line, so ⁻10 is *less than* 2. Write ⁻10 < 2.

23. ⁻3 is to the *right* of ⁻6 on the number line, so ⁻3 is *greater than* ⁻6. Write ⁻3 > ⁻6.

25. ⁻10 is to the *left* of ⁻2 on the number line, so ⁻10 is *less than* ⁻2. Write ⁻10 < ⁻2.

27. 0 is to the *right* of ⁻8 on the number line, so 0 is *greater than* ⁻8. Write 0 > ⁻8.

29. 10 is to the *right* of ⁻2 on the number line, so 10 is *greater than* ⁻2. Write 10 > ⁻2.

31. ⁻4 is to the *left* of 4 on the number line, so ⁻4 is *less than* 4. Write ⁻4 < 4.

33. |15| = 15 because the distance from 0 to 15 on the number line is 15 spaces.

35. |⁻3| = 3 because the distance between 0 and ⁻3 on the number line is 3 spaces.

37. |0| = 0 because the distance from 0 to 0 on the number line is 0 spaces.

39. |200| = 200 because the distance between 0 and 200 on the number line is 200 spaces.

41. |⁻75| = 75 because the distance between 0 and ⁻75 on the number line is 75 spaces.

43. |⁻8042| = 8042 because the distance between 0 and ⁻8042 on the number line is 8042 spaces.

45. Graph ⁻1.5 as A, 0.5 as B, ⁻1 as C, and 0 as D.

46. From Exercise 45, in order from lowest to highest: ⁻1.5, ⁻1, 0, 0.5

47. A: ⁻1.5 is in the Below ⁻1 range. This patient may be at risk.

B: 0.5 is in the Above 0 range. This patient is above normal.

C: ⁻1 is in the 0 to ⁻1 range. This patient is normal.

D: 0 is in the 0 to ⁻1 range. This patient is normal.

48. (a) A patient who did not understand the importance of the negative sign would think the interpretation of ⁻1.5 was "above normal" (range Above 0) and wouldn't get treatment.

(b) For Patient D's score of 0, the sign plays no role. Zero is neither positive nor negative.

1.3 Adding Integers

1.3 Margin Exercises

1. (a) ⁻2 + ⁻2 = ⁻4

(b) 2 + 2 = 4

(c) ⁻10 + ⁻1 = ⁻11

Chapter 1 Introduction to Algebra: Integers

(d) $10 + 1 = 11$

(e) $^-3 + ^-7 = ^-10$

(f) $3 + 7 = 10$

2. (a) $^-6 + ^-6$ Adding *like* signed integers

   Step 1: $|^-6| = 6; |^-6| = 6$; Add $6 + 6 = 12$

   Step 2: Both numbers are negative, so the sum is negative.
   $$^-6 + ^-6 = ^-12$$

   (b) $9 + 7$ Adding *like* signed integers

   Step 1: $|9| = 9; |7| = 7$; Add $9 + 7 = 16$

   Step 2: Both numbers are positive, so the sum is positive.
   $$9 + 7 = 16$$

   (c) $^-5 + ^-10$ Adding *like* signed integers

   Step 1: $|^-5| = 5; |^-10| = 10$; Add $5 + 10 = 15$

   Step 2: Both numbers are negative, so the sum is negative.
   $$^-5 + ^-10 = ^-15$$

   (d) $^-12 + ^-4$ Adding *like* signed integers

   Step 1: $|^-12| = 12; |^-4| = 4$; Add $12 + 4 = 16$

   Step 2: Both numbers are negative, so the sum is negative.
   $$^-12 + ^-4 = ^-16$$

   (e) $13 + 2$ Adding *like* signed integers

   Step 1: $|13| = 13; |2| = 2$; Add $13 + 2 = 15$

   Step 2: Both numbers are positive, so the sum is positive.
   $$13 + 2 = 15$$

3. (a) $^-3 + 7$ Adding *unlike* signed integers

   Step 1: $|^-3| = 3; |7| = 7$; Subtract $7 - 3 = 4$

   Step 2: 7 has the larger absolute value and is positive, so the sum is positive.
   $$^-3 + 7 = ^+4 \text{ or } 4$$

   (b) $6 + ^-12$ Adding *unlike* signed integers

   Step 1: $|6| = 6; |^-12| = 12$; Subtract $12 - 6 = 6$

   Step 2: $^-12$ has the larger absolute value and is negative, so the sum is negative.
   $$6 + ^-12 = ^-6$$

   (c) $12 + ^-7$ Adding *unlike* signed integers

   Step 1: $|12| = 12; |^-7| = 7$; Subtract $12 - 7 = 5$

   Step 2: 12 has the larger absolute value and is positive, so the sum is positive.
   $$12 + ^-7 = ^+5 \text{ or } 5$$

   (d) $^-10 + 2$ Adding *unlike* signed integers

   Step 1: $|^-10| = 10; |2| = 2$; Subtract $10 - 2 = 8$

   Step 2: $^-10$ has the larger absolute value and is negative, so the sum is negative.
   $$^-10 + 2 = ^-8$$

   (e) $5 + ^-9$ Adding *unlike* signed integers

   Step 1: $|5| = 5; |^-9| = 9$; Subtract $9 - 5 = 4$

   Step 2: $^-9$ has the larger absolute value and is negative, so the sum is negative.
   $$5 + ^-9 = ^-4$$

   (f) $^-8 + 9$ Adding *unlike* signed integers

   Step 1: $|^-8| = 8; |9| = 9$; Subtract $9 - 8 = 1$

   Step 2: 9 has the larger absolute value and is positive, so the sum is positive.
   $$^-8 + 9 = ^+1 \text{ or } 1$$

4. (a) Starting temperature in the morning is $^-15$ degrees. A rise of 21 degrees implies a positive number. A drop of 10 degrees implies a negative number.

   $^-15 + 21 + ^-10$ *Add left to right.*
   $= 6 + ^-10$
   $= ^-4$

   The new temperature is 4 degrees below zero or $^-4$ degrees.

   (b) The beginning balance is $60. Checks imply negative numbers and deposits imply positive numbers.

   $60 + ^-20 + ^-75 + 85$ *Add left to right.*
   $= 40 + ^-75 + 85$
   $= ^-35 + 85$
   $= 50$

   His account balance is $50.

5. (a) $175 + 25 = 25 + 175$
   Both sums are 200.

   (b) $7 + ^-37 = ^-37 + 7$
   Both sums are $^-30$.

   (c) $^-16 + 16 = 16 + ^-16$
   Both sums are 0.

   (d) $^-9 + ^-41 = ^-41 + ^-9$
   Both sums are $^-50$.

1.3 Adding Integers

6. (a) $^-12 + 12 + {^-19} = (^-12 + 12) + {^-19}$
$= 0 + {^-19}$
$= {^-19}$

(b) $31 + 75 + {^-75} = 31 + (75 + {^-75})$
$= 31 + 0$
$= 31$

(c) $16 + {^-1} + {^-9} = 16 + ({^-1} + {^-9})$
$= 16 + {^-10}$
$= 6$

(d) $^-8 + 5 + {^-25} = {^-8} + (5 + {^-25})$
$= {^-8} + {^-20}$
$= {^-28}$

1.3 Section Exercises

1. $^-2 + 5 = {^+3}$ or 3

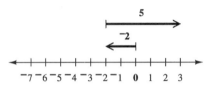

3. $^-5 + {^-2} = {^-7}$

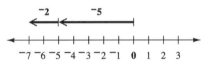

5. $3 + {^-4} = {^-1}$

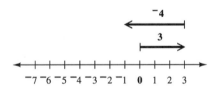

7. (a) $^-5 + {^-5} = {^-10}$ Adding *like* signed integers

Step 1: Add the absolute values.

$|{^-5}| = 5$

Add $5 + 5$ to get 10.

Step 2: Both integers are negative, so the sum is negative.

$^-5 + {^-5} = {^-10}$

(b) $5 + 5 = 10$ Adding *like* signed integers
Both addends are positive, so the sum is positive.

9. (a) $7 + 5 = 12$ Adding *like* signed integers
Both addends are positive, so the sum is positive.

(b) $^-7 + {^-5} = {^-12}$ Adding *like* signed integers

Step 1: Add the absolute values.

$|{^-7}| = 7; |{^-5}| = 5$

Add $7 + 5$ to get 12.

Step 2: Both integers are negative, so the sum is negative.

$^-7 + {^-5} = {^-12}$

11. (a) $^-25 + {^-25} = {^-50}$ Adding *like* signed integers

Step 1: Add the absolute values.

$|{^-25}| = 25$

Add $25 + 25$ to get 50.

Step 2: Both integers are negative, so the sum is negative.

$^-25 + {^-25} = {^-50}$

(b) $25 + 25 = 50$ Adding *like* signed integers
Both addends are positive, so the sum is positive.

13. (a) $48 + 110 = 158$ Adding *like* signed integers
Both addends are positive, so the sum is positive.

(b) $^-48 + {^-110} = {^-158}$ Adding *like* signed integers

Step 1: Add the absolute values.

$|{^-48}| = 48; |{^-110}| = 110$

Add $48 + 110$ to get 158.

Step 2: Both numbers are negative, so the sum is negative.

$^-48 + {^-110} = {^-158}$

15. The absolute values are the same in each pair of answers, so the only difference in the sums is the common sign.

17. (a) $^-6 + 8$ Adding *unlike* signed integers

Step 1: $|{^-6}| = 6; |8| = 8$

Subtract $8 - 6$ to get 2.

Step 2: 8 has the larger absolute value and is positive, so the sum is positive.

$^-6 + 8 = {^+2}$ or 2

(b) $6 + {^-8}$ Adding *unlike* signed integers

Step 1: $|6| = 6; |{^-8}| = 8$

Subtract $8 - 6$ to get 2.

Step 2: $^-8$ has the larger absolute value and is negative, so the sum is negative.

$6 + {^-8} = {^-2}$

19. **(a)** $^-9 + 2$ Adding *unlike* signed integers

Step 1: $|^-9| = 9; |2| = 2$

Subtract $9 - 2$ to get 7.

Step 2: $^-9$ has the larger absolute value and is negative, so the sum is negative.

$$^-9 + 2 = \,^-7$$

(b) $9 + \,^-2$ Adding *unlike* signed integers

Step 1: $|9| = 9; |^-2| = 2$

Subtract $9 - 2$ to get 7.

Step 2: 9 has the larger absolute value and is positive, so the sum is positive.

$$9 + \,^-2 = \,^+7 \text{ or } 7$$

21. **(a)** $20 + \,^-25$ Adding *unlike* signed integers

Step 1: $|20| = 20; |^-25| = 25$

Subtract $25 - 20$ to get 5.

Step 2: $^-25$ has the larger absolute value and is negative, so the sum is negative.

$$20 + \,^-25 = \,^-5$$

(b) $^-20 + 25$ Adding *unlike* signed integers

Step 1: $|^-20| = 20; |25| = 25$

Subtract $25 - 20$ to get 5.

Step 2: 25 has the larger absolute value and is positive, so the sum is positive.

$$^-20 + 25 = \,^+5 \text{ or } 5$$

23. **(a)** $200 + \,^-50$ Adding *unlike* signed integers

Step 1: $|200| = 200; |^-50| = 50$

Subtract $200 - 50$ to get 150.

Step 2: 200 has the larger absolute value and is positive, so the sum is positive.

$$200 + \,^-50 = \,^+150 \text{ or } 150$$

(b) $^-200 + 50$ Adding *unlike* signed integers

Step 1: $|^-200| = 200; |50| = 50$

Subtract $200 - 50$ to get 150.

Step 2: $^-200$ has the larger absolute value and is negative, so the sum is negative.

$$^-200 + 50 = \,^-150$$

25. Each pair of answers differs only in the sign of the answer. This occurs because the signs of the addends are reversed.

27. $^-8 + 5$ Adding *unlike* signed integers

Step 1: $|^-8| = 8; |5| = 5$

Subtract $8 - 5$ to get 3.

Step 2: $^-8$ has the larger absolute value and is negative, so the sum is negative.

$$^-8 + 5 = \,^-3$$

29. $^-1 + 8$ Adding *unlike* signed integers

Step 1: $|^-1| = 1; |8| = 8$

Subtract $8 - 1$ to get 7.

Step 2: 8 has the larger absolute value and is positive, so the sum is positive.

$$^-1 + 8 = \,^+7 \text{ or } 7$$

31. $^-2 + \,^-5$ Adding *like* signed integers

Step 1: $|^-2| = 2; |^-5| = 5$

Add $2 + 5$ to get 7.

Step 2: Both integers are negative, so the sum is negative.

$$^-2 + \,^-5 = \,^-7$$

33. $6 + \,^-5$ Adding *unlike* signed integers

Step 1: $|6| = 6; |^-5| = 5$

Subtract $6 - 5$ to get 1.

Step 2: 6 has the larger absolute value and is positive, so the sum is positive.

$$6 + \,^-5 = \,^+1 \text{ or } 1$$

35. $4 + \,^-12$ Adding *unlike* signed integers

Step 1: $|4| = 4; |^-12| = 12$

Subtract $12 - 4$ to get 8.

Step 2: $^-12$ has the larger absolute value and is negative, so the sum is negative.

$$4 + \,^-12 = \,^-8$$

37. $^-10 + \,^-10$ Adding *like* signed integers

Step 1: $|^-10| = 10; |^-10| = 10$

Add $10 + 10$ to get 20.

Step 2: Both integers are negative, so the sum is negative.

$$^-10 + \,^-10 = \,^-20$$

39. $^-17 + 0 = \,^-17$

Adding zero to any number leaves the number unchanged.

41. $1 + {}^-23$ Adding *unlike* signed integers

Step 1: $|1| = 1; |{}^-23| = 23$

Subtract $23 - 1$ to get 22.

Step 2: ${}^-23$ has the larger absolute value and is negative, so the sum is negative.

$$1 + {}^-23 = {}^-22$$

43. ${}^-2 + {}^-12 + {}^-5$ *Add left to right.*
$= {}^-14 + {}^-5$
$= {}^-19$

45. $8 + 6 + {}^-8$ *Commute addends.*
$= 8 + {}^-8 + 6$ *Add left to right.*
$= 0 + 6$
$= 6$

47. ${}^-7 + 6 + {}^-4$ *Add left to right.*
$= {}^-1 + {}^-4$
$= {}^-5$

49. ${}^-3 + {}^-11 + 14$ *Add left to right.*
$= {}^-14 + 14$
$= 0$

51. $10 + {}^-6 + {}^-3 + 4$ *Add left to right.*
$= 4 + {}^-3 + 4$
$= 1 + 4$
$= 5$

53. ${}^-7 + 28 + {}^-56 + 3$ *Add left to right.*
$= 21 + {}^-56 + 3$
$= {}^-35 + 3$
$= {}^-32$

55. "Yards gained" are positive $({}^+13)$, and "yards lost" are negative $({}^-17)$.

$13 + {}^-17 = {}^-4$ yards

The team lost 4 yards.

57. The overdrawn amount is negative $({}^-\$62)$, and the deposit amount is positive $({}^+\$50)$.

${}^-\$62 + \$50 = {}^-\$12$

Nick is $12 overdrawn.

59. $88 stolen implies a loss of money or ${}^-\$88$.

Jay received $35 back implies a gain of money or ${}^+\$35$.

${}^-\$88 + \$35 = {}^-\$53$

Jay's net loss was $53.

61. Jeff: ${}^-20 + 75 + {}^-55$ *Add left to right.*
$= 55 + {}^-55$
$= 0$ points

Terry: $42 + {}^-15 + 20$ *Add left to right.*
$= 27 + 20$
$= 47$ points

63. ${}^-2 + 0 + 5 + {}^-5$ *Add left to right.*
$= {}^-2 + 5 + {}^-5$
$= 3 + {}^-5$
$= {}^-2$

Angela lost 2 pounds.

65. $3 + {}^-2 + {}^-2 + 3$ *Add left to right.*
$= 1 + {}^-2 + 3$
$= {}^-1 + 3$
$= 2$

Brittany gained 2 pounds.

67. $\underbrace{{}^-18 + {}^-5}_{{}^-23} = \underbrace{{}^-5 + {}^-18}_{{}^-23}$ *Commutative property*

Both sums are ${}^-23$.

69. $\underbrace{{}^-4 + 15}_{{}^+11} = \underbrace{15 + {}^-4}_{{}^+11}$ *Commutative property*

Both sums are ${}^+11$ or 11.

71. $6 + {}^-14 + 14$

Option 1: $(6 + {}^-14) + 14 = {}^-8 + 14$
$= 6$

Option 2: $6 + ({}^-14 + 14) = 6 + 0$
$= 6$

Option 2 is easier.

73. ${}^-14 + {}^-6 + {}^-7$

Option 1: $({}^-14 + {}^-6) + {}^-7 = {}^-20 + {}^-7$
$= {}^-27$

Option 2: ${}^-14 + ({}^-6 + {}^-7) = {}^-14 + {}^-13$
$= {}^-27$

Option 1 might seem easier.

75. Answers will vary. Some possibilities are:
${}^-6 + 0 = {}^-6; 10 + 0 = 10; 0 + 3 = 3$

77. Be sure to use the *negative* key as opposed to the *subtraction* key.

${}^-7081 + 2965 = {}^-4116$

79. ${}^-179 + {}^-61 + 8926 = 8686$

81. $86 + {}^-99{,}000 + 0 + 2837 = {}^-96{,}077$

Chapter 1 Introduction to Algebra: Integers

1.4 Subtracting Integers

1.4 Margin Exercises

1. (a) The opposite of 5 is $^-5$. $5 + {^-5} = 0$
   (b) The opposite of 48 is $^-48$. $48 + {^-48} = 0$
   (c) The opposite of 0 is 0. $0 + 0 = 0$
   (d) The opposite of $^-1$ is 1. $^-1 + 1 = 0$
   (e) The opposite of $^-24$ is 24. $^-24 + 24 = 0$

2. (a) $^-6 - 5$ *Change subtraction to addition. Change 5 to $^-5$.*
   $= {^-6} + {^-5}$
   $= {^-11}$

   (b) $3 - {^-10}$ *Change subtraction to addition. Change $^-10$ to $^+10$.*
   $= 3 + {^+10}$
   $= 13$

   (c) $^-8 - {^-2}$ *Change subtraction to addition. Change $^-2$ to $^+2$.*
   $= {^-8} + {^+2}$
   $= {^-6}$

   (d) $0 - 10$ *Change subtraction to addition. Change 10 to $^-10$.*
   $= 0 + {^-10}$
   $= {^-10}$

   (e) $^-4 - {^-12}$ *Change subtraction to addition. Change $^-12$ to $^+12$.*
   $= {^-4} + {^+12}$
   $= 8$

   (f) $9 - 7$ *Change subtraction to addition. Change 7 to $^-7$.*
   $= 9 + {^-7}$
   $= 2$

3. (a) $6 - 7 + {^-3}$ *Change subtraction to addition. Change 7 to $^-7$.*
   $= 6 + {^-7} + {^-3}$ *Add left to right.*
   $= {^-1} + {^-3}$
   $= {^-4}$

   (b) $^-2 + {^-3} - {^-5}$ *Change subtraction to addition. Change $^-5$ to $^+5$.*
   $= {^-2} + {^-3} + {^+5}$ *Add left to right.*
   $= {^-5} + 5$
   $= 0$

   (c) $7 - 7 - 7$ *Change all subtractions to additions. Change 7 to $^-7$.*
   $= 7 + {^-7} + {^-7}$ *Add left to right.*
   $= 0 + {^-7}$
   $= {^-7}$

   (d) $^-3 - 9 + 4 - {^-20} = {^-3} + {^-9} + 4 + {^+20}$
   $= {^-12} + 4 + {^+20}$
   $= {^-8} + {^+20}$
   $= 12$

1.4 Section Exercises

1. The opposite of 6 is $^-6$. $6 + {^-6} = 0$

3. The opposite of $^-13$ is 13. $^-13 + 13 = 0$

5. The opposite of 0 is 0. $0 + 0 = 0$

7. $19 - 5$ *Change subtraction to addition. Change 5 to $^-5$.*
   $= 19 + {^-5}$
   $= 14$

9. $10 - 12$ *Change subtraction to addition. Change 12 to $^-12$.*
   $= 10 + {^-12}$
   $= {^-2}$

11. $7 - 19$ *Change subtraction to addition. Change 19 to $^-19$.*
    $= 7 + {^-19}$
    $= {^-12}$

13. $^-15 - 10$ *Change subtraction to addition. Change 10 to $^-10$.*
    $= {^-15} + {^-10}$
    $= {^-25}$

15. $^-9 - 14$ *Change subtraction to addition. Change 14 to $^-14$.*
    $= {^-9} + {^-14}$
    $= {^-23}$

17. $^-3 - {^-8}$ *Change subtraction to addition. Change $^-8$ to $^+8$.*
    $= {^-3} + {^+8}$
    $= 5$

19. $6 - {^-14}$ *Change subtraction to addition. Change $^-14$ to $^+14$.*
    $= 6 + {^+14}$
    $= 20$

21. $1 - {^-10}$ *Change subtraction to addition. Change $^-10$ to $^+10$.*
    $= 1 + {^+10}$
    $= 11$

1.4 Subtracting Integers

23. $^-30 - 30$ *Change subtraction to addition. Change 30 to $^-30$.*
$= {}^-30 + {}^-30$
$= {}^-60$

25. $^-16 - {}^-16$ *Change subtraction to addition. Change $^-16$ to $^+16$.*
$= {}^-16 + {}^+16$
$= 0$

27. $13 - 13$ *Change subtraction to addition. Change 13 to $^-13$.*
$= 13 + {}^-13$
$= 0$

29. $0 - 6$ *Change subtraction to addition. Change 6 to $^-6$.*
$= 0 + {}^-6$
$= {}^-6$

31. **(a)** $3 - {}^-5$ *Change subtraction to addition. Change $^-5$ to $^+5$.*
$= 3 + {}^+5$
$= 8$

(b) $3 - 5$ *Change subtraction to addition. Change 5 to $^-5$.*
$= 3 + {}^-5$
$= {}^-2$

(c) $^-3 - {}^-5$ *Change subtraction to addition. Change $^-5$ to $^+5$.*
$= {}^-3 + {}^+5$
$= 2$

(d) $^-3 - 5$ *Change subtraction to addition. Change 5 to $^-5$.*
$= {}^-3 + {}^-5$
$= {}^-8$

33. **(a)** $4 - 7$ *Change subtraction to addition. Change 7 to $^-7$.*
$= 4 + {}^-7$
$= {}^-3$

(b) $4 - {}^-7$ *Change subtraction to addition. Change $^-7$ to $^+7$.*
$= 4 + {}^+7$
$= 11$

(c) $^-4 - 7$ *Change subtraction to addition. Change 7 to $^-7$.*
$= {}^-4 + {}^-7$
$= {}^-11$

(d) $^-4 - {}^-7$ *Change subtraction to addition. Change $^-7$ to $^+7$.*
$= {}^-4 + {}^+7$
$= 3$

35. $^-2 - 2 - 2$ *Change all subtractions to additions. Change 2 to $^-2$.*
$= {}^-2 + {}^-2 + {}^-2$ *Add left to right.*
$= {}^-4 + {}^-2$
$= {}^-6$

37. $9 - 6 - 3 - 5$ *Change all subtractions to additions. Change 6 to $^-6$, 3 to $^-3$, and 5 to $^-5$.*
$= 9 + {}^-6 + {}^-3 + {}^-5$ *Add left to right.*
$= 3 + {}^-3 + {}^-5$
$= 0 + {}^-5$
$= {}^-5$

39. $3 - {}^-3 - 10 - {}^-7$ *Change all subtractions to additions. Change $^-3$ to $^+3$, 10 to $^-10$, and $^-7$ to $^+7$.*
$= 3 + {}^+3 + {}^-10 + {}^+7$ *Add left to right.*
$= 6 + {}^-10 + {}^+7$
$= {}^-4 + {}^+7$
$= 3$

41. $^-2 + {}^-11 - {}^-3$ *Change subtraction to addition. Change $^-3$ to $^+3$.*
$= {}^-2 + {}^-11 + {}^+3$ *Add left to right.*
$= {}^-13 + {}^+3$
$= {}^-10$

43. $4 - {}^-13 + {}^-5$ *Change subtraction to addition. Change $^-13$ to $^+13$.*
$= 4 + {}^+13 + {}^-5$ *Add left to right.*
$= 17 + {}^-5$
$= 12$

45. $6 + 0 - 12 + 1$ *Change subtraction to addition. Change 12 to $^-12$.*
$= 6 + 0 + {}^-12 + 1$ *Add left to right.*
$= 6 + {}^-12 + 1$
$= {}^-6 + 1$
$= {}^-5$

47. **(a)** The 30°F column and the 10 mph wind row intersect at 21°F. The difference between the actual temperature and the wind chill temperature is $30 - 21 = 30 + {}^-21 = 9$ degrees.

(b) The 15°F column and the 15 mph wind row intersect at 0°F. The difference between the actual temperature and the wind chill temperature is $15 - 0 = 15$ degrees.

(c) The 5°F column and the 25 mph wind row intersect at $^-17$°F. The difference between the actual temperature and the wind chill temperature is $5 - {^-17} = 5 + {^+17} = 22$ degrees.

(d) The $^-10$°F column and the 35 mph wind row intersect at $^-41$°F. The difference between the actual temperature and the wind chill temperature is $^-10 - {^-41} = {^-10} + {^+41} = 31$ degrees.

49. The student forgot to change 6 to its opposite, $^-6$.

Correct Method:
$^-6 - 6$ *Change subtraction to addition. Change 6 to $^-6$.*
$= {^-6} + {^-6}$ *Add.*
$= {^-12}$

51. $^-2 + {^-11} + |{^-2}|$ *$|{^-2}| = 2$ because the distance from 0 to $^-2$ is 2 units.*
$= {^-2} + {^-11} + 2$ *Add left to right.*
$= {^-13} + 2$
$= {^-11}$

53. $0 - |{^-7} + 2|$ *Simplify the sum within the absolute value bars first.*
$= 0 - |{^-5}|$ *$|{^-5}| = 5$ because the distance from 0 to $^-5$ is 5 units.*
$= 0 - 5$ *Change subtraction to addition. Change 5 to $^-5$.*
$= 0 + {^-5}$ *Add.*
$= {^-5}$

55. $^-3 - ({^-2} + 4) + {^-5}$ *Simplify the sum within the parentheses first.*
$= {^-3} - 2 + {^-5}$ *Change subtraction to addition and change 2 to $^-2$.*
$= {^-3} + {^-2} + {^-5}$ *Add left to right.*
$= {^-5} + {^-5}$
$= {^-10}$

57. $^-3 - 5 = {^-3} + {^-5} = {^-8}$
$5 - {^-3} = 5 + 3 = 8$

$^-4 - {^-3} = {^-4} + 3 = {^-1}$
$^-3 - {^-4} = {^-3} + 4 = 1$

Subtraction is *not* commutative; the absolute value of the answer is the same, but the sign changes.

58. Subtracting 0 from a number does *not* change the number. For example, $^-5 - 0 = {^-5}$. But subtracting a number from 0 *does* change the number to its opposite. For example, $0 - {^-5} = 5$.

1.5 Problem Solving: Rounding and Estimating

1.5 Margin Exercises

1. **(a)** $^-746$ (nearest ten)
Draw a line under the 4. $^-74\underline{6}$
$^-746$ is closer to $^-75\underline{0}$.

(b) 2412 (nearest thousand)
Draw a line under the leading 2. $\underline{2}412$
2412 is closer to $\underline{2}000$.

(c) $^-89{,}512$ (nearest hundred)
Draw a line under the 5. $^-89{,}\underline{5}12$
$^-89{,}512$ is closer to $^-89{,}\underline{5}00$.

(d) 546,325 (nearest ten thousand)
Draw a line under the 4. $54\underline{6}{,}325$
546,325 is closer to $55\underline{0}{,}000$.

2. **(a)** $3\underline{4} \approx 30$
Because the next digit to the right of the underlined place is 4 or less, do not change the digit in the underlined place. Change all digits to the right of the underlined place to zeros.
Note: The symbol "$\approx$" means "approximately equal to."

(b) $^-6\underline{1} \approx {^-60}$
Because the next digit to the right of the underlined place is 4 or less, do not change the digit in the underlined place. Change all digits to the right of the underlined place to zeros.

(c) $^-6\underline{8}3 \approx {^-680}$
Because the next digit to the right of the underlined place is 4 or less, do not change the digit in the underlined place. Change all digits to the right of the underlined place to zeros.

(d) $17\underline{9}2 \approx 1790$
Because the next digit to the right of the underlined place is 4 or less, do not change the digit in the underlined place. Change all digits to the right of the underlined place to zeros.

3. **(a)** $\underline{1}725 \approx 2000$
Because the next digit to the right of the underlined place is 5 or more, add 1 to the digit in the underlined place. Change all digits to the right of the underlined place to zeros.

(b) $^-6{\underline{5}}11 \approx {}^-7000$
Because the next digit to the right of the underlined place is 5 or more, add 1 to the digit in the underlined place. Change all digits to the right of the underlined place to zeros.

(c) $5\underline{8},829 \approx 59,000$
Because the next digit to the right of the underlined place is 5 or more, add 1 to the digit in the underlined place. Change all digits to the right of the underlined place to zeros.

(d) $^-8\underline{3},904 \approx {}^-84,000$
Because the next digit to the right of the underlined place is 5 or more, add 1 to the digit in the underlined place. Change all digits to the right of the underlined place to zeros.

4. (a) $^-60\underline{3}6 \approx {}^-6040$
Underline the tens place. Next digit is 5 or more. Tens place changes. Add 1 to 3. Change all digits to the right of the underlined place to zeros.

(b) $31,\underline{9}68 \approx 32,000$
Underline the hundreds place. Next digit is 5 or more. Hundreds place changes. Add 1 to 9. Write 0 and carry 1 into the thousands place. Change all digits to the right of the underlined place to zero.

(c) $^-7\underline{3},077 \approx {}^-73,000$
Underline the thousands place. Next digit is 4 or less. Leave 3 as 3. Change all digits to the right of the underlined place to zeros.

(d) $\underline{4}952 \approx 5000$
Underline the thousands place. Next digit is 5 or more. Thousands place changes. Add 1 to 4. Change all digits to the right of the underlined place to zeros.

(e) $85,\underline{9}49 \approx 85,900$
Underline the hundreds place. Next digit is 4 or less. Leave 9 as 9. Change all digits to the right of the underlined place to zeros.

(f) $4\underline{0},387 \approx 40,000$
Underline the thousands place. Next digit is 4 or less. Leave 0 as 0. Change all digits to the right of the underlined place to zeros.

5. (a) $^-\underline{1}4,679 \approx {}^-10,000$
Underline the ten thousands place. Next digit is 4 or less. Leave 1 as 1. Change all digits to the right of the underlined place to zeros.

(b) $72\underline{4},518,715 \approx 725,000,000$
Underline the millions place. Next digit is 5 or more. Add 1 to 4. Change all digits to the right of the underlined place to zeros.

(c) $^-4\underline{9},900,700 \approx {}^-50,000,000$
Underline the millions place. Next digit is 5 or more. Add 1 to 9. Write 0 and carry 1 to the ten millions place. Change all digits to the right of the underlined place to zeros.

(d) $\underline{3}06,779,000 \approx 300,000,000$
Underline the hundred millions place. Next digit is 4 or less. Leave 3 as 3. Change all digits to the right of the underlined place to zeros.

6. (a) $^-\underline{9}4 \approx {}^-90$
Underline the first digit. Next digit is 4 or less. Leave 9 as 9. Change 4 to 0.

(b) $\underline{5}08 \approx 500$
Underline the first digit. Next digit is 4 or less. Leave 5 as 5. Change 8 to 0.

(c) $^-\underline{2}522 \approx {}^-3000$
Underline the first digit. Next digit is 5 or more. Add 1 to 2. Change all digits to the right of the underlined place to zeros.

(d) $\underline{9}700 \approx 10,000$
Underline the first digit. Next digit is 5 or more. Add 1 to 9. Write 0 and carry 1 to the ten thousands place. Change all digits to the right of the underlined place to zeros.

(e) $\underline{6}1,888 \approx 60,000$
Underline the first digit. Next digit is 4 or less. Leave 6 as 6. Change all digits to the right to zeros.

(f) $^-\underline{9}63,369 \approx {}^-1,000,000$
Underline the first digit. Next digit is 5 or more. Add 1 to 9. Write 0 and carry 1 to the millions place. Change all digits to the right to zeros.

7. "Overdrawn" implies a negative number, $^-\$3881$
"Deposit" implies a positive number, $^+\$2090$

Estimate: $^-\$3\underline{8}81 \approx {}^-\$4000$
$^+\$\underline{2}090 \approx {}^+\$2000$
Balance $= {}^-4000 + 2000 = {}^-2000$
Approximately $2000 overdrawn.

Exact: Balance $= {}^-\$3881 + \$2090 = {}^-\$1791$
Pao Xiong is overdrawn by $1791.

The estimate of $2000 overdrawn is fairly close to the exact amount.

1.5 Section Exercises

1. $6\underline{2}5 \approx 630$ (nearest ten)
Next digit is 5 or more. Tens place changes. Add 1 to 2. Change 5 to 0.

3. $^-10\underline{8}3 \approx {}^-1080$ (nearest ten)
Next digit is 4 or less. Tens place remains 8. Change 3 to 0.

Chapter 1 Introduction to Algebra: Integers

5. $7\underline{8}62 \approx 7900$ (nearest hundred)
Next digit is 5 or more. Hundreds place changes. Add 1 to 8. Change 6 and 2 to 0.

7. $^{-}86,\underline{8}13 \approx {}^{-}86,800$ (nearest hundred)
Next digit is 4 or less. Hundreds place remains 8. Change 1 and 3 to 0.

9. $42,\underline{4}95 \approx 42,500$ (nearest hundred)
Next digit is 5 or more. Hundreds place changes. Add 1 to 4. Change 9 and 5 to 0.

11. $^{-}5\underline{9}96 \approx {}^{-}6000$ (nearest hundred)
Next digit is 5 or more. Hundreds place changes. Add 1 to 9 and carry 1 to thousands. Change 9 and 6 to 0.

13. $^{-}7\underline{8},499 \approx {}^{-}78,000$ (nearest thousand)
Next digit is 4 or less. Thousands place remains 8. Change 4, 9, and 9 to 0.

15. $\underline{5}847 \approx 6000$ (nearest thousand)
Next digit is 5 or more. Thousands place changes. Add 1 to 5. Change 8, 4, and 7 to 0.

17. $5\underline{9}5,008 \approx 600,000$ (nearest ten-thousand)
Next digit is 5 or more. Ten-thousands place changes. Add 1 to 9 and carry 1 to hundred thousands. Change 5 and 8 to 0.

19. $^{-}\underline{8},906,422 \approx {}^{-}9,000,000$ (nearest million)
Next digit is 5 or more. Millions place changes. Add 1 to 8. Change other digits to 0.

21. $13\underline{9},610,000 \approx 140,000,000$ (nearest million)
Next digit is 5 or more. Millions place changes. Add 1 to 9. Carry one to ten-millions. Change 6 and 1 to zeros.

23. $19,\underline{9}51,880,500 \approx 20,000,000,000$ (nearest hundred-million)
Next digit is 5 or more. Hundred-millions place changes. Add 1 to 9. Write 0 and regroup 1 to the ten-billions place. All digits to the right of the underlined place change to 0.

25. $\underline{8},608,200,000 \approx 9,000,000,000$ (nearest billion)
Next digit is 5 or more. Billions place changes. Add 1 to 8. All digits to the right of the underlined place change to 0.

27. $\underline{3}1,500 \approx 30,000$ miles
Next digit is 4 or less. Leave 3 as 3. Change 1 and 5 to 0. 31,500 is closer to 30,000 than 40,000.

29. $^{-}\underline{5}6 \approx {}^{-}60$ degrees
Next digit is 5 or more. Change 5 to 6 and change 6 to 0. $^{-}56$ is closer to $^{-}60$ than $^{-}50$.

31. $\$\underline{9}942 \approx \$10,000$
Next digit is 5 or more. Add 1 to 9 and carry 1 to the ten-thousands place. Change 9, 4, and 2 to 0. $9942 is closer to $10,000 than $9000.

33. $\underline{6}0,950,000 \approx 60,000,000$ Americans
Next digit is 4 or less. Leave 6 as 6. Change 9 and 5 to 0. 60,950,000 is closer to 60,000,000 than 70,000,000.

35. $^{-}\underline{2}55 \approx {}^{-}300$ feet
Next digit is 5 or more. Change 2 to 3. Change 5s to 0. $^{-}255$ is closer to $^{-}300$ than $^{-}200$.

37. $\underline{6}70,053 \approx 700,000$ people in Alaska
Next digit is 5 or more. Change 6 to 7. Change all other digits to 0. 670,053 is closer to 700,000 than 600,000.

$\underline{3}6,457,549 \approx 40,000,000$ people in California
Next digit is 5 or more. Change 3 to 4. Change all other digits to 0. 36,457,549 is closer to 40,000,000 than 30,000,000.

39. Answers will vary but should mention looking only at the second digit, rounding first digit up when second digit is 5 or more, leaving first digit unchanged when second digit is 4 or less. Examples will vary. Some possibilities are $\underline{2}7 \approx 30, \underline{6}41 \approx 600$.

41. $^{-}42 + 89$
$^{-}\underline{4}2$ is closer to $^{-}40$ than $^{-}50$.
$\underline{8}9$ is closer to 90 than 80.

Estimate: $^{-}40 + 90 = 50$;
Exact: $^{-}42 + 89 = 47$

43. $16 + {}^{-}97$
$\underline{1}6$ is closer to 20 than 10.
$^{-}\underline{9}7$ is closer to $^{-}100$ than $^{-}90$.

Estimate: $20 + {}^{-}100 = {}^{-}80$;
Exact: $16 + {}^{-}97 = {}^{-}81$

45. $^{-}273 + {}^{-}399$
$^{-}\underline{2}73$ is closer to $^{-}300$ than $^{-}200$.
$^{-}\underline{3}99$ is closer to $^{-}400$ than $^{-}300$.

Estimate: $^{-}300 + {}^{-}400 = {}^{-}700$;
Exact: $^{-}273 + {}^{-}399 = {}^{-}672$

47. $3081 + 6826$
$\underline{3}081$ is closer to 3000 than 4000.
$\underline{6}826$ is closer to 7000 than 6000.

Estimate: $3000 + 7000 = 10,000$;
Exact: $3081 + 6826 = 9907$

49. 23 − 81   *Change subtraction to addition.*
23 + ⁻81   *Change 81 to ⁻81.*

2̲3 is closer to 20 than 30.
⁻8̲1 is closer to ⁻80 than ⁻90.

Estimate: 20 + ⁻80 = ⁻60;
Exact: 23 − 81 = 23 + ⁻81 = ⁻58

51. ⁻39 − 39   *Change subtraction to addition.*
⁻39 + ⁻39   *Change 39 to ⁻39.*

⁻3̲9 is closer to ⁻40 than ⁻30.

Estimate: ⁻40 + ⁻40 = ⁻80;
Exact: ⁻39 − 39 = ⁻39 + ⁻39 = ⁻78

53. ⁻106 + 34 − ⁻72   *Change subtraction to*
addition of the opposite.
⁻106 + 34 + ⁺72

⁻1̲06 ≈ ⁻100; 3̲4 ≈ 30; 7̲2 ≈ 70

Estimate: ⁻100 + 30 + ⁺70 = 0;
Exact:
⁻106 + 34 − ⁻72 = ⁻106 + 34 + ⁺72 = 0

55. Already raised: $52,882 ≈ $50,000
Amount needed: $78,650 ≈ $80,000
Amount that still needs to be collected:

Estimate: 80,000 − 50,000 = $30,000;
Exact: 78,650 − 52,882 = $25,768

57. Estimate Dorene's expenses.
Rent: $8̲45 ≈ $800
Food: $3̲25 ≈ $300
Childcare: $3̲65 ≈ $400
Transportation: $1̲82 ≈ $200
Other: $2̲40 ≈ $200

Estimate: Dorene's total expenses:
$800 + $300 + $400 + $200 + $200 = $1900.
Estimate Dorene's monthly take home pay.
$2̲120 ≈ $2000
Subtract Dorene's expenses from her take home
pay to estimate her monthly savings.
$2000 − $1900 = $100.

Exact:
Total Expenses = $845 + $325 + $365
+ $182 + $240
= $1957
Monthly savings = $2120 − $1957 = $163.

59. The final temperature equals the initial
temperature plus the two increases.

⁻102 ≈ ⁻100; 37 ≈ 40; 52 ≈ 50

Estimate: ⁻100 + 40 + 50 = ⁻10 degrees
Exact: ⁻102 + 37 + 52 = ⁻13 degrees

61. 4̲12 ≈ 400 doors
1̲47 ≈ 100 windows
Total number of doors and windows:
Estimate: 400 + 100 = 500 doors and windows
Exact: 412 + 147 = 559 doors and windows

1.6   Multiplying Integers

1.6   Margin Exercises

1. **(a)** $\underline{100} \times \underline{6} = \underline{600}$
    Factor Factor Product

    Equivalent forms: 100 · 6 = 600
    100(6) = 600
    (100)(6) = 600

    (b) $\underline{7} \times \underline{12} = \underline{84}$
    Factor Factor Product

    Equivalent forms: 7 · 12 = 84
    7(12) = 84
    (7)(12) = 84

2. **(a)** 7(⁻2) = ⁻14
    (*different* signs, product is *negative*)

    (b) ⁻5 · ⁻5 = 25
    (*same* signs, product is *positive*)

    (c) ⁻1(14) = ⁻14
    (*different* signs, product is *negative*)

    (d) 10 · 6 = 60
    (*same* signs, product is *positive*)

    (e) (⁻4)(⁻9) = 36
    (*same* signs, product is *positive*)

3. **(a)** 5 · (⁻10 · 2)   Work within parentheses first.
    = 5 · ⁻20   (*different* signs,
    product is *negative*)
    = ⁻100   (*different* signs,
    product is *negative*)

    (b) ⁻1 · 8 · ⁻5   No parentheses.
    Multiply from left to right.
    = ⁻8 · ⁻5   (*different* signs,
    product is *negative*)
    = 40   (*same* signs,
    product is *positive*)

    (c) ⁻3 · ⁻2 · ⁻4   Multiply from left to right.
    = 6 · ⁻4   (*same* signs,
    product is *positive*)
    = ⁻24   (*different* signs,
    product is *negative*)

14 Chapter 1 Introduction to Algebra: Integers

(d) $^-2 \cdot (7 \cdot {}^-3)$ Work within parentheses first.
$= {}^-2 \cdot {}^-21$ (*different* signs, product is *negative*)
$= 42$ (*same* signs, product is *positive*)

(e) $({}^-1)({}^-1)({}^-1)$ Multiply from left to right.
$= 1({}^-1)$ (*same* signs, product is *positive*)
$= {}^-1$ (*different* signs, product is *negative*)

4. (a) $819 \cdot 0 = 0$; multiplication property of 0

(b) $1({}^-90) = {}^-90$; multiplication property of 1

(c) $25 \cdot 1 = 25$; multiplication property of 1

(d) $(0)({}^-75) = 0$; multiplication property of 0

5. (a) $(3 \cdot 3) \cdot 2 = 3 \cdot (3 \cdot 2)$
$9 \cdot 2 = 3 \cdot 6$
$18 = 18$
Associative property of multiplication

(b) $11 \cdot 8 = 8 \cdot 11$
$88 = 88$
Commutative property of multiplication

(c) $2 \cdot {}^-15 = {}^-15 \cdot 2$
$^-30 = {}^-30$
Commutative property of multiplication

(d) $^-4 \cdot (2 \cdot 5) = ({}^-4 \cdot 2) \cdot 5$
$^-4 \cdot 10 = {}^-8 \cdot 5$
$^-40 = {}^-40$
Associative property of multiplication

6. (a) $3(8 + 7) = 3 \cdot 8 + 3 \cdot 7$
$3(15) = 24 + 21$
$45 = 45$ Both results are 45.

(b) $10(6 + {}^-9) = 10 \cdot 6 + 10 \cdot {}^-9$
$10({}^-3) = 60 + {}^-90$
$^-30 = {}^-30$ Both results are $^-30$.

(c) $^-6(4 + 4) = {}^-6 \cdot 4 + {}^-6 \cdot 4$
$^-6(8) = {}^-24 + {}^-24$
$^-48 = {}^-48$ Both results are $^-48$.

7. 27,095 fans rounds to 30,000 fans and 81 games rounds to 80 games.

Total attendance for the season:

Estimate: $30{,}000(80) = 2{,}400{,}000$ fans

Exact: $27{,}095(81) = 2{,}194{,}695$ fans

1.6 Section Exercises

1. (a) $9 \cdot 7 = 63$ Factors have the *same* sign, so the product is *positive*.

(b) $^-9 \cdot {}^-7 = 63$ Factors have the *same* sign, so the product is *positive*.

(c) $^-9 \cdot 7 = {}^-63$ Factors have *different* signs, so the product is *negative*.

(d) $9 \cdot {}^-7 = {}^-63$ Factors have *different* signs, so the product is *negative*.

3. (a) $7({}^-8) = {}^-56$ Factors have *different* signs, so the product is *negative*.

(b) $^-7(8) = {}^-56$ Factors have *different* signs, so the product is *negative*.

(c) $7(8) = 56$ Factors have the *same* sign, so the product is *positive*.

(d) $^-7({}^-8) = 56$ Factors have the *same* sign, so the product is *positive*.

5. $^-5 \cdot 7 = {}^-35$ (*different* signs, product is *negative*)

7. $({}^-5)(9) = {}^-45$
(*different* signs, product is *negative*)

9. $3({}^-6) = {}^-18$ (*different* signs, product is *negative*)

11. $10({}^-5) = {}^-50$
(*different* signs, product is *negative*)

13. $({}^-1)(40) = {}^-40$
(*different* signs, product is *negative*)

15. $^-56 \cdot 1 = {}^-56$; multiplication property of 1

17. $^-8({}^-4) = 32$ (*same* signs, product is *positive*)

19. $11 \cdot 7 = 77$ (*same* signs, product is *positive*)

21. $25 \cdot 0 = 0$; multiplication property of 0

23. $^-19({}^-7) = 133$ (*same* signs, product is *positive*)

25. $^-13({}^-1) = 13$ (*same* signs, product is *positive*)

27. $(0)({}^-25) = 0$; multiplication property of 0

29. $^-4 \cdot {}^-6 \cdot 2$ Multiply from left to right.
$= 24 \cdot 2$
$= 48$

31. $({}^-4)({}^-2)({}^-7)$ Multiply from left to right.
$= 8({}^-7)$
$= {}^-56$

33. $5({}^-8)(4)$ Multiply from left to right.
$= {}^-40(4)$
$= {}^-160$

35. $({}^-3)(\underline{5}) = {}^-15$ (negative product, different signs)

37. $\underline{^-3} \cdot 10 = {}^-30$ (negative product, different signs)

39. $^-17 = 17(\underline{^-1})$ (negative product, different signs)

41. $(\underline{0})({}^-350) = 0$ (multiplication property of 0)

1.6 Multiplying Integers

43. $5 \cdot {}^-4 \cdot \underline{5} = {}^-100$ because
${}^-20 \cdot 5 = {}^-100$ (negative product, different signs)

45. $({}^-\underline{4})({}^-5)({}^-2) = {}^-40$ because
$({}^-4)(10) = {}^-40$ (negative product, different signs)

47. Commutative property: changing the order of the factors does not change the product.
Associative property: changing the *grouping* of the factors does not change the product.
Examples will vary. Some possibilities are
${}^-4 \cdot 7 = 7 \cdot {}^-4 = {}^-28$,
$(2 \cdot 5) \cdot {}^-8 = 2 \cdot (5 \cdot {}^-8) = {}^-80$.

49. Examples will vary. Some possibilities are:

(a) $6 \cdot {}^-1 = {}^-6$; $2 \cdot {}^-1 = {}^-2$; $15 \cdot {}^-1 = {}^-15$

(b) ${}^-6 \cdot {}^-1 = 6$; ${}^-2 \cdot {}^-1 = 2$; ${}^-15 \cdot {}^-1 = 15$

The result of multiplying any nonzero number times ${}^-1$ is the number with the opposite sign.

50.
${}^-2 \cdot {}^-2 = 4$
${}^-2 \cdot {}^-2 \cdot {}^-2 = {}^-8$
${}^-2 \cdot {}^-2 \cdot {}^-2 \cdot {}^-2 = 16$
${}^-2 \cdot {}^-2 \cdot {}^-2 \cdot {}^-2 \cdot {}^-2 = {}^-32$

The absolute value doubles each time and the sign changes. The next three products are ${}^-2 \cdot {}^-32 = 64$, ${}^-2 \cdot 64 = {}^-128$, and ${}^-2 \cdot {}^-128 = 256$.

51. $9({}^-3 + 5)$ rewritten by using the distributive property is $9 \cdot {}^-3 + 9 \cdot 5$.

$9({}^-3 + 5) = 9 \cdot {}^-3 + 9 \cdot 5$
$9(2) = {}^-27 + 45$
$18 = 18$

53. $25 \cdot 8$ rewritten by using the commutative property is $8 \cdot 25$.

$25 \cdot 8 = 8 \cdot 25$
$200 = 200$

55. ${}^-3 \cdot (2 \cdot 5)$ rewritten using the associative property is $({}^-3 \cdot 2) \cdot 5$.

${}^-3 \cdot (2 \cdot 5) = ({}^-3 \cdot 2) \cdot 5$
$\phantom{{}^-3 \cdot}{}^-3 \cdot 10 = {}^-6 \cdot 5$
$\phantom{{}^-3 \cdot 10}{}^-30 = {}^-30$

57. Income: $\$324 \approx \$300$
52 weeks ≈ 50

Estimate: $\$300 \cdot 50 = \$15,000$
Exact: $\$324 \cdot 52 = \$16,848$

59. Monthly loss: ${}^-\$9950 \approx {}^-\$10,000$
12 months ≈ 10

Estimate: ${}^-\$10,000 \cdot 10 = {}^-\$100,000$
Exact: ${}^-\$9950 \cdot 12 = {}^-\$119,400$

61. Tuition: $\$182$ per credit $\approx \$200$ per credit
13 credits ≈ 10 credits

Estimate: $\$200 \cdot 10 = \$2000$
Exact: $\$182 \cdot 13 = \$2366$

63. Hours: $24 \approx 20$
365 days ≈ 400

Estimate: $20 \cdot 400 = 8000$ hours
Exact: $24 \cdot 365 = 8760$ hours

65. ${}^-8 \cdot |{}^-8 \cdot 8|$ Multiply the factors within the absolute value (different signs, negative product).
$= {}^-8 \cdot |{}^-64|$ ${}^-64$ is 64 units from 0, so $|{}^-64| = 64$.
$= {}^-8 \cdot 64$ Multiply.
$= {}^-512$ (different signs, negative product)

67. $({}^-37)({}^-1)(85)(0) = 0$;
multiplication property of 0

69. $|6 - 7| \cdot {}^-355,299$ Subtract within the absolute value first.
$= |6 + {}^-7| \cdot {}^-355,299$
$= |{}^-1| \cdot {}^-355,299$ ${}^-1$ is 1 unit from 0, so $|{}^-1| = 1$
$= 1 \cdot {}^-355,299$ multiplication property of 1
$= {}^-355,299$

71. The charge for each cat's shots will be $\$24 + \$29 = \$53$.
The total for all four cats will be four times that, plus the office visit charge.

$4 \cdot \$53 + \$35$ Do the multiplication first.
$= \$212 + \$35$
$= \$247$

73. The temperature drops 3 degrees (${}^-3$ degrees) for every 1000 feet climbed into the air. An altitude of 24,000 feet would require 24 increases of 1000 feet each.

${}^-3 \cdot 24 = {}^-72$ degrees

The temperature at 24,000 feet is $50 + {}^-72 = {}^-22$ degrees.

75. Points possible on tests:
$6(100) = 600$ points
Bonus points possible on tests:
$6(4) = 24$ points
Points possible on quizzes:
$8(6) = 48$ points
Points possible on homework assignments:
$20(5) = 100$ points
Total points possible:
$600 + 24 + 48 + 100 = 772$ points

1.7 Dividing Integers

1.7 Margin Exercises

1. **(a)** $\dfrac{40}{-8} = {}^-5$
   (*different* signs, quotient is *negative*)

   (b) $\dfrac{49}{7} = 7$
   (*same* signs, quotient is *positive*)

   (c) $\dfrac{-32}{4} = {}^-8$
   (*different* signs, quotient is *negative*)

   (d) $\dfrac{-10}{-10} = 1$
   (*same* signs, quotient is *positive*)

   (e) ${}^-81 \div 9 = {}^-9$
   (*different* signs, quotient is *negative*)

   (f) ${}^-100 \div {}^-50 = 2$
   (*same* signs, quotient is *positive*)

2. **(a)** $\dfrac{-12}{0}$ is undefined; division by 0 is undefined.

   (b) $\dfrac{0}{39} = 0$;
   0 divided by any nonzero number is 0.

   (c) $\dfrac{-9}{1} = {}^-9$;
   any number divided by 1 is the number.

   (d) $\dfrac{21}{21} = 1$;
   any nonzero number divided by itself is 1.

3. **(a)** $60 \div {}^-3({}^-5)$ Work from left to right.
   $= {}^-20({}^-5)$
   $= 100$

   (b) ${}^-6({}^-16 \div {}^-8) \cdot 2$ Work inside parentheses first.
   $= {}^-6(2) \cdot 2$ Now work from left to right.
   $= {}^-12 \cdot 2$
   $= {}^-24$

   (c) ${}^-8(10) \div 4({}^-3) \div {}^-6$ Work left to right.
   $= {}^-80 \div 4({}^-3) \div {}^-6$
   $= {}^-20({}^-3) \div {}^-6$
   $= 60 \div {}^-6$
   $= {}^-10$

   (d) $56 \div {}^-8 \div {}^-1$ No parentheses.
   $= {}^-7 \div {}^-1$ Work left to right.
   $= 7$

4. A loss of money implies a negative number:
   ${}^-\$2724$ in stocks rounds to ${}^-\$3000$.
   12 months rounds to 10 months.
   To find an average, use division.

   Estimate: ${}^-\$3000 \div 10 = {}^-\$300$ each month
   Exact: ${}^-\$2724 \div 12 = {}^-\$227$ each month

5. **(a)** 116 cookies are separated into packages of 12 each.

   ```
        9
   12)1 1 6
      1 0 8
      -----
          8
   ```

   They will have 9 packages of a dozen cookies each, and there will be 8 cookies left over for them to eat.

   (b) 249 senior citizens separated into buses with a 44-person limit per bus.

   ```
        5
   44)2 4 9
      2 2 0
      -----
        2 9
   ```

   If Coreen dispatches 5 buses, then 29 senior citizens will not have transportation to the game. She will need to dispatch 6 buses.

1.7 Section Exercises

1. **(a)** $14 \div 2 = 7$ (*same* signs, quotient is *positive*)

   (b) ${}^-14 \div {}^-2 = 7$
   (*same* signs, quotient is *positive*)

   (c) $14 \div {}^-2 = {}^-7$
   (*different* signs, quotient is *negative*)

   (d) ${}^-14 \div 2 = {}^-7$
   (*different* signs, quotient is *negative*)

3. **(a)** ${}^-42 \div 6 = {}^-7$
   (*different* signs, quotient is *negative*)

   (b) ${}^-42 \div {}^-6 = 7$
   (*same* signs, quotient is *positive*)

   (c) $42 \div {}^-6 = {}^-7$
   (*different* signs, quotient is *negative*)

(d) $42 \div 6 = 7$ (*same* signs, quotient is *positive*)

5. (a) $\dfrac{35}{35} = 1$; any nonzero number divided by itself is 1.

(b) $\dfrac{35}{1} = 35$; any number divided by 1 is the number.

(c) $\dfrac{^-13}{1} = {^-}13$; any number divided by 1 is the number.

(d) $\dfrac{^-13}{^-13} = 1$; any nonzero number divided by itself is 1.

7. (a) $\dfrac{0}{50} = 0$; zero divided by any nonzero number is 0.

(b) $\dfrac{50}{0}$ is undefined; division by zero is undefined.

(c) $\dfrac{^-11}{0}$ is undefined; division by zero is undefined.

(d) $\dfrac{0}{^-11} = 0$; zero divided by any nonzero number is 0.

9. $\dfrac{^-8}{2} = {^-}4$ (*different* signs, quotient is *negative*)

11. $\dfrac{21}{^-7} = {^-}3$ (*different* signs, quotient is *negative*)

13. $\dfrac{^-54}{^-9} = 6$ (*same* signs, quotient is *positive*)

15. $\dfrac{55}{^-5} = {^-}11$ (*different* signs, quotient is *negative*)

17. $\dfrac{^-28}{0}$ is undefined. Division by zero is undefined.

19. $\dfrac{14}{^-1} = {^-}14$ (*different* signs, quotient is *negative*)

21. $\dfrac{^-20}{^-2} = 10$ (*same* signs, quotient is *positive*)

23. $\dfrac{^-48}{^-12} = 4$ (*same* signs, quotient is *positive*)

25. $\dfrac{^-18}{18} = {^-}1$ (*different* signs, quotient is *negative*)

27. $\dfrac{0}{^-9} = 0$; zero divided by any nonzero number is 0.

29. $\dfrac{^-573}{^-3} = 191$ (*same* signs, quotient is *positive*)

31. $\dfrac{163{,}672}{^-328} = {^-}499$
(*different* signs, quotient is *negative*)

33. ${^-}60 \div 10 \div {^-}3$ No parentheses, so start at the left.
$= {^-}6 \div {^-}3$
$= 2$

35. ${^-}64 \div {^-}8 \div {^-}2$ No parentheses, so start at the left.
$= 8 \div {^-}2$
$= {^-}4$

37. $100 \div {^-}5({^-}2)$ Start at the left.
$= {^-}20({^-}2)$
$= 40$

39. $48 \div 3 \cdot (12 \div {^-}4)$ Start inside the parentheses. Now work from the left.
$= 48 \div 3 \cdot {^-}3$
$= 16 \cdot {^-}3$
$= {^-}48$

41. ${^-}5 \div {^-}5({^-}10) \div {^-}2$ Start at the left.
$= 1({^-}10) \div {^-}2$
$= {^-}10 \div {^-}2$
$= 5$

43. $64 \cdot 0 \div {^-}8(10)$ Start at the left.
$= 0 \div {^-}8(10)$
$= 0(10)$
$= 0$

45. $2 \div 1 = 2$ but $1 \div 2 = 0.5$, so division is not commutative.

46. $\underbrace{(12 \div 6)}_{2 \div 2} \div 2 \qquad 12 \div \underbrace{(6 \div 2)}_{12 \div 3}$
$\qquad\quad 1 \qquad\quad \text{different} \qquad\quad 4$
$\qquad\qquad\qquad \text{quotients}$

Division is not associative.

47. Similar: If the signs match, the result is positive. If the signs are different, the result is negative. Different: Multiplication is commutative, division is not. You can multiply by 0, but dividing by 0 is undefined.

48. Examples will vary. The properties are: Any nonzero number divided by itself is 1 (example: $4 \div 4 = 1$). Any number divided by 1 is the number (example: ${^-}7 \div 1 = {^-}7$). Division by 0 is undefined (example: ${^-}5 \div 0$ is undefined). Zero divided by any other number (except 0) is 0 (example: $0 \div 9 = 0$).

49. Examples will vary.

(a) $\frac{^-6}{^-1} = 6$; $\frac{^-2}{^-1} = 2$; $\frac{^-15}{^-1} = 15$

(b) $\frac{6}{^-1} = ^-6$; $\frac{2}{^-1} = ^-2$; $\frac{15}{^-1} = ^-15$

When dividing by $^-1$, change the sign of the dividend to its opposite to get the quotient.

50. Division is not commutative. $\frac{0}{^-3} = 0$ because $0 \cdot ^-3 = 0$. But $\frac{^-3}{0}$ is undefined because when $\frac{^-3}{0} = ?$ is rewritten as $? \cdot 0 = ^-3$, no number can replace ? and make a true statement.

51. Depth below sea level implies a negative, $^-35{,}836$ feet. Use division to find the size of each step. $^-35{,}836 \approx ^-40{,}000$; $17 \approx 20$

Estimate: $^-40{,}000 \div 20 = ^-2000$ feet
Exact: $^-35{,}836 \div 17 = ^-2108$ feet

53. Overdrawn implies a negative, $^-\$238$. Transfer of money into the account, implies a positive, $\$450$. $^-238 \approx ^-200$; $450 \approx 500$

Estimate: $^-200 + 500 = \$300$
Exact: $^-238 + 450 = \$212$

55. The number of non-foggy days equals the total number of days in a year minus the number of foggy days. $365 \approx 400$; $106 \approx 100$

Estimate: $400 - 100 = 300$ days
Exact: $365 - 106 = 259$ days

57. Descending implies a negative, $^-730$ feet each minute. $^-730 \approx ^-700$
Because the plane took $37 \approx 40$ minutes to land, use multiplication to find how far the plane descended.

Estimate: $^-700 \cdot 40 = ^-28{,}000$ feet
Exact: $^-730 \cdot 37 = ^-27{,}010$ feet

59. Use division to find how many miles were covered in each hour. $315 \approx 300$; $5 \approx 5$

Estimate: $300 \div 5 = 60$ miles
Exact: $315 \div 5 = 63$ miles

61. To find the average, add all the scores and divide by the number of scores.
Sum of scores: $143 + 190 + 162 + 177 = 672$
Number of scores: 4 scores given

$$\text{Average score} = \frac{672}{4} = 168$$

His average score was 168.

63. Calculate the total weight from the data on the back, then use subtraction to find the difference between that and the figure on the front.

Total weight from back:
$13 \cdot 40$ grams $= 520$ grams

Difference from front: $520 - 510 = 10$ grams

The back claims 10 more grams than the front.

65. The $\$302$ already in Stephanie's account and her $\$347$ paycheck are positives. The money she paid for day care, $\$116$, and rent, $\$548$, are negatives.

$$\$302 + ^-\$116 + ^-\$548 + \$347 = ^-\$15$$

Stephanie's balance is $^-\$15$. She is overdrawn by $\$15$.

67. Use division to convert minutes to hours.

$$\frac{1000 \text{ minutes}}{60 \text{ minutes per hour}}$$

$$\begin{array}{r} 16 \\ 60\overline{)1000} \\ \underline{60} \\ 400 \\ \underline{360} \\ 40 \end{array}$$

A new subscriber will receive 1000 minutes, which is 16 hours, with 40 minutes left over.

69. Use division to find the number of rooms.

$$\frac{163 \text{ people}}{5 \text{ people per room}}$$

$$\begin{array}{r} 32 \\ 5\overline{)163} \\ \underline{15} \\ 13 \\ \underline{10} \\ 3 \end{array}$$

32 rooms will be full, and that leaves 3 people. So 33 rooms are needed, with space for 2 people $(5 - 3 = 2)$ unused.

71. $|^-8| \div ^-4 \cdot |^-5| \cdot |1|$ Simplify the absolute values first.

$= 8 \div ^-4 \cdot 5 \cdot 1$ No parentheses, so start from the left.

$= ^-2 \cdot 5 \cdot 1$

$= ^-10 \cdot 1$

$= ^-10$

73. $^-6(^-8) \div (^-5 - ^-5)$ Start inside the parentheses.

$= ^-6 \cdot ^-8 \div 0$ Multiply and divide from left to right.

$= 48 \div 0$ Division by zero is undefined.

75. Start by entering 1 000 000 000.
Divide by 60. (≈ 16,666,667 minutes)
Divide by 60. (≈ 277,778 hours)
Divide by 24. (≈ 11,574 days)
Divide by 365. (≈ 31.7 years)
31.70979198 rounds to 32 years to receive one billion dollars.

Summary Exercises on Operations with Integers

1. $\quad 2 - 8 \quad$ Add the opposite.
$= 2 + {}^-8$
$= {}^-6$

3. $\quad {}^-14 - {}^-7 \quad$ Add the opposite.
$= {}^-14 + 7$
$= {}^-7$

5. $\quad {}^-9({}^-7) \quad$ (*same* signs, product is *positive*)
$= 63$

7. $\quad (1)({}^-56) \quad$ Multiplication property of 1
$= {}^-56$

9. $\quad 5 - {}^-7 \quad$ Add the opposite.
$= 5 + 7$
$= 12$

11. $\quad {}^-18 + 5 = {}^-13$

13. $\quad {}^-40 - {}^-40 \quad$ Add the opposite.
$= {}^-40 + {}^+40 \quad$ Addition of opposites is 0.
$= 0$

15. $\quad 8({}^-6) \quad$ (*different* signs, product is *negative*)
$= {}^-48$

17. $\quad {}^-5(10) \quad$ (*different* signs, product is *negative*)
$= {}^-50$

19. $\quad 0 - 14 \quad$ Add the opposite.
$= 0 + {}^-14 \quad$ Addition property of 0.
$= {}^-14$

21. $\quad {}^-13 + 13 \quad$ Addition of opposites is 0.
$= 0$

23. $\quad 20 - 50 \quad$ Add the opposite.
$= 20 + {}^-50$
$= {}^-30$

25. $\quad ({}^-4)({}^-6)(2) \quad$ Multiply from left to right.
$= (24)(2)$
$= 48$

27. $\quad {}^-60 \div 10 \div {}^-3 \quad$ Divide from left to right.
$= {}^-6 \div {}^-3$
$= 2$

29. $\quad 64(0) \div {}^-8 \quad$ Multiply.
$= 0 \div {}^-8 \quad$ Divide.
$= 0$

31. $\quad {}^-9 + 8 + {}^-2 \quad$ Add from left to right.
$= {}^-1 + {}^-2$
$= {}^-3$

33. $\quad 8 + 6 + {}^-8 \quad$ Add from left to right.
$= 14 + {}^-8$
$= 6$

35. $\quad {}^-25 \div {}^-1 \div {}^-5 \quad$ Divide from left to right.
$= 25 \div {}^-5$
$= {}^-5$

37. $\quad {}^-72 \div {}^-9 \div {}^-4 \quad$ Divide from left to right.
$= 8 \div {}^-4$
$= {}^-2$

39. $\quad 9 - 6 - 3 - 5 \quad$ Add the opposite.
$= 9 + {}^-6 + {}^-3 + {}^-5 \quad$ Add from left to right.
$= 3 + {}^-3 + {}^-5$
$= 0 + {}^-5$
$= {}^-5$

41. $\quad {}^-1(9732)({}^-1)({}^-1) \quad$ Multiply from left to right.
$= ({}^-9732)({}^-1)({}^-1)$
$= (9732)({}^-1)$
$= {}^-9732$

43. $\quad {}^-10 - 4 + 0 + 18 \quad$ Add the opposite.
$= {}^-10 + {}^-4 + 0 + 18 \quad$ Add from left to right.
$= {}^-14 + 0 + 18$
$= {}^-14 + 18$
$= 4$

45. $\quad 5 - |{}^-3| + 3 \quad$ Absolute value first
$= 5 - 3 + 3 \quad$ Add the opposite.
$= 5 + {}^-3 + 3 \quad$ Add from left to right.
$= 2 + 3$
$= 5$

47. $\quad {}^-3 - ({}^-2 + 4) - 5 \quad$ Parentheses first
$= {}^-3 - 2 - 5 \quad$ Add the opposites.
$= {}^-3 + {}^-2 + {}^-5 \quad$ Add from left to right.
$= {}^-5 + {}^-5$
$= {}^-10$

49. **(a)** If zero is divided by a nonzero number, the quotient is 0.

(b) If any number is multiplied by 0, the product is 0.

(c) If a nonzero number is divided by itself, the quotient is 1.

1.8 Exponents and Order of Operations

1.8 Margin Exercises

1. **(a)** $3 \cdot 3 \cdot 3 \cdot 3 = 3^4$
   is read "3 to the fourth power."

   (b) $6 \cdot 6 = 6^2$
   is read "6 squared" or "6 to the second power."

   (c) $9 = 9^1$
   is read "9 to the first power."

   (d) $(2)(2)(2)(2)(2)(2) = (2)^6 = 2^6$
   is read "2 to the sixth power."

2. **(a)** $(^-2)^3 = (^-2)(^-2)(^-2)$
   $= 4(^-2)$
   $= {}^-8$

   (b) $(^-6)^2 = (^-6)(^-6)$
   $= 36$

   (c) $2^4(^-3)^2 = \underbrace{(2)(2)(2)(2)}\ \underbrace{(^-3)(^-3)}$
   $ = (16) \cdot (9)$
   $ = 144$

   (d) $3^3 \cdot (^-4)^2 = \underbrace{(3)(3)(3)}\ \underbrace{(^-4)(^-4)}$
   $ = (27) \cdot (16)$
   $ = 432$

3. **(a)** $^-9 + {}^-15 - 3$ Change subtraction to addition. Change 3 to $^-3$.
   $= {}^-9 + {}^-15 + {}^-3$ Add from left to right.
   $= {}^-24 + {}^-3$
   $= {}^-27$

   (b) $^-4 - 2 + {}^-6$ Change subtraction to addition. Change 2 to $^-2$.
   $= {}^-4 + {}^-2 + {}^-6$ Add from left to right.
   $= {}^-6 + {}^-6$
   $= {}^-12$

   (c) $3(^-4) \div {}^-6$ Multiply and divide left to right.
   $= {}^-12 \div {}^-6$
   $= 2$

   (d) $^-18 \div 9(^-4)$ Multiply and divide left to right.
   $= {}^-2(^-4)$
   $= 8$

4. **(a)** $8 + 6(14 \div 2)$ Parentheses first
   $= 8 + 6(7)$ Multiply.
   $= 8 + 42$ Addition last
   $= 50$

   (b) $4(1) + 8(9 - 2)$ Parentheses first
   $= 4(1) + 8(7)$ Multiply.
   $= 4 + 56$ Addition last
   $= 60$

   (c) $3(5 + 1) + 20 \div 4$ Parentheses first
   $= 3(6) + 20 \div 4$ Multiply and divide left to right.
   $= 18 + 20 \div 4$
   $= 18 + 5$ Addition last
   $= 23$

5. **(a)** $2 + 40 \div (^-5 + 3)$ Parentheses first
   $= 2 + 40 \div {}^-2$ Divide.
   $= 2 + {}^-20$ Addition last
   $= {}^-18$

   (b) $^-5(5) - (15 + 5)$ Parentheses first
   $= {}^-5(5) - 20$ Multiply.
   $= {}^-25 - 20$ Change subtraction to addition. Change 20 to its opposite.
   $= {}^-25 + {}^-20$ Addition last
   $= {}^-45$

   (c) $(^-24 \div 2) + (15 - 3)$ Parentheses first
   $= {}^-12 + 12$ Addition last
   $= 0$

   (d) $^-3(2 - 8) - 5(4 - 3)$ Parentheses first
   $= {}^-3(^-6) - 5(1)$ Multiply left to right.
   $= 18 - 5(1)$
   $= 18 - 5$ Change subtraction to addition. Change 5 to its opposite.
   $= 18 + {}^-5$ Addition last
   $= 13$

   (e) $3(3) - (10 \cdot 3) \div 5$ Parentheses first
   $= 3(3) - 30 \div 5$ Multiply and divide left to right.
   $= 9 - 30 \div 5$
   $= 9 - 6$
   $= 9 + {}^-6$
   $= 3$

   (f) $6 - (2 + 7) \div (^-4 + 1)$ Parentheses first
   $= 6 - 9 \div {}^-3$ Divide.
   $= 6 - {}^-3$
   $= 6 + {}^+3$
   $= 9$

1.8 Exponents and Order of Operations

6. **(a)** $2^3 - 3^2$ Apply exponents.
$= 8 - 9$ $2^3 = 2 \cdot 2 \cdot 2;\ 3^2 = 3 \cdot 3$
$= 8 + {}^-9$ Change subtraction to addition. Change 9 to $^-9$.
$= {}^-1$ Add.

(b) $6^2 \div ({}^-4)({}^-3)$ Apply exponent.
$= 36 \div ({}^-4)({}^-3)$ Work left to right.
$= {}^-9({}^-3)$
$= 27$

(c) $({}^-4)^2 - 3^2(5-2)$ Parentheses first
$= ({}^-4)^2 - 3^2(3)$ Apply exponents.
     $({}^-4)^2 = ({}^-4)({}^-4) = 16$
     $3^2 = 3 \cdot 3 = 9$
$= 16 - 9(3)$ Multiply.
$= 16 - 27$ Change subtraction to addition. Change 27 to $^-27$.
$= 16 + {}^-27$ Add.
$= {}^-11$

(d) $({}^-3)^3 + (3-9)^2$ Parentheses first
     $3 - 9 = 3 + {}^-9 = {}^-6$
$= ({}^-3)^3 + ({}^-6)^2$ Apply exponents.
     $({}^-3)^3 = ({}^-3)({}^-3)({}^-3) = {}^-27$
     $({}^-6)^2 = ({}^-6)({}^-6) = 36$
$= {}^-27 + 36$ Add.
$= 9$

7. **(a)** Numerator:
$= {}^-3(2^3)$ Exponent first
$= {}^-3(8)$ Multiply.
$= {}^-24$

Denominator:
$= {}^-10 - 6 + 8$ Change subtraction to addition. Change 6 to $^-6$.
$= {}^-10 + {}^-6 + 8$ Add from left to right.
$= {}^-16 + 8$
$= {}^-8$

Last step is division: $\dfrac{{}^-24}{{}^-8} = 3$

(b) Numerator:
$= ({}^-10)({}^-5)$
$= 50$

Denominator:
$= {}^-6 \div 3(5)$ Multiply and divide left to right.
$= {}^-2(5)$
$= {}^-10$

Last step is division: $\dfrac{50}{{}^-10} = {}^-5$

(c) Numerator:
$= 6 + 18 \div ({}^-2)$ Divide before adding.
$= 6 + {}^-9$
$= {}^-3$

Denominator:
$= (1 - 10) \div 3$ Parentheses first
$= {}^-9 \div 3$
$= {}^-3$

Last step is division: $\dfrac{{}^-3}{{}^-3} = 1$

(d) Numerator:
$= 6^2 - 3^2(4)$ Exponents first
$= 36 - 9(4)$ Multiply before subtracting.
$= 36 - 36$
$= 0$

Denominator:
$= 5 + (3 - 7)^2$ Parentheses first
     $3 - 7 = 3 + {}^-7 = {}^-4$
$= 5 + ({}^-4)^2$ Exponent next
     $({}^-4)^2 = ({}^-4)({}^-4) = 16$
$= 5 + 16$ Addition last
$= 21$

Last step is division: $\dfrac{0}{21} = 0$

1.8 Section Exercises

1. Exponential Form: 4^3
Factored Form: $4 \cdot 4 \cdot 4$
Simplified: 64
Read as: 4 cubed or 4 to the third power

3. Exponential Form: 2^7
Factored Form: $2 \cdot 2 \cdot 2 \cdot 2 \cdot 2 \cdot 2 \cdot 2$
Simplified: 128
Read as: 2 to the seventh power

5. Exponential Form: 5^4
Factored Form: $5 \cdot 5 \cdot 5 \cdot 5$
Simplified: 625
Read as: 5 to the fourth power

7. Exponential Form: 7^2
Factored Form: $7 \cdot 7$
Simplified: 49
Read as: 7 squared

22 Chapter 1 Introduction to Algebra: Integers

9. Exponential Form: 10^1
Factored Form: 10
Simplified: 10
Read as: 10 to the first power

11. (a) $10^1 = 10$

(b) $10^2 = 10 \cdot 10 = 100$

(c) $10^3 = 10 \cdot 10 \cdot 10$
$= 100 \cdot 10$
$= 1000$

(d) $10^4 = 10 \cdot 10 \cdot 10 \cdot 10$
$= 100 \cdot 10 \cdot 10$
$= 1000 \cdot 10$
$= 10{,}000$

13. (a) $4^1 = 4$

(b) $4^2 = 4 \cdot 4 = 16$

(c) $4^3 = 4 \cdot 4 \cdot 4$
$= 16 \cdot 4$
$= 64$

(d) $4^4 = 4 \cdot 4 \cdot 4 \cdot 4$
$= 16 \cdot 4 \cdot 4$
$= 64 \cdot 4$
$= 256$

15. 5^{10} on the calculator is 9,765,625.

17. 2^{12} on the calculator is 4096.

19. $(^-2)^2 = (^-2)(^-2)$
$= 4$

21. $(^-5)^2 = (^-5)(^-5)$
$= 25$

23. $(^-4)^3 = (^-4)(^-4)(^-4)$
$= 16(^-4)$
$= ^-64$

25. $(^-3)^4 = (^-3)(^-3)(^-3)(^-3)$
$= 9(^-3)(^-3)$
$= ^-27(^-3)$
$= 81$

27. $(^-10)^3 = (^-10)(^-10)(^-10)$
$= 100(^-10)$
$= ^-1000$

29. $1^4 = 1$;   1 times itself any number
of times equals 1.

31. $3^3 \cdot 2^2$ Apply the exponents first.
$3^3 = 3 \cdot 3 \cdot 3 = 27$
$2^2 = 2 \cdot 2 = 4$
$= 27 \cdot 4$
$= 108$

33. $2^3(^-5)^2$ Apply the exponents first.
$2^3 = 2 \cdot 2 \cdot 2 = 8$
$(^-5)^2 = ^-5 \cdot ^-5 = 25$
$= 8(25)$
$= 200$

35. $6^1(^-5)^3$ Apply the exponents first.
$6^1 = 6$
$(^-5)^3 = (^-5)(^-5)(^-5) = ^-125$
$= 6(^-125)$
$= ^-750$

37. $(^-2)(^-2)^4$ Apply the exponent first.
$(^-2)^4 = (^-2)(^-2)(^-2)(^-2) = 16$
$= ^-2(16)$
$= ^-32$

39. $(^-2)^2 = \underline{4}$ $(^-2)^6 = \underline{64}$
$(^-2)^3 = \underline{^-8}$ $(^-2)^7 = \underline{^-128}$
$(^-2)^4 = \underline{16}$ $(^-2)^8 = \underline{256}$
$(^-2)^5 = \underline{^-32}$ $(^-2)^9 = \underline{^-512}$

(a) When a negative number is raised to an even power, the answer is positive; when raised to an odd power, the answer is negative.

(b) 15 is odd, so the sign of $(^-2)^{15}$ is negative. 24 is even, so the sign of $(^-2)^{24}$ is positive.

41. $12 \div 6(^-3)$ Divide.
$= 2(^-3)$ Multiply.
$= ^-6$

43. $^-1 + 15 - 7 - 7$ Add left to right.
$= 14 - 7 - 7$ Add the opposite.
$= 14 + ^-7 + ^-7$ Add left to right.
$= 7 + ^-7$ Add.
$= 0$

45. $10 - 7^2$ Exponent: $7^2 = 7 \cdot 7$
$= 10 - 49$ Add the opposite.
$= 10 + ^-49$ Add.
$= ^-39$

47. $2 - ^-5 + 3^2$ Exponent: $3^2 = 3 \cdot 3$
$= 2 - ^-5 + 9$ Add the opposite.
$= 2 + (^+5) + 9$ Add from left to right.
$= 7 + 9$ Add.
$= 16$

1.8 Exponents and Order of Operations

49. $3 + 5(6 - 2)$ Add the opposite.
$= 3 + 5(6 + {}^-2)$ Parentheses
$= 3 + 5(4)$ Multiply.
$= 3 + 20$ Add.
$= 23$

51. ${}^-7 + 6(8 - 14)$ Add the opposite.
$= {}^-7 + 6(8 + {}^-14)$ Parentheses
$= {}^-7 + 6({}^-6)$ Multiply.
$= {}^-7 + {}^-36$ Add.
$= {}^-43$

53. $2({}^-3 + 5) - (9 - 12)$ Add the opposite.
$= 2({}^-3 + 5) - (9 + {}^-12)$ Parentheses
$= 2(2) - ({}^-3)$ Multiply.
$= 4 - ({}^-3)$ Add the opposite.
$= 4 + {}^+3$ Add.
$= 7$

55. ${}^-5(7 - 13) \div {}^-10$ Add the opposite.
$= {}^-5(7 + {}^-13) \div {}^-10$ Parentheses
$= {}^-5({}^-6) \div {}^-10$ Multiply.
$= 30 \div {}^-10$ Divide.
$= {}^-3$

57. $9 \div ({}^-3)^2 + {}^-1$ Exponent first
$= 9 \div 9 + {}^-1$ Divide.
$= 1 + {}^-1$ Add.
$= 0$

59. $2 - {}^-5({}^-2)^3$ Exponent first
$= 2 - {}^-5({}^-8)$ Multiply.
$= 2 - 40$ Add the opposite.
$= 2 + {}^-40$ Add.
$= {}^-38$

61. ${}^-2({}^-7) + 3(9)$ Multiply.
$= 14 + 3(9)$ Multiply.
$= 14 + 27$ Add.
$= 41$

63. $30 \div {}^-5 - 36 \div {}^-9$ Divide.
$= {}^-6 - 36 \div {}^-9$ Divide.
$= {}^-6 - {}^-4$ Add the opposite.
$= {}^-6 + {}^+4$ Add.
$= {}^-2$

65. $2(5) - 3(4) + 5(3)$ Multiply.
$= 10 - 3(4) + 5(3)$ Multiply.
$= 10 - 12 + 5(3)$ Multiply.
$= 10 - 12 + 15$ Add the opposite.
$= 10 + {}^-12 + 15$ Add.
$= {}^-2 + 15$ Add.
$= 13$

67. $4(3^2) + 7(3 + 9) - {}^-6$ Parentheses first
$= 4(3^2) + 7(12) - {}^-6$ Exponent next
$= 4(9) + 7(12) - {}^-6$ Multiply.
$= 36 + 7(12) - {}^-6$ Multiply.
$= 36 + 84 - {}^-6$ Add the opposite.
$= 36 + 84 + ({}^+6)$ Add.
$= 120 + 6$ Add.
$= 126$

69. $({}^-4)^2 \cdot (7 - 9)^2 \div 2^3$ Add the opposite.
$= ({}^-4)^2 \cdot (7 + {}^-9)^2 \div 2^3$ Parentheses first
$= ({}^-4)^2 \cdot ({}^-2)^2 \div 2^3$ Exponents next
$= 16 \cdot 4 \div 8$ Multiply.
$= 64 \div 8$ Divide.
$= 8$

71. $\dfrac{{}^-1 + 5^2 - {}^-3}{{}^-6 - 9 + 12}$

Numerator:
${}^-1 + 5^2 - {}^-3$ Exponents first
$= {}^-1 + 25 - {}^-3$ Add the opposite.
$= {}^-1 + 25 + ({}^+3)$ Add left to right.
$= 24 + 3$
$= 27$

Denominator:
${}^-6 - 9 + 12$ Add the opposite.
$= {}^-6 + {}^-9 + 12$ Add left to right.
$= {}^-15 + 12$
$= {}^-3$

Last step is division: $\dfrac{27}{-3} = {}^-9$

73. $\dfrac{{}^-2(4^2) - 4(6 - 2)}{{}^-4(8 - 13) \div {}^-5}$

Numerator:
${}^-2(4^2) - 4(6 - 2)$ Parentheses first
$= {}^-2(4^2) - 4(4)$ Exponent
$= {}^-2(16) - 4(4)$ Multiply from left to right.
$= {}^-32 - 16$ Add the opposite.
$= {}^-32 + {}^-16$ Add.
$= {}^-48$

Denominator:
${}^-4(8 - 13) \div {}^-5$ Add the opposite.
$= {}^-4(8 + {}^-13) \div {}^-5$ Parentheses first
$= {}^-4({}^-5) \div {}^-5$ Multiply.
$= 20 \div {}^-5$ Divide.
$= {}^-4$

Last step is division: $\dfrac{{}^-48}{-4} = 12$

75. $\dfrac{2^3 \cdot (^-2 - 5) + 4(^-1)}{4 + 5(^-6 \cdot 2) + (5 \cdot 11)}$

Numerator:
$2^3 \cdot (^-2 - 5) + 4(^-1)$ Add the opposite.
$= 2^3 \cdot (^-2 + ^-5) + 4(^-1)$ Parentheses
$= 2^3 \cdot (^-7) + 4(^-1)$ Exponent
$= 8 \cdot (^-7) + 4(^-1)$ Multiply.
$= ^-56 + 4(^-1)$ Multiply.
$= ^-56 + ^-4$ Add.
$= ^-60$

Denominator:
$4 + 5(^-6 \cdot 2) + (5 \cdot 11)$ Parentheses
$= 4 + 5(^-12) + (5 \cdot 11)$ Parentheses
$= 4 + 5(^-12) + 55$ Multiply.
$= 4 + ^-60 + 55$ Add left to right.
$= ^-56 + 55$
$= ^-1$

Last step is division: $\dfrac{^-60}{^-1} = 60$

77. $5^2(9 - 11)(^-3)(^-3)^3$ Add the opposite.
$= 5^2(9 + ^-11)(^-3)(^-3)^3$ Parentheses first
$= 5^2(^-2)(^-3)(^-3)^3$ Exponents
$= 25(^-2)(^-3)(^-27)$ Mult. left to right.
$= ^-50(^-3)(^-27)$
$= 150(^-27)$
$= ^-4050$

79. $|^-12| \div 4 + 2 \cdot |(^-2)^3| \div 4$ Work inside the absolute value signs first.
$= |^-12| \div 4 + 2 \cdot |^-8| \div 4$ Absolute values
$= 12 \div 4 + 2 \cdot 8 \div 4$ Divide.
$= 3 + 2 \cdot 8 \div 4$ Multiply.
$= 3 + 16 \div 4$ Divide.
$= 3 + 4$ Add.
$= 7$

81. $\dfrac{^-9 + 18 \div ^-3(^-6)}{32 - 4(12) \div 3(2)}$

Numerator:
$^-9 + 18 \div ^-3(^-6)$ Divide.
$= ^-9 + ^-6(^-6)$ Multiply.
$= ^-9 + 36$ Add.
$= 27$

Denominator:
$32 - 4(12) \div 3(2)$ Multiply.
$= 32 - 48 \div 3(2)$ Divide.
$= 32 - 16(2)$ Multiply.
$= 32 - 32$ Add the opposite.
$= 32 + ^-32$ Add.
$= 0$

Last step is division: $\dfrac{27}{0}$ is undefined.

Chapter 1 Review Exercises

1. The whole numbers are: 86; 0; 35,600

2. 806 in words: eight hundred six

3. 319,012 in words: three hundred nineteen thousand, twelve

4. 60,003,200 in words: sixty million, three thousand, two hundred

5. 15,749,000,000,006 in words: fifteen trillion, seven hundred forty-nine billion, six

6. Five hundred four thousand, one hundred
   The first group name is *thousand*, so you need to fill *two groups* of three digits.
   $\underline{5}\,\underline{0}\,\underline{4},\underline{1}\,\underline{0}\,\underline{0} = 504{,}100$

7. Six hundred twenty million, eighty thousand
   The first group name is *million*, so you need to fill *three groups* of three digits.
   $\underline{6}\,\underline{2}\,\underline{0},\underline{0}\,\underline{8}\,\underline{0},\underline{0}\,\underline{0}\,\underline{0} = 620{,}080{,}000$

8. Ninety-nine billion, seven million, three hundred fifty-six
   The first group name is *billion,* so you need to fill *four groups* of three digits.
   $\underline{0}\,\underline{9}\,\underline{9},\underline{0}\,\underline{0}\,\underline{7},\underline{0}\,\underline{0}\,\underline{0},\underline{3}\,\underline{5}\,\underline{6} = 99{,}007{,}000{,}356$

9. Graph $^-3\tfrac{1}{2}, 2, ^-5, 0$

   ◄—•—┼•—┼—┼—•—┼—•—┼—┼—┼—►
   $^-5\ ^-4\ ^-3\ ^-2\ ^-1\ 0\ 1\ 2\ 3\ 4\ 5$

10. 0 is to the *right* of $^-4$ on the number line, so 0 is *greater than* $^-4$. Write $0 > ^-4$.

11. $^-3$ is to the *left* of $^-1$ on the number line, so $^-3$ is *less than* $^-1$. Write $^-3 < ^-1$.

12. 2 is to the *right* of $^-2$ on the number line, so 2 is *greater than* $^-2$. Write $2 > ^-2$.

13. $^-2$ is to the *left* of 1 on the number line, so $^-2$ is *less than* 1. Write $^-2 < 1$.

14. $|^-5| = 5$ because the distance from 0 to $^-5$ on the number line is 5 spaces.

15. $|9| = 9$ because the distance from 0 to 9 on the number line is 9 spaces.

16. $|0| = 0$ because the distance from 0 to 0 on the number line is 0 spaces.

17. $|{}^-125| = 125$ because the distance from 0 to ${}^-125$ on the number line is 125 spaces.

18. ${}^-9 + 8 = {}^-1$

Add *unlike* signed integers.

$|{}^-9| = 9$; $|8| = 8$; Subtract $9 - 8$ to get 1.

${}^-9$ has the larger absolute value and is negative, so the sum is negative.

19. ${}^-8 + {}^-5 = {}^-13$

Add *like* signed integers.

$|{}^-8| = 8$; $|{}^-5| = 5$; Add $8 + 5$ to get 13.

Both numbers are negative, so the sum is negative.

20. $16 + {}^-19 = {}^-3$

Add *unlike* signed integers.

$|16| = 16$; $|{}^-19| = 19$; Subtract $19 - 16$ to get 3.

${}^-19$ has the larger absolute value and is negative, so the sum is negative.

21. ${}^-4 + 4 = 0$

Addition of opposites is always zero.

22. $6 + {}^-5 = {}^+1$ or 1

23. ${}^-12 + {}^-12 = {}^-24$

24. $0 + {}^-7 = {}^-7$

25. ${}^-16 + 19 = {}^+3$ or 3

26. $9 + {}^-4 + {}^-8 + 3$ *Add from left to right.*
$= 5 + {}^-8 + 3$
$= {}^-3 + 3$
$= 0$

27. ${}^-11 + {}^-7 + 5 + {}^-4$ *Add from left to right.*
$= {}^-18 + 5 + {}^-4$
$= {}^-13 + {}^-4$
$= {}^-17$

28. The opposite of ${}^-5$ is 5. ${}^-5 + 5 = 0$

29. The opposite of 18 is ${}^-18$. $18 + {}^-18 = 0$

30. $5 - 12$ *Change subtraction to addition. Change 12 to ${}^-12$.*
$= 5 + {}^-12$
$= {}^-7$

31. $24 - 7$ *Add the opposite.*
$= 24 + {}^-7$
$= 17$

32. ${}^-12 - 4$ *Add the opposite.*
$= {}^-12 + {}^-4$
$= {}^-16$

33. $4 - {}^-9$ *Add the opposite.*
$= 4 + {}^+9$
$= 13$

34. ${}^-12 - {}^-30$ *Add the opposite.*
$= {}^-12 + {}^+30$
$= 18$

35. ${}^-8 - 14$ *Add the opposite.*
$= {}^-8 + {}^-14$
$= {}^-22$

36. ${}^-6 - {}^-6$ *Add the opposite.*
$= {}^-6 + {}^+6$
$= 0$

37. ${}^-10 - 10$ *Add the opposite.*
$= {}^-10 + {}^-10$
$= {}^-20$

38. ${}^-8 - {}^-7$ *Add the opposite.*
$= {}^-8 + {}^+7$
$= {}^-1$

39. $0 - 3$ *Add the opposite.*
$= 0 + {}^-3$
$= {}^-3$

40. $1 - {}^-13$ *Add the opposite.*
$= 1 + {}^+13$
$= 14$

41. $15 - 0$ *Add the opposite.*
$= 15 + 0$
$= 15$

42. $3 - 12 - 7$ *Add the opposites.*
$= 3 + {}^-12 + {}^-7$ *Add left to right.*
$= {}^-9 + {}^-7$
$= {}^-16$

43. ${}^-7 - {}^-3 + 7$ *Add the opposite.*
$= {}^-7 + {}^+3 + 7$
$= {}^-7 + 7 + {}^+3$ *Commutative property*
$= 0 + {}^+3$
$= 3$

44. $4 + {}^-2 - 0 - 10$ *Add the opposites.*
$= 4 + {}^-2 + 0 + {}^-10$ *Add left to right.*
$= 2 + 0 + {}^-10$
$= 2 + {}^-10$
$= {}^-8$

45. ${}^-12 - 12 + 20 - {}^-4$ *Add the opposites.*
$= {}^-12 + {}^-12 + 20 + {}^+4$ *Add left to right.*
$= {}^-24 + 20 + 4$
$= {}^-4 + 4$
$= 0$

46. $2\underline{0}5 \approx 210$

Underline the tens place. The next digit is 5 or more. Add 1 to 0. Change 5 to 0.

47. $5\underline{9},499 \approx 59,000$

Underline the thousands place. The next digit is 4 or less. Leave 9 as 9. Change all digits to the right of the underlined place to zeros.

48. $8\underline{5},066,000 \approx 85,000,000$

Underline the millions place. The next digit is 4 or less. Leave 5 as 5. Change all digits to the right of the underlined place to zeros.

49. ${}^-2\underline{9}63 \approx {}^-3000$

Underline the hundreds place. The next digit is 5 or more. Add 1 to 9. Write 0 and carry the one to the thousands place. Change all digits to the right of the underlined place to zeros.

50. ${}^-7,0\underline{6}3,885 \approx {}^-7,060,000$

Underline the ten-thousands place. The next digit is 4 or less. Leave 6 as 6. Change all digits to the right of the underlined place to zeros.

51. $39\underline{9},712 \approx 400,000$

Underline the thousands place. The next digit is 5 or more. Add 1 to 9. Write 0 and carry the one to the ten-thousands place: $9 + 1 = 10$. Write 0 and carry the one to the hundred-thousands place: $3 + 1 = 4$. Change all digits to the right of the underlined place to zeros.

52. Weight loss implies a negative number:

$${}^-1\underline{9}7 \text{ pounds} \approx {}^-200 \text{ pounds}$$

Underline the first digit. The next digit is 5 or more. Add 1 to 1. Change all digits to the right of the underlined place to zeros.

53. Below sea level implies a negative number:

$${}^-\underline{1}312 \approx {}^-1000 \text{ feet}$$

Underline the first digit. The next digit is 4 or less. Leave 1 as 1. Change all digits to the right of the underlined place to zeros.

54. $\underline{3}62,000,000$ directories $\approx 400,000,000$ directories

Underline the first digit. The next digit is 5 or more. Add 1 to 3. Change all digits to the right of the underlined place to zeros.

55. $\underline{9},402,000,000$ people $\approx 9,000,000,000$ people

Underline the first digit. The next digit is 4 or less. Leave 9 as 9. Change all digits to the right of the underlined place to zeros.

56. ${}^-6(9) = {}^-54$ (*different* signs, product is *negative*)

57. $({}^-7)({}^-8) = 56$ (*same* signs, product is *positive*)

58. $10({}^-10) = {}^-100$
(*different* signs, product is *negative*)

59. ${}^-45 \cdot 0 = 0$; multiplication property of 0

60. ${}^-1({}^-24) = 24$ (*same* signs, product is *positive*)

61. $17 \cdot 1 = 17$; multiplication property of 1

62. $4({}^-12) = {}^-48$
(*different* signs, product is *negative*)

63. $({}^-5)({}^-25) = 125$ (*same* signs, product is *positive*)

64. ${}^-3({}^-4)({}^-3)$ *Multiply from left to right.*
$= 12 \cdot {}^-3$
$= {}^-36$

65. ${}^-5(2)({}^-5)$ *Multiply from left to right.*
$= {}^-10 \cdot {}^-5$
$= 50$

66. $({}^-8)({}^-1)({}^-9)$ *Multiply from left to right.*
$= 8({}^-9)$
$= {}^-72$

67. $\dfrac{{}^-63}{{}^-7} = 9$ (*same* signs, quotient is *positive*)

68. $\dfrac{70}{{}^-10} = {}^-7$ (*different* signs, quotient is *negative*)

69. $\dfrac{{}^-15}{0}$ is undefined. Division by zero is undefined.

70. ${}^-100 \div {}^-20 = 5$ (*same* signs, quotient is *positive*)

71. $18 \div {}^-1 = {}^-18$
(*different* signs, quotient is *negative*)

72. $\dfrac{0}{12} = 0$; 0 divided by any nonzero number is 0.

73. $\dfrac{^{-}30}{^{-}2} = 15$ (*same* signs, quotient is *positive*)

74. $\dfrac{^{-}35}{35} = {}^{-}1$ (*different* signs, quotient is *negative*)

75. ${}^{-}40 \div {}^{-}4 \div {}^{-}2$ *Divide from left to right.*
$= 10 \div {}^{-}2$
$= {}^{-}5$

76. ${}^{-}18 \div 3({}^{-}3)$ *Divide.*
$= {}^{-}6({}^{-}3)$ *Multiply.*
$= 18$

77. $0 \div {}^{-}10(5) \div 5$ *Divide.*
$= 0(5) \div 5$ *Multiply.*
$= 0 \div 5$ *Divide.*
$= 0$

78. 1250 hours are separated into 8-hour days.

$$\begin{array}{r} 156 \\ 8\overline{)1250} \\ \underline{8} \\ 45 \\ \underline{40} \\ 50 \\ \underline{48} \\ 2 \end{array}$$

It took 156 work days of 8 hours each, plus 2 extra hours.

79. $10^4 = 10 \cdot 10 \cdot 10 \cdot 10 = 10{,}000$

80. $2^5 = 2 \cdot 2 \cdot 2 \cdot 2 \cdot 2 = 32$

81. $3^3 = 3 \cdot 3 \cdot 3 = 27$

82. $({}^{-}4)^2 = ({}^{-}4)({}^{-}4) = 16$

83. $({}^{-}5)^3 = ({}^{-}5)({}^{-}5)({}^{-}5)$
$= (25)({}^{-}5)$
$= {}^{-}125$

84. $8^1 = 8$

85. $6^2 \cdot 3^2$ *Exponents*
$= 36 \cdot 9$
$= 324$

86. $5^2({}^{-}2)^3$ *Exponents*
$= 25({}^{-}8)$
$= {}^{-}200$

87. ${}^{-}30 \div 6 - 4(5)$ *Divide.*
$= {}^{-}5 - 4(5)$ *Multiply.*
$= {}^{-}5 - 20$ *Add the opposite.*
$= {}^{-}5 + {}^{-}20$ *Add.*
$= {}^{-}25$

88. $6 + 8(2 - 3)$ *Parentheses first*
$\;\; 2 - 3 = 2 + {}^{-}3$
$= 6 + 8({}^{-}1)$ *Multiply.*
$= 6 + {}^{-}8$ *Add.*
$= {}^{-}2$

89. $16 \div 4^2 + ({}^{-}6 + 9)^2$ *Parentheses first*
$= 16 \div 4^2 + (3)^2$ *Exponents*
$= 16 \div 16 + 9$ *Divide.*
$= 1 + 9$ *Add.*
$= 10$

90. ${}^{-}3(4) - 2(5) + 3({}^{-}2)$ *Multiply left to right.*
$= {}^{-}12 - 10 + {}^{-}6$ *Add the opposite.*
$= {}^{-}12 + {}^{-}10 + {}^{-}6$ *Add left to right.*
$= {}^{-}22 + {}^{-}6$
$= {}^{-}28$

91. $\dfrac{{}^{-}10 + 3^2 - {}^{-}9}{3 - 10 - 1}$

Numerator:
${}^{-}10 + 3^2 - {}^{-}9$ *Exponent*
$= {}^{-}10 + 9 - {}^{-}9$ *Add the opposite.*
$= {}^{-}10 + 9 + {}^{+}9$ *Add from left to right.*
$= {}^{-}1 + {}^{+}9$
$= 8$

Denominator:
$3 - 10 - 1$ *Add the opposites.*
$= 3 + {}^{-}10 + {}^{-}1$ *Add from left to right.*
$= {}^{-}7 + {}^{-}1$
$= {}^{-}8$

Last step is division: $\dfrac{8}{^{-}8} = {}^{-}1$

92. $\dfrac{{}^{-}1(1 - 3)^3 + 12 \div 4}{{}^{-}5 + 24 \div 8 \cdot 2(6 - 6) + 5}$

Numerator:
${}^{-}1(1 - 3)^3 + 12 \div 4$ *Parentheses first*
$\phantom{= {}^{-}1(1 - 3)^3 + 12 \div 4}\;\; 1 - 3 = 1 + {}^{-}3 = {}^{-}2$
$= {}^{-}1({}^{-}2)^3 + 12 \div 4$ *Exponent*
$= {}^{-}1({}^{-}8) + 12 \div 4$ *Multiply.*
$= 8 + 12 \div 4$ *Divide.*
$= 8 + 3$ *Add.*
$= 11$

continued

Denominator:
$$-5 + 24 \div 8 \cdot 2(6-6) + 5 \quad \text{Parentheses first}$$
$$= -5 + 24 \div 8 \cdot 2(0) + 5 \quad \text{Divide.}$$
$$= -5 + 3 \cdot 2(0) + 5 \quad \text{Multiply.}$$
$$= -5 + 6(0) + 5 \quad \text{Multiply.}$$
$$= -5 + 0 + 5 \quad \text{Add.}$$
$$= -5 + 5 \quad \text{Add.}$$
$$= 0$$

Last step is division: $\frac{11}{0}$ is undefined.

93. [1.3] Associative property of addition
94. [1.6] Commutative property of multiplication
95. [1.3] Addition property of 0
96. [1.6] Multiplication property of 0
97. [1.6] Distributive property
98. [1.6] Associative property of multiplication
99. [1.6] 192 rounds to 200.
    $11,900 rounds to $10,000.
    Estimate: $10,000 \cdot 200 = \$2,000,000$ total value
    Exact: $11,900 \cdot 192 = \$2,284,800$ total value
100. [1.3] Account balance of $185 rounds to $200.
    The deposit of $428 rounds to $400.
    The check for $706 rounds to $700.
    Estimate: $\$200 + \$400 + {}^-\$700 = {}^-\$100$
    Exact: $\$185 + \$428 + {}^-\$706 = {}^-\$93$
101. [1.7] 22 gallons rounds to 20 gallons.
    880 miles rounds to 900 miles.
    Divide miles by gallons to get miles per gallon.
    Estimate: $900 \div 20 = 45$ miles for each gallon
    Exact: $880 \div 22 = 40$ miles for each gallon
102. [1.6] 19 calculators rounds to 20.
    12 modems rounds to 10.
    $39 rounds to $40. $85 rounds to $90.
    Estimate:
    $(\$40 \cdot 20) + (\$90 \cdot 10) = \$800 + \$900 = \$1700$
    Exact:
    $(\$39 \cdot 19) + (\$85 \cdot 12) = \$741 + \$1020 = \$1761$
103. [1.3] Expenses imply negative numbers.
    Jan. $\$2400 + {}^-\$3100 = {}^-\$700$ (loss)
    Feb. $\$1900 + {}^-\$2000 = {}^-\$100$ (loss)
    Mar. $\$2500 + {}^-\$1800 = \$700$ (profit)
    Apr. $\$2300 + {}^-\$1400 = \$900$ (profit)
    May $\$1600 + {}^-\$1600 = \$0$ (neither)
    June $\$1900 + {}^-\$1200 = \$700$ (profit)
104. [1.3] January had the greatest loss.
    April had the greatest profit.

105. [1.7] To find her average monthly income, add her income from each month and divide by the number of months.
$$\frac{\$2400 + \$1900 + \$2500 + \$2300 + \$1600 + \$1900}{6}$$
$$= \frac{\$12,600}{6} = \$2100$$

106. [1.7] To find her average monthly expenses, add her expenses from each month and divide by the number of months.
$$\frac{\$3100 + \$2000 + \$1800 + \$1400 + \$1600 + \$1200}{6}$$
$$= \frac{\$11,100}{6} = \$1850$$

Chapter 1 Test

1. 20,008,307 in words: twenty million, eight thousand, three hundred seven

2. Thirty billion, seven hundred thousand, five

   The first group name is *billion*, so you need to fill *four groups* of three digits.

   $\underline{0\,3\,0},\underline{0\,0\,0},\underline{7\,0\,0},\underline{0\,0\,5} = 30,000,700,005$

3. Graph 3, $^-2$, 0, $^-\frac{1}{2}$

   $\xleftarrow{\quad\bullet\quad\bullet\bullet\quad\bullet\quad}\rightarrow$
   $^-3\ ^-2\ ^-1\ \ 0\ \ 1\ \ 2\ \ 3$

4. 0 is to the *right* of $^-3$ on the number line, so 0 is *greater than* $^-3$. Write $0 > {}^-3$.

   $^-2$ is to the *left* of $^-1$ on the number line, so $^-2$ is *less than* $^-1$. Write $^-2 < {}^-1$.

5. $|10| = 10$ because the distance from 0 to 10 on the number line is 10 spaces.

   $|{}^-14| = 14$ because the distance from 0 to $^-14$ on the number line is 14 spaces.

6. $3 - 9$ Add the opposite.
   $= 3 + {}^-9$
   $= {}^-6$

7. $^-12 + 7 = {}^-5$

8. $\frac{^-28}{^-4} = 7$ (*same* signs, quotient is *positive*)

9. $^-1(40) = {}^-40$
   (*different* signs, product is *negative*)

10. $^-5 - {}^-15$ Change subtraction to addition.
    $= {}^-5 + {}^+15$ Add.
    $= 10$

11. $({}^-8)({}^-8) = 64$ (*same* signs, product is *positive*)

12. $^-25 + {}^-25$ Add.
    $= {}^-50$

13. $\frac{17}{0}$ is undefined.

14. $^-30 - 30$ *Change subtraction to addition.*
    $= {}^-30 + {}^-30$ *Add.*
    $= {}^-60$

15. $\frac{50}{-10} = {}^-5$ (*different* signs, quotient is *negative*)

16. $5({}^-9) = {}^-45$ (*different* signs, product is *negative*)

17. $0 - {}^-6$ *Change subtraction to addition.*
    $= 0 + {}^+6$ *Addition property of zero*
    $= 6$

18. $^-35 \div 7(-5)$ *Divide.*
    $= {}^-5({}^-5)$ *Multiply.*
    $= 25$

19. $^-15 - {}^-8 + 7$ *Change subtraction to addition.*
    $= {}^-15 + {}^+8 + 7$ *Add from left to right.*
    $= {}^-7 + 7$
    $= 0$

20. $3 - 7({}^-2) - 8$ *Multiply.*
    $= 3 - {}^-14 - 8$ *Change subtraction to addition.*
    $= 3 + {}^+14 + {}^-8$ *Add from left to right.*
    $= 17 + {}^-8$
    $= 9$

21. $({}^-4)^2 \cdot 2^3$ *Exponents*
    $= 16 \cdot 8$
    $= 128$

22. $\frac{5^2 - 3^2}{(4)({}^-2)}$

    Numerator:
    $5^2 - 3^2$ *Exponents*
    $= 25 - 9$ *Add the opposite.*
    $= 25 + {}^-9$ *Add.*
    $= 16$

    Denominator:
    $(4)({}^-2)$ *Multiply.*
    $= {}^-8$

    Last step is division: $\frac{16}{-8} = {}^-2$

23. $^-2({}^-4 + 10) + 5(4)$ *Parentheses*
    $= {}^-2(6) + 5(4)$ *Multiply.*
    $= {}^-12 + 5(4)$ *Multiply.*
    $= {}^-12 + 20$ *Add.*
    $= 8$

24. $^-3 + ({}^-7 - {}^-10) + 4(6 - 10)$ *Parentheses*
    $= {}^-3 + ({}^-7 + {}^+10) + 4(6 + {}^-10)$
    $= {}^-3 + 3 + 4({}^-4)$ *Multiply.*
    $= {}^-3 + 3 + {}^-16$ *Add.*
    $= 0 + {}^-16$ *Add.*
    $= {}^-16$

25. An exponent shows how many times to use a factor in repeated multiplication. Examples will vary. Some possibilities are
    $(2)^4 = 2 \cdot 2 \cdot 2 \cdot 2 = 16$; $({}^-3)^2 = ({}^-3)({}^-3) = 9$.

26. Commutative property: changing the *order* of addends does not change the sum.

    One possible example: $2 + 5 = 5 + 2$

    Associative property: changing the *grouping* of addends does not change the sum.

    One possible example:
    $({}^-1 + 4) + 2 = {}^-1 + (4 + 2)$

27. 8$\underline{5}$1 ≈ 900

    Underline the hundreds place. The next digit is 5 or more. Add 1 to 8. Change all digits to the right of the underlined place to zeros.

28. 36,4$\underline{2}$0,498,725 ≈ 36,420,000,000

    Underline the millions place. The next digit is 4 or less. Leave 0 as 0. Change all digits to the right of the underlined place to zeros.

29. 34$\underline{9}$,812 ≈ 350,000

    Underline the thousands place. The next digit is 5 or more. Add 1 to 9. Write 0 and carry 1 to the ten-thousands place. Change all digits to the right of the underlined place to zeros.

30. $184 account balance rounds to $200.
    The $293 deposit rounds to $300.
    The $506 tuition check rounds to $500.

    Balance in her account:
    Estimate: $200 + $300 + ${}^-$500 = $0
    Exact: $184 + $293 + ${}^-$506 = ${}^-$29

31. Loss of 1140 yards rounds to 1000 yds.
    12 games rounds to 10 games.

    Average loss in each game:
    Estimate: ${}^-1000 \div 10 = {}^-100$ yards
    Exact: ${}^-1140 \div 12 = {}^-95$ yards

32. Cereal 1: 220 calories rounds to 200
Cereal 2: 110 calories rounds to 100
31 days rounds to 30

Estimate of calories saved by eating Cereal 2:
$200 - 100 = 100$ calories per day
$30 \cdot 100 = 3000$ calories in the month

Exact calories saved by eating Cereal 2:
$220 - 110 = 110$ calories per day
$31 \cdot 110 = 3410$ calories in the month

33. The difference between the high and low temperatures on Mars is:

$^{-}10 - {^{-}100}$ *Add the opposite.*
$= {^{-}10} + {^{+}100}$
$= 90$ degrees difference

34. 1276 books separated into cartons holding 48 books each.

$$\begin{array}{r} 26 \\ 48\overline{\smash{)}1276} \\ \underline{96} \\ 316 \\ \underline{288} \\ 28 \end{array}$$

Anthony will need 27 cartons to ship all of the books. He will have 26 full boxes and one box with only 28 books in it.

CHAPTER 2 UNDERSTANDING VARIABLES AND SOLVING EQUATIONS

2.1 Introduction to Variables

2.1 Margin Exercises

1. The expression is $c + 15$.

   The variable c represents the class limit.
   The constant is 15.

2. **(a)** Evaluate the expression $c + 3$ when c is 25.

   $c + 3$ Replace c with 25.
   $25 + 3$

   28 books must be ordered.

   (b) Evaluate the expression $c + 3$ when c is 60.

   $c + 3$ Replace c with 60.
   $60 + 3$

   63 books must be ordered.

3. **(a)** Evaluate the expression $4s$ when s is 3 feet.

   $4s$ Replace s with 3 feet.
   $4 \cdot 3$ feet
   12 feet

   The perimeter of the square table is 12 feet.

   (b) Evaluate the expression $4s$ when s is 7 miles.

   $4s$ Replace s with 7 miles.
   $4 \cdot 7$ miles
   28 miles

   The perimeter of the square park is 28 miles.

4. Evaluate the expression $100 + \dfrac{a}{2}$ when a is 40.

   $100 + \dfrac{a}{2}$ Replace a with 40.
   $100 + \dfrac{40}{2}$ Divide.
   $100 + 20$ Add.
   120

   The approximate systolic blood pressure is 120.

5. **(a)** Evaluate the expression $\dfrac{t}{g}$ when t is 532 and g is 4.

   $\dfrac{t}{g}$ Replace t with 532 and g with 4.
   $\dfrac{532}{4}$ Divide.
   133

   The average score is 133.

   (b)

Value of x	Value of y	Expression $x - y$
100	5	$100 - 5$ is 95
3	7	$3 - 7$ is $^-4$
8	0	$8 - 0$ is 8

6. **(a)** Multiplying any number a by 0 gives a product of 0.

   $a \cdot 0 = 0$ or $0 \cdot a = 0$

   (b) Changing the grouping of addends (a, b, c) does not change the sum.

   $(a + b) + c = a + (b + c)$

7. **(a)** x^5 can be written as $x \cdot x \cdot x \cdot x \cdot x$

   (b) $4a^2b^2$ can be written as $4 \cdot a \cdot a \cdot b \cdot b$

   (c) $^-10xy^3$ can be written as $^-10 \cdot x \cdot y \cdot y \cdot y$

   (d) s^4tu^2 can be written as $s \cdot s \cdot s \cdot s \cdot t \cdot u \cdot u$

8. **(a)** y^3 means

   $y \cdot y \cdot y$ Replace y with $^-5$.
   $^-5 \cdot {^-5} \cdot {^-5}$ Multiply left to right.
   $25 \cdot {^-5}$
   $^-125$

   (b) r^2s^2 means

   $r \cdot r \cdot s \cdot s$ Replace r with 6 and s with 3.
   $6 \cdot 6 \cdot 3 \cdot 3$ Multiply left to right.
   $36 \cdot 3 \cdot 3$
   $108 \cdot 3$
   324

   (c) $10xy^2$ means

   $10 \cdot x \cdot y \cdot y$ Replace x with 4 and y with $^-3$.
   $10 \cdot 4 \cdot {^-3} \cdot {^-3}$ Multiply left to right.
   $40 \cdot {^-3} \cdot {^-3}$
   $^-120 \cdot {^-3}$
   360

   (d) $^-3c^4$ means

   $^-3 \cdot c \cdot c \cdot c \cdot c$ Replace c with 2.
   $^-3 \cdot 2 \cdot 2 \cdot 2 \cdot 2$ Multiply left to right.
   $^-6 \cdot 2 \cdot 2 \cdot 2$
   $^-12 \cdot 2 \cdot 2$
   $^-24 \cdot 2$
   $^-48$

2.1 Section Exercises

1. $c + 4$ c is the variable.
   4 is the constant.

3. $^-3 + m$ m is the variable.
$^-3$ is the constant.

5. $5h$ h is the variable.
5 is the coefficient.

7. $2c - 10$ c is the variable.
2 is the coefficient.
10 is the constant.

9. $x - y$ Both x and y are variables.

11. $^-6g + 9$ g is the variable.
$^-6$ is the coefficient.
9 is the constant.

13. Expression (rule) for ordering robes: $g + 10$

   (a) Evaluate the expression when there are 654 graduates.

   $g + 10$ Replace g with 654.
   $\underline{654 + 10}$ Follow the rule and add.
   664 robes must be ordered.

   (b) Evaluate the expression when there are 208 graduates.

   $g + 10$ Replace g with 208.
   $\underline{208 + 10}$ Follow the rule and add.
   218 robes must be ordered.

   (c) Evaluate the expression when there are 95 graduates.

   $g + 10$ Replace g with 95.
   $\underline{95 + 10}$ Follow the rule and add.
   105 robes must be ordered.

15. Expression (rule) for finding perimeter of equilateral triangle of side length s: $3s$

   (a) Evaluate the expression when s, the side length, is 11 inches.

   $3s$ Replace s with 11.
   $\underline{3 \cdot 11}$ Follow the rule and multiply.
   33 inches is the perimeter.

   (b) Evaluate the expression when s, the side length, is 3 feet.

   $3s$ Replace s with 3.
   $\underline{3 \cdot 3}$ Follow the rule and multiply.
   9 feet is the perimeter.

17. Expression (rule) for ordering brushes: $3c - 5$

   (a) Evaluate the expression when c, the class size, is 12.

   $3c - 5$ Replace c with 12.
   $\underline{3 \cdot 12} - 5$ Multiply before subtracting.
   $\underline{36 - 5}$
   31 brushes must be ordered.

   (b) Evaluate the expression when c, the class size, is 16.

   $3c - 5$ Replace c with 16.
   $\underline{3 \cdot 16} - 5$ Multiply before subtracting.
   $\underline{48 - 5}$
   43 brushes must be ordered.

19. Expression (rule) for average test score, where p is the total points and t is the number of tests: p/t

   (a) Evaluate the expression when p, the total points, is 332 and t, the number of tests, is 4.

   $\dfrac{p}{t}$ Replace p with 332 and t with 4.
   $\dfrac{332}{4}$ Follow the rule and divide.
   83 points is the average test score.

   (b) Evaluate the expression when p, the total points, is 637 and t, the number of tests, is 7.

   $\dfrac{p}{t}$ Replace p with 637 and t with 7.
   $\dfrac{637}{7}$ Follow the rule and divide.
   91 points is the average test score.

21.

Value of x	Expression $x + x + x + x$	Expression $4x$
12	$12 + 12 + 12 + 12$ is 48	$4 \cdot 12$ is 48
0	$0 + 0 + 0 + 0$ is 0	$4 \cdot 0$ is 0
$^-5$	$^-5 + ^-5 + ^-5 + ^-5$ is $^-20$	$4 \cdot ^-5$ is $^-20$

23.

Value of x	Value of y	Expression $^-2x + y$
$^-4$	5	$^-2(^-4) + 5$ is $8 + 5$, or 13
$^-6$	$^-2$	$^-2(^-6) + ^-2$ is $12 + ^-2$, or 10
0	$^-8$	$^-2(0) + ^-8$ is $0 + ^-8$, or $^-8$

25. A variable is a letter that represents the part of a rule that varies or changes depending on the situation. An expression expresses, or tells, the rule for doing something. For example, $c + 5$ is an expression, and c is the variable.

27. Let b represent "a number." Multiplying a number by 1 leaves the number unchanged.

$$b \cdot 1 = b \quad \text{or} \quad 1 \cdot b = b$$

29. Let b represent "any number." Any number divided by 0 is undefined.

$$\frac{b}{0} \text{ is undefined} \quad \text{or} \quad b \div 0 \text{ is undefined.}$$

31. c^6 written without exponents is

$$c \cdot c \cdot c \cdot c \cdot c \cdot c$$

33. $x^4 y^3$ written without exponents is

$$x \cdot x \cdot x \cdot x \cdot y \cdot y \cdot y$$

35. $^-3a^3b$ can be written as $^-3 \cdot a \cdot a \cdot a \cdot b$. The exponent 3 applies only to the base a.

37. $9xy^2$ can be written as $9 \cdot x \cdot y \cdot y$. The exponent 2 applies only to the base y.

39. $^-2c^5 d$ can be written as $^-2 \cdot c \cdot c \cdot c \cdot c \cdot c \cdot d$. The exponent 5 applies only to the base c.

41. $a^3 b c^2$ can be written as $a \cdot a \cdot a \cdot b \cdot c \cdot c$. The exponent 3 applies only to the base a. The exponent 2 applies only to the base c.

43. Evaluate t^2 when t is $^-4$.

t^2 means
$t \cdot t$ Replace t with $^-4$.
$\underline{^-4 \cdot {}^-4}$ Multiply.
16

45. Evaluate rs^3 when r is $^-3$ and s is 2.

rs^3 means
$r \cdot s \cdot s \cdot s$ Replace r with $^-3$ and s with 2.
$\underline{^-3 \cdot 2} \cdot 2 \cdot 2$ Multiply left to right.
$\underline{^-6 \cdot 2} \cdot 2$
$\underline{^-12 \cdot 2}$
$^-24$

47. Evaluate $3rs$ when r is $^-3$ and s is 2.

$3rs$ means
$3 \cdot r \cdot s$ Replace r with $^-3$ and s with 2.
$\underline{3 \cdot {}^-3} \cdot 2$ Multiply left to right.
$\underline{^-9 \cdot 2}$
$^-18$

49. Evaluate $^-2s^2 t^2$ when s is 2 and t is $^-4$.

$^-2s^2 t^2$ means
$^-2 \cdot s \cdot s \cdot t \cdot t$ Replace s with 2 and t with $^-4$.
$\underline{^-2 \cdot 2} \cdot 2 \cdot {}^-4 \cdot {}^-4$ Multiply left to right.
$\underline{^-4 \cdot 2} \cdot {}^-4 \cdot {}^-4$
$\underline{^-8 \cdot {}^-4} \cdot {}^-4$
$\underline{32 \cdot {}^-4}$
$^-128$

51. Evaluate $r^2 s^5 t^3$ when r is $^-3$, s is 2, and t is $^-4$, using a calculator.

$r^2 s^5 t^3$ Replace r with $^-3$, s with 2, and t with $^-4$.
$\underline{(^-3)^2 (2)^5 (^-4)^3}$ Use the y^x key.
$\underline{(9)(32)(^-64)}$ Multiply left to right.
$\underline{(288)(^-64)}$
$^-18{,}432$

53. Evaluate $^-10 r^5 s^7$ when r is $^-3$ and s is 2, using a calculator.

$^-10 r^5 s^7$ Replace r with $^-3$ and s with 2.
$^-10 \underline{(^-3)^5} \underline{(2)^7}$ Use the y^x key.
$\underline{^-10(^-243)} (128)$ Multiply left to right.
$\underline{2430(128)}$
$311{,}040$

55. Evaluate $|xy| + |xyz|$ when x is 4, y is $^-2$, and z is $^-6$.

$|xy| + |xyz|$ Replace x with 4, y with $^-2$, and z with $^-6$.
$|\underline{4 \cdot {}^-2}| + |\underline{4 \cdot {}^-2} \cdot {}^-6|$ Multiply left to right within the absolute value signs.
$|{}^-8| + |\underline{^-8 \cdot {}^-6}|$
$|{}^-8| + |48|$ Evaluate the absolute values.
$\underline{8 + 48}$ Add.
56

57. Evaluate $\dfrac{z^2}{-3y+z}$ when z is $^-6$ and y is $^-2$.

$\dfrac{z^2}{-3y+z}$ Replace z with $^-6$ and y with $^-2$.

$\dfrac{(^-6)^2}{-3(^-2)+^-6}$ Follow the order of operations.

$\dfrac{36}{0}$ Numerator: $(^-6)^2 = ^-6 \cdot ^-6 = 36$
Denominator: $^-3(^-2) + ^-6 = 6 + ^-6$

Undefined   Division by 0 is undefined.

59. (a) Evaluate $\dfrac{s}{5}$ when s is 15 seconds.

$\dfrac{s}{5}$ Replace s with 15.

$\dfrac{15}{5}$ Divide.

3 miles

(b) Evaluate $\dfrac{s}{5}$ when s is 10 seconds.

$\dfrac{s}{5}$ Replace s with 10.

$\dfrac{10}{5}$ Divide.

2 miles

(c) Evaluate $\dfrac{s}{5}$ when s is 5 seconds.

$\dfrac{s}{5}$ Replace s with 5.

$\dfrac{5}{5}$ Divide.

1 mile

60. (a) Using part (c) of Exercise 59, the distance covered in $2\tfrac{1}{2}$ seconds is half of the distance covered in 5 seconds, or $\tfrac{1}{2}$ mile.

(b) Using part (a) of Exercise 59, the time to cover $1\tfrac{1}{2}$ miles is half the time to cover 3 miles, or $7\tfrac{1}{2}$ seconds. Or, using parts (b) and (c), find the number halfway between 5 seconds and 10 seconds

(c) Using parts (a) and (b) of Exercise 59, find the number halfway between 10 seconds and 15 seconds; that is $12\tfrac{1}{2}$ seconds.

2.2 Simplifying Expressions

2.2 Margin Exercises

1. (a) $3b^2 + {}^-3b + 3 + b^3 + b$

The like terms are ^-3b and b since the variable parts match; both are b.

The coefficients are $^-3$ and 1.

(b) $^-4xy + 4x^2y + {}^-4xy^2 + {}^-4 + 4$

The like terms are the constants, $^-4$ and 4. There are no variable parts.

(c) $5r^2 + 2r + {}^-2r^2 + 5 + 5r^3$

The like terms are $5r^2$ and $^-2r^2$ since the variable parts match; both are r^2.

The coefficients are 5 and $^-2$.

(d) $^-10 + {}^-x + {}^-10x + {}^-x^2 + {}^-10y$

The like terms are ^-x and ^-10x since the variable parts match; both are x.

The coefficients are $^-1$ and $^-10$.

2. (a) $10b + 4b + 10b$ These are like terms.
$(10 + 4 + 10)b$ Add the coefficients.
$24b$ The variable part, b, stays the same.

(b) $y^3 + 8y^3$ These are like terms.
$1y^3 + 8y^3$ Rewrite y^3 as $1y^3$.
$(1 + 8)y^3$ Add the coefficients.
$9y^3$ The variable part, y^3, stays the same.

(c) $^-7n - n$ These are like terms.
$^-7n - 1n$ Rewrite n as $1n$.
$^-7n + {}^-1n$ Change to addition.
$(^-7 + {}^-1)n$ Add the coefficients.
^-8n The variable part, n, stays the same.

(d) $3c - 5c - 4c$ These are like terms.
$3c + {}^-5c + {}^-4c$ Change to addition.
$(3 + {}^-5 + {}^-4)c$ Add the coefficients.
^-6c The variable part, c, stays the same.

(e) $^-9xy + xy$ These are like terms.
$^-9xy + 1xy$ Rewrite xy as $1xy$.
$(^-9 + 1)xy$ Add the coefficients.
^-8xy The variable part, xy, stays the same.

(f) $^-4p^2 - 3p^2 + 8p^2$ These are like terms.
$^-4p^2 + {}^-3p^2 + 8p^2$ Change to addition.
$(^-4 + {}^-3 + 8)p^2$ Add the coefficients.
$1p^2$ or p^2 The variable part, p^2, stays the same.

2.2 Simplifying Expressions

(g) $ab - ab$ — These are like terms.
$1ab - 1ab$ — Rewrite ab as $1ab$.
$1ab + {}^-1ab$ — Change to addition.
$(1 + {}^-1)ab$ — Add the coefficients.
$0ab$ — Zero times anything is zero.
0

3. (a) $3b^2 + 4d^2 + 7b^2$ — Rewrite using the commutative property.
$3b^2 + 7b^2 + 4d^2$ — Combine $3b^2 + 7b^2$.
$(3 + 7)b^2 + 4d^2$ — Add the coefficients.
$10b^2 + 4d^2$

(b) $4a + b - 6a + b$ — Change to addition.
$4a + b + {}^-6a + b$ — Rewrite using the commutative property.
$4a + {}^-6a + b + b$ — Rewrite b as $1b$.
$4a + {}^-6a + 1b + 1b$
$(4 + {}^-6)a + (1 + 1)b$ — Add the coefficients of like terms.
${}^-2a + 2b$

(c) ${}^-6x + 5 + 6x + 2$ — Rewrite using the commutative property.
${}^-6x + 6x + 5 + 2$ — Add the coefficients of like terms.
$({}^-6 + 6)x + (5 + 2)$
$0x + 7$ — Zero times anything is zero.
$0 + 7$
7

(d) $2y - 7 - y + 7$ — Change to addition.
$2y + {}^-7 + {}^-y + 7$ — Rewrite ${}^-y$ as ${}^-1y$.
$2y + {}^-7 + {}^-1y + 7$ — Rewrite using the commutative property.
$2y + {}^-1y + {}^-7 + 7$ — Add the coefficients of like terms.
$(2 + {}^-1)y + ({}^-7 + 7)$
$1y + 0$
$1y$ or y

(e) ${}^-3x - 5 + 12 + 10x$ — Change to addition.
${}^-3x + {}^-5 + 12 + 10x$ — Rewrite using the commutative property.
${}^-3x + 10x + {}^-5 + 12$ — Add the coefficients of like terms.
$({}^-3 + 10)x + ({}^-5 + 12)$
$7x + 7$

4. (a) $7(4c)$ can be written as
$\underbrace{(7 \cdot 4)} \cdot c$
$\underbrace{28 \cdot c}$
$28c$

(b) ${}^-3(5y^3)$ can be written as
$\underbrace{({}^-3 \cdot 5)} \cdot y^3$
$\underbrace{{}^-15 \cdot y^3}$
${}^-15y^3$

(c) $20({}^-2a)$ can be written as
$\underbrace{(20 \cdot {}^-2)} \cdot a$
$\underbrace{{}^-40 \cdot a}$
${}^-40a$

(d) ${}^-10({}^-x)$ — Rewrite ${}^-x$ as ${}^-1x$.
${}^-10({}^-1x)$ can be written as
$\underbrace{({}^-10 \cdot {}^-1)} \cdot x$
$\underbrace{10 \cdot x}$
$10x$

5. (a) $7(a + 10)$ can be written as
$7 \cdot a + \underbrace{7 \cdot 10}$
$7a + 70$

(b) $3(x - 3)$ can be written as
$3 \cdot x - \underbrace{3 \cdot 3}$
$3x - 9$

(c) $4(2y + 6)$ can be written as
$4 \cdot 2y + \underbrace{4 \cdot 6}$
$\underbrace{4 \cdot 2} \cdot y + 24$
$\underbrace{8 \cdot y} + 24$
$8y + 24$

(d) ${}^-5(3b + 2)$
$\underbrace{{}^-5 \cdot 3b} + \underbrace{{}^-5 \cdot 2}$
${}^-15b + {}^-10$ — Change addition to subtraction.
${}^-15b - 10$

(e) ${}^-8(c + 4)$
$\underbrace{{}^-8 \cdot c} + \underbrace{{}^-8 \cdot 4}$
${}^-8c + {}^-32$ — Change addition to subtraction.
${}^-8c - 32$

6. (a) ${}^-4 + 5(y + 1)$ — Distributive property
${}^-4 + 5 \cdot y + 5 \cdot 1$
${}^-4 + 5y + 5$ — Rewrite using the commutative property.
$\underbrace{{}^-4 + 5} + 5y$ — Combine constants.
$1 + 5y$ or $5y + 1$

(b) $2(3w + 4) - 5$ Distributive property
$\underline{2 \cdot 3w} + \underline{2 \cdot 4} - 5$ Multiply.
$6w + \underbrace{8 - 5}$ Combine constants.
$6w + 3$

(c) $5(6x - 2) + 3x$ Distributive property
$\underline{5 \cdot 6x} - \underline{5 \cdot 2} + 3x$
$30x - 10 + 3x$ Change to addition.
$30x + {}^-10 + 3x$ Rewrite using the commutative property.
$30x + 3x + {}^-10$
$(30 + 3)x + {}^-10$ Add the coefficients of like terms.
$33x + {}^-10$ or $33x - 10$

(d) $21 + 7(a^2 - 3)$ Distributive property
$21 + 7 \cdot a^2 - \underline{7 \cdot 3}$
$21 + 7a^2 - 21$ Change to addition.
$21 + 7a^2 + {}^-21$ Rewrite using the commutative property.
$\underbrace{21 + {}^-21} + 7a^2$ Combine constants.
$\underbrace{0 + 7a^2}$
$7a^2$

(e) ${}^-y + 3(2y + 5) - 18$ Distributive property
${}^-y + \underline{3 \cdot 2y} + \underline{3 \cdot 5} - 18$ Rewrite ${}^-y$ as ${}^-1y$.
${}^-1y + 6y + 15 + {}^-18$ Change to addition.
$\underbrace{({}^-1 + 6)}y + (15 + {}^-18)$ Add the coefficients of like terms.
$5y + {}^-3$ or $5y - 3$

2.2 Section Exercises

1. $2b^2$ and b^2 are the only like terms in the expression. The variable parts match; both are b^2. The coefficients are 2 and 1.

3. ${}^-xy$ and $2xy$ are the like terms in the expression. The variable parts match; both are xy. The coefficients are ${}^-1$ and 2.

5. 7, 3, and ${}^-4$ are like terms. There are no variable parts; constants are considered like terms.

7. $6r + 6r$ These are like terms.
   Add the coefficients.
$(6 + 6)r$
$12r$ The variable part, r, stays the same.

9. $x^2 + 5x^2$ These are like terms.
   Rewrite x^2 as $1x^2$.
$1x^2 + 5x^2$ Add the coefficients.
$(1 + 5)x^2$
$6x^2$ The variable part, x^2, stays the same.

11. $p - 5p$ These are like terms.
   Rewrite p as $1p$.
$1p - 5p$ Change to addition.
$1p + {}^-5p$ Add the coefficients.
$(1 + {}^-5)p$
${}^-4p$ The variable part, p, stays the same.

13. ${}^-2a^3 - a^3$ These are like terms.
   Rewrite a^3 as $1a^3$.
${}^-2a^3 - 1a^3$ Change to addition.
${}^-2a^3 + {}^-1a^3$ Add the coefficients.
$({}^-2 + {}^-1)a^3$
${}^-3a^3$ The variable part, a^3, stays the same.

15. $\underline{c - c}$
0 Any number minus itself is 0.

17. $9xy + xy - 9xy$ These are like terms.
   Rewrite xy as $1xy$.
$9xy + 1xy - 9xy$ Change to addition.
$9xy + 1xy + {}^-9xy$ Add the coefficients.
$(9 + 1 + {}^-9)xy$
$1xy$ or xy The variable part, xy, stays the same.

19. $5t^4 + 7t^4 - 6t^4$ These are like terms.
   Change to addition.
$5t^4 + 7t^4 + {}^-6t^4$ Add the coefficients.
$(5 + 7 + {}^-6)t^4$
$6t^4$ The variable part, t^4, stays the same.

21. $y^2 + y^2 + y^2 + y^2$ These are like terms.
   Write in the understood coefficient of 1.
$1y^2 + 1y^2 + 1y^2 + 1y^2$
$(1 + 1 + 1 + 1)y^2$ Add the coefficients.
$4y^2$ The variable part, y^2, stays the same.

2.2 Simplifying Expressions

23. $^-x - 6x - x$ — These are like terms. Rewrite ^-x as ^-1x and x as $1x$.
$^-1x - 6x - 1x$ — Change to addition.
$^-1x + ^-6x + ^-1x$ — Add the coefficients.
$(^-1 + ^-6 + ^-1)x$
^-8x — The variable part, x, stays the same.

25. $8a + 4b + 4a$ — Use the commutative property to rewrite the expression so that like terms are next to each other.
$8a + 4a + 4b$ — Add the coefficients of like terms.
$(8 + 4)a + 4b$
$12a + 4b$ — The variable part, a, stays the same.

27. $6 + 8 + 7rs$ — Use the commutative property to put the constants at the end.
$7rs + 6 + 8$ — Add the coefficients of like terms.
$7rs + 14$ — The only like terms are constants.

29. $a + ab^2 + ab^2$ — Write in the understood coefficient of 1.
$1a + 1ab^2 + 1ab^2$ — Add the coefficients of like terms.
$1a + (1 + 1)ab^2$
$1a + 2ab^2$ — The variable part, ab^2, stays the same.
or
$a + 2ab^2$

31. $6x + y - 8x + y$ — Write in the understood coefficients of 1. Change to addition. Rewrite using the commutative property. Add the coefficients of like terms.
$6x + 1y + ^-8x + 1y$
$6x + ^-8x + 1y + 1y$
$\underbrace{(6 + ^-8)}x + \underbrace{(1 + 1)}y$
$^-2x \quad + \quad ^+2y$
$^-2x + 2y$

33. $8b^2 - a^2 - b^2 + a^2$ — Write in the understood coefficient of 1.
$8b^2 - 1a^2 - 1b^2 + 1a^2$ — Change to addition.
$8b^2 + ^-1a^2 + ^-1b^2 + 1a^2$ — Rewrite using the commutative property.
$8b^2 + ^-1b^2 + ^-1a^2 + 1a^2$ — Add the coefficients of like terms.
$\underbrace{(8 + ^-1)}b^2 + \underbrace{(^-1 + 1)}a^2$
$7b^2 \quad + \quad 0 \cdot a^2$
$7b^2 + 0$
$7b^2$

35. $^-x^3 + 3x - 3x^2 + 2$
There are no like terms. The expression cannot be simplified.

37. $^-9r + 6t - s - 5r + s + t - 6t + 5s - r$
Write in the understood coefficient of 1.
Change to addition.
$^-9r + 6t + ^-1s + ^-5r + 1s + 1t + ^-6t + 5s + ^-1r$
Rewrite using the commutative property.
$^-9r + ^-5r + ^-1r + ^-1s + 1s + 5s + 6t + 1t + ^-6t$
Add the coefficients of like terms.
$\underbrace{(^-9 + ^-5 + ^-1)}r + \underbrace{(^-1 + 1 + 5)}s + \underbrace{(6 + 1 + ^-6)}t$
$^-15r \quad + \quad ^+5s \quad + \quad ^+1t$
$^-15r + 5s + t$

39. By using the associative property, we can write $3(10a)$ as
$$(3 \cdot 10) \cdot a = 30 \cdot a = 30a.$$
So, $3(10a)$ simplifies to $30a$.

41. By using the associative property, we can write $^-4(2x^2)$ as
$$(^-4 \cdot 2) \cdot x^2 = ^-8 \cdot x^2 = ^-8x^2.$$
So, $^-4(2x^2)$ simplifies to $^-8x^2$.

43. By using the associative property, we can write $5(^-4y^3)$ as
$$(5 \cdot ^-4) \cdot y^3 = ^-20 \cdot y^3 = ^-20y^3.$$
So, $5(^-4y^3)$ simplifies to $^-20y^3$.

45. By using the associative property, we can write $^-9(^-2cd)$ as
$$(^-9 \cdot ^-2) \cdot c \cdot d = 18 \cdot c \cdot d = 18cd.$$
So, $^-9(^-2cd)$ simplifies to $18cd$.

47. By using the associative property, we can write $7(3a^2bc)$ as
$$(7 \cdot 3) \cdot a^2 \cdot b \cdot c = 21 \cdot a^2 \cdot b \cdot c = 21a^2bc.$$
So, $7(3a^2bc)$ simplifies to $21a^2bc$.

Chapter 2 Understanding Variables and Solving Equations

49. $^-12(^-w)$ Write in the understood coefficient of $^-1$.
$^-12(^-1w)$ Rewrite using the associative property.
$(^-12 \cdot {}^-1)w$
$12 \cdot w$
$12w$

51. $6(b+6)$ Distributive property
$6 \cdot b + 6 \cdot 6$
$6b + 36$

53. $7(x-1)$ Distributive property
$7 \cdot x - 7 \cdot 1$
$7x - 7$

55. $3(7t+1)$ Distributive property
$3 \cdot 7t + 3 \cdot 1$
$21t + 3$

57. $^-2(5r+3)$ Distributive property
$^-2 \cdot 5r + {}^-2 \cdot 3$
$^-10r + {}^-6$ Change addition to subtraction of the opposite.
$^-10r - 6$

59. $^-9(k+4)$ Distributive property
$^-9 \cdot k + {}^-9 \cdot 4$
$^-9k + {}^-36$ Change addition to subtraction of the opposite.
$^-9k - 36$

61. $50(m-6)$ Distributive property
$50 \cdot m - 50 \cdot 6$
$50m - 300$

63. $10 + 2(4y+3)$ Distributive property
$10 + 2 \cdot 4y + 2 \cdot 3$
$10 + 8y + 6$ Rewrite using the commutative property.
$8y + 10 + 6$ Combine like terms.
$8y + 16$

65. $6(a^2 - 2) + 15$ Distributive property
$6 \cdot a^2 - 6 \cdot 2 + 15$
$6a^2 - 12 + 15$ Combine like terms.
$6a^2 + 3$

67. $2 + 9(m-4)$ Distributive property
$2 + 9 \cdot m - 9 \cdot 4$
$2 + 9m - 36$ Change to addition.
$2 + 9m + {}^-36$ Rewrite using the commutative property.
$9m + 2 + {}^-36$ Add the coefficients of like terms.
$9m + {}^-34$ Change addition to subtraction of the opposite.
$9m - 34$

69. $^-5(k+5) + 5k$ Distributive property
$^-5 \cdot k + {}^-5 \cdot 5 + 5k$
$^-5k + {}^-25 + 5k$ Rewrite using the commutative property.
$^-5k + 5k + {}^-25$ Add the coefficients of like terms.
$(^-5 + 5)k + {}^-25$
$0k + {}^-25$ Zero times any number is 0.
$0 + {}^-25$ Zero added to any number is the number
$^-25$

71. $4(6x - 3) + 12$ Distributive property
$4 \cdot 6x - 4 \cdot 3 + 12$
$24x - 12 + 12$ Change to addition.
$24x + {}^-12 + 12$ Combine like terms. Any number plus its opposite is 0.
$24x + 0$
$24x$

73. $5 + 2(3n + 4) - n$ Distributive property
$5 + 2 \cdot 3n + 2 \cdot 4 - n$ Rewrite n as $1n$.
$5 + 6n + 8 - 1n$ Change to addition.
$5 + 6n + 8 + {}^-1n$ Rewrite using the commutative property.
$5 + 8 + 6n + {}^-1n$ Add the coefficients of like terms.
$(5 + 8) + (6 + {}^-1)n$
$13 + 5n$ or $5n + 13$

75. $^-p + 6(2p - 1) + 5$ Distributive property
$^-p + 6 \cdot 2p - 6 \cdot 1 + 5$

$^-p + 12p - 6 + 5$ Rewrite ^-p as ^-1p. Change to addition. Add the coefficients of like terms.

$^-1p + 12p + {}^-6 + 5$

$(^-1 + 12)p + (^-6 + 5)$

$11p + {}^-1$ Change addition to subtraction of the opposite.

$11p - 1$

77. A simplified expression still has variables, but is written in a simpler way. When evaluating an expression, the variables are all replaced by specific numbers and the final result is a numerical answer.

79. Like terms have matching variable parts, that is, matching letters and exponents. The coefficients do not have to match. Examples will vary. Possible examples: In $^-6x + 9 + x$, the terms ^-6x and x are like terms. In $4k + 3 - 8k^2 + 10$, the terms 3 and 10 are like terms.

81. $\underbrace{{}^-2x + 7x}_{5x} + 8$
$5x + 8$

Keep the variable part unchanged when combining like terms.

83. $^-4(3y) - 5 + 2(5y + 7)$ Distributive property
$^-4 \cdot 3y - 5 + 2 \cdot 5y + 2 \cdot 7$

$^-12y - 5 + 10y + 14$ Change subtraction to adding the opposite. Group like terms and add the coefficients.

$^-12y + {}^-5 + 10y + 14$

$\underbrace{{}^-12y + 10y}_{{}^-2y} + \underbrace{{}^-5 + 14}_{9}$
$^-2y + 9$

85. $^-10 + 4(^-3b + 3) + 2(6b - 1)$
Distributive property
$^-10 + 4 \cdot {}^-3b + 4 \cdot 3 + 2 \cdot 6b - 2 \cdot 1$
$^-10 + {}^-12b + 12 + 12b - 2$
Change to addition.
$^-10 + {}^-12b + 12 + 12b + {}^-2$
Group like terms and add the coefficients.
$\underbrace{{}^-12b + 12b}_{0b} + \underbrace{{}^-10 + 12 + {}^-2}_{0}$
0

87. $^-5(^-x + 2) + 8(^-x) + 3(^-2x - 2) + 16$
Distributive property
$^-5 \cdot {}^-x + {}^-5 \cdot 2 + 8 \cdot {}^-x + 3 \cdot {}^-2x - 3 \cdot 2 + 16$
$5x + {}^-10 + {}^-8x + {}^-6x - 6 + 16$
Change to addition.
$5x + {}^-10 + {}^-8x + {}^-6x + {}^-6 + 16$
Group like terms and add the coefficients.
$\underbrace{5x + {}^-8x + {}^-6x}_{{}^-9x} + \underbrace{{}^-10 + {}^-6 + 16}_{0}$
^-9x

Summary Exercises on Variables and Expressions

1. $^-10 - m$ m is the variable.
$^-10$ is the constant.

3. $6 + 4x$ x is the variable.
4 is the coefficient.
6 is the constant.

5. Expression (rule) for finding the total cost of a car with down payment d, monthly payment m, and number of payments t: $d + mt$

(a) Evaluate the expression when the down payment is $3000, the monthly payment is $280, and the number of payments is 36.

$d + mt$ Replace d with $3000, m with $280, and t with 36.

$\$3000 + \$280 \cdot 36$ Multiply before adding.

$\$3000 + \$10{,}080$

$13,080 is the total cost of the car.

(b) Evaluate the expression when the down payment is $1750, the monthly payment is $429, and the number of payments is 48.

$d + mt$ Replace d with $1750, m with $429, and t with 48.

$\$1750 + \$429 \cdot 48$ Multiply before adding.

$\$1750 + \$20{,}592$

$22,342 is the total cost of the car.

7. b^3cd written without exponents is
$b \cdot b \cdot b \cdot c \cdot d$

9. $w^4 = w \cdot w \cdot w \cdot w$

Replace w with 5.

$\underbrace{5 \cdot 5}_{25} \cdot 5 \cdot 5$ Multiply left to right.

$\underbrace{25 \cdot 5}_{125} \cdot 5$

$\underbrace{125 \cdot 5}_{625}$

11. yz^2

    Replace y with $^-6$ and z with 0.
    If 0 is multiplied by any number, the result is 0. Thus, there is no need to make any calculations since the result is 0.

13. $x^3 = x \cdot x \cdot x$

    Replace x with $^-2$.

    $\underline{^-2 \cdot {}^-2} \cdot {}^-2$ Multiply left to right.
    $\underline{4 \cdot {}^-2}$
    ${}^-8$

15. $3xy^2 = 3 \cdot x \cdot y \cdot y$

    Replace x with $^-2$ and y with $^-6$.

    $\underline{3 \cdot {}^-2} \cdot {}^-6 \cdot {}^-6$ Multiply left to right.
    $\underline{{}^-6 \cdot {}^-6} \cdot {}^-6$
    $\underline{36 \cdot {}^-6}$
    ${}^-216$

17. $^-7wx^4y^3$ Use a calculator.
    Replace w with 5, x with $^-2$, and y with $^-6$.
    $^-7(5)(^-2)^4(^-6)^3 = ^-35(16)(^-216) = 120{,}960$

19. $^-3x - 5 + 12 + 10x = {}^-3x + 10x + {}^-5 + 12$
    $ = ({}^-3 + 10)x + ({}^-5 + 12)$
    $ = 7x + 7$

21. $^-9xy + 9xy = ({}^-9 + 9)xy$
    $ = 0xy$
    $ = 0$

23. $3f - 5f - 4f = 3f + {}^-5f + {}^-4f$
    $ = (3 + {}^-5 + {}^-4)f$
    $ = {}^-6f$

25. $^-a - 6b - a = {}^-a + {}^-6b + {}^-a$
    $ = {}^-a + {}^-a + {}^-6b$
    $ = {}^-1a + {}^-1a + {}^-6b$
    $ = ({}^-1 + {}^-1) \cdot a + {}^-6b$
    $ = {}^-2a + {}^-6b$
     or $^-2a - 6b$

27. $5r^2 + 2r - 2r^2 + 5r^3$
    $= 5r^3 + 5r^2 + {}^-2r^2 + 2r$
    $= 5r^3 + (5 + {}^-2)r^2 + 2r$
    $= 5r^3 + 3r^2 + 2r$

29. $^-3(m + 3) + 3m = {}^-3 \cdot m + {}^-3 \cdot 3 + 3m$
    $ = {}^-3m + 3m + {}^-9$
    $ = ({}^-3 + 3) \cdot m + {}^-9$
    $ = 0m + {}^-9$
    $ = 0 + {}^-9$
    $ = {}^-9$

31. $2 + 12(3x - 1) = 2 + 12 \cdot 3x - 12 \cdot 1$
    $ = 2 + (12 \cdot 3) \cdot x - 12$
    $ = 2 + {}^-12 + 36x$
    $ = {}^-10 + 36x$
     or $36x - 10$

33. **(a)** Simplifying the expression correctly:

    $6(n + 2) = 6 \cdot n + 6 \cdot 2$
    $ = 6n + 12$

    The student forgot to multiply $6 \cdot 2$.

    (b) Simplifying the expression correctly:

    $^-5(^-4a) = ({}^-5 \cdot {}^-4) \cdot a$
    $ = 20a$

    Two negative factors give a *positive* product.

    (c) Simplifying the expression correctly:

    $3y + 2y - 10 = (3 + 2)y - 10$
    $ = 5y - 10$

    Keep the variable part unchanged; that is, adding ys to ys gives an answer with ys, not y^2s.

2.3 Solving Equations Using Addition

2.3 Margin Exercises

1. **(a)** $c + 15 = 80$ Given equation
    $95 + 15 \stackrel{?}{=} 80$ Replace c with 95.
    $110 \neq 80$
    No, 95 is not the solution.
    $65 + 15 \stackrel{?}{=} 80$ Replace c with 65.
    $80 = 80$
    Yes, 65 is the solution. (No need to check 70)

    (b) $28 = c - 4$ Given equation
    $28 \stackrel{?}{=} 20 - 4$ Replace c with 20.
    $28 \neq 16$
    No, 20 is not the solution.
    $28 \stackrel{?}{=} 24 - 4$ Replace c with 24.
    $28 \neq 20$
    No, 24 is not the solution.
    $28 \stackrel{?}{=} 32 - 4$ Replace c with 32.
    $28 = 28$
    Yes, 32 is the solution.

2. **(a)** Solve $12 = y + 5$.

To get y by itself add the opposite of 5, $^-5$, to both sides.

$$\begin{aligned} 12 &= y + 5 \\ ^-5 & ^-5 \\ \hline 7 &= y + 0 \\ 7 &= y \quad \text{The solution is 7.} \end{aligned}$$

Check: $12 = y + 5$ Replace y with 7.
$12 = 7 + 5$
$12 = 12$ Balances

(b) Solve $b - 2 = {^-6}$.

Change to addition.

$b + {^-2} = {^-6}$

To get b by itself add the opposite of $^-2$, $^+2$, to both sides.

$$\begin{aligned} b + {^-2} &= {^-6} \\ {^+2} & {^+2} \\ \hline b + 0 &= {^-4} \\ b &= {^-4} \quad \text{The solution is } {^-4}. \end{aligned}$$

Check: $b - 2 = {^-6}$ Replace b with $^-4$.
$^-4 - 2 = {^-6}$
$^-4 + {^-2} = {^-6}$
$^-6 = {^-6}$ Balances

3. **(a)** $2 - 8 = k - 2$

Rewrite both sides by changing subtraction to addition. Combine like terms.

$2 + {^-8} = k + {^-2}$
$^-6 = k + {^-2}$

To get k by itself add the opposite of $^-2$, $^+2$, to both sides.

$$\begin{aligned} ^-6 &= k + {^-2} \\ {^+2} & {^+2} \\ \hline ^-4 &= k \quad \text{The solution is } {^-4}. \end{aligned}$$

Check: $2 - 8 = k - 2$ Replace k with $^-4$.
$2 - 8 = {^-4} - 2$
$2 + {^-8} = {^-4} + {^-2}$
$^-6 = {^-6}$ Balances

(b) $4r + 1 - 3r = {^-8} + 11$

Change to addition.

$4r + 1 + {^-3r} = {^-8} + 11$

Rewrite the left side by using the commutative property.

$$\begin{aligned} 4r + {^-3r} + 1 &= {^-8} + 11 \quad \text{Combine like terms.} \\ 1r + 1 &= 3 \quad \text{To get } r \text{ by itself add} \\ ^-1 ^-1 & {^-1} \text{ to both sides.} \\ \hline 1r + 0 &= 2 \\ 1r &= 2 \\ \text{or} \quad r &= 2 \quad \text{The solution is 2.} \end{aligned}$$

Check:
$4r + 1 - 3r = {^-8} + 11$ Replace r with 2.
$4 \cdot 2 + 1 - 3 \cdot 2 = {^-8} + 11$
$8 + 1 - 6 = 3$
$9 - 6 = 3$
$3 = 3$ Balances

2.3 Section Exercises

1. $n - 50 = 8$ Given equation
$58 - 50 \stackrel{?}{=} 8$ Replace n with 58.
$8 = 8$
Yes, 58 is the solution.
(No need to check 42 and 60.)

3. $^-6 = y + 10$ Given equation
$^-6 \stackrel{?}{=} {^-4} + 10$ Replace y with $^-4$.
$^-6 \neq 6$
No, $^-4$ is not the solution.
$^-6 \stackrel{?}{=} {^-16} + 10$ Replace y with $^-16$.
$^-6 = {^-6}$
Yes, $^-16$ is the solution.

5. $t + 12 = 0$ Given equation
$0 + 12 \stackrel{?}{=} 0$ Replace t with 0.
$12 \neq 0$
No, 0 is not the solution.
$^-12 + 12 \stackrel{?}{=} 0$ Replace t with $^-12$.
$0 = 0$
Yes, $^-12$ is the solution.

7. $p + 5 = 9$ Add the opposite of 5, $^-5$, to both sides.
$$\begin{aligned} ^-5 & ^-5 \\ \hline p + 0 &= 4 \\ p &= 4 \quad \text{The solution is 4.} \end{aligned}$$

Check: $p + 5 = 9$ Replace p with 4.
$4 + 5 = 9$
$9 = 9$ Balances

9. $8 = r - 2$ Change to addition.
$8 = r + {}^-2$ Add the opposite of $^-2$, 2, to both sides.
$$\frac{2\ \ 2}{10 = r + 0}$$
$10 = r$ The solution is 10.

Check: $8 = r - 2$ Replace r with 10.
$8 = 10 - 2$
$8 = 8$ Balances

11. $^-5 = n + 3$ Add the opposite of 3, $^-3$, to both sides.
$$\frac{{}^-3\ \ {}^-3}{{}^-8 = n + 0}$$
$^-8 = n$ The solution is $^-8$.

Check: $^-5 = n + 3$ Replace n with $^-8$.
$^-5 = {}^-8 + 3$
$^-5 = {}^-5$ Balances

13. $^-4 + k = 14$ Add the opposite of $^-4$, 4, to both sides.
$$\frac{4\ \ 4}{0 + k = 18}$$
$k = 18$ The solution is 18.

Check: $^-4 + k = 14$ Replace k with 18.
$^-4 + 18 = 14$
$14 = 14$ Balances

15. $y - 6 = 0$ Change to addition.
$y + {}^-6 = 0$ Add the opposite of $^-6$, 6, to both sides.
$$\frac{\phantom{y+{}^-6=}6\ \ 6}{y + 0 = 6}$$
$y = 6$ The solution is 6.

Check: $y - 6 = 0$ Replace y with 6.
$6 - 6 = 0$ Change to addition.
$6 + {}^-6 = 0$
$0 = 0$ Balances

17. $7 = r + 13$ Add the opposite of 13, $^-13$, to both sides.
$$\frac{{}^-13\ \ {}^-13}{{}^-6 = r + 0}$$
$^-6 = r$ The solution is $^-6$.

Check: $7 = r + 13$ Replace r with $^-6$.
$7 = {}^-6 + 13$
$7 = 7$ Balances

19. $x - 12 = {}^-1$ Change to addition.
$x + {}^-12 = {}^-1$ Add the opposite of $^-12$, 12, to both sides.
$$\frac{\phantom{x+{}^-12=}12\ \ 12}{x + 0 = 11}$$
$x = 11$ The solution is 11.

Check: $x - 12 = {}^-1$ Replace x with 11.
$11 - 12 = {}^-1$ Change to addition.
$11 + {}^-12 = {}^-1$
$^-1 = {}^-1$ Balances

21. $^-5 = {}^-2 + t$ Add the opposite of $^-2$, 2, to both sides.
$$\frac{2\phantom{={}^-2+t}\ \ 2}{{}^-3 = 0 + t}$$
$^-3 = t$ The solution is $^-3$.

Check: $^-5 = {}^-2 + t$ Replace t with $^-3$.
$^-5 = {}^-2 + {}^-3$
$^-5 = {}^-5$ Balances

23. The given solution is $^-2$.

Check: $z - 5 = 3$ Replace z with $^-2$.
$^-2 - 5 = 3$ Change to addition.
$^-2 + {}^-5 = 3$
$^-7 \neq 3$ Does not balance

Correct solution:
$z - 5 = 3$ Change to addition.
$z + {}^-5 = 3$ Add the opposite of $^-5$, 5, to both sides.
$$\frac{\phantom{z+{}^-5=}5\ \ 5}{z + 0 = 8}$$
$z = 8$ The solution is 8.

Check: $z - 5 = 3$ Replace z with 8.
$8 - 5 = 3$ Change to addition.
$8 + {}^-5 = 3$
$3 = 3$ Balances

25. The given solution is $^-18$.

Check: $7 + x = {}^-11$ Replace x with $^-18$.
$7 + {}^-18 = {}^-11$
$^-11 = {}^-11$ Balances

$^-18$ is the correct solution.

27. The given solution is 10.

Check: $^-10 = ^-10 + b$ Replace b with 10.
$^-10 = ^-10 + 10$
$^-10 \neq 0$ Does not balance

Correct solution:

$^-10 = ^-10 + b$ Add the opposite of $^-10$, 10, to both sides.
$\underline{10 10}$
$0 = 0 + b$
$0 = b$ The solution is 0.

Check: $^-10 = ^-10 + b$ Replace b with 0.
$^-10 = ^-10 + 0$
$^-10 = ^-10$ Balances

29. $c - 4 = ^-8 + 10$ Simplify the right side.
$c - 4 = 2$ Change to addition.
$c + ^-4 = 2$ Add 4 to both sides.
$\underline{4 4}$
$c + 0 = 6$
$c = 6$ The solution is 6.

Check: $c - 4 = ^-8 + 10$ Replace c with 6.
$6 - 4 = ^-8 + 10$
$2 = 2$ Balances

31. $^-1 + 4 = y - 2$ Simplify the left side.
$3 = y - 2$ Change to addition.
$3 = y + ^-2$ Add 2 to both sides.
$\underline{2 2}$
$5 = y + 0$
$5 = y$ The solution is 5.

Check: $^-1 + 4 = y - 2$ Replace y with 5.
$^-1 + 4 = 5 - 2$ Change to addition.
$^-1 + 4 = 5 + ^-2$
$3 = 3$ Balances

33. $10 + b = ^-14 - 6$ Change to addition.
$10 + b = ^-14 + ^-6$ Add.
$10 + b = ^-20$ Add $^-10$
$\underline{^-10 ^-10}$ to both sides.
$0 + b = ^-30$
$b = ^-30$ The solution is $^-30$.

Check:
$10 + b = ^-14 - 6$ Replace b with $^-30$.
$10 + ^-30 = ^-14 + ^-6$
$^-20 = ^-20$ Balances

35. $t - 2 = 3 - 5$ Change to addition.
$t + ^-2 = 3 + ^-5$ Simplify the right side.
$t + ^-2 = ^-2$ Add 2 to both sides.
$\underline{2 2}$
$t + 0 = 0$
$t = 0$ The solution is 0.

Check: $t - 2 = 3 - 5$ Replace t with 0.
$0 - 2 = 3 - 5$ Change to addition.
$0 + ^-2 = 3 + ^-5$
$^-2 = ^-2$ Balances

37. $10z - 9z = ^-15 + 8$ Change to addition.
$10z + ^-9z = ^-15 + 8$ Combine like terms.
$1z = ^-7$ $1z$ is the same as z.
$z = ^-7$ The solution is $^-7$.

Check:
$10z - 9z = ^-15 + 8$ Replace z with $^-7$.
$10 \cdot ^-7 - 9 \cdot ^-7 = ^-15 + 8$
$^-70 - ^-63 = ^-7$ Change to addition.
$^-70 + 63 = ^-7$
$^-7 = ^-7$ Balances

39. $^-5w + 2 + 6w = ^-4 + 9$ Rearrange and combine like terms.
$\underbrace{^-5w + 6w} + 2 = \underbrace{^-4 + 9}$
$1w + 2 = 5$ Add $^-2$ to both sides.
$\underline{^-2 ^-2}$
$1w + 0 = 3$
$w = 3$ The solution is 3.

Check:
$^-5w + 2 + 6w = ^-4 + 9$ Replace w with 3.
$^-5 \cdot 3 + 2 + 6 \cdot 3 = ^-4 + 9$
$^-15 + 2 + 18 = ^-4 + 9$
$5 = 5$ Balances

41. $^-3 - 3 = 4 - 3x + 4x$ Change to addition.
$\underbrace{^-3 + ^-3} = 4 + \underbrace{^-3x + 4x}$ Combine like terms.
$^-6 = 4 + 1x$ Add $^-4$ to both sides.
$\underline{^-4 ^-4}$
$^-10 = 0 + 1x$
$^-10 = 1x$ $1x$ is the same as x.
$^-10 = x$ The solution is $^-10$.

43. $^-3 + 7 - 4 = ^-2a + 3a$ Change to addition.
$^-3 + 7 + ^-4 = ^-2a + 3a$ Combine like terms.
$0 = 1a$
$0 = a$ The solution is 0.

45.
$y - 75 = {}^-100$ Change to addition.
$y + {}^-75 = {}^-100$ Add 75 to both sides.
        75   75
$\overline{}$
$y + 0 = {}^-25$
$ y = {}^-25$ The solution is ${}^-25$.

47. ${}^-x + 3 + 2x = 18$ Rearrange and combine like terms.
$1x + 3 = 18$ Add ${}^-3$ to both sides.
    ${}^-3$ ${}^-3$
$\overline{}$
$1x + 0 = 15$
$x = 15$ The solution is 15.

49. $82 = {}^-31 + k$ Add 31 to both sides.
   31   31
$\overline{}$
$113 = 0 + k$
$113 = k$ The solution is 113.

51. ${}^-2 + 11 = 2b - 9 - b$ Change to addition.
${}^-2 + 11 = 2b + {}^-9 + {}^-b$ Rearrange and combine like terms.
$9 = 1b + {}^-9$ Add 9 to both sides.
$9 9$
$\overline{}$
$18 = 1b + 0$
$18 = b$ The solution is 18.

53. $r - 6 = 7 - 10 - 8$ Change to addition.
$r + {}^-6 = 7 + {}^-10 + {}^-8$ Combine like terms.
$r + {}^-6 = {}^-11$ Add 6 to both sides.
   6   6
$\overline{}$
$r + 0 = {}^-5$
$r = {}^-5$ The solution is ${}^-5$.

55. ${}^-14 = n + 91$ Add ${}^-91$ to both sides.
  ${}^-91$ ${}^-91$
$\overline{}$
${}^-105 = n + 0$
${}^-105 = n$ The solution is ${}^-105$.

57. ${}^-9 + 9 = 5 + h$ Combine like terms.
$0 = 5 + h$ Add ${}^-5$ to both sides.
  ${}^-5$ ${}^-5$
$\overline{}$
${}^-5 = 0 + h$
${}^-5 = h$ The solution is ${}^-5$.

59. No, the solution is ${}^-14$, the number used to replace x in the original equation.

61. $g + 10 = 305$ Add the opposite of 10, ${}^-10$, to both sides.
  ${}^-10$ ${}^-10$
$\overline{}$
$g + 0 = 295$
$g = 295$

There were 295 graduates this year.

63. $92 = c + 37$ Add ${}^-37$ to both sides.
  ${}^-37$ ${}^-37$
$\overline{}$
$55 = c + 0$
$55 = c$

When the temperature is 92 degrees, a field cricket chirps 55 times (in 15 seconds).

65. $p - 65 = 45$ Change to addition.
$p + {}^-65 = 45$ Add 65 to both sides.
  65   65
$\overline{}$
$p + 0 = 110$
$p = 110$

Ernesto's parking fees average $110 per month in winter.

67. ${}^-17 - 1 + 26 - 38 = {}^-3 - m - 8 + 2m$
Change all subtractions to additions.
Write the understood coefficient of 1.
${}^-17 + {}^-1 + 26 + {}^-38 = {}^-3 + {}^-1m + {}^-8 + 2m$
Use the commutative property to group like terms on the right side.
${}^-17 + {}^-1 + 26 + {}^-38 = {}^-3 + {}^-8 + {}^-1m + 2m$
Combine like terms on each side.
${}^-30 = {}^-11 + 1m$
To get m by itself, add the opposite of ${}^-11$, 11, to both sides.
${}^-30 = {}^-11 + 1m$
  11   11
$\overline{}$
${}^-19 = 1m$
${}^-19 = m$ The solution is ${}^-19$.

69. ${}^-6x + 2x + 6 + 5x = |0 - 9| - |{}^-6 + 5|$
Change subtraction within absolute value to addition and rearrange the terms.
${}^-6x + 2x + 5x + 6 = |0 + {}^-9| - |{}^-6 + 5|$
Simplify inside absolute value signs. Collect like terms.
$1x + 6 = |{}^-9| - |{}^-1|$
Evaluate absolute values.
$1x + 6 = 9 - 1$
Change to addition.
$1x + 6 = 9 + {}^-1$
$1x + 6 = 8$ Add ${}^-6$ to both sides.
  ${}^-6$ ${}^-6$
$\overline{}$
$1x + 0 = 2$
$x = 2$ The solution is 2.

71. (a) Equations will vary. Some possibilities are:

$n - 1 = {}^-3$ Change to addition.
$n + {}^-1 = {}^-3$ Add 1 to both sides.
$\underline{1} \quad \underline{1}$
$n + 0 = {}^-2$
$n = {}^-2$ The solution is ${}^-2$.

$8 = x + 10$ Add the opposite of 10, ${}^-10$, to both sides.
$\underline{{}^-10} \quad \underline{{}^-10}$
${}^-2 = x + 0$
${}^-2 = x$ The solution is ${}^-2$.

(b) Equations will vary. Some possibilities are:

$y + 6 = 6$ Add the opposite of 6, ${}^-6$, to both sides.
$\underline{{}^-6} \quad \underline{{}^-6}$
$y + 0 = 0$
$y = 0$ The solution is 0.

${}^-5 = {}^-5 + b$ Add the opposite of ${}^-5$, 5, to both sides.
$\underline{5} \quad \underline{5}$
$0 = 0 + b$
$0 = b$ The solution is 0.

72. (a) $x + 1 = 1\frac{1}{2}$ Add the opposite of 1, ${}^-1$, to both sides.
$\underline{{}^-1} \quad \underline{{}^-1}$
$x + 0 = \frac{1}{2}$
$x = \frac{1}{2}$ The solution is $\frac{1}{2}$.

(b) $\frac{1}{4} = y - 1$ Change to addition.
$\frac{1}{4} = y + {}^-1$ Add the opposite of ${}^-1$, 1, to both sides.
$\underline{1} \quad \underline{1}$
$1\frac{1}{4} = y + 0$
$1\frac{1}{4} = y$ or $y = \frac{5}{4}$ The solution is $\frac{5}{4}$.

(c) $\$2.50 + n = \$3.35$ Add the opposite of $\$2.50$, ${}^-\$2.50$, to both sides.
$\underline{{}^-\$2.50} \quad \underline{{}^-\$2.50}$
$\$0 + n = \$0.85$
$n = \$0.85$ The solution is $\$0.85$.

(d) Equations will vary. Some possibilities are:

$a - \$7.32 = \$9.16$ The solution is $\$16.48$.
$5c - \$11.20 = 4c - \$2.00$ The solution is $\$9.20$..

2.4 Solving Equations Using Division

2.4 Margin Exercises

1. (a) Solve $4s = 44$.

Use division to undo multiplication. Divide *both* sides by the coefficient of the variable, which is 4.

$$\frac{4s}{4} = \frac{44}{4}$$
$$s = 11$$

The solution is 11.

Check: $4s = 44$ Replace s with 11.
$4 \cdot 11 = 44$
$44 = 44$ Balances

(b) $27 = {}^-9p$ Divide *both* sides by ${}^-9$.
$$\frac{27}{{}^-9} = \frac{{}^-9p}{{}^-9}$$
$${}^-3 = p$$

The solution is ${}^-3$.

Check: $27 = {}^-9p$ Replace p with ${}^-3$.
$27 = {}^-9 \cdot {}^-3$
$27 = 27$ Balances

(c) ${}^-40 = {}^-5x$ Divide *both* sides by ${}^-5$.
$$\frac{{}^-40}{{}^-5} = \frac{{}^-5x}{{}^-5}$$
$$8 = x$$

The solution is 8.

Check: ${}^-40 = {}^-5x$ Replace x with 8.
${}^-40 = {}^-5 \cdot 8$
${}^-40 = {}^-40$ Balances

(d) $7t = {}^-70$ Divide *both* sides by 7.
$$\frac{7t}{7} = \frac{{}^-70}{7}$$
$$t = {}^-10$$

The solution is ${}^-10$.

Check: $7t = {}^-70$ Replace t with ${}^-10$.
$7 \cdot {}^-10 = {}^-70$
${}^-70 = {}^-70$ Balances

2. (a) ${}^-28 = {}^-6n + 10n$ Combine like terms.
${}^-28 = 4n$ Divide *both* sides by 4.
$$\frac{{}^-28}{4} = \frac{4n}{4}$$
${}^-7 = n$ The solution is ${}^-7$.

Check:
${}^-28 = {}^-6n + 10n$ Replace n with ${}^-7$.
${}^-28 = {}^-6 \cdot {}^-7 + 10 \cdot {}^-7$
${}^-28 = 42 + {}^-70$
${}^-28 = {}^-28$ Balances

(b) $\quad p - 14p = {}^-2 + 18 - 3 \quad$ Change to addition. Rewrite p as $1p$.
$\quad\quad 1p + {}^-14p = {}^-2 + 18 + {}^-3 \quad$ Combine like terms.
$\quad\quad\quad {}^-13p = 13 \quad$ Divide *both* sides by ${}^-13$.
$\quad\quad\quad \dfrac{{}^-13p}{{}^-13} = \dfrac{13}{{}^-13}$
$\quad\quad\quad\quad p = {}^-1 \quad$ The solution is ${}^-1$.

Check:
$\quad p - 14p = {}^-2 + 18 - 3 \quad$ Replace p with ${}^-1$.
$\quad {}^-1 - 14 \cdot {}^-1 = {}^-2 + 18 - 3$
$\quad\quad {}^-1 - {}^-14 = 16 - 3$
$\quad\quad {}^-1 + 14 = 13$
$\quad\quad\quad 13 = 13 \quad$ Balances

3. (a) $\quad {}^-k = {}^-12 \quad$ Write in the understood ${}^-1$ as the coefficient of k.
$\quad\quad {}^-1k = {}^-12 \quad$ Divide *both* sides by ${}^-1$.
$\quad\quad \dfrac{{}^-1k}{{}^-1} = \dfrac{{}^-12}{{}^-1}$
$\quad\quad\quad k = 12 \quad$ The solution is 12.

Check: $\quad {}^-k = {}^-12$
$\quad\quad {}^-1k = {}^-12 \quad$ Replace k with 12.
$\quad\quad {}^-1 \cdot 12 = {}^-12$
$\quad\quad\quad {}^-12 = {}^-12 \quad$ Balances

(b) $\quad 7 = {}^-t \quad$ Write ${}^-t$ as ${}^-1t$.
$\quad 7 = {}^-1t \quad$ Divide *both* sides by ${}^-1$.
$\quad \dfrac{7}{{}^-1} = \dfrac{{}^-1t}{{}^-1}$
$\quad {}^-7 = t \quad$ The solution is ${}^-7$.

Check: $\quad 7 = {}^-t$
$\quad\quad 7 = {}^-1t \quad$ Replace t with ${}^-7$.
$\quad\quad 7 = {}^-1 \cdot {}^-7$
$\quad\quad 7 = 7 \quad$ Balances

(c) $\quad {}^-m = {}^-20 \quad$ Write ${}^-m$ as ${}^-1m$.
$\quad {}^-1m = {}^-20 \quad$ Divide *both* sides by ${}^-1$.
$\quad \dfrac{{}^-1m}{{}^-1} = \dfrac{{}^-20}{{}^-1}$
$\quad\quad m = 20 \quad$ The solution is 20.

Check: $\quad {}^-m = {}^-20$
$\quad\quad {}^-1m = {}^-20 \quad$ Replace m with 20.
$\quad\quad {}^-1 \cdot 20 = {}^-20$
$\quad\quad {}^-20 = {}^-20 \quad$ Balances

2.4 Section Exercises

1. $\quad 6z = 12 \quad$ Divide *both* sides by 6.
$\quad \dfrac{6z}{6} = \dfrac{12}{6}$
$\quad z = 2 \quad$ The solution is 2.

Check: $\quad 6z = 12 \quad$ Replace z with 2.
$\quad\quad 6 \cdot 2 = 12$
$\quad\quad 12 = 12 \quad$ Balances

3. $\quad 48 = 12r \quad$ Divide *both* sides by 12.
$\quad \dfrac{48}{12} = \dfrac{12r}{12}$
$\quad 4 = r \quad$ The solution is 4.

Check: $\quad 48 = 12r \quad$ Replace r with 4.
$\quad\quad 48 = 12 \cdot 4$
$\quad\quad 48 = 48 \quad$ Balances

5. $\quad 3y = 0 \quad$ Divide *both* sides by 3.
$\quad \dfrac{3y}{3} = \dfrac{0}{3}$
$\quad y = 0 \quad$ The solution is 0.

Check: $\quad 3y = 0 \quad$ Replace y with 0.
$\quad\quad 3 \cdot 0 = 0$
$\quad\quad 0 = 0 \quad$ Balances

7. $\quad {}^-7k = 70 \quad$ Divide *both* sides by ${}^-7$.
$\quad \dfrac{{}^-7k}{{}^-7} = \dfrac{70}{{}^-7}$
$\quad k = {}^-10 \quad$ The solution is ${}^-10$.

Check: $\quad {}^-7k = 70 \quad$ Replace k with ${}^-10$.
$\quad\quad {}^-7 \cdot {}^-10 = 70$
$\quad\quad 70 = 70 \quad$ Balances

9. $\quad {}^-54 = {}^-9r \quad$ Divide *both* sides by ${}^-9$.
$\quad \dfrac{{}^-54}{{}^-9} = \dfrac{{}^-9r}{{}^-9}$
$\quad 6 = r \quad$ The solution is 6.

Check: $\quad {}^-54 = {}^-9r \quad$ Replace r with 6.
$\quad\quad {}^-54 = {}^-9 \cdot 6$
$\quad\quad {}^-54 = {}^-54 \quad$ Balances

11. $\quad {}^-25 = 5b \quad$ Divide *both* sides by 5.
$\quad \dfrac{{}^-25}{5} = \dfrac{5b}{5}$
$\quad {}^-5 = b \quad$ The solution is ${}^-5$.

Check: $\quad {}^-25 = 5b \quad$ Replace b with ${}^-5$.
$\quad\quad {}^-25 = 5 \cdot {}^-5$
$\quad\quad {}^-25 = {}^-25 \quad$ Balances

13. $\quad 2r = {}^-7 + 13 \quad$ Combine like terms.
$\quad 2r = 6 \quad$ Divide *both* sides by 2.
$\quad \dfrac{2r}{2} = \dfrac{6}{2}$
$\quad r = 3 \quad$ The solution is 3.

Check: $\quad 2r = {}^-7 + 13 \quad$ Replace r with 3.
$\quad\quad 2 \cdot 3 = {}^-7 + 13$
$\quad\quad 6 = 6 \quad$ Balances

2.4 Solving Equations Using Division 47

15.
$$-12 = 5p - p \quad \text{Change to addition.}$$
$$-12 = 5p + {}^-p \quad \text{Rewrite } {}^-p \text{ as } {}^-1p.$$
$$-12 = 5p + {}^-1p \quad \text{Combine like terms.}$$
$$-12 = 4p \quad \text{Divide both sides by 4.}$$
$$\frac{-12}{4} = \frac{4p}{4}$$
$$-3 = p \quad \text{The solution is } {}^-3.$$

Check:
$$-12 = 5p - p \quad \text{Replace } p \text{ with } {}^-3.$$
$$-12 = 5 \cdot {}^-3 - {}^-3$$
$$-12 = {}^-15 - {}^-3 \quad \text{Change to addition.}$$
$$-12 = {}^-15 + {}^+3$$
$$-12 = {}^-12 \quad \text{Balances}$$

17.
$$3 - 28 = 5a \quad \text{Change to addition.}$$
$$3 + {}^-28 = 5a \quad \text{Combine like terms.}$$
$$\frac{-25}{5} = \frac{5a}{5} \quad \text{Divide both sides by 5.}$$
$$-5 = a \quad \text{The solution is } {}^-5.$$

19.
$$x - 9x = 80 \quad \text{Change to addition.}$$
$$x + {}^-9x = 80 \quad \text{Rewrite } x \text{ as } 1x.$$
$$1x + {}^-9x = 80 \quad \text{Combine like terms.}$$
$$-8x = 80 \quad \text{Divide both sides by } {}^-8.$$
$$\frac{-8x}{-8} = \frac{80}{-8}$$
$$x = {}^-10 \quad \text{The solution is } {}^-10.$$

21.
$$13 - 13 = 2w - w \quad \text{Change to addition.}$$
$$13 + {}^-13 = 2w + {}^-w \quad \text{Rewrite } {}^-w \text{ as } {}^-1w.$$
$$13 + {}^-13 = 2w + {}^-1w \quad \text{Combine like terms.}$$
$$0 = 1w \quad 1w \text{ is the same as } w.$$
$$0 = w \quad \text{The solution is 0.}$$

23.
$$3t + 9t = 20 - 10 + 26 \quad \text{Change to addition.}$$
$$3t + 9t = 20 + {}^-10 + 26 \quad \text{Combine like terms.}$$
$$12t = 36 \quad \text{Divide both sides by 12.}$$
$$\frac{12t}{12} = \frac{36}{12}$$
$$t = 3 \quad \text{The solution is 3.}$$

25.
$$0 = {}^-9t \quad \text{Divide both sides by } {}^-9.$$
$$\frac{0}{-9} = \frac{-9t}{-9}$$
$$0 = t \quad \text{The solution is 0.}$$

27.
$$-14m + 8m = 6 - 60 \quad \text{Change to addition.}$$
$$-14m + 8m = 6 + {}^-60 \quad \text{Combine like terms.}$$
$$-6m = {}^-54 \quad \text{Divide both sides by } {}^-6.$$
$$\frac{-6m}{-6} = \frac{-54}{-6}$$
$$m = 9 \quad \text{The solution is 9.}$$

29.
$$100 - 96 = 31y - 35y \quad \text{Change to addition.}$$
$$100 + {}^-96 = 31y + {}^-35y \quad \text{Combine like terms.}$$
$$4 = {}^-4y \quad \text{Divide both sides by } {}^-4.$$
$$\frac{4}{-4} = \frac{-4y}{-4}$$
$$-1 = y \quad \text{The solution is } {}^-1.$$

31.
$$3(2z) = {}^-30 \quad \text{To multiply on the left, use the associative property.}$$
$$(3 \cdot 2) \cdot z = {}^-30$$
$$6z = {}^-30 \quad \text{Divide both sides by 6.}$$
$$\frac{6z}{6} = \frac{-30}{6}$$
$$z = {}^-5 \quad \text{The solution is } {}^-5.$$

33.
$$50 = {}^-5(5p) \quad \text{To multiply on the right, use the associative property.}$$
$$50 = ({}^-5 \cdot 5) \cdot p$$
$$50 = {}^-25p \quad \text{Divide both sides by } {}^-25.$$
$$\frac{50}{-25} = \frac{-25p}{-25}$$
$$-2 = p \quad \text{The solution is } {}^-2.$$

35.
$$-2({}^-4k) = 56 \quad \text{Associative property}$$
$$({}^-2 \cdot {}^-4) \cdot k = 56$$
$$8k = 56 \quad \text{Divide both sides by 8.}$$
$$\frac{8k}{8} = \frac{56}{8}$$
$$k = 7 \quad \text{The solution is 7.}$$

37.
$$-90 = {}^-10({}^-3b) \quad \text{Associative property}$$
$$-90 = ({}^-10 \cdot {}^-3) \cdot b$$
$$-90 = 30b \quad \text{Divide both sides by 30.}$$
$$\frac{-90}{30} = \frac{30b}{30}$$
$$-3 = b \quad \text{The solution is } {}^-3.$$

39.
$$-x = 32 \quad \text{Write in the understood } {}^-1.$$
$$-1x = 32 \quad \text{Divide both sides by } {}^-1.$$
$$\frac{-1x}{-1} = \frac{32}{-1}$$
$$x = {}^-32 \quad \text{The solution is } {}^-32.$$

41.
$$-2 = {}^-w \quad \text{Write in the understood } {}^-1.$$
$$-2 = {}^-1w \quad \text{Divide both sides by } {}^-1.$$
$$\frac{-2}{-1} = \frac{-1w}{-1}$$
$$2 = w \quad \text{The solution is 2.}$$

43.
$$-n = {}^-50 \quad \text{Write in the understood } {}^-1.$$
$$-1n = {}^-50 \quad \text{Divide both sides by } {}^-1.$$
$$\frac{-1n}{-1} = \frac{-50}{-1}$$
$$n = 50 \quad \text{The solution is 50.}$$

45.
$10 = {}^-p$ Write in the understood $^-1$.
$10 = {}^-1p$ Divide *both* sides by $^-1$.
$\dfrac{10}{{}^-1} = \dfrac{{}^-1p}{{}^-1}$
$^-10 = p$ The solution is $^-10$.

47. Each solution is the opposite of the number in the equation. So the rule is: When you change the sign of the variable from negative to positive, then change the number in the equation to its opposite. In $^-x = 5$, the opposite of 5 is $^-5$, so $x = {}^-5$.

49. Divide by the coefficient of x, which is 3, *not* by the opposite of 3.

$3x = 16 - 1$
$3x = 15$
$\dfrac{3x}{3} = \dfrac{15}{3}$
$x = 5$ The correct solution is 5.

51. $3s = 45$ Divide *both* sides by 3.
$\dfrac{3s}{3} = \dfrac{45}{3}$
$s = 15$

The length of one side is 15 feet.

53. $120 = 5s$ Divide *both* sides by 5.
$\dfrac{120}{5} = \dfrac{5s}{5}$
$24 = s$

The length of one side is 24 meters.

55. $89 - 116 = {}^-4({}^-4y) - 9(2y) + y$

Use the associative property to simplify the products.

$89 - 116 = ({}^-4 \cdot {}^-4) \cdot y - (9 \cdot 2) \cdot y + y$
$89 - 116 = 16y - 18y + y$

Change to addition. Combine like terms.

$89 + {}^-116 = 16y + {}^-18y + y$
$^-27 = {}^-1y$

Divide both sides by the coefficient, $^-1$, to get y by itself.

$\dfrac{{}^-27}{{}^-1} = \dfrac{{}^-1y}{{}^-1}$
$27 = y$ The solution is 27.

57. $^-37(14x) + 28(21x) = |72 - 72| + |{}^-166 + 96|$

Simplify within the absolute values.

$^-37(14x) + 28(21x) = |0| + |{}^-70|$

Simplify the absolute values.

$^-37(14x) + 28(21x) = 0 + 70$

Use the associative property to multiply on the left.

$({}^-37 \cdot 14) \cdot x + (28 \cdot 21) \cdot x = 0 + 70$
$^-518x + 588x = 70$

Combine like terms.

$70x = 70$

Divide both sides by the coefficient, 70, to get x by itself.

$\dfrac{70x}{70} = \dfrac{70}{70}$
$x = 1$ The solution is 1.

2.5 Solving Equations with Several Steps

2.5 Margin Exercises

1. (a) $2r + 7 = 13$ To get $2r$ by itself, add $^-7$ to both sides.
$ \dfrac{{}^-7 {}^-7}{2r + 0 = 6}$
$2r = 6$ To solve for r, divide both sides by the coefficient, 2.
$\dfrac{2r}{2} = \dfrac{6}{2}$
$r = 3$ The solution is 3.

Check: $2r + 7 = 13$
$2 \cdot 3 + 7 = 13$
$6 + 7 = 13$
$13 = 13$ Balances

(b) $20 = 6y - 4$ Change to addition.
$20 = 6y + {}^-4$ Add 4 to both sides.
$ \dfrac{4 4}{24 = 6y + 0}$
$24 = 6y$ Divide both sides by 6.
$\dfrac{24}{6} = \dfrac{6y}{6}$
$4 = y$ The solution is 4.

Check: $20 = 6y - 4$
$20 = 6 \cdot 4 - 4$
$20 = 24 - 4$
$20 = 20$ Balances

2.5 Solving Equations with Several Steps

(c) ${}^-10z - 9 = 11$ Change to addition.
${}^-10z + {}^-9 = 11$ Add 9 to both sides.
$\underline{9 9}$
${}^-10z + 0 = 20$
${}^-10z = 20$ Divide both sides
 by ${}^-10$.
$\dfrac{{}^-10z}{{}^-10} = \dfrac{20}{{}^-10}$
$z = {}^-2$ The solution is ${}^-2$.

Check: ${}^-10z - 9 = 11$
${}^-10 \cdot {}^-2 - 9 = 11$
$20 - 9 = 11$
$11 = 11$ Balances

2. (a) Solve with the variable on *left* side.

$3y - 1 = 2y + 7$ Add ${}^-2y$ to both sides.
$\underline{{}^-2y {}^-2y}$
$1y - 1 = 0 + 7$ Change to addition.
$1y + {}^-1 = 7$ Add 1 to both sides.
$\underline{1 1}$
$1y + 0 = 8$
$1y = 8$
 or $y = 8$ The solution is 8.

Solve with the variable on the *right* side.

$3y - 1 = 2y + 7$ Add ${}^-3y$ to both sides.
$\underline{{}^-3y {}^-3y}$
$0 - 1 = {}^-1y + 7$
${}^-1 = {}^-1y + 7$ Add ${}^-7$ to both sides.
$\underline{{}^-7 {}^-7}$
${}^-8 = {}^-1y + 0$
${}^-8 = {}^-1y$ Divide both sides by ${}^-1$.
$\dfrac{{}^-8}{{}^-1} = \dfrac{{}^-1y}{{}^-1}$
$8 = y$ The solution is 8.

(b) Solve with the variable on *left* side.

$3p - 2 = p - 6$ Rewrite p as $1p$.
$3p - 2 = 1p - 6$ Add ${}^-1p$ to both sides.
$\underline{{}^-1p {}^-1p}$
$2p - 2 = 0 - 6$ Change to addition.
$2p + {}^-2 = {}^-6$ Add 2 to both sides.
$\underline{2 2}$
$2p + 0 = {}^-4$
$2p = {}^-4$ Divide both sides by 2.
$\dfrac{2p}{2} = \dfrac{{}^-4}{2}$
$p = {}^-2$ The solution is ${}^-2$.

Solve with the variable on the *right* side.

$3p - 2 = p - 6$ Rewrite p as $1p$.
$3p - 2 = 1p - 6$ Add ${}^-3p$ to both sides.
$\underline{{}^-3p {}^-3p}$
$0 - 2 = {}^-2p - 6$
${}^-2 = {}^-2p + {}^-6$ Add 6 to both sides.
$\underline{6 6}$
$4 = {}^-2p + 0$
$4 = {}^-2p$ Divide both sides by ${}^-2$.
$\dfrac{4}{{}^-2} = \dfrac{{}^-2p}{{}^-2}$
${}^-2 = p$ The solution is ${}^-2$.

3. (a) ${}^-12 = 4(y - 1)$ Distribute on the right.
${}^-12 = 4 \cdot y - 4 \cdot 1$
${}^-12 = 4y - 4$ Change to addition.
${}^-12 = 4y + {}^-4$ Add 4 to both sides.
$\underline{4 4}$
${}^-8 = 4y + 0$
${}^-8 = 4y$ Divide both sides by 4.
$\dfrac{{}^-8}{4} = \dfrac{4y}{4}$
${}^-2 = y$ The solution is ${}^-2$.

Check: ${}^-12 = 4(y - 1)$ Replace y with ${}^-2$.
${}^-12 = 4({}^-2 - 1)$
${}^-12 = 4({}^-3)$
${}^-12 = {}^-12$ Balances

(b) $5(m + 4) = 20$ Distribute on the left.
$5 \cdot m + 5 \cdot 4 = 20$
$5m + 20 = 20$ Add ${}^-20$ to both sides.
$\underline{{}^-20 {}^-20}$
$5m + 0 = 0$
$5m = 0$ Divide both sides by 5.
$\dfrac{5m}{5} = \dfrac{0}{5}$
$m = 0$ The solution is 0.

Check: $5(m + 4) = 20$ Replace m with 0.
$5(0 + 4) = 20$
$5(4) = 20$
$20 = 20$ Balances

Chapter 2 Understanding Variables and Solving Equations

(c) $\quad 6(t-2) = 18 \quad$ Distribute on the left.
$\quad 6 \cdot t - 6 \cdot 2 = 18$
$\quad\quad 6t - 12 = 18 \quad$ Change to addition.
$\quad\quad 6t + {}^{-}12 = 18 \quad$ Add 12 to both sides.
$\quad\quad\quad\quad\quad 12 \quad\quad 12$
$\quad\quad\quad \overline{6t + 0 = 30}$
$\quad\quad\quad\quad 6t = 30 \quad$ Divide both sides by 6.
$\quad\quad\quad\quad \dfrac{6t}{6} = \dfrac{30}{6}$
$\quad\quad\quad\quad\quad t = 5 \quad$ The solution is 5.

Check: $\quad 6(t-2) = 18 \quad$ Replace t with 5.
$\quad\quad\quad\quad 6(5-2) = 18$
$\quad\quad\quad\quad\quad 6(3) = 18$
$\quad\quad\quad\quad\quad 18 = 18 \quad$ Balances

4. (a)
$\quad 3(b+7) = 2b - 1 \quad$ Distribute on the left.
$\quad 3 \cdot b + 3 \cdot 7 = 2b - 1$
$\quad\quad 3b + 21 = 2b - 1 \quad$ Variables to the left.
$\quad\quad\quad {}^{-}2b \quad\quad\quad {}^{-}2b \quad$ Add ${}^{-}2b$ to both sides.
$\quad\quad\quad \overline{1b + 21 = 0 - 1}$
$\quad\quad\quad 1b + 21 = {}^{-}1 \quad$ Add ${}^{-}21$ to both sides.
$\quad\quad\quad\quad {}^{-}21 \quad\quad {}^{-}21$
$\quad\quad\quad \overline{1b + 0 = {}^{-}22}$
$\quad\quad\quad\quad 1b = {}^{-}22$
$\quad\quad\text{or}\quad b = {}^{-}22 \quad$ The solution is ${}^{-}22$.

Check: $\quad 3(b+7) = 2b - 1$
$\quad\quad\quad 3({}^{-}22 + 7) = 2 \cdot {}^{-}22 - 1$
$\quad\quad\quad\quad 3({}^{-}15) = {}^{-}44 - 1$
$\quad\quad\quad\quad\quad {}^{-}45 = {}^{-}45 \quad$ Balances

(b) $\quad 6 - 2n = 14 + 4(n-5) \quad$ Distribute.
$\quad 6 - 2n = 14 + 4 \cdot n - 4 \cdot 5$
$\quad 6 - 2n = 14 + 4n - 20 \quad$ Add the opposite.
$\quad 6 + {}^{-}2n = 14 + 4n + {}^{-}20 \quad$ Combine like terms.
$\quad 6 + {}^{-}2n = {}^{-}6 + 4n \quad$ Add $2n$ to both sides.
$\quad\quad 2n \quad\quad\quad 2n$
$\quad \overline{6 + 0 = {}^{-}6 + 6n}$
$\quad\quad 6 = {}^{-}6 + 6n \quad$ Add 6 to both sides.
$\quad\quad 6 \quad\quad 6$
$\quad \overline{12 = 0 + 6n}$
$\quad\quad 12 = 6n \quad$ Divide both sides by 6.
$\quad\quad \dfrac{12}{6} = \dfrac{6n}{6}$
$\quad\quad\quad 2 = n \quad$ The solution is 2.

Check: $\quad 6 - 2n = 14 + 4(n-5)$
$\quad\quad\quad 6 - 2 \cdot 2 = 14 + 4(2-5)$
$\quad\quad\quad 6 - 4 = 14 + 4({}^{-}3)$
$\quad\quad\quad 2 = 14 + {}^{-}12$
$\quad\quad\quad 2 = 2 \quad$ Balances

2.5 Section Exercises

1. $\quad 7p + 5 = 12 \quad$ To get $7p$ by itself, add
$\quad\quad {}^{-}5 \quad\quad {}^{-}5 \quad {}^{-}5$ to both sides.
$\quad \overline{7p + 0 = 7}$
$\quad\quad 7p = 7 \quad$ Divide both sides by 7.
$\quad\quad \dfrac{7p}{7} = \dfrac{7}{7}$
$\quad\quad\quad p = 1 \quad$ The solution is 1.

Check: $\quad 7p + 5 = 12$
$\quad\quad\quad 7(1) + 5 = 12$
$\quad\quad\quad 7 + 5 = 12$
$\quad\quad\quad 12 = 12 \quad$ Balances

3. $\quad 2 = 8y - 6 \quad$ Change to addition.
$\quad\quad 2 = 8y + {}^{-}6 \quad$ To get $8y$ by itself, add
$\quad\quad 6 \quad\quad\quad 6 \quad$ 6 to both sides.
$\quad \overline{8 = 8y + 0} \quad$ Divide both sides by 8.
$\quad\quad \dfrac{8}{8} = \dfrac{8y}{8}$
$\quad\quad\quad 1 = y \quad$ The solution is 1.

Check: $\quad 2 = 8y - 6$
$\quad\quad\quad 2 = 8(1) - 6$
$\quad\quad\quad 2 = 8 - 6$
$\quad\quad\quad 2 = 2 \quad$ Balances

5. $\quad {}^{-}3m + 1 = 1 \quad$ To get ${}^{-}3m$ by itself, add
$\quad\quad {}^{-}1 \quad\quad {}^{-}1 \quad {}^{-}1$ to both sides.
$\quad \overline{{}^{-}3m + 0 = 0}$
$\quad\quad {}^{-}3m = 0 \quad$ Divide both sides by ${}^{-}3$.
$\quad\quad \dfrac{{}^{-}3m}{{}^{-}3} = \dfrac{0}{{}^{-}3}$
$\quad\quad\quad m = 0 \quad$ The solution is 0.

Check: $\quad {}^{-}3m + 1 = 1$
$\quad\quad\quad {}^{-}3(0) + 1 = 1$
$\quad\quad\quad 0 + 1 = 1$
$\quad\quad\quad 1 = 1 \quad$ Balances

7. $\quad 28 = {}^{-}9a + 10 \quad$ To get ${}^{-}9a$ by itself, add
$\quad\quad {}^{-}10 \quad\quad\quad {}^{-}10 \quad {}^{-}10$ to both sides.
$\quad \overline{18 = {}^{-}9a + 0}$
$\quad\quad 18 = {}^{-}9a \quad$ Divide both sides by ${}^{-}9$.
$\quad\quad \dfrac{18}{{}^{-}9} = \dfrac{{}^{-}9a}{{}^{-}9}$
$\quad\quad\quad {}^{-}2 = a \quad$ The solution is ${}^{-}2$.

2.5 Solving Equations with Several Steps

Check: $28 = {}^-9a + 10$
$28 = {}^-9({}^-2) + 10$
$28 = 18 + 10$
$28 = 28$ Balances

9. ${}^-5x - 4 = 16$ Change to addition.
${}^-5x + {}^-4 = 16$ To get ${}^-5x$ by itself, add
$\phantom{{}^-5x +} 4 4$ 4 to both sides.
${}^-5x + 0 = 20$
${}^-5x = 20$ Divide both sides by ${}^-5$.
$\dfrac{{}^-5x}{{}^-5} = \dfrac{20}{{}^-5}$
$x = {}^-4$ The solution is ${}^-4$.

Check: ${}^-5x - 4 = 16$
${}^-5({}^-4) - 4 = 16$
$20 - 4 = 16$
$16 = 16$ Balances

11. Solve with the variable on the *left* side.

$6p - 2 = 4p + 6$ Change to addition.
$6p + {}^-2 = 4p + 6$ Add ${}^-4p$ to both sides.
${}^-4p \phantom{= {}}{}^-4p$
$2p + {}^-2 = 0 + 6$
$2p + {}^-2 = 6$ Add 2 to both sides.
$\phantom{2p + {}^-}2 \phantom{= {}}2$
$2p + 0 = 8$
$2p = 8$ Divide both sides by 2.
$\dfrac{2p}{2} = \dfrac{8}{2}$
$p = 4$ The solution is 4.

Solve with the variable on the *right* side.

$6p - 2 = 4p + 6$ Change to addition.
$6p + {}^-2 = 4p + 6$ Add ${}^-6p$ to both sides.
${}^-6p \phantom{= {}}{}^-6p$
$0 + {}^-2 = {}^-2p + 6$
${}^-2 = {}^-2p + 6$ Add ${}^-6$ to both sides.
${}^-6 \phantom{= {}}{}^-6$
${}^-8 = {}^-2p + 0$
${}^-8 = {}^-2p$ Divide both sides by ${}^-2$.
$\dfrac{{}^-8}{{}^-2} = \dfrac{{}^-2p}{{}^-2}$
$4 = p$ The solution is 4.

Check: $6p - 2 = 4p + 6$
$6(4) - 2 = 4(4) + 6$
$24 - 2 = 16 + 6$
$22 = 22$ Balances

13. Solve with the variable on the *left* side.

${}^-2k - 6 = 6k + 10$ Change to addition.
${}^-2k + {}^-6 = 6k + 10$ Add ${}^-6k$ to both sides.
${}^-6k \phantom{= {}}{}^-6k$
${}^-8k + {}^-6 = 0 + 10$
${}^-8k + {}^-6 = 10$ Add 6 to both sides.
$\phantom{{}^-8k + {}^-}6 \phantom{= {}}6$
${}^-8k + 0 = 16$ Divide both sides by ${}^-8$.
$\dfrac{{}^-8k}{{}^-8} = \dfrac{16}{{}^-8}$
$k = {}^-2$ The solution is ${}^-2$.

Solve with the variable on the *right* side.

${}^-2k - 6 = 6k + 10$ Change to addition.
${}^-2k + {}^-6 = 6k + 10$ Add $2k$ to both sides.
$2k \phantom{= {}}2k$
$0 + {}^-6 = 8k + 10$
${}^-6 = 8k + 10$ Add ${}^-10$ to both sides.
${}^-10 \phantom{= {}}{}^-10$
${}^-16 = 8k + 0$ Divide both sides by 8.
$\dfrac{{}^-16}{8} = \dfrac{8k}{8}$
${}^-2 = k$ The solution is ${}^-2$.

Check: ${}^-2k - 6 = 6k + 10$
${}^-2({}^-2) - 6 = 6({}^-2) + 10$
$4 + {}^-6 = {}^-12 + 10$
${}^-2 = {}^-2$ Balances

15. Solve with the variable on the *left* side.

${}^-18 + 7a = 2a + 7$ Add ${}^-2a$ to both sides.
${}^-2a \phantom{= {}}{}^-2a$
${}^-18 + 5a = 0 + 7$
${}^-18 + 5a = 7$ Add 18 to both sides.
$18 \phantom{= {}}18$
$0 + 5a = 25$
$5a = 25$ Divide both sides by 5.
$\dfrac{5a}{5} = \dfrac{25}{5}$
$a = 5$ The solution is 5.

continued

Chapter 2 Understanding Variables and Solving Equations

Solve with the variable on the *right* side.

$$-18 + 7a = 2a + 7 \quad \text{Add } {}^-7a \text{ to both sides.}$$
$$\underline{\quad -7a \qquad -7a \quad}$$
$$-18 + 0 = -5a + 7$$
$$-18 = -5a + 7 \quad \text{Add } {}^-7 \text{ to both sides.}$$
$$\underline{\quad -7 \qquad\qquad -7 \quad}$$
$$-25 = -5a + 0$$
$$-25 = -5a \quad \text{Divide both sides by } {}^-5.$$
$$\frac{-25}{-5} = \frac{-5a}{-5}$$
$$5 = a \quad \text{The solution is 5.}$$

Check: $-18 + 7a = 2a + 7$
$$-18 + 7(5) = 2(5) + 7$$
$$-18 + 35 = 10 + 7$$
$$17 = 17 \quad \text{Balances}$$

17. $8(w - 2) = 32$ Distribute.
$8w - 16 = 32$ Change to addition.
$8w + {}^-16 = 32$ Add 16 to both sides.
$\underline{\qquad\quad 16 \quad\; 16}$
$8w + 0 = 48$
$8w = 48$ Divide both sides by 8.
$\frac{8w}{8} = \frac{48}{8}$
$w = 6$ The solution is 6.

19. $-10 = 2(y + 4)$ Distribute.
$-10 = 2y + 8$ Add -8 to both sides.
$\underline{\;\;-8 \qquad\quad -8}$
$-18 = 2y + 0$
$-18 = 2y$ Divide both sides by 2.
$\frac{-18}{2} = \frac{2y}{2}$
$-9 = y$ The solution is -9.

21. $-4(t + 2) = 12$ Distribute.
$-4t + {}^-8 = 12$ Add 8 to both sides.
$\underline{\qquad\quad 8 \quad\; 8}$
$-4t + 0 = 20$
$-4t = 20$ Divide both sides by -4.
$\frac{-4t}{-4} = \frac{20}{-4}$
$t = -5$ The solution is -5.

23. $6(x - 5) = -30$ Distribute.
$6x - 30 = -30$ Change to addition.
$6x + {}^-30 = -30$ Add 30 to both sides.
$\underline{\qquad\quad 30 \qquad 30}$
$6x + 0 = 0$
$6x = 0$ Divide both sides by 6.
$\frac{6x}{6} = \frac{0}{6}$
$x = 0$ The solution is 0.

25. $-12 = 12(h - 2)$ Distribute.
$-12 = 12h - 24$ Change to addition.
$-12 = 12h + {}^-24$ Add 24 to both sides.
$\underline{\;\; 24 \qquad\qquad 24}$
$12 = 12h + 0$
$12 = 12h$ Divide both sides by 12.
$\frac{12}{12} = \frac{12h}{12}$
$1 = h$ The solution is 1.

27. $0 = -2(y + 2)$ Distribute.
$0 = -2y + {}^-4$ Add 4 to both sides.
$\underline{\; 4 \qquad\qquad 4}$
$4 = -2y + 0$
$4 = -2y$ Divide both sides by -2.
$\frac{4}{-2} = \frac{-2y}{-2}$
$-2 = y$ The solution is -2.

29. $6m + 18 = 0$ To get $6m$ by itself, add -18 to both sides.
$\underline{\quad\;\; -18 \;\; -18}$
$6m + 0 = -18$
$6m = -18$ Divide both sides by 6.
$\frac{6m}{6} = \frac{-18}{6}$
$m = -3$ The solution is -3.

31. $6 = 9w - 12$ Change to addition.
$6 = 9w + {}^-12$ Add 12 to both sides.
$\underline{\; 12 \qquad\quad 12}$
$18 = 9w + 0$
$18 = 9w$ Divide both sides by 9.
$\frac{18}{9} = \frac{9w}{9}$
$2 = w$ The solution is 2.

33. $5x = 3x + 10$ Add $-3x$ to both sides to get the variable term on one side.
$\underline{\;\; -3x \;\; -3x}$
$2x = 0 + 10$
$2x = 10$ Divide both sides by 2.
$\frac{2x}{2} = \frac{10}{2}$
$x = 5$ The solution is 5.

2.5 Solving Equations with Several Steps

35.
$2a + 11 = 8a - 7$ Change to addition.
$2a + 11 = 8a + {}^-7$ Add ${}^-2a$ to both sides.
$\underline{{}^-2a {}^-2a}$
$0 + 11 = 6a + {}^-7$
$11 = 6a + {}^-7$ Add 7 to both sides.
$\underline{7 7}$
$18 = 6a + 0$
$18 = 6a$ Divide both sides by 6.
$\dfrac{18}{6} = \dfrac{6a}{6}$
$3 = a$ The solution is 3.

37.
$7 - 5b = 28 + 2b$ Change to addition.
$7 + {}^-5b = 28 + 2b$ Add $5b$ to both sides.
$\underline{5b 5b}$
$7 + 0 = 28 + 7b$
$7 = 28 + 7b$ Add ${}^-28$ to both sides.
$\underline{{}^-28 {}^-28}$
${}^-21 = 0 + 7b$
${}^-21 = 7b$ Divide both sides by 7.
$\dfrac{{}^-21}{7} = \dfrac{7b}{7}$
${}^-3 = b$ The solution is ${}^-3$.

39.
${}^-20 + 2k = k - 4k$ Change to addition.
${}^-20 + 2k = k + {}^-4k$ Combine like terms.
${}^-20 + 2k = {}^-3k$ Add ${}^-2k$ to both sides.
$\underline{{}^-2k {}^-2k}$
${}^-20 + 0 = {}^-5k$
${}^-20 = {}^-5k$ Divide both sides by ${}^-5$.
$\dfrac{{}^-20}{{}^-5} = \dfrac{{}^-5k}{{}^-5}$
$4 = k$ The solution is 4.

41.
$10(c - 6) + 4 = 2 + c - 58$ Distribute.
$10c - 60 + 4 = 2 + c - 58$ Change to addition.
$10c + {}^-60 + 4 = 2 + c + {}^-58$ Group like terms.
$10c + {}^-60 + 4 = 2 + {}^-58 + c$ Combine like terms.
$10c + {}^-56 = {}^-56 + c$ Add ${}^-c$ to both sides.
$\underline{{}^-c {}^-c}$
$9c + {}^-56 = {}^-56 + 0$
$9c + {}^-56 = {}^-56$ Add 56 to both sides.
$\underline{56 56}$
$9c + 0 = 0$ Divide both sides by 9.
$\dfrac{9c}{9} = \dfrac{0}{9}$
$c = 0$ The solution is 0.

43.
${}^-18 + 13y + 3 = 3(5y - 1) - 2$ Distribute.
${}^-18 + 13y + 3 = 15y - 3 - 2$ Add the opposites.
${}^-18 + 13y + 3 = 15y + {}^-3 + {}^-2$ Group like terms.
$13y + {}^-18 + 3 = 15y + {}^-3 + {}^-2$ Combine like terms.
$13y + {}^-15 = 15y + {}^-5$ Add ${}^-13y$ to both sides.
$\underline{{}^-13y {}^-13y}$
$0 + {}^-15 = 2y + {}^-5$
${}^-15 = 2y + {}^-5$ Add 5 to both sides.
$\underline{5 5}$
${}^-10 = 2y + 0$
${}^-10 = 2y$ Divide both sides by 2.
$\dfrac{{}^-10}{2} = \dfrac{2y}{2}$
${}^-5 = y$

The solution is ${}^-5$.

45.
$6 - 4n + 3n = 20 - 35$ Change to addition.
$6 + {}^-4n + 3n = 20 + {}^-35$ Combine like terms.
$6 + {}^-1n = {}^-15$ Add ${}^-6$ to both sides.
$\underline{{}^-6 {}^-6}$
$0 + {}^-1n = {}^-21$ Divide both sides by ${}^-1$.
$\dfrac{{}^-1n}{{}^-1} = \dfrac{{}^-21}{{}^-1}$
$n = 21$ The solution is 21.

47.
$6(c - 2) = 7(c - 6)$ Distribute.
$6c - 12 = 7c - 42$ Change to addition.
$6c + {}^-12 = 7c + {}^-42$ Add ${}^-6c$ to both sides.
$\underline{{}^-6c {}^-6c}$
$0 + {}^-12 = 1c + {}^-42$
${}^-12 = 1c + {}^-42$ Add 42 to both sides.
$\underline{42 42}$
$30 = 1c + 0$
$30 = c$ The solution is 30.

49.
$$-5(2p+2) - 7 = 3(2p+5) \quad \text{Distribute.}$$
$$-10p + -10 - 7 = 6p + 15 \quad \text{Change to addition.}$$
$$-10p + -10 + -7 = 6p + 15 \quad \text{Combine like terms.}$$
$$-10p + -17 = 6p + 15 \quad \text{Add } -6p \text{ to both}$$
$$\underline{-6p -6p} \quad \text{sides.}$$
$$-16p + -17 = 0 + 15$$
$$-16p + -17 = 15 \quad \text{Add 17 to both}$$
$$\underline{17 17} \quad \text{sides.}$$
$$-16p + 0 = 32 \quad \text{Divide both sides by } -16.$$
$$\frac{-16p}{-16} = \frac{32}{-16}$$
$$p = -2 \quad \text{The solution is } -2.$$

51.
$$-6b - 4b + 7b = 10 - b + 3b \quad \text{Change to addition.}$$
$$-6b + -4b + 7b = 10 + -b + 3b \quad \text{Combine like terms.}$$
$$-3b = 10 + 2b \quad \text{Add } -2b \text{ to}$$
$$\underline{-2b -2b} \quad \text{both sides.}$$
$$-5b = 10 + 0$$
$$-5b = 10 \quad \text{Divide both sides by } -5.$$
$$\frac{-5b}{-5} = \frac{10}{-5}$$
$$b = -2 \quad \text{The solution is } -2.$$

53. The series of steps may vary. One possibility is:

$$-2t - 10 = 3t + 5 \quad \text{Change to addition.}$$
$$-2t + -10 = 3t + 5 \quad \text{Add } 2t \text{ to both sides}$$
$$\underline{2t 2t} \quad \text{(addition property).}$$
$$0 + -10 = 5t + 5 \quad \text{Add } -5 \text{ to both sides}$$
$$\underline{-5 -5} \quad \text{(addition property).}$$
$$\frac{-15}{5} = \frac{5t}{5} \quad \text{Divide both sides by 5 (division property).}$$
$$-3 = t$$

The solution is -3.

55. Check:
$$-8 + 4a = 2a + 2$$
$$-8 + 4(3) = 2(3) + 2$$
$$-8 + 12 = 6 + 2$$
$$4 \neq 8$$

The check does not balance, so 3 is not the correct solution. The student added $-2a$ to -8 on the left side, instead of adding $-2a$ to $4a$. The correct solution, obtained using $-8 + 2a = 2$, $2a = 10$, is $a = 5$.

57. (a) It must be negative, because the sum of two positive numbers is always positive.

(b) The sum of x and a positive number is negative, so x must be negative.

58. (a) It must be positive, because the sum of two negative numbers is always negative.

(b) The sum of d and a negative number is positive, so d must be positive.

59. (a) It must be positive. When the signs are the same, the product is positive, and when the signs are different, the product is negative.

(b) The product of n and a negative number is negative, so n must be positive.

60. (a) It must be negative also. When the signs are different, the product is negative, and when the signs match, the product is positive.

(b) The product of y and a negative number is positive, so y must be negative.

Chapter 2 Review Exercises

1. (a) In the expression $-3 + 4k$, k is the variable, 4 is the coefficient, and -3 is the constant term.

(b) The term that has 20 as the constant term and -9 as the coefficient is $-9y + 20$.

2. (a) Evaluate $4c + 10$ when c is 15.
$$4c + 10 \quad \text{Replace } c \text{ with } 15.$$
$$4 \cdot 15 + 10$$
$$60 + 10$$
$$70 \quad \text{Order 70 test tubes.}$$

(b) Evaluate $4c + 10$ when c is 24.
$$4c + 10 \quad \text{Replace } c \text{ with } 24.$$
$$4 \cdot 24 + 10$$
$$96 + 10$$
$$106 \quad \text{Order 106 test tubes.}$$

3. (a) $x^2 y^4$ means $x \cdot x \cdot y \cdot y \cdot y \cdot y$

(b) $5ab^3$ means $5 \cdot a \cdot b \cdot b \cdot b$

4. (a) n^2 means

$n \cdot n \quad$ Replace n with -3.

$-3 \cdot -3$

9

(b) n^3 means

$n \cdot n \cdot n \quad$ Replace n with -3.

$-3 \cdot -3 \cdot -3$

$9 \cdot -3$

-27

(c) $^-4mp^2$ means

$^-4 \cdot m \cdot p \cdot p$ Replace m with 2 and p with 4.

$\underbrace{^-4 \cdot 2} \cdot 4 \cdot 4$
$\underbrace{^-8 \cdot 4} \cdot 4$
$\underbrace{^-32 \cdot 4}$
$^-128$

(d) $5m^4n^2$ means

$5 \cdot m \cdot m \cdot m \cdot m \cdot n \cdot n$ Replace m with 2 and n with $^-3$.

$\underbrace{5 \cdot 2} \cdot 2 \cdot 2 \cdot 2 \cdot ^-3 \cdot ^-3$
$\underbrace{10 \cdot 2} \cdot 2 \cdot 2 \cdot ^-3 \cdot ^-3$
$\underbrace{20 \cdot 2} \cdot 2 \cdot ^-3 \cdot ^-3$
$\underbrace{40 \cdot 2} \cdot ^-3 \cdot ^-3$
$\underbrace{80 \cdot ^-3} \cdot ^-3$
$\underbrace{^-240 \cdot ^-3}$
720

5. $ab + ab^2 + 2ab$
$1ab + ab^2 + 2ab$ Combine like terms.
$3ab + ab^2$ or $ab^2 + 3ab$

6. $^-3x + 2y - x - 7$ Rewrite x as $1x$.
$^-3x + 2y - 1x - 7$ Change to addition.
$^-3x + 2y + ^-1x + ^-7$ Combine like terms.
$^-4x + 2y + ^-7$ or $^-4x + 2y - 7$

7. $^-8(^-2g^3)$ Associative property
$(^-8 \cdot ^-2) \cdot g^3$
$16 \cdot g^3$
$16g^3$

8. $4(3r^2t)$ Associative property
$(4 \cdot 3) \cdot r^2t$
$12 \cdot r^2t$
$12r^2t$

9. $5(k + 2)$ Distribute.
$5 \cdot k + 5 \cdot 2$
$5k + 10$

10. $^-2(3b + 4)$ Distribute.
$^-2 \cdot 3b + ^-2 \cdot 4$
$^-6b + ^-8$ or $^-6b - 8$

11. $3(2y - 4) + 12$ Distribute.
$3 \cdot 2y - 3 \cdot 4 + 12$
$6y - 12 + 12$
$6y + ^-12 + 12$
$6y + 0$
$6y$

12. $^-4 + 6(4x + 1) - 4x$ Distribute.
$^-4 + 24x + 6 - 4x$
$^-4 + 24x + 6 + ^-4x$
$2 + 20x$ or $20x + 2$

13. Expressions will vary. One possibility is $6a^3 + a^2 + 3a - 6$.

14. $16 + n = 5$ Add $^-16$ to both sides.
$\underline{^-16} \quad \underline{^-16}$
$0 + n = ^-11$
$n = ^-11$ The solution is $^-11$.

Check: $16 + n = 5$
$16 + ^-11 = 5$
$5 = 5$ Balances

15. $^-4 + 2 = 2a - 6 - a$
$^-4 + 2 = 2a + ^-6 + ^-1a$
$^-2 = 1a + ^-6$
$\underline{6} \quad \underline{6}$
$4 = 1a + 0$
$4 = a$

The solution is 4.

Check: $^-4 + 2 = 2a - 6 - a$
$^-4 + 2 = 2(4) - 6 - 4$
$^-2 = 8 + ^-6 + ^-4$
$^-2 = 2 + ^-4$
$^-2 = ^-2$ Balances

16. $48 = ^-6m$ Divide both sides by $^-6$.
$\dfrac{48}{^-6} = \dfrac{^-6m}{^-6}$
$^-8 = m$ The solution is $^-8$.

17. $k - 5k = ^-40$
$1k - 5k = ^-40$
$1k + ^-5k = ^-40$ Combine like terms.
$^-4k = ^-40$ Divide both sides by $^-4$.
$\dfrac{^-4k}{^-4} = \dfrac{^-40}{^-4}$
$k = 10$ The solution is 10.

18. $\underbrace{^-17 + 11 + 6} = 7t$
$0 = 7t$ Divide both sides by 7.
$\dfrac{0}{7} = \dfrac{7t}{7}$
$0 = t$ The solution is 0.

19. $^-2p + 5p = 3 - 21$
$^-2p + 5p = 3 + ^-21$
$3p = ^-18$ Divide both sides by 3.
$\dfrac{3p}{3} = \dfrac{^-18}{3}$
$p = ^-6$ The solution is $^-6$.

20. $^-30 = 3(^-5r)$
$^-30 = ^-15r$ Divide both sides by $^-15$.
$\dfrac{^-30}{^-15} = \dfrac{^-15r}{^-15}$
$2 = r$ The solution is 2.

21. $12 = ^-h$
$12 = ^-1h$
$\dfrac{12}{^-1} = \dfrac{^-1h}{^-1}$
$^-12 = h$ The solution is $^-12$.

22. $12w - 4 = 8w + 12$
$12w + ^-4 = 8w + 12$
$\underline{ ^-8w \underline{^-8w}}$
$4w + ^-4 = 0 + 12$
$4w + ^-4 = 12$ Add 4 to both sides.
$\underline{4 \underline{4}}$
$4w + 0 = 16$
$4w = 16$ Divide both sides by 4.
$\dfrac{4w}{4} = \dfrac{16}{4}$
$w = 4$ The solution is 4.

23. $0 = ^-4(c + 2)$ Distribute.
$0 = ^-4 \cdot c + ^-4 \cdot 2$
$0 = ^-4c + ^-8$
$\underline{8 \underline{8}}$ Add 8 to both sides.
$8 = ^-4c + 0$
$8 = ^-4c$ Divide both sides by $^-4$.
$\dfrac{8}{^-4} = \dfrac{^-4c}{^-4}$
$^-2 = c$ The solution is $^-2$.

24. $34 = 2n + 4$ Add $^-4$ to both sides.
$\underline{^-4 ^-4}$
$30 = 2n + 0$
$30 = 2n$ Divide both sides by 2.
$\dfrac{30}{2} = \dfrac{2n}{2}$
$15 = n$

The number of employees is 15.

25. [2.5] $12 + 7a = 4a - 3$ Add ^-4a to both sides.
$\underline{ ^-4a ^-4a}$
$12 + 3a = 0 - 3$
$12 + 3a = ^-3$
$\underline{^-12 ^-12}$
$0 + 3a = ^-15$ Divide both sides by 3.
$\dfrac{3a}{3} = \dfrac{^-15}{3}$
$a = ^-5$ The solution is $^-5$.

26. [2.5] $^-2(p - 3) = ^-14$ Distribute.
$^-2p + 6 = ^-14$ Add $^-6$ to both sides.
$\underline{ ^-6 ^-6}$
$^-2p + 0 = ^-20$
$\dfrac{^-2p}{^-2} = \dfrac{^-20}{^-2}$
$p = 10$ The solution is 10.

27. [2.5] $10y = 6y + 20$ Add ^-6y to both sides.
$\underline{^-6y ^-6y}$
$4y = 0 + 20$
$\dfrac{4y}{4} = \dfrac{20}{4}$
$y = 5$ The solution is 5.

28. [2.5] $2m - 7m = 5 - 20$ Add the opposites.
$2m + ^-7m = 5 + ^-20$ Combine like terms.
$^-5m = ^-15$ Divide both sides by $^-5$.
$\dfrac{^-5m}{^-5} = \dfrac{^-15}{^-5}$
$m = 3$ The solution is 3.

29. [2.5] $20 = 3x - 7$
$20 = 3x + ^-7$
$\underline{7 7}$
$27 = 3x + 0$
$\dfrac{27}{3} = \dfrac{3x}{3}$
$9 = x$ The solution is 9.

30. [2.5]
$$\begin{aligned} b + 6 &= 3b - 8 \\ {}^-3b & {}^-3b \\ \hline {}^-2b + 6 &= 0 - 8 \\ {}^-2b + 6 &= {}^-8 \\ {}^-6 & {}^-6 \\ \hline {}^-2b + 0 &= {}^-14 \\ \dfrac{{}^-2b}{{}^-2} &= \dfrac{{}^-14}{{}^-2} \\ b &= 7 \end{aligned}$$ The solution is 7.

31. [2.3]
$$\begin{aligned} z + 3 &= 0 \\ {}^-3 & {}^-3 \\ \hline z + 0 &= {}^-3 \\ z &= {}^-3 \end{aligned}$$ The solution is $^-3$.

32. [2.5]
$$\begin{aligned} 3(2n - 1) &= 3(n + 3) \quad \text{Distribute.} \\ 6n - 3 &= 3n + 9 \\ {}^-3n & {}^-3n \\ \hline 3n - 3 &= 0 + 9 \\ 3n - 3 &= 9 \quad \text{Add 3 to both sides.} \\ 3 & 3 \\ \hline 3n + 0 &= 12 \\ \dfrac{3n}{3} &= \dfrac{12}{3} \\ n &= 4 \end{aligned}$$ The solution is 4.

33. [2.5]
$$\begin{aligned} {}^-4 + 46 &= 7({}^-3t + 6) \quad \text{Distribute.} \\ {}^-4 + 46 &= {}^-21t + 42 \\ 42 &= {}^-21t + 42 \quad \text{Add } {}^-42 \text{ to both sides.} \\ {}^-42 & {}^-42 \\ \hline 0 &= {}^-21t + 0 \\ \dfrac{0}{{}^-21} &= \dfrac{{}^-21t}{{}^-21} \\ 0 &= t \end{aligned}$$ The solution is 0.

34. [2.5]
$$\begin{aligned} 6 + 10d - 19 &= 2(3d + 4) - 1 \quad \text{Distribute.} \\ 6 + 10d + {}^-19 &= 6d + 8 - 1 \\ {}^-13 + 10d &= 6d + 7 \\ {}^-6d & {}^-6d \\ \hline {}^-13 + 4d &= 0 + 7 \\ {}^-13 + 4d &= 7 \quad \text{Add 13 to both sides.} \\ 13 & 13 \\ \hline 0 + 4d &= 20 \\ \dfrac{4d}{4} &= \dfrac{20}{4} \\ d &= 5 \end{aligned}$$

The solution is 5.

35. [2.5]
$$\begin{aligned} {}^-4(3b + 9) &= 24 + 3(2b - 8) \quad \text{Distribute.} \\ {}^-12b - 36 &= 24 + 6b - 24 \\ {}^-12b + {}^-36 &= 24 + 6b + {}^-24 \\ {}^-12b + {}^-36 &= 6b \quad \text{Add } 12b \text{ to both sides.} \\ 12b & 12b \\ \hline 0 + {}^-36 &= 18b \\ \dfrac{{}^-36}{18} &= \dfrac{18b}{18} \\ {}^-2 &= b \end{aligned}$$

The solution is $^-2$.

Chapter 2 Test

1. In the expression $^-7w + 6$, $^-7$ is the coefficient, w is the variable, and 6 is the constant term.

2. Evaluate the expression $3a + 2c$ when a is 45 and c is 21.

   $3a + 2c$
   $3 \cdot 45 + 2 \cdot 21$
   $135 + 42$
   177

   Buy 177 hot dogs.

3. $x^5 y^3$ means $x \cdot x \cdot x \cdot x \cdot x \cdot y \cdot y \cdot y$

4. $4ab^4$ means $4 \cdot a \cdot b \cdot b \cdot b \cdot b$

5. $^-2s^2 t$ means

   $^-2 \cdot s \cdot s \cdot t$ Replace s with $^-5$ and t with 4.
   $^-2 \cdot {}^-5 \cdot {}^-5 \cdot 4$
   $\underbrace{10 \cdot {}^-5} \cdot 4$
   $\underbrace{{}^-50 \cdot 4}$
   $^-200$

6. $3w^3 - 8w^3 + w^3$
   $3w^3 - 8w^3 + 1w^3$
   $\underbrace{3w^3 + {}^-8w^3} + 1w^3$
   $\underbrace{{}^-5w^3 + 1w^3}$
   $^-4w^3$

7. $xy - xy$
   $1xy - 1xy$
   $(1 - 1)xy$
   $0xy$
   0

Chapter 2 Understanding Variables and Solving Equations

8. $^-6c - 5 + 7c + 5$
    $^-6c + {}^-5 + 7c + 5$
    $\underbrace{{}^-6c + 7c}\ +\ \underbrace{{}^-5 + 5}$
    $\ \ \ 1c\quad +\quad 0$
    $\ \ \ 1c\ \ \text{or}\ \ c$

9. $3m^2 - 3m + 3mn$
    There are no like terms.
    The expression cannot be simplified.

10. $^-10(4b^2)$ Associative property of multiplication
    $(^-10 \cdot 4) \cdot b^2$
    $^-40b^2$

11. $^-5(^-3k)$ Associative property of multiplication
    $(^-5 \cdot {}^-3) \cdot k$
    $15k$

12. $7(3t + 4)$ Distributive property
    $7(3t) + 7(4)$
    $21t + 28$

13. $^-4(a + 6)$ Distributive property
    $^-4 \cdot a + {}^-4 \cdot 6$
    $^-4a + {}^-24$
    $^-4a - 24$

14. $^-8 + 6(x - 2) + 5$
    $^-8 + 6x - 12 + 5$
    $^-8 + 6x + {}^-12 + 5$ Combine like terms.
    $6x + {}^-15\ \ \text{or}\ \ 6x - 15$

15. $^-9b - c - 3 + 9 + 2c$
    $^-9b - 1c - 3 + 9 + 2c$
    $^-9b + {}^-1c + {}^-3 + 9 + 2c$ Combine like terms.
    $^-9b + c + 6$

16. $^-4 = x - 9$ **Check:** $^-4 = x - 9$
    $\ \ 9\quad\quad\ \ 9$ $^-4 = 5 - 9$
    $\overline{\ \ 5\ = x + 0}$ $^-4 = {}^-4$
    $\ \ 5\ = x$ Balances

    The solution is 5.

17. $^-7w = 77$ **Check:** $^-7w = 77$
    $\dfrac{^-7w}{^-7} = \dfrac{77}{^-7}$ $^-7 \cdot {}^-11 = 77$
    $w = {}^-11$ $77 = 77$
    Balances

    The solution is $^-11$.

18. $^-p = 14$ **Check:** $^-p = 14$
    $^-1p = 14$ $^-1p = 14$
    $\dfrac{^-1p}{^-1} = \dfrac{14}{^-1}$ $^-1 \cdot {}^-14 = 14$
    $p = {}^-14$ $14 = 14$
    Balances

    The solution is $^-14$.

19. $^-15 = {}^-3(a + 2)$ **Check:** $^-15 = {}^-3(a + 2)$
    $^-15 = {}^-3a - 6$ $^-15 = {}^-3(3 + 2)$
    $\ \ \ \ 6\quad\quad\ \ \ \ 6$ $^-15 = {}^-3(5)$
    $\overline{^-9 = {}^-3a}$ $^-15 = {}^-15$
    $\dfrac{^-9}{^-3} = \dfrac{^-3a}{^-3}$ Balances
    $3 = a$

    The solution is 3.

20. $6n + 8 - 5n\ =\ ^-4 + 4$
    $6n + 8 + {}^-5n\ =\ 0$
    $n + 8\ =\ 0$
    $\quad\ \ {}^-8\quad\quad\ \ {}^-8$
    $\overline{\quad\ \ n\ =\ {}^-8}$

    The solution is $^-8$.

21. $5 - 20 = 2m - 3m$
    $5 + {}^-20 = 2m + {}^-3m$
    $^-15 = {}^-1m$
    $\dfrac{^-15}{^-1} = \dfrac{^-1m}{^-1}$
    $15 = m$

    The solution is 15.

22. $^-2x + 2\ =\ 5x + 9$ Add $2x$ to both sides.
    $\ \ 2x\quad\quad\ \ \ 2x$
    $\overline{\ \ \ \ \ 2\ =\ 7x + 9}$
    $\ \ \ ^-9\quad\quad\ {}^-9$
    $\overline{\ \ \ ^-7\ =\ 7x}$
    $\dfrac{^-7}{7} = \dfrac{7x}{7}$
    $^-1 = x$

    The solution is $^-1$.

23. $3m - 5\ =\ 7m - 13$
    $^-3m\quad\quad\ {}^-3m$
    $\overline{0 - 5\ =\ 4m - 13}$
    $^-5\ =\ 4m - 13$ Add 13 to both sides.
    $\ \ 13\quad\quad\ \ \ 13$
    $\overline{\ \ \dfrac{8}{4}\ =\ \dfrac{4m}{4}}$
    $2\ =\ m$

    The solution is 2.

24. $\quad 2 + 7b - 44 = {}^-3b + 12 + 9b$
$\quad\quad\quad 7b - 42 = 6b + 12$
$\quad\quad\quad\quad\underline{{}^-6b \quad\quad\quad {}^-6b}$ Add 42 to both sides.
$\quad\quad\quad 1b - 42 = 12$
$\quad\quad\quad\quad\underline{\quad\quad 42 \quad\quad 42}$
$\quad\quad\quad\quad\quad 1b = 54$
$\quad\quad\quad\quad\quad\; b = 54$

The solution is 54.

25. $\quad 3c - 24 = 6(c - 4)\quad$ Distribute.
$\quad\quad 3c - 24 = 6c - 24$
$\quad\quad\underline{{}^-3c \quad\quad\quad {}^-3c}$ Add 24 to both sides.
$\quad\quad\quad {}^-24 = 3c - 24$
$\quad\quad\quad\underline{\quad 24 \quad\quad\quad 24}$
$\quad\quad\quad\quad 0 = 3c$
$\quad\quad\quad\quad \dfrac{0}{3} = \dfrac{3c}{3}$
$\quad\quad\quad\quad 0 = c$

The solution is 0.

26. Equations will vary. Two possibilities are $x - 5 = {}^-9$ and ${}^-24 = 6y$.

Solving:

$\quad x - 5 = {}^-9 \quad\quad\quad {}^-24 = 6y$
$\quad\underline{{}^{+}5 \quad\quad {}^{+}5} \quad\quad\quad \dfrac{{}^-24}{6} = \dfrac{6y}{6}$
$\quad\quad x = {}^-4 \quad\quad\quad\quad {}^-4 = y$

Cumulative Review Exercises (Chapters 1–2)

1. 306,000,004,210 in words is three hundred six billion, four thousand, two hundred ten.

2. eight hundred million, sixty-six thousand: 800,066,000

3. (a) ${}^-3$ lies to the *right* of ${}^-10$ on the number line, so ${}^-3 > {}^-10$.

   (b) ${}^-1$ lies to the *left* of 0 on the number line, so ${}^-1 < 0$.

4. (a) ${}^-6 + 2 = 2 + {}^-6$

   Commutative property of addition:

   Changing the order of the addends does not change the sum.

   (b) $0 \cdot 25 = 0$

   Multiplication property of zero: Multiplying any number by 0 gives a product of 0.

   (c) $5({}^-6 + 4) = 5 \cdot {}^-6 + 5 \cdot 4$

   Distributive property:

   Multiplication distributes over addition.

5. (a) $9047 \approx 9000$

   Underline the hundreds place: 9$\underline{0}$47

   The next digit is 4 or less, so leave 0 as 0. Change 4 and 7 to 0.

   (b) $289{,}610 \approx 290{,}000$

   Underline the thousands place: 28$\underline{9}$,610

   The next digit is 5 or more, so add 1 to 9, write the 0 and add 1 to the ten-thousands place. Change 6 and 1 to 0.

6. $\quad 0 - 8 \quad$ Change to addition.
   $\quad = 0 + {}^-8$
   $\quad = {}^-8$

7. $\quad |{}^-6| + |4| \quad {}^-6$ is 6 units from 0.
   $\quad\quad\quad\quad\quad\;\; 4$ is 4 units from 0.
   $\quad = 6 + 4$
   $\quad = 10$

8. $\quad {}^-3({}^-10) \quad$ Same sign, positive product
   $\quad = 30$

9. $\quad ({}^-5)^2$
   $\quad = {}^-5 \cdot {}^-5 \quad$ Same sign, positive product
   $\quad = 25$

10. $\quad \dfrac{{}^-42}{{}^-6} \quad$ Same sign, positive quotient
    $\quad = 7$

11. $\quad {}^-19 + 19 \quad$ Addition of a number and its opposite is zero.
    $\quad = 0$

12. $\quad ({}^-4)^3 \quad$ Exponent
    $\quad \underbrace{{}^-4 \cdot {}^-4} \cdot {}^-4 \quad$ Multiply left to right.
    $\quad\;\; \underbrace{16 \cdot {}^-4}$
    $\quad\quad\; {}^-64$

13. $\dfrac{{}^-14}{0}$ is undefined. Division by 0 is undefined.

14. $\quad {}^-5 \cdot 12 \quad$ Different signs, negative product
    $\quad = {}^-60$

15. $\quad {}^-20 - 20 \quad$ Change to addition.
    $\quad = {}^-20 + {}^-20$
    $\quad = {}^-40$

16. $\dfrac{45}{{}^-5} = {}^-9 \quad$ Different signs, negative quotient

17. $^-50 + 25 = ^-25$

18. $^-10 + 6(4-7)$ Distribute.
$^-10 + \underbrace{6\cdot 4} - \underbrace{6\cdot 7}$
$^-10 + 24 - 42$
$\underbrace{^-10 + 24} + ^-42$
$\underbrace{14 + ^-42}$
$^-28$

19. $\dfrac{^-20 - 3(^-5) + 16}{(^-4)^2 - 3^3}$

Numerator:
$^-20 - 3(^-5) + 16$ Multiply.
$^-20 - ^-15 + 16$ Change to addition.
$\underbrace{^-20 + 15} + 16$ Add left to right.
$\underbrace{^-5 + 16}$
11

Denominator:
$(^-4)^2 - 3^3$
$\underbrace{(^-4)(^-4)} - \underbrace{3\cdot 3\cdot 3}$
$16 \quad - \quad 27$
$\underbrace{16 + ^-27}$
$^-11$

Last step is division: $\dfrac{11}{^-11} = ^-1$

20. 22 days rounds to 20.
616 miles rounds to 600.
Average distance "per" day implies division.

Estimate: $\dfrac{600 \text{ miles}}{20 \text{ days}} = 30$ miles per day

Exact: $\dfrac{616 \text{ miles}}{22 \text{ days}} = 28$ miles per day

The average distance the tiger traveled each day was 28 miles.

21. $^-48$ degrees rounds to $^-50$.
"Rise" of 23 degrees rounds to 20.

A start temperature of $^-48$ degrees followed by a rise of 23 degrees implies addition.

Estimate: $^-50 + 20 = ^-30$ degrees
Exact: $^-48 + 23 = ^-25$ degrees

The daytime temperature was $^-25$ degrees.

22. 52 shares rounds to 50.
$2132 rounds to $2000.
$8 stays $8 (it's a single digit number).

Each stock dropped in value by $8 and Doug owned 52 shares. Multiply to find out how much money he lost. Then, subtract this amount from the original total value.

Estimate: $\$2000 - (50\cdot 8) = \$1600$
Exact: $\$2132 - (52\cdot 8) = \$1716$

His shares are now worth $1716.

23. $758 rounds to $800.
$45 rounds to $50.
12 months (in one year) rounds to 10.

Estimate: $10(\$800 + \$50)$
$10(\$850) = \$8500$

Exact: $12(\$758 + \$45) = 12(\$803) = \$9636$

She will spend $9636 for rent and parking in one year.

24. $^-4ab^3c^2$ means $^-4 \cdot a \cdot b \cdot b \cdot b \cdot c \cdot c$

25. $3xy^3$ means

$3\cdot x\cdot y\cdot y\cdot y$ Replace x with $^-5$ and y with $^-2$.
$3\cdot ^-5\cdot ^-2\cdot ^-2\cdot ^-2$ Multiply left to right.
$\underbrace{^-15\cdot ^-2}\cdot ^-2\cdot ^-2$
$\underbrace{30\cdot ^-2}\cdot ^-2$
$\underbrace{^-60\cdot ^-2}$
120

26. $3h - 7h + 5h$ Change to addition.
$\underbrace{3h + ^-7h} + 5h$ Combine like terms.
$\underbrace{^-4h + 5h}$
$1h$ or h

27. $c^2d - c^2d$ Write with the understood coefficients of 1.
$= 1c^2d - 1c^2d$ Change to addition.
$= 1c^2d + ^-1c^2d$ Combine like terms.
$= (1 + ^-1)c^2d$
$= 0 \cdot c^2d$
$= 0$

28. $4n^2 - 4n + 6 - 8 + n^2$
$4n^2 + ^-4n + 6 + ^-8 + n^2$
$\underbrace{4n^2 + n^2} + ^-4n + \underbrace{6 + ^-8}$
$5n^2 \quad + ^-4n + \quad ^-2$
or $5n^2 - 4n - 2$

29. $^-10(3b^2)$ Associative property
$\underbrace{(^-10 \cdot 3)}b^2$
$^-30b^2$

30. $7(4p - 4)$ Distribute.
$\underbrace{7 \cdot 4p} - \underbrace{7 \cdot 4}$
$28p - 28$

31. $3 + 5(^-2w^2 - 3) + w^2$
$3 + ^-10w^2 - 15 + w^2$
$3 + ^-10w^2 + ^-15 + w^2$
$^-9w^2 + ^-12$ or $^-9w^2 - 12$

32. $3x = x - 8$ **Check:** $3x = x - 8$
$\dfrac{^-x \quad\quad ^-x}{2x = 0 - 8}$ $3(^-4) = ^-4 - 8$
$2x = ^-8$ $^-12 = ^-4 + ^-8$
$\dfrac{2x}{2} = \dfrac{^-8}{2}$ $^-12 = ^-12$
$x = ^-4$ Balances

The solution is $^-4$.

33. $^-44 = ^-2 + 7y$ **Check:** $^-44 = ^-2 + 7y$
$\dfrac{\quad 2 \quad\quad 2}{^-42 = 0 + 7y}$ $^-44 = ^-2 + 7(^-6)$
$^-42 = 7y$ $^-44 = ^-2 + ^-42$
$\dfrac{^-42}{7} = \dfrac{7y}{7}$ $^-44 = ^-44$
$^-6 = y$ Balances

The solution is $^-6$.

34. $2k - 5k = ^-21$ **Check:** $2k - 5k = ^-21$
$2k + ^-5k = ^-21$ $2(7) - 5(7) = ^-21$
$^-3k = ^-21$ $14 - 35 = ^-21$
$\dfrac{^-3k}{^-3} = \dfrac{^-21}{^-3}$ $14 + ^-35 = ^-21$
$k = 7$ $^-21 = ^-21$
                  Balances

The solution is 7.

35. $m - 6 = ^-2m + 6$
$\dfrac{2m \quad\quad\quad 2m}{3m - 6 = 0 + 6}$
$3m - 6 = 6$
$\dfrac{\quad 6 \quad\quad 6}{3m + 0 = 12}$
$\dfrac{3m}{3} = \dfrac{12}{3}$
$m = 4$ The solution is 4.

Check: $m - 6 = ^-2m + 6$
$4 - 6 = ^-2(4) + 6$
$4 + ^-6 = ^-8 + 6$
$^-2 = ^-2$ Balances

36. $4 - 4x = 18 + 10x$ Add $4x$ to both sides.
$\dfrac{\quad 4x \quad\quad\quad 4x}{4 + 0 = 18 + 14x}$
$4 = 18 + 14x$
$\dfrac{^-18 \quad\quad ^-18}{^-14 = 0 + 14x}$
$\dfrac{^-14}{14} = \dfrac{14x}{14}$
$^-1 = x$ The solution is $^-1$.

37. $18 = ^-r$
$18 = ^-1r$
$\dfrac{18}{^-1} = \dfrac{^-1r}{^-1}$
$^-18 = r$ The solution is $^-18$.

38. $^-8b - 11 + 7b = b - 1$
$^-1b - 11 = 1b - 1$ Add $1b$ to both sides.
$\dfrac{\quad 1b \quad\quad\quad 1b}{0b - 11 = 2b - 1}$
$^-11 = 2b - 1$ Add 1 to both sides.
$\dfrac{\quad 1 \quad\quad\quad 1}{^-10 = 2b + 0}$
$\dfrac{^-10}{2} = \dfrac{2b}{2}$
$^-5 = b$

The solution is $^-5$.

39. $^-2(t + 1) = 4(1 - 2t)$
$^-2t + ^-2 = 4 - 8t$ Add 2 to both sides.
$\dfrac{\quad 2 \quad\quad\quad 2}{^-2t + 0 = 6 - 8t}$
$^-2t = 6 - 8t$ Add $8t$ to both sides.
$\dfrac{\quad 8t \quad\quad\quad 8t}{6t = 6 + 0}$
$\dfrac{6t}{6} = \dfrac{6}{6}$
$t = 1$

The solution is 1.

40.
$$5 + 6y - 23 = 5(2y + 8) - 10$$
$$5 + 6y + {}^-23 = 10y + 40 + {}^-10$$
$$6y + {}^-18 = 10y + 30 \quad \text{Add 18 to both sides.}$$
$$\underline{18 \qquad 18}$$
$$6y + 0 = 10y + 48$$
$$6y = 10y + 48$$
$$\underline{{}^-10y \qquad {}^-10y}$$
$${}^-4y = 0 + 48$$
$$\frac{{}^-4y}{{}^-4} = \frac{48}{{}^-4}$$
$$y = {}^-12$$

The solution is $^-12$.

CHAPTER 3 SOLVING APPLICATION PROBLEMS

3.1 Problem Solving: Perimeter

3.1 Margin Exercises

1. **(a)** $P = 4s$ Perimeter formula for a square
   $P = 4 \cdot 20$ in.   Replace s, side length, with 20 in.
   $P = 80$ in.

   (b)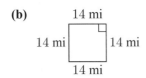

   $P = 4s$ Perimeter formula for a square
   $P = 4 \cdot 14$ mi   Replace s with 14 mi.
   $P = 56$ mi

2. **(a)** $P = 4s$ Perimeter formula for a square
   28 in. $= 4s$ Replace P, perimeter, with 28 in.
   $\dfrac{28 \text{ in.}}{4} = \dfrac{4s}{4}$ Divide both sides by 4.
   7 in. $= s$

   The length of one side is 7 inches.

   7 in. [square with 7 in. on each side]

   $P = 7$ in. $+ 7$ in. $+ 7$ in. $+ 7$ in. $= 28$ in.

   (b) $P = 4s$ Perimeter formula for a square
   100 ft $= 4s$ Replace P with 100 ft.
   $\dfrac{100 \text{ ft}}{4} = \dfrac{4s}{4}$ Divide both sides by 4.
   25 ft $= s$

   The length of one side is 25 feet.

   25 ft [square with 25 ft on each side]

   $P = 25$ ft $+ 25$ ft $+ 25$ ft $+ 25$ ft $= 100$ ft

   (c) $P = 4s$ Perimeter formula for a square
   64 cm $= 4s$ Replace P with 64 cm.
   $\dfrac{64 \text{ cm}}{4} = \dfrac{4s}{4}$ Divide both sides by 4.
   16 cm $= s$

   The length of one side is 16 cm.

   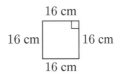

   $P = 16$ cm $+ 16$ cm $+ 16$ cm $+ 16$ cm $= 64$ cm

3. **(a)** $P = 2l + 2w$ Perimeter formula for a rectangle
   $P = 2 \cdot 17$ cm $+ 2 \cdot 10$ cm   Replace l with 17 cm and w with 10 cm.
   $P = 34$ cm $+ 20$ cm   Multiply first.
   $P = 54$ cm   Add last.

   The perimeter is 54 cm.

   Check:
   $P = 17$ cm $+ 17$ cm $+ 10$ cm $+ 10$ cm $= 54$ cm

   (b)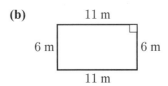

   $P = 2l + 2w$ Perimeter formula for a rectangle
   $P = 2 \cdot 11$ m $+ 2 \cdot 6$ m   Replace l with 11 m and w with 6 m.
   $P = 22$ m $+ 12$ m   Multiply first.
   $P = 34$ m   Add last.

   The perimeter is 34 m.

   Check:
   $P = 11$ m $+ 11$ m $+ 6$ m $+ 6$ m $= 34$ m

   (c) $P = 2l + 2w$
   $P = 2 \cdot 25$ ft $+ 2 \cdot 12$ ft   Replace l with 25 ft and w with 12 ft.
   $P = 50$ ft $+ 24$ ft   Multiply first.
   $P = 74$ ft   Add last.

   The perimeter is 74 ft.

   Check:
   $P = 25$ ft $+ 25$ ft $+ 12$ ft $+ 12$ ft $= 74$ ft

64 Chapter 3 Solving Application Problems

4. (a)
$$P = 2l + 2w \quad \text{Replace } P \text{ with 36 in. and } w \text{ with 8 in.}$$
$$36 \text{ in.} = 2l + 2 \cdot 8 \text{ in.} \quad \text{Multiply.}$$
$$36 \text{ in.} = 2l + 16 \text{ in.} \quad \text{To get } 2l \text{ by itself, add}$$
$$\underline{^-16 \text{ in.} \quad\quad ^-16 \text{ in.}} \quad ^-16 \text{ in. to both sides.}$$
$$20 \text{ in.} = 2l + 0$$
$$\frac{20 \text{ in.}}{2} = \frac{2l}{2} \quad \text{Divide both sides by 2.}$$
$$10 \text{ in.} = l$$

The length is 10 in.

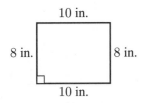

$P = 10 \text{ in.} + 10 \text{ in.} + 8 \text{ in.} + 8 \text{ in.} = 36 \text{ in.}$

(b)
$$P = 2l + 2w \quad \text{Replace } P \text{ with 32 cm and } w \text{ with 4 cm.}$$
$$32 \text{ cm} = 2l + 2 \cdot 4 \text{ cm} \quad \text{Multiply.}$$
$$32 \text{ cm} = 2l + 8 \text{ cm} \quad \text{To get } 2l \text{ by itself,}$$
$$\underline{^-8 \text{ cm} \quad\quad ^-8 \text{ cm}} \quad \text{add } ^-8 \text{ cm both sides.}$$
$$24 \text{ cm} = 2l + 0$$
$$\frac{24 \text{ cm}}{2} = \frac{2l}{2} \quad \text{Divide both sides by 2.}$$
$$12 \text{ cm} = l$$

The length is 12 cm.

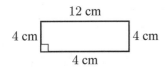

$P = 12 \text{ cm} + 12 \text{ cm} + 4 \text{ cm} + 4 \text{ cm} = 32 \text{ cm}$

(c)
$$P = 2l + 2w \quad \text{Replace } P \text{ with 14 ft and } l \text{ with 6 ft.}$$
$$14 \text{ ft} = 2 \cdot 6 \text{ ft} + 2w \quad \text{Multiply.}$$
$$14 \text{ ft} = 12 \text{ ft} + 2w \quad \text{To get } 2w \text{ by itself,}$$
$$\underline{^-12 \text{ ft} \quad\quad ^-12 \text{ ft}} \quad \text{add } ^-12 \text{ ft to both sides.}$$
$$2 \text{ ft} = 0 + 2w$$
$$\frac{2 \text{ ft}}{2} = \frac{2w}{2} \quad \text{Divide both sides by 2.}$$
$$1 \text{ ft} = w$$

The width is 1 ft.

$P = 6 \text{ ft} + 6 \text{ ft} + 1 \text{ ft} + 1 \text{ ft} = 14 \text{ ft}$

5. (a) $P = 15 \text{ m} + 27 \text{ m} + 15 \text{ m} + 27 \text{ m}$
$P = 84 \text{ m}$

(b) $P = 5 \text{ ft} + 4 \text{ ft} + 5 \text{ ft} + 4 \text{ ft}$
$P = 18 \text{ ft}$

6. (a) $P = 31 \text{ mm} + 16 \text{ mm} + 25 \text{ mm}$
$P = 72 \text{ mm}$

(b)

$P = 5 \text{ in.} + 5 \text{ in.} + 5 \text{ in.}$
$P = 15 \text{ in.}$

7. "Fencing needed to go *around* the flower bed" implies perimeter.

$P = 6 \text{ m} + 2 \text{ m} + 4 \text{ m} + 3 \text{ m} + 5 \text{ m}$
$P = 20 \text{ m}$

20 m of fencing are needed.

3.1 Section Exercises

1. The perimeter formula for a square is $P = 4s$.

$P = 4s \quad \text{Replace } s \text{ with 9 cm.}$
$P = 4 \cdot 9 \text{ cm}$
$P = 36 \text{ cm}$

The perimeter of the square is 36 cm.

3. The perimeter formula for a square is $P = 4s$.

$P = 4s \quad \text{Replace } s \text{ with 25 in.}$
$P = 4 \cdot 25 \text{ in.}$
$P = 100 \text{ in.}$

The perimeter of the square is 100 in.

5. A square park measuring 1 mile (mi) on each side

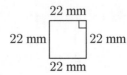

$P = 4s \quad \text{Replace } s \text{ with 1 mi.}$
$P = 4 \cdot 1 \text{ mi}$
$P = 4 \text{ mi}$

The perimeter of the park is 4 miles.

7. A 22 mm square postage stamp

$P = 4s$ Replace s with 22 mm.
$P = 4 \cdot 22$ mm
$P = 88$ mm

The perimeter of the stamp is 88 mm.

9. The perimeter of a square is 120 ft.

   $P = 4s$ Replace P with 120 ft.
   120 ft $= 4s$ Divide both sides by 4.
   $\dfrac{120 \text{ ft}}{4} = \dfrac{4s}{4}$
   30 ft $= s$

   The length of one side of the square is 30 ft.

11. The perimeter of a square is 4 millimeters (mm).

    $P = 4s$ Replace P with 4 mm.
    4 mm $= 4s$ Divide both sides by 4.
    $\dfrac{4 \text{ mm}}{4} = \dfrac{4s}{4}$
    1 mm $= s$

    The length of one side of the square is 1 mm.

13. The perimeter of a square parking lot is 92 yards (yd).

    $P = 4s$ Replace P with 92 yd.
    92 yd $= 4s$ Divide both sides by 4.
    $\dfrac{92 \text{ yd}}{4} = \dfrac{4s}{4}$
    23 yd $= s$

    The length of one side of the parking lot is 23 yards.

15. The perimeter of a square closet is 8 feet (ft).

    $P = 4s$ Replace P with 8 ft.
    8 ft $= 4s$ Divide both sides by 4.
    $\dfrac{8 \text{ ft}}{4} = \dfrac{4s}{4}$
    2 ft $= s$

    The length of one side of the closet is 2 ft.

17. $P = 2l + 2w$ Replace l with 8 yd and w with 6 yd.
    $P = 2 \cdot 8$ yd $+ 2 \cdot 6$ yd   Multiply first.
    $P = 16$ yd $+ 12$ yd   Add last.
    $P = 28$ yd

    The perimeter of the rectangle is 28 yd.

    Check: $P = 8$ yd $+ 8$ yd $+ 6$ yd $+ 6$ yd
    $P = 28$ yd

19. $P = 2l + 2w$ Replace l with 25 cm and w with 10 cm.
    $P = 2 \cdot 25$ cm $+ 2 \cdot 10$ cm   Multiply first.
    $P = 50$ cm $+ 20$ cm   Add last.
    $P = 70$ cm

    The perimeter of the rectangle is 70 cm.

    Check: $P = 25$ cm $+ 25$ cm $+ 10$ cm $+ 10$ cm
    $P = 70$ cm

21.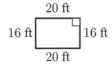

    $P = 2l + 2w$ Replace l with 20 ft and w with 16 ft.
    $P = 2 \cdot 20$ ft $+ 2 \cdot 16$ ft   Multiply first.
    $P = 40$ ft $+ 32$ ft   Add last.
    $P = 72$ ft

    The perimeter of the rectangular living room is 72 ft.

23.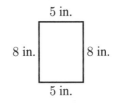

    $P = 2l + 2w$ Replace l with 8 in. and w with 5 in.
    $P = 2 \cdot 8$ in. $+ 2 \cdot 5$ in.   Multiply first.
    $P = 16$ in. $+ 10$ in.   Add last.
    $P = 26$ in.

    The perimeter of the rectangular piece of paper is 26 inches.

25. $P = 2l + 2w$ Replace P with 30 cm and w with 6 cm.
    30 cm $= 2l + 2 \cdot 6$ cm   Multiply on the right.
    $\begin{aligned}30 \text{ cm} &= 2l + 12 \text{ cm} \\ {}^-12 \text{ cm} & \quad\quad {}^-12 \text{ cm}\end{aligned}$ Add $^-12$ cm to both sides.
    18 cm $= 2l + 0$ cm
    $\dfrac{18 \text{ cm}}{2} = \dfrac{2l}{2}$ Divide both sides by 2.
    9 cm $= l$

    The length is 9 cm.

    Check:

    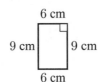

continued

66 Chapter 3 Solving Application Problems

$P = 9 \text{ cm} + 9 \text{ cm} + 6 \text{ cm} + 6 \text{ cm}$
$P = 30 \text{ cm}$

30 cm matches the original perimeter, so 9 cm is the correct length.

27. $P = 2l + 2w$ Replace P with 10 mi and l with 4 mi.
$10 \text{ mi} = 2 \cdot 4 \text{ mi} + 2w$ Multiply on the right.
$10 \text{ mi} = 8 \text{ mi} + 2w$ Add $^-8$ mi to both sides.
$\underline{^-8 \text{ mi} \qquad ^-8 \text{ mi}}$
$2 \text{ mi} = 0 \text{ mi} + 2w$
$\dfrac{2 \text{ mi}}{2} = \dfrac{2w}{2}$ Divide both sides by 2.
$1 \text{ mi} = w$

The width is 1 mi.

Check:

$P = 4 \text{ mi} + 4 \text{ mi} + 1 \text{ mi} + 1 \text{ mi}$
$P = 10 \text{ mi}$

10 mi matches the original perimeter, so 1 mi is the correct width.

29. $P = 2l + 2w$ Replace l with 6 ft and P with 16 ft.
$16 \text{ ft} = 2 \cdot 6 \text{ ft} + 2w$ Multiply on the right.
$16 \text{ ft} = 12 \text{ ft} + 2w$ Add $^-12$ ft to both sides.
$\underline{^-12 \text{ ft} \qquad ^-12 \text{ ft}}$
$4 \text{ ft} = 0 \text{ ft} + 2w$
$\dfrac{4 \text{ ft}}{2} = \dfrac{2w}{2}$ Divide both sides by 2.
$2 \text{ ft} = w$

The width of the table is 2 ft.

Check:

$P = 6 \text{ ft} + 6 \text{ ft} + 2 \text{ ft} + 2 \text{ ft}$
$P = 16 \text{ ft}$

31. $P = 2l + 2w$ Replace w with 1 m and P with 6 m.
$6 \text{ m} = 2l + 2 \cdot 1 \text{ m}$ Multiply on the right.
$6 \text{ m} = 2l + 2 \text{ m}$ Add $^-2$ m to both sides.
$\underline{^-2 \text{ m} \qquad ^-2 \text{ m}}$
$4 \text{ m} = 2l + 0 \text{ m}$
$\dfrac{4 \text{ m}}{2} = \dfrac{2}{2}l$ Divide both sides by 2.
$2 \text{ m} = l$

The length of the door is 2 m.

Check:

$P = 2 \text{ m} + 2 \text{ m} + 1 \text{ m} + 1 \text{ m}$
$P = 6 \text{ m}$

33. Add all four sides to find the perimeter.
$P = 58 \text{ m} + 58 \text{ m} + 46 \text{ m} + 46 \text{ m}$
$P = 208 \text{ m}$

The perimeter is 208 m.

35. Add all four sides to find the perimeter.
$P = 100 \text{ ft} + 60 \text{ ft} + 100 \text{ ft} + 60 \text{ ft}$
$P = 320 \text{ ft}$

The perimeter is 320 ft.

37. Add all three sides to find the perimeter of the triangle.
$P = 12 \text{ mm} + 26 \text{ mm} + 16 \text{ mm}$
$P = 54 \text{ mm}$

The perimeter is 54 mm.

39. Add all six sides to find the perimeter.
$P = 4 \text{ ft} + 12 \text{ ft} + 12 \text{ ft} + 3 \text{ ft} + 8 \text{ ft} + 9 \text{ ft}$
$P = 48 \text{ ft}$

The perimeter is 48 ft.

41. Add all six sides to find the perimeter.
$P = 13 \text{ in.} + 8 \text{ in.} + 18 \text{ in.} + 13 \text{ in.} + 18 \text{ in.} + 8 \text{ in.}$
$P = 78 \text{ in.}$

The perimeter is 78 in.

43. Add all five sides to find the perimeter.
$P = 34 \text{ m} + 22 \text{ m} + 20 \text{ m} + 22 \text{ m} + 27 \text{ m}$
$P = 125 \text{ m}$

The perimeter is 125 m.

45. The perimeter is 115 cm.
$P = 10 \text{ cm} + 10 \text{ cm} + 30 \text{ cm} + 25 \text{ cm} + ?$
$P = 75 \text{ cm} + ?$

Because the perimeter is 115 cm, replace P with 115 cm.

continued

115 cm = 75 cm + ?
115 cm = 75 cm + ?   Add ⁻75 cm to both sides.
⁻75 cm   ⁻75 cm
―――――――――――――――――――
40 cm = 0 cm + ?
40 cm = ?

The length of the unknown side is 40 cm.

47. The perimeter is 78 in.

78 in. = 15 in. + 6 in. + 6 in. + 6 in. + 6 in.
         + 6 in. + 9 in. + ? + 6 in. + 6 in.
78 in. = 66 in. + ?
⁻66 in.   ⁻66 in.
―――――――――――――――――
12 in. = ?

The length of the unknown side is 12 in.

49. (a) Sketches will vary.

(b) Formula for perimeter of an equilateral triangle is $P = 3s$, where s is the length of one side.

(c) The formula will not work for other kinds of triangles because the sides will have different lengths.

51. Use $d = rt$ with $r = 70$ miles per hour.

(a) In $t = 2$ hours, you will travel $d = 70 \cdot 2 = 140$ miles.

(b) In $t = 5$ hours, you will travel $d = 70 \cdot 5 = 350$ miles.

(c) In $t = 8$ hours, you will travel $d = 70 \cdot 8 = 560$ miles.

52. Use $d = rt$ with $r = 35$ miles per hour.

(a) In $t = 2$ hours, you will travel $d = 35 \cdot 2 = 70$ miles.

(b) In $t = 5$ hours, you will travel $d = 35 \cdot 5 = 175$ miles.

(c) In $t = 8$ hours, you will travel $d = 35 \cdot 8 = 280$ miles.

(d) The rate is half of 70 miles per hour, so in each case the distance traveled will be half as far. Divide each result in Exercise 51 by 2.

53. Use $rt = d$ with $d = 3000$ miles and r in miles per hour. In each case, divide by r.

(a) $60t = 3000$
$\dfrac{60t}{60} = \dfrac{3000}{60}$
$t = 50$ hours

(b) $50t = 3000$
$\dfrac{50t}{50} = \dfrac{3000}{50}$
$t = 60$ hours

(c) $20t = 3000$
$\dfrac{20t}{20} = \dfrac{3000}{20}$
$t = 150$ hours

54. Use $rt = d$. In each case, divide by t.

(a) $r \cdot 11 = 671$
$\dfrac{11r}{11} = \dfrac{671}{11}$
$r = 61$ miles per hour

(b) $r \cdot 27 = 1539$
$\dfrac{27r}{27} = \dfrac{1539}{27}$
$r = 57$ miles per hour

(c) $r \cdot 16 = 1040$
$\dfrac{16r}{16} = \dfrac{1040}{16}$
$r = 65$ miles per hour

3.2 Problem Solving: Area

3.2 Margin Exercises

1. (a) $A = l \cdot w$ Replace l with 9 ft and w with 4 ft.
$A = 9 \text{ ft} \cdot 4 \text{ ft}$
$A = 36 \text{ ft}^2$

The area of the rectangle is 36 ft².

(b) 20 yd, 35 yd

$A = l \cdot w$ Replace l with 35 yd and w with 20 yd.
$A = 35 \text{ yd} \cdot 20 \text{ yd}$
$A = 700 \text{ yd}^2$

The area of the rectangle is 700 yd².

(c) 2 m, 3 m

$A = l \cdot w$ Replace l with 3 m and w with 2 m.
$A = 3 \text{ m} \cdot 2 \text{ m}$
$A = 6 \text{ m}^2$

The area of the patio is 6 m².

2. (a) $A = l \cdot w$ — Replace l with 6 cm and A with 12 cm².

$12 \text{ cm}^2 = 6 \text{ cm} \cdot w$

$\dfrac{12 \text{ cm} \cdot \text{cm}}{6 \text{ cm}} = \dfrac{6 \text{ cm} \cdot w}{6 \text{ cm}}$ — To get w by itself, divide both sides by 6 cm.

$2 \text{ cm} = w$

The width of the slide is 2 cm.

2 cm
6 m

Check: $A = l \cdot w$
$A = 6 \text{ cm} \cdot 2 \text{ cm}$
$A = 12 \text{ cm}^2$

(b) $A = l \cdot w$ — Replace w with 10 ft and A with 160 ft².

$160 \text{ ft}^2 = l \cdot 10 \text{ ft}$

$\dfrac{160 \text{ ft} \cdot \text{ft}}{10 \text{ ft}} = \dfrac{l \cdot 10 \text{ ft}}{10 \text{ ft}}$ — To get l by itself, divide both sides by 10 ft.

$16 \text{ ft} = l$

The length of the lot is 16 ft.

10 ft
16 ft

Check: $A = l \cdot w$
$A = 16 \text{ ft} \cdot 10 \text{ ft}$
$A = 160 \text{ ft}^2$

(c) $A = l \cdot w$ — Replace A with 93 m² and l with 31 m.

$93 \text{ m}^2 = 31 \text{ m} \cdot w$

$\dfrac{93 \text{ m} \cdot \text{m}}{31 \text{ m}} = \dfrac{31 \text{ m} \cdot w}{31 \text{ m}}$ — To get w by itself, divide both sides by 31 m.

$3 \text{ m} = w$

The width of the floor is 3 m.

3 m
31 m

Check: $A = l \cdot w$
$A = 31 \text{ m} \cdot 3 \text{ m}$
$A = 93 \text{ m}^2$

3. (a) 12 in. / 12 in.

$A = s^2$
$A = s \cdot s$
$A = 12 \text{ in.} \cdot 12 \text{ in.}$
$A = 144 \text{ in.}^2$

The area of the fabric is 144 in.².

(b) 7 mi / 7 mi

$A = s^2$
$A = s \cdot s$
$A = 7 \text{ mi} \cdot 7 \text{ mi}$
$A = 49 \text{ mi}^2$

The area of the township is 49 mi².

(c) 20 mm / 20 mm

$A = s^2$
$A = s \cdot s$
$A = 20 \text{ mm} \cdot 20 \text{ mm}$
$A = 400 \text{ mm}^2$

The area of the earring is 400 mm².

4. (a) $A = s^2$ — Replace A with 16 mi².

$16 \text{ mi}^2 = s^2$
$16 \text{ mi}^2 = s \cdot s$
$16 \text{ mi}^2 = 4 \text{ mi} \cdot 4 \text{ mi}$

The length of one side of the nature center is 4 mi.

(b) $A = s^2$ — Replace A with 100 m².

$100 \text{ m}^2 = s^2$
$100 \text{ m}^2 = s \cdot s$
$100 \text{ m}^2 = 10 \text{ m} \cdot 10 \text{ m}$

The length of one side of the floor is 10 m.

(c) $A = s^2$ — Replace A with 81 in.².

$81 \text{ in.}^2 = s^2$
$81 \text{ in.}^2 = s \cdot s$
$81 \text{ in.}^2 = 9 \text{ in.} \cdot 9 \text{ in.}$

The length of one side of the clock face is 9 inches.

3.2 Problem Solving: Area

5. (a) *Note*: The base is 50 ft and the height is 42 ft. The height forms a 90° angle with the base.

$A = b \cdot h$ Replace b with 50 ft and h with 42 ft.
$A = 50 \text{ ft} \cdot 42 \text{ ft}$
$A = 2100 \text{ ft}^2$

(b) *Note*: The base is 18 in. and the height is 10 in. The height forms a 90° angle with the base.

$A = b \cdot h$
$A = 18 \text{ in.} \cdot 10 \text{ in.}$
$A = 180 \text{ in.}^2$

(c) $A = b \cdot h$ Replace b with 8 cm and h with 1 cm.
$A = 8 \text{ cm} \cdot 1 \text{ cm}$
$A = 8 \text{ cm}^2$

6. (a) $A = b \cdot h$ Replace A with 140 in.² and b with 14 in.

$140 \text{ in.}^2 = 14 \text{ in.} \cdot h$
$\dfrac{140 \text{ in.} \cdot \text{in.}}{14 \text{ in.}} = \dfrac{14 \text{ in.} \cdot h}{14 \text{ in.}}$ Divide both sides by 14 in.
$10 \text{ in.} = h$

The height is 10 inches.

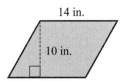

Check: $A = b \cdot h$
$A = 14 \text{ in.} \cdot 10 \text{ in.}$
$A = 140 \text{ in.}^2$

(b) $A = b \cdot h$ Replace A with 4 yd² and h with 1 yd.

$4 \text{ yd}^2 = b \cdot 1 \text{ yd}$
$\dfrac{4 \text{ yd} \cdot \text{yd}}{1 \text{ yd}} = \dfrac{b \cdot 1 \text{ yd}}{1 \text{ yd}}$ Divide both sides by 1 yd.
$4 \text{ yd} = b$

The base is 4 yd.

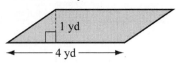

Check: $A = b \cdot h$
$A = 4 \text{ yd} \cdot 1 \text{ yd}$
$A = 4 \text{ yd}^2$

7. The sod will *cover* the playground, so you need to find the area of the playground.
The playground is rectangular. Use the formula for the area of a rectangle.

$A = l \cdot w$ Replace l with 22 yd and w with 16 yd.
$A = 22 \text{ yd} \cdot 16 \text{ yd}$
$A = 352 \text{ yd}^2$

The area is 352 yd², so the neighbors need to buy 352 yd² of sod. The cost of the sod is $3 per square yard, which means $3 for 1 square yard.

Cost = $3 \cdot 352$
 = $1056

The neighbors will spend $1056 on sod.

3.2 Section Exercises

1. The figure is a rectangle.

$A = l \cdot w$ Replace l with 11 ft and w with 7 ft.
$A = 11 \text{ ft} \cdot 7 \text{ ft}$
$A = 77 \text{ ft}^2$

The area is 77 ft², or 77 square feet.

3. The figure is a square.

$A = s^2$
$A = s \cdot s$ Remember, s^2 means $s \cdot s$.
$A = 10 \text{ m} \cdot 10 \text{ m}$ Replace s with 10 m.
$A = 100 \text{ m}^2$

The area is 100 m², or 100 square meters.

5. The figure is a parallelogram.

$A = bh$

Turn your book sideways to identify that the height is 25 mm and the base is 31 mm. Replace b with 31 mm and h with 25 mm.

$A = 31 \text{ mm} \cdot 25 \text{ mm}$
$A = 775 \text{ mm}^2$

The area is 775 mm², or 775 square millimeters.

7. The figure is a square.

$A = s^2$
$A = s \cdot s$ Remember, s^2 means $s \cdot s$.
$A = 6 \text{ in.} \cdot 6 \text{ in.}$ Replace s with 6 in.
$A = 36 \text{ in.}^2$

The area is 36 in.², or 36 square inches.

9.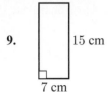

$A = l \cdot w$ Replace l with 15 cm and w with 7 cm.
$A = 15 \text{ cm} \cdot 7 \text{ cm}$
$A = 105 \text{ cm}^2$

The area of the calculator is 105 cm².

11.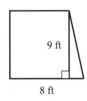

Area of a parallelogram.

$A = b \cdot h$ Replace b with 8 ft and h with 9 ft.
$A = 8 \text{ ft} \cdot 9 \text{ ft}$
$A = 72 \text{ ft}^2$

The area of the parallelogram is 72 ft².

13.

25 mi
25 mi

$A = s^2$ Area formula for a square
$A = s \cdot s$ Replace s with 25 mi.
$A = 25 \text{ mi} \cdot 25 \text{ mi}$
$A = 625 \text{ mi}^2$

The fire burned an area of 625 mi².

15.

1 m
1 m

$A = s^2$ Area of a square
$A = s \cdot s$ Replace s with 1 m.
$A = 1 \text{ m} \cdot 1 \text{ m}$
$A = 1 \text{ m}^2$

The area is 1 m².

17. $A = l \cdot w$ Replace A with 18 ft², and w with 3 ft.
  Divide both sides by 3 ft.
  $18 \text{ ft}^2 = l \cdot 3 \text{ ft}$
  $\dfrac{18 \text{ ft} \cdot \cancel{\text{ft}}}{3 \cancel{\text{ft}}} = \dfrac{l \cdot 3 \text{ ft}}{3 \text{ ft}}$
  $6 \text{ ft} = l$

The length of the desk is 6 ft.

6 ft
3 ft 3 ft
6 ft

Check: $A = l \cdot w$
  $A = 6 \text{ ft} \cdot 3 \text{ ft}$
  $A = 18 \text{ ft}^2$

19. $A = l \cdot w$ Replace A with 7200 yd², and l with 90 yd.
  Divide both sides by 90 yd.
  $7200 \text{ yd}^2 = 90 \text{ yd} \cdot w$
  $\dfrac{7200 \text{ yd} \cdot \cancel{\text{yd}}}{90 \cancel{\text{yd}}} = \dfrac{90 \text{ yd} \cdot w}{90 \text{ yd}}$
  $80 \text{ yd} = w$

The width of the parking lot is 80 yd.

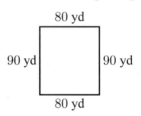

Check: $A = l \cdot w$
  $A = 90 \text{ yd} \cdot 80 \text{ yd}$
  $A = 7200 \text{ yd}^2$

21. $A = l \cdot w$ Replace A with 154 in.², and w with 11 in.
  Divide both sides by 11 in.
  $154 \text{ in.}^2 = l \cdot 11 \text{ in.}$
  $\dfrac{154 \text{ in.} \cdot \cancel{\text{in.}}}{11 \cancel{\text{in.}}} = \dfrac{l \cdot 11 \text{ in.}}{11 \text{ in.}}$
  $14 \text{ in.} = l$

The length of the photo is 14 inches.

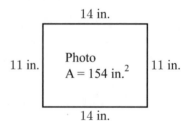

Check: $A = l \cdot w$
  $A = 14 \text{ in.} \cdot 11 \text{ in.}$
  $A = 154 \text{ in.}^2$

23. $A = s^2$ Area formula for a square
    $A = s \cdot s$ Replace A with 36 m².
    $36 \text{ m}^2 = s \cdot s$ By inspection, what number times itself is 36?
    $36 \text{ m}^2 = 6 \text{ m} \cdot 6 \text{ m}$ $6 \cdot 6$ is 36, so $6 \text{ m} \cdot 6 \text{ m}$ is 36 m².

    The length of one side of the floor is 6 m.

25. $A = s^2$ Area formula for a square
    $A = s \cdot s$ Replace A with 4 ft².
    $4 \text{ ft}^2 = s \cdot s$ By inspection, what number times itself is 4?
    $4 \text{ ft}^2 = 2 \text{ ft} \cdot 2 \text{ ft}$ $2 \cdot 2$ is 4, so $2 \text{ ft} \cdot 2 \text{ ft}$ is 4 ft².

    The length of one side of the sign is 2 ft.

27. Use the area formula for a parallelogram. Replace A with 500 cm² and b with 25 cm.

    $A = b \cdot h$
    $500 \text{ cm}^2 = 25 \text{ cm} \cdot h$ Divide both sides by 25 cm.
    $$\frac{500 \text{ cm} \cdot \cancel{\text{cm}}}{25 \cancel{\text{cm}}} = \frac{25 \text{ cm} \cdot h}{25 \text{ cm}}$$
    $20 \text{ cm} = h$

    The height is 20 cm.

    Check: $A = b \cdot h$
    $A = 25 \text{ cm} \cdot 20 \text{ cm}$
    $A = 500 \text{ cm}^2$

29. Use the area formula for a parallelogram. Replace A with 221 in.² and h with 13 in.

    $A = b \cdot h$
    $221 \text{ in.}^2 = b \cdot 13 \text{ in.}$ Divide both sides by 13 in.
    $$\frac{221 \text{ in.} \cdot \cancel{\text{in.}}}{13 \cancel{\text{in.}}} = \frac{b \cdot 13 \text{ in.}}{13 \text{ in.}}$$
    $17 \text{ in.} = b$

    The base is 17 inches.

    Check: $A = b \cdot h$
    $A = 17 \text{ in.} \cdot 13 \text{ in.}$
    $A = 221 \text{ in.}^2$

31. Use the area formula for a parallelogram. Replace A with 9 m² and b with 9 m.

    $A = b \cdot h$
    $9 \text{ m}^2 = 9 \text{ m} \cdot h$ Divide both sides by 9 m.
    $$\frac{9 \text{ m} \cdot \cancel{\text{m}}}{9 \cancel{\text{m}}} = \frac{9 \text{ m} \cdot h}{9 \text{ m}}$$
    $1 \text{ m} = h$

    The height is 1 m.

    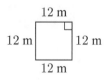

    Check: $A = b \cdot h$
    $A = 9 \text{ m} \cdot 1 \text{ m}$
    $A = 9 \text{ m}^2$

33. The height of the parallelogram is not part of the perimeter. Also, square units are used for area, not perimeter.

    $P = 25 \text{ cm} + 25 \text{ cm} + 25 \text{ cm} + 25 \text{ cm}$
    $P = 100 \text{ cm}$

35. The figure is a square.

    $P = 4s = 4 \cdot 45 \text{ in.} = 180 \text{ in.}$
    $A = s^2 = s \cdot s = 45 \text{ in.} \cdot 45 \text{ in.} = 2025 \text{ in.}^2$

37. The figure is a rectangle with base 9 ft and height 39 ft.

    $P = 39 \text{ ft} + 39 \text{ ft} + 9 \text{ ft} + 9 \text{ ft}$
    $P = 96 \text{ ft}$

    $A = b \cdot h$
    $A = 9 \text{ ft} \cdot 39 \text{ ft}$
    $A = 351 \text{ ft}^2$

39. The figure is a parallelogram with base 18 cm and height 10 cm.

    $P = 18 \text{ cm} + 12 \text{ cm} + 18 \text{ cm} + 12 \text{ cm}$
    $P = 60 \text{ cm}$

    $A = b \cdot h$
    $A = 18 \text{ cm} \cdot 10 \text{ cm}$
    $A = 180 \text{ cm}^2$

41. The mat is a square.

    $P = 4 \cdot s$ $A = s \cdot s$
    $P = 4 \cdot 12 \text{ m}$ $A = 12 \text{ m} \cdot 12 \text{ m}$
    $P = 48 \text{ m}$ $A = 144 \text{ m}^2$

43.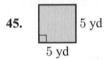

Tyra is decorating the top edges of her walls, so find the perimeter of her ceiling.

$P = 2 \cdot l + 2 \cdot w$ Perimeter of a rectangle.

$P = 2 \cdot 5 \text{ m} + 2 \cdot 4 \text{ m}$ Replace l with 5 m and w with 4 m.

$P = 18 \text{ m}$

The strip costs $6 per meter. To find the cost of 18 meters, multiply $6 · 18 to get $108.

To have the top edges of her walls decorated, Tyra will spend $108.

45. [figure: square, 5 yd by 5 yd]

Mr. and Mrs. Gomez are *covering* the bedroom floor, so find the area.

$A = s^2$ Area of a square
$A = s \cdot s$
$A = 5 \text{ yd} \cdot 5 \text{ yd}$
$A = 25 \text{ yd}^2$

The carpet cost is $23 per square yard. To find the cost for 25 square yards, multiply $23 · 25 to get $575.

The cost of padding and installation is $6 per square yard. To find the cost for 25 square yards, multiply $6 · 25 to get $150.

To have the bedroom carpeted, Mr. and Mrs. Gomez will spend $575 + $150, or $725 total.

47. The football field is a rectangle with length 100 yards and area 5300 yd².

$A = l \cdot w$ Area formula for a rectangle.

$5300 \text{ yd}^2 = 100 \text{ yd} \cdot w$ Replace A with 5300 yd² and l with 100 yd.

$\dfrac{5300 \text{ yd} \cdot \text{yd}}{100 \text{ yd}} = \dfrac{100 \text{ yd} \cdot w}{100 \text{ yd}}$ Divide both sides by 100 yd.

$53 \text{ yd} = w$

The width of the field is 53 yards.

49. $P = 4 \cdot s$ $A = s \cdot s$
     $= 4 \cdot 13 \text{ ft}$ $= 13 \text{ ft} \cdot 13 \text{ ft}$
     $= 52 \text{ ft}$ $= 169 \text{ ft}^2$

$\dfrac{169 \text{ ft}^2}{8 \text{ campers}} = 21.125 \text{ ft}^2$ for each camper

$\approx 21 \text{ ft}^2$ for each camper

51. Since perimeter $P = 2 \cdot l + 2 \cdot w$, if $w = 1$ and $P = 12$, then $2 \cdot l = 12 - 2 \cdot 1 = 10$ and $l = 5$. Similarly, if $w = 2$, $l = 4$, and if $w = 3$, $l = 3$. So there are only three possibilities for whole number plots with perimeter 12 ft and $l \geq w$.

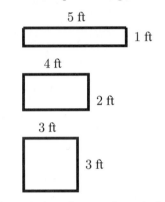

52. **(a)** Use the area formula for a rectangle, $A = l \cdot w$.

l	w	$A = l \cdot w$
5 ft	1 ft	5 ft²
4 ft	2 ft	8 ft²
3 ft	3 ft	9 ft²

(b) The table in part (a) shows that the 3 ft by 3 ft plot has the greatest area.

53. As in Exercise 51, we let w equal a whole number and find l.

w	$2 \cdot w$	$2l = 16 - 2 \cdot w$	l
1	2	14	7
2	4	12	6
3	6	10	5
4	8	8	4

So there are only four possibilities for whole number plots with perimeter 16 ft.

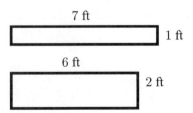

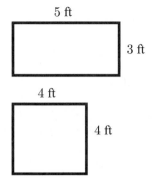

54. (a) Use the area formula for a rectangle, $A = l \cdot w$.

w	l	$A = l \cdot w$
1 ft	7 ft	7 ft^2
2 ft	6 ft	12 ft^2
3 ft	5 ft	15 ft^2
4 ft	4 ft	16 ft^2

(b) Based on Exercises 52 (a) and 54 (a), square plots have the greatest area.

Summary Exercises on Perimeter and Area

1. The figure is a rectangle with base 3 m and height 13 m.

$P = 13\text{ m} + 13\text{ m} + 3\text{ m} + 3\text{ m}$
$P = 32\text{ m}$

$A = b \cdot h$
$A = 3\text{ m} \cdot 13\text{ m}$
$A = 39\text{ m}^2$

3. The figure is a parallelogram with base 7 yd and height 8 yd.

$P = 10\text{ yd} + 10\text{ yd} + 7\text{ yd} + 7\text{ yd}$
$P = 34\text{ yd}$

$A = b \cdot h$
$A = 7\text{ yd} \cdot 8\text{ yd}$
$A = 56\text{ yd}^2$

5. The figure is a square.

$P = 4s = 4 \cdot 9\text{ in.} = 36\text{ in.}$
$A = s^2 = s \cdot s = 9\text{ in.} \cdot 9\text{ in.} = 81\text{ in.}^2$

7. The figure is a rectangle with length 9 ft and width 4 ft.

$P = 9\text{ ft} + 9\text{ ft} + 4\text{ ft} + 4\text{ ft}$
$P = 26\text{ ft}$

$A = b \cdot h$
$A = 9\text{ ft} \cdot 4\text{ ft}$
$A = 36\text{ ft}^2$

9.
$P = 2l + 2w$ Replace P with 16 ft and l with 7 ft.
$16\text{ ft} = 2 \cdot 7\text{ ft} + 2w$ Multiply.
$16\text{ ft} = 14\text{ ft} + 2w$ To get $2w$ by itself,
$\underline{\phantom{16\text{ ft} =}{}^-14\text{ ft} {}^-14\text{ ft}}$ add $^-14$ ft to both sides.
$2\text{ ft} = 0 + 2w$
$\dfrac{2\text{ ft}}{2} = \dfrac{2w}{2}$ Divide both sides by 2.
$1\text{ ft} = w$

The width is 1 ft.

Check: $P = 7\text{ ft} + 7\text{ ft} + 1\text{ ft} + 1\text{ ft} = 16\text{ ft}$

11.
$A = s^2$ Replace A with 36 in.^{2}.
$36\text{ in.}^2 = s^2$
$36\text{ in.}^2 = s \cdot s$
$36\text{ in.}^2 = 6\text{ in.} \cdot 6\text{ in.}$

The length of one side of the photograph is 6 in.

Check: $A = s \cdot s$
$A = 6\text{ in.} \cdot 6\text{ in.}$
$A = 36\text{ in.}^2$

13.
$P = 4s$ Perimeter formula for a square
$64\text{ cm} = 4s$ Replace P with 64 cm.
$\dfrac{64\text{ cm}}{4} = \dfrac{4s}{4}$ Divide both sides by 4.
$16\text{ cm} = s$

The length of one side of the map is 16 cm.

Check:
$P = 16\text{ cm} + 16\text{ cm} + 16\text{ cm} + 16\text{ cm} = 64\text{ cm}$

15. The figure is a triangle.

$P = 168\text{ m} + 168\text{ m} + 168\text{ m}$
$P = 504\text{ m}$

504 meters of fencing are needed.

17.
$$P = 2l + 2w \quad \text{Replace } w \text{ with 12 in.}$$
$$\text{and } P \text{ with 42 in.}$$
$$42 \text{ in.} = 2l + 2 \cdot 12 \text{ in.} \quad \text{Multiply on the right.}$$
$$42 \text{ in.} = 2l + 24 \text{ in.} \quad \text{Add } ^-24 \text{ in. to both sides.}$$
$$\underline{^-24 \text{ in.} \qquad \qquad ^-24 \text{ in.}}$$
$$18 \text{ in.} = 2l + 0 \text{ in.}$$
$$\frac{18 \text{ in.}}{2} = \frac{2l}{2} \quad \text{Divide both sides by 2.}$$
$$9 \text{ in.} = l$$

The height of the screen is 9 inches.

19. Each flag is rectangular in shape.
$$P = 3 \text{ ft} + 5 \text{ ft} + 3 \text{ ft} + 5 \text{ ft}$$
$$P = 16 \text{ ft}$$
$$A = l \cdot w$$
$$A = 5 \text{ ft} \cdot 3 \text{ ft}$$
$$A = 15 \text{ ft}^2$$

To make seven flags, complete with binding, we will need $7(15 \text{ ft}^2) = 105 \text{ ft}^2$ of fabric and $7(16 \text{ ft}) = 112 \text{ ft}$ of binding.

3.3 Solving Application Problems with One Unknown Quantity

3.3 Margin Exercises

1. (a) 15 less than a number
$$x - 15$$

   (b) 12 more than a number
$$x + 12 \quad \text{or} \quad 12 + x$$

   (c) A number increased by 13
$$x + 13 \quad \text{or} \quad 13 + x$$

   (d) A number minus 8
$$x - 8$$

   (e) 10 plus a number
$$10 + x \quad \text{or} \quad x + 10$$

   (f) A number subtracted from 6
$$6 - x$$

   (g) 6 subtracted from a number
$$x - 6$$

2. (a) Double a number
$$2x$$

   (b) The product of $^-8$ and a number
$$^-8x$$

   (c) The quotient of 15 and a number
$$\frac{15}{x}$$

   (d) $30 - 5 \cdot x$ or $30 - 5x$

   Note: $5x - 30$ would be "30 subtracted from 5 times a number."

3. (a) Let x represent the unknown number.

   $\underbrace{\text{3 times a number}}_{3x} \; \underbrace{\text{is added to}}_{+} \; \underbrace{4}_{4} \; \underbrace{\text{the result is}}_{=} \; \underbrace{19}_{19}$

   $3x + 4 = 19$ To get $3x$ by itself, add $^-4$ to both sides.
   $$\underline{ \;^-4 \qquad ^-4}$$
   $$3x + 0 = 15$$
   $$\frac{3x}{3} = \frac{15}{3} \quad \text{Divide both sides by 3.}$$
   $$x = 5$$

   Check: 3 times 5 $[= 15]$ is added to 4 $[= 19]$. The result is 19, so 5 is the number.

   (b) Let x represent the unknown number.

   $\underbrace{\text{6 times a number}}_{6x} \; \underbrace{\text{7 is subtracted from}}_{-\;\;7} \; \underbrace{\text{the result is}}_{=} \; \underbrace{^-25}_{^-25}$

   $6x - 7 = ^-25$ Add 7 to both sides.
   $$\underline{ +7 \qquad +7}$$
   $$6x + 0 = ^-18$$
   $$\frac{6x}{6} = \frac{^-18}{6} \quad \text{Divide both sides by 6.}$$
   $$x = ^-3$$

   Check: 6 times $^-3$ $[= ^-18]$ minus 7 $[= ^-25]$ is $^-25$, so $^-3$ is the number.

4. *Step 1*
   The problem is about the number of people on a bus.

   Unknown: number of people who got on at the first stop

   Known: 3 got on; 5 got on; 10 got off; 4 people still on the bus

   Step 2(a)
   Let p be the number of people who got on at the first stop.

   Step 3
   $$p + 3 + 5 - 10 = 4$$

3.3 Solving Application Problems with One Unknown Quantity

Step 4
$$\begin{aligned} p + 3 + 5 - 10 &= 4 \\ p + 8 + {}^-10 &= 4 \\ p + {}^-2 &= 4 \\ \underline{+ 2 + 2} & \\ p + 0 &= 6 \\ p &= 6 \end{aligned}$$

Step 5
6 people got on at the first stop.

Step 6
First stop: 6 got on
Second stop: 3 got on $6 + 3 = 9$ on
Third stop: 5 got on $9 + 5 = 14$ on
Fourth stop: 10 got on $14 - 10 = 4$ on
4 people are left. **matches**

5. *Step 1*
    The problem is about money being donated and spent by a college.

    Unknown: amount of money given by each donor.

    Known: Each donor gave the "same" dollar amount. College gave out $1250, $900, and $850. $250 was left over.

    Step 2(a)
    Let m be the amount of money donated by each donor.

    Step 3
    $\underbrace{\text{5 equal donations}}_{5 \cdot m} - \underbrace{\text{money spent on scholarships}}_{(1250 + 900 + 850)} = \underbrace{\text{money left over}}_{250}$

    Step 4
    $$\begin{aligned} 5m - (1250 + 900 + 850) &= 250 \\ 5m - 3000 &= 250 \\ 5m + ({}^-3000) &= 250 \\ \underline{+ 3000 + 3000} & \\ 5m + 0 &= 3250 \\ \frac{5m}{5} &= \frac{3250}{5} \\ m &= 650 \end{aligned}$$

    Step 5
    Each donor gave $650.

    Step 6
    5 donors each gave $650, so $5 \cdot \$650 = \$3250$ donated

    College gave out $1250, $900, $850, so $\$3250 - \$1250 - \$900 - \$850 = \$250$.
    College had $250 left. **matches**

6. *Step 1*
    The problem is about money being donated by two people.

    Unknown: LuAnn's donation

    Known: Susan donated $10 more than twice what LuAnn donated. Susan donated $22.

    Step 2(a)
    Let d represent LuAnn's donation.

    Step 3
    $\underbrace{\text{Susan's donation}}_{22} \underset{\downarrow}{\text{ is }} \underset{=}{} \underset{\downarrow}{\text{ \$10 }} \underset{10}{} \underset{\downarrow}{\text{ more than }} \underset{+}{} \underset{\downarrow}{\text{ twice }} \underset{2\cdot}{} \underbrace{\text{LuAnn's donation}}_{d}$

    Step 4
    $$\begin{aligned} 22 &= 10 + 2d \\ \underline{{}^-10 {}^-10} & \\ 12 &= 0 + 2d \\ \frac{12}{2} &= \frac{2d}{2} \\ 6 &= d \end{aligned}$$

    Step 5
    LuAnn donated $6.

    Step 6
    $10 more than twice $6 is
    $\$10 + (2 \cdot \$6) = \$10 + \$12 = \$22$.

    Susan donated $22.

3.3 Section Exercises

1. 14 plus a number

    $14 + x$ or $x + 14$

3. $^-5$ added to a number

    $^-5 + x$ or $x + {}^-5$

5. 20 minus a number

    $20 - x$

7. 9 less than a number

    $x - 9$

9. Subtract 4 from a number

    $x - 4$

11. $^-6$ times a number

    ^-6x

13. Double a number

    $2x$

15. A number divided by 2

    $\dfrac{x}{2}$

17. Twice a number added to 8

$$8 + 2x \quad \text{or} \quad 2x + 8$$

19. 10 fewer than seven times a number

$$7x - 10$$

21. The sum of twice a number and the number

$$2x + x \quad \text{or} \quad x + 2x$$

23. Let n represent the unknown number.

four times a number [$\downarrow$ $4n$] decreased by 2 [$\downarrow$ -2] result [$\downarrow$ $=$] is 26. [$\downarrow$ 26]

$$4n - 2 = 26$$
$$4n - 2 + 2 = 26 + 2$$
$$4n = 28$$
$$\frac{4n}{4} = \frac{28}{4}$$
$$n = 7$$

The number is 7.

Check: Four times 7 [$= 28$] is decreased by 2 [$= 26$]. *True*

25. Let n represent the unknown number.

Twice a number [$\downarrow$ $2n$] added to [$\downarrow$ $+$] the number [$\downarrow$ n] is [$\downarrow$ $=$] $^-15$. [$\downarrow$ $^-15$]

$$2n + n = {}^-15$$
$$3n = {}^-15$$
$$\frac{3n}{3} = \frac{{}^-15}{3}$$
$$n = {}^-5$$

The number is $^-5$.

Check: Twice $^-5$ [$= {}^-10$] is added to $^-5$ [$= {}^-15$]. *True*

27. Let n represent the unknown number.

Product of a number and 5 [$\downarrow$ $5n$] increased by [$\downarrow$ $+$] 12 [$\downarrow$ 12] the result is [$\downarrow$ $=$] 7 times the number. [$\downarrow$ $7n$]

$$5n + 12 = 7n$$
$$5n + 12 + {}^-5n = 7n + {}^-5n$$
$$12 = 2n$$
$$\frac{12}{2} = \frac{2n}{2}$$
$$6 = n$$

The number is 6.

Check: The product of 6 and 5 [$= 30$] is increased by 12 [$= 42$] is seven times 6 [$= 42$]. *True*

29. Let n represent the unknown number.

30 subtract [$\downarrow$ $30 -$] 3 times a number [$\downarrow$ $3n$] is [$\downarrow$ $=$] 2 plus [$\downarrow$ $2 +$] the number [$\downarrow$ n]

$$30 - 3n = 2 + n$$
$$30 - 3n + 3n = 2 + n + 3n$$
$$30 = 2 + 4n$$
$$30 + {}^-2 = 2 + 4n + {}^-2$$
$$28 = 4n$$
$$\frac{28}{4} = \frac{4n}{4}$$
$$7 = n$$

The number is 7.

Check: Three times 7 [$= 21$] is subtracted from 30 [$= 9$] is 2 plus 7 [$= 9$]. *True*

31. *Step 1*
The problem is about Ricardo's weight.

Unknown: his original weight

Known: He gained 15 pounds, lost 28, and then regained 5 pounds to weigh 177.

Step 2(a)
Let w represent Ricardo's original weight.

Step 3

weight at beginning [$\downarrow$ w] + pounds gained [$\downarrow$ 15] − pounds lost [$\downarrow$ 28] + pounds regained [$\downarrow$ 5] = final weight [$\downarrow$ 177]

Step 4
$$w + 15 - 28 + 5 = 177 \quad \text{Combine like terms.}$$
$$w - 8 = 177$$
$$w = 185$$

Step 5
He weighed 185 pounds originally.

Step 6
$$185 + 15 - 28 + 5 = 177$$

The correct answer is 185 pounds.

33. *Step 1*
The problem is about the number of cookies in the cookie jar.

Unknown: number of cookies the children ate

Known: 18 cookies at the start, three dozen cookies put in the jar, and 49 cookies at the end

Step 2(a)
Let c represent the number of cookies the children ate.

Step 3

cookies at the start	−	cookies eaten	+	cookies added	=	ended up with 49 cookies
↓		↓		↓		↓
18	−	c	+	36	=	49

Step 4

$18 - c + 36 = 49$ Write the understood $1c$. Change subtraction to adding the opposite.

$$54 + (^-1c) = 49$$
$$\underline{^-54 \qquad\qquad ^-54}$$
$$0 + (^-1c) = ^-5$$
$$0 + (^-1c) = ^-5$$
$$\frac{^-1c}{^-1} = \frac{^-5}{^-1}$$
$$c = 5$$

Step 5
The children ate 5 cookies.

Step 6
There were 18 cookies at the start. The children ate 5 cookies, so that left $18 - 5 = 13$ cookies in the jar.

Three dozen cookies were added, so there were $13 + 36 = 49$ cookies.

The correct answer is 5 cookies.

35. *Step 1*
The problem is about boxes of pens.

Unknown: the number of pens in each box

Known: The bookstore started with 6 boxes of pens. The bookstore sold 32 pens, then 35 pens, and 5 pens were left.

Step 2(a)
Let p represent the number of pens in each box.

Step 3

total number of boxes	·	number of pens in each box	−	32	−	35	=	5
↓		↓		↓		↓		↓
6	·	p	−	32	−	35	=	5

Step 4
$$6p - 32 - 35 = 5$$
$$6p - 67 = 5$$
$$\underline{^+67 \qquad\qquad ^+67}$$
$$6p + 0 = 72$$
$$\frac{6p}{6} = \frac{72}{6}$$
$$p = 12$$

Step 5
There were 12 pens in each box.

Step 6
The bookstore ordered 6 boxes of red pens each containing 12 pens. $6(12) = 72$ pens
32 were sold: $72 - 32 = 40$
35 were sold: $40 - 35 = 5$
5 pens were left on the shelf.

The correct answer is 12 pens per box.

37. *Step 1*
The problem is about the bank account of a music club.

Unknown: the dues paid by each member

Known: 14 total members, earned $340, spent $575, and account is overdrawn by $25

Step 2(a)
Let d represent the dues paid by each member.

Step 3

total dues paid by 14 members	+	earned	−	spent	=	ended up with an overdrawn bank account
↓		↓		↓		↓
$14 \cdot d$	+	340	−	575	=	$^-25$

Step 4
$$14d + 340 - 575 = ^-25$$
$$14d + 340 + (^-575) = ^-25$$
$$14d + (^-235) = ^-25$$
$$\underline{^+235 \qquad\qquad ^+235}$$
$$14d + 0 = 210$$
$$\frac{14d}{14} = \frac{210}{14}$$
$$d = 15$$

Step 5
Each member paid $15 in dues.

Chapter 3 Solving Application Problems

Step 6
14 members paid dues of $15 each:
$14(\$15) = \$210$ in the bank account.

Earned $340: $210 + $340 = $550 in the account.

Spent $575: $550 − $575 = ⁻$25 in the account.

The account was overdrawn by $25.

The correct answer is $15 dues.

39. *Step 1*
The problem is about Tamu's age.

Unknown: how old Tamu is right now

Known: Tamu's age equals the number you get after multiplying his age by 4 and then subtracting 75.

Step 2(a)
Let a represent Tamu's age.

Step 3

Tamu's age	=	his age multiplied by 4	subtract	75
↓		↓	↓	↓
a	=	$4a$	−	75

Step 4
$$a = 4a - 75$$
$$\underline{{}^-a \qquad\qquad {}^-a}$$
$$0 = 3a - 75$$
$$\underline{{}^+75 \qquad\qquad {}^+75}$$
$$75 = 3a$$
$$\frac{75}{3} = \frac{3a}{3}$$
$$25 = a$$

Step 5
Tamu is 25 years old.

Step 6
75 subtracted from 4 times 25 :
$4(25) - 75 = 100 - 75 = 25$

The correct solution is 25 years old.

41. *Step 1*
The problem is about spending money for clothes.

Unknown: amount spent on clothes by Brenda

Known: Consuelo spent $3 less than twice the amount that Brenda spent, Consuelo spent $81

Step 2(a)
Let m represent the amount of money that Brenda spent.

Step 3

Consuelo spent	=	$3 less than twice the amount that Brenda spent
↓		↓
81	=	$2m - 3$

Step 4
$$81 = 2m - 3$$
$$81 = 2m + (^-3)$$
$$\underline{{}^+3 \qquad\qquad {}^+3}$$
$$\frac{84}{2} = \frac{2m}{2}$$
$$42 = m$$

Step 5
Brenda spent $42 on clothes.

Step 6
Consuelo spent $3 less than twice the amount that Brenda spent, or
$2 \cdot \$42 + (^-\$3) = \$84 - (\$3) = \$81$.

The correct answer is $42.

43. *Step 1*
The problem concerns bags of candy.

Unknown: the number of pieces of candy in each bag

Known: There were originally 5 bags. She gave 3 pieces of candy each to 48 children. Afterwards 1 bag remained.

Step 2(a)
Let b represent the number of pieces in each bag.

Step 3

total amount of candy at the beginning	subtract	total amount of candy given out	results in	amount of candy in one bag
↓	↓	↓	↓	↓
$5 \cdot b$	−	$3 \cdot 48$	=	b

Step 4
$$5b - 3 \cdot 48 = b$$
$$5b - 144 = b$$
$$\begin{array}{r} +144 \\ \hline 5b + 0 \end{array} = \begin{array}{r} +144 \\ \hline b + 144 \end{array}$$
$$\begin{array}{r} {}^-b \\ \hline 4b \end{array} = \begin{array}{r} {}^-b \\ \hline 144 \end{array}$$
$$\frac{4b}{4} = \frac{144}{4}$$
$$b = 36$$

Step 5
There were 36 pieces of candy in each bag.

Step 6
5 bags of 36 pieces of candy: $5 \cdot 36 = 180$
3 pieces were handed out to each of 48 children:
$3 \cdot 48 = 144$
What remained: $180 - 144 = 36$
Equals the amount in one bag: 36

The correct answer is 36.

45. *Step 1*
The problem is about the recommended daily iron intake.

Unknown: the recommended daily iron intake for an infant

Known: The recommended daily iron intake for an adult female is 4 mg less than twice that for an infant. The amount for an adult female is 18 mg.

Step 2(a)
Let d represent the daily amount for an infant.

Step 3

the amount for an adult female	is	twice	the amount for an infant	less	4 mg
↓	↓	↓	↓	↓	↓
18	=	2 ·	d	−	4

Step 4
$$18 = 2d - 4$$
$$\begin{array}{r} +4 \\ \hline 22 \end{array} = \begin{array}{r} +4 \\ \hline 2d \end{array}$$
$$\frac{22}{2} = \frac{2d}{2}$$
$$11 = d$$

Step 5
The daily recommended iron intake for an infant is 11 mg.

Step 6
4 less than twice 11: $2 \cdot 11 - 4 = 18$
Equals the amount for an adult female: 18

The correct answer is 11 mg.

3.4 Solving Application Problems with Two Unknown Quantities

3.4 Margin Exercises

1. *Step 1*
The problem is about the amount made by Keonda and her daughter.

Unknowns: how much Keonda made; how much the daughter made

Known: Keonda made $12 more than her daughter. Together they made $182.

Step 2(b)
There are two unknowns. You know the least about the money made by the daughter, so let m be the money made by the daughter.

Since Keonda made $12 more, $m + 12$ represents Keonda's earnings.

Step 3

Daughter's money	+	Keonda's money	=	total money
↓	↓	↓	↓	↓
m	+	$m + 12$	=	182

Step 4
$$m + m + 12 = 182$$
$$2m + 12 = 182$$
$$\begin{array}{r} {}^-12 \\ \hline 2m + 0 \end{array} = \begin{array}{r} {}^-12 \\ \hline 170 \end{array}$$
$$\frac{2m}{2} = \frac{170}{2}$$
$$m = 85$$

Step 5
The daughter made $85.
Keonda made $85 + $12 = $97.

Step 6
$97 is $12 more than $85 and the sum of their earnings is $85 + $97 = $182 (matches).

2. *Step 1*
The problem is about putting fishing line on two reels.

Unknowns: lengths of the two lines

Known: The length of one line is 25 yd less than the length of the other line. The total length is 175 yd.

Step 2(b)
There are two unknowns. Let r represent the length of the line on the first reel, then $r - 25$ is the length of the line on the second reel.

Step 3

length of first line	+	length of second line	=	total length
↓	↓	↓	↓	↓
r	+	$r - 25$	=	175

Step 4

$$\begin{aligned} r + (r-25) &= 175 \\ 2r - 25 &= 175 \\ +25 &\quad +25 \\ \hline 2r &= 200 \\ \frac{2r}{2} &= \frac{200}{2} \\ r &= 100 \end{aligned}$$

Step 5
The first reel had 100 yards of line on it and the second reel had $100 - 25 = 75$ yards of line on it.

Step 6
75 yd is 25 yd less than 100 yd and $100 \text{ yd} + 75 \text{ yd} = 175 \text{ yd}$.

3. *Step 1*
The problem is about a rectangular garden plot.

Unknowns: garden length and width

Known: The length is 3 yd longer than the width. The perimeter of the garden is 22 yd.

Step 2(b)
There are two unknowns. You know the least about the width, so let w represent the width. The length is 3 yd more than the width, so let $w + 3$ represent the length.

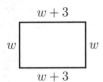

Step 3
$P = 2l + 2w$
$22 = 2 \cdot (w + 3) + 2w$

Step 4
$$\begin{aligned} 22 &= 2(w+3) + 2w \\ 22 &= 2 \cdot w + 2 \cdot 3 + 2w \\ 22 &= 2w + 6 + 2w \\ 22 &= 4w + 6 \\ -6 &\quad -6 \\ \hline 16 &= 4w + 0 \\ \frac{16}{4} &= \frac{4w}{4} \\ 4 &= w \end{aligned}$$

Step 5
The width is 4 yd.
The length is $4 + 3 = 7$ yd.

Step 6
7 yd is 3 yd more than 4 yd.
$P = 2 \cdot 7 \text{ yd} + 2 \cdot 4 \text{ yd}$
$P = 14 \text{ yd} + 8 \text{ yd}$
$ = 22 \text{ yd}$ (matches)

3.4 Section Exercises

1. *Step 1*
The problem is about someone's age.

Unknowns: how old I am; how old my sister is

Known: My sister is 9 years older than I am. The sum of our ages is 51.

Step 2(b)
Let a represent how old I am. Since my sister is 9 years older, let $a + 9$ represent her age.

Step 3

My age	+	sister's age	=	sum of ages
↓		↓		↓
a	+	$a + 9$	=	51

Step 4

$$\begin{aligned} a + a + 9 &= 51 \\ 2a + 9 &= 51 \\ -9 &\quad -9 \\ \hline 2a + 0 &= 42 \\ \frac{2a}{2} &= \frac{42}{2} \\ a &= 21 \end{aligned}$$

Step 5
I am 21 years old, and my sister is $21 + 9 = 30$ years old.

Step 6
30 is 9 more than 21 and the sum of our ages is $21 + 30 = 51$.

3. *Step 1*
The problem concerns a couple's earnings.

Unknowns: how much each earned

Known: Lien earned $1500 more than her husband. Together they earned $37,500.

Step 2(b)
Let m represent the husband's earnings. Then Lien earned $m + 1500$.

Step 3

Husband's earnings	+	Lien's earnings	=	amount earned together
↓		↓		↓
m	+	$m + 1500$	=	$37{,}500$

Step 4
$$m + m + 1500 = 37{,}500$$
$$2m + 1500 = 37{,}500$$
$$\underline{{}^-1500 \quad {}^-1500}$$
$$2m + 0 = 36{,}000$$
$$\frac{2m}{2} = \frac{36{,}000}{2}$$
$$m = 18{,}000$$
$$m + 1500 = 18{,}000 + 1500$$
$$= 19{,}500$$

Step 5
Lien earned $19,500 and her husband earned $18,000.

Step 6
$19,500 is $1500 more than $18,000 and the sum of $18,000 and $19,500 is $37,500.

5. *Step 1*
The problem is about the price of a computer and the price of a printer.

Unknowns: the computer's price and the printer's price

Known: The computer's price is five times the printer's price. The total paid is $1320.

Step 2(b)
Let m represent the price of the printer. The price of the computer can be represented by $5 \cdot m$, or $5m$.

Step 3
The total price for both is $1320.

printer price	+	computer price	=	total price
↓		↓		↓
m	+	$5m$	=	1320

Step 4
$$m + 5m = 1320$$
$$\frac{6m}{6} = \frac{1320}{6}$$
$$m = 220$$

Step 5
The printer's price is $220. The computer's price is $5 \cdot \$220 = \$1100$.

Step 6
Five times $220 is $1100 and the sum of $220 and $1100 is $1320.

7. *Step 1*
The problem concerns cutting a board into two pieces.

Unknowns: the length of each piece of board

Known: The board had a total length of 78 cm. One piece is 10 cm longer than the other.

Step 2(b)
Let x represent the shorter length. Then the longer length is $x + 10$ cm.

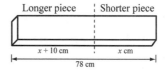

Step 3
The sum of the lengths is 78 cm, so
$x + (x + 10) = 78$.

Step 4
$$x + x + 10 = 78$$
$$2x + 10 = 78$$
$$\underline{{}^-10 \quad {}^-10}$$
$$2x + 0 = 68$$
$$\frac{2x}{2} = \frac{68}{2}$$
$$x = 34$$

Step 5
The shorter length is $x = 34$ cm, so the longer length is $34 + 10 = 44$ cm.

Step 6
44 cm is 10 cm longer than 34 cm, and the sum of 34 cm and 44 cm is 78 cm.

9. *Step 1*
The problem is about the lengths of two pieces of wire.

Unknowns: the length of each piece

Known: One piece is 7 ft shorter than the other. The wire was 31 ft long before it was cut into two pieces.

Step 2(b)
Let x represent the length of the long piece. Then $x - 7$ represents the piece that is 7 ft shorter.

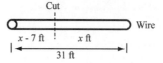

Step 3
The lengths of the two pieces together should be 31, so $x + (x - 7) = 31$.

Step 4
$$x + x - 7 = 31$$
$$2x + (^-7) = 31$$
$$7 \quad 7$$
$$\overline{2x + 0 = 38}$$
$$2x = 38$$
$$x = 19$$

Step 5
The long piece is 19 ft. The short piece is $19 - 7 = 12$ ft.

Step 6
12 ft is 7 ft shorter than 19 ft and the sum of 19 ft and 12 ft is 31 ft.

11. *Step 1*
The problem is about the number of Senators and Representatives in Congress.

Unknowns: The number of Senators and the number of Representatives

Known: The number of Representatives is 65 less than 5 times the number of Senators. The total number of Representatives and Senators is 535.

Step 2(b)
Let s represent the number of Senators. Then the number of Representatives is $5s - 65$.

Step 3
Since the total number in both houses of Congress equals 535, we have $s + 5s - 65 = 535$.

Step 4
$$s + 5s - 65 = 535$$
$$6s - 65 = 535$$
$$+ 65 \quad\quad + 65$$
$$\overline{6s + 0 = 600}$$
$$\frac{6s}{6} = \frac{600}{6}$$
$$s = 100$$

Step 5
The number of Senators is 100 and the number of Representatives is
$5s - 65 = 5(100) - 65 = 435$.

Step 6
Since $5(100) - 65 = 435$ and $100 + 435 = 535$, this answer is correct.

13. *Step 1*
The problem is about cutting a fence into parts.

Unknown: The length of each part

Known: Two parts are of equal length. The third part is 25 m longer than each of the other parts. The fence is 706 m long.

Step 2(b)
Let x represent the length of one of the two equal parts. Then the third part is $x + 25$ m long.

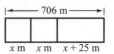

Step 3
Since the sum of the parts is 706 m,
$x + x + (x + 25) = 706$.

Step 4
$$x + x + x + 25 = 706$$
$$3x + 25 = 706$$
$$-25 \quad\quad -25$$
$$\overline{3x + 0 = 681}$$
$$\frac{3x}{3} = \frac{681}{3}$$
$$x = 227$$

Step 5
The two equal parts are each 227 m long, so the longer part is $x + 25 = 227 + 25 = 252$ m in length.

Step 6
Since $252 - 227 = 25$ and
$227 + 227 + 252 = 706$, this answer is correct.

15. *Step 1*
The problem is about the length l of a rectangle.

Unknown: Rectangle's length

Known: Rectangle's width and perimeter

Step 2(a)
Let x represent the length.

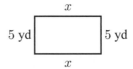

Step 3
$P = 2l + 2w \quad\quad$ Perimeter $= 48$ yd
$48 = 2 \cdot x + 2 \cdot 5$

Step 4
$$48 = 2x + 10$$
$$-10 \quad\quad -10$$
$$\overline{38 = 2x + 0}$$
$$\frac{38}{2} = \frac{2x}{2}$$
$$19 = x$$

Step 5
The rectangle's length is 19 yd.

Step 6
$P = 2l + 2w$
$P = 2 \cdot 19 + 2 \cdot 5$
$P = 38 + 10$
$P = 48$

17. *Step 1*
The problem is about the dimensions of a rectangular dog pen.

Unknowns: The length and width of the dog pen

Known: The perimeter is 36 ft. The length is twice the width.

Step 2(b)
Let w represent the width. Then the length equals $2w$.

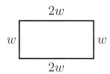

Step 3
$P = 2l + 2w$ Perimeter = 36 ft
$36 = 2 \cdot 2w + 2w$

Step 4
$36 = 4w + 2w$
$36 = 6w$
$\dfrac{36}{6} = \dfrac{6w}{6}$
$6 = w$

Step 5
The width is 6 ft, so the length is $2w = 2 \cdot 6 = 12$ ft.

Step 6
Since 12 is twice 6 and $2 \cdot 12 + 2 \cdot 6 = 24 + 12 = 36$, this answer is correct.

19. *Step 1*
The problem is about finding the length and the width of a jewelry box.

Unknowns: The length and width of the box

Known: The length is 3 in. more than twice the width. The perimeter is 36 in.

Step 2(b)
Let w represent the width. Then, since the length is 3 in. more than twice the width, use $2w + 3$ to represent the length.

Step 3
$2l + 2w = P$
$2 \cdot (2w + 3) + 2 \cdot w = 36$

Step 4
$2(2w + 3) + 2w = 36$
$4w + 6 + 2w = 36$
$4w + 2w + 6 = 36$
$6w + 6 = 36$
$\,^{-}6 \,^{-}6$
$\dfrac{6w}{6} = \dfrac{30}{6}$
$w = 5$

Step 5
The width is 5 inches. The length is $2(5) + 3 = 13$ inches.

Step 6
The length of 13 inches is 3 inches longer than twice the width of 5 inches.

$P = 2l + 2w$
$P = 2 \cdot 13 + 2 \cdot 5$
$P = 26 + 10$
$P = 36$ in.

21. *Step 1*
The problem is about a framed photograph.

Unknowns: The outside perimeter and the total area of the photograph and frame

Known: The photo inside the frame is a rectangle with length 10 in. and width 8 in. The frame is 2 in. wide.

Step 2(b)
Let P represent the outside perimeter and A represent the total area of the photograph and frame. Note that $P = 2l + 2w$ and $A = lw$, where $l = 10 + 2 \cdot 2 = 14$ and $w = 8 + 2 \cdot 2 = 12$ are the length and the width of the frame, respectively.

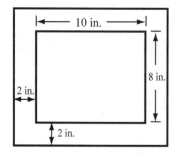

84 Chapter 3 Solving Application Problems

Step 3
$P = 2(14) + 2(12)$ and
$A = 14 \cdot 12$

Step 4
$P = 2(14) + 2(12)$ $A = 14 \cdot 12 = 168$ in.^{2}
$ = 28 + 24$
$ = 52$ in.

Step 5
The outside perimeter is 52 in. and the total area of the photograph and frame is 168 in.^{2}.

Step 6
$P = 12$ in. $+ 14$ in. $+ 12$ in. $+ 14$ in. $= 52$ in.
$A = 12 \cdot 14 = 168$ in.^{2}.

Chapter 3 Review Exercises

1. The figure is a square.
   $P = 4s$
   $P = 4 \cdot 28$ cm
   $P = 112$ cm

2. The figure is a rectangle.
   $P = 2 \cdot l + 2 \cdot w$
   $P = 2 \cdot 8$ mi $+ 2 \cdot 3$ mi
   $P = 16$ mi $+ 6$ mi
   $P = 22$ mi

3. The figure is a parallelogram.
   $P =$ sum of all four sides
   $P = 7$ yd $+ 14$ yd $+ 7$ yd $+ 14$ yd
   $P = 42$ yd

4. $P =$ sum of all sides
   $P = 26$ m $+ 44$ m $+ 14$ m $+ 20$ m $+ 12$ m $+ 24$ m
   $P = 140$ m

5. $P = 4s$ Replace P with 12 ft.
   12 ft $= 4s$
   $\dfrac{12 \text{ ft}}{4} = \dfrac{4s}{4}$
   3 ft $= s$

   One side of the table is 3 ft.

6. $P = 2l + 2w$ Replace P with 128 yd and w with 31 yd.
   128 yd $= 2l + 2 \cdot 31$ yd
   128 yd $= 2l + 62$ yd   To get $2l$ by itself, add $^-62$ yd to both sides.
   $\dfrac{^-62 \text{ yd} \qquad ^-62 \text{ yd}}{66 \text{ yd} = 2l + 0}$
   $\dfrac{66 \text{ yd}}{2} = \dfrac{2l}{2}$ To get l by itself, divide both sides by 2.
   33 yd $= l$

   The length of the playground is 33 yd.

7. $P = 2l + 2w$ Replace P with 72 in. and l with 21 in.
   72 in. $= 2 \cdot 21$ in. $+ 2w$
   72 in. $= 42$ in. $+ 2w$ Add $^-42$ in. to both sides.
   $\dfrac{^-42 \text{ in.} \qquad ^-42 \text{ in.}}{30 \text{ in.} = 0 + 2w}$
   $\dfrac{30 \text{ in.}}{2} = \dfrac{2w}{2}$ Divide both sides by 2.
   15 in. $= w$

   The width of the painting is 15 inches.

8.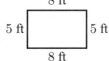

   $A = l \cdot w$ The tablecloth is a rectangle.
   $A = 8$ ft $\cdot 5$ ft
   $A = 40$ ft^2

   The area of the tablecloth is 40 ft^2.

9.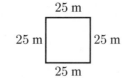

   $A = s^2$
   $A = s \cdot s$
   $A = 25$ m $\cdot 25$ m
   $A = 625$ m^2

   The area of the dance floor is 625 m^2.

10.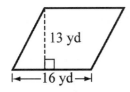

$A = b \cdot h$
$A = 16 \text{ yd} \cdot 13 \text{ yd}$
$A = 208 \text{ yd}^2$

The area of the lot is 208 yd².

11.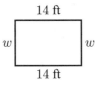

$A = l \cdot w$
$126 \text{ ft}^2 = 14 \text{ ft} \cdot w$
$\dfrac{126 \text{ ft} \cdot \cancel{\text{ft}}}{14 \cancel{\text{ft}}} = \dfrac{14 \text{ ft} \cdot w}{14 \text{ ft}}$
$9 \text{ ft} = w$

The width of the patio is 9 ft.

12.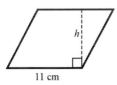

$A = b \cdot h$ Replace A with 88 cm² and b with 11 cm.

$88 \text{ cm}^2 = 11 \text{ cm} \cdot h$
$\dfrac{88 \text{ cm} \cdot \cancel{\text{cm}}}{11 \cancel{\text{cm}}} = \dfrac{11 \text{ cm} \cdot h}{11 \text{ cm}}$
$8 \text{ cm} = h$

The height is 8 cm.

13.

$A = s^2$ Replace A with 100 mi² and solve by inspection.

$100 \text{ mi}^2 = s^2$
$100 \text{ mi}^2 = s \cdot s$
$100 \text{ mi}^2 = 10 \text{ mi} \cdot 10 \text{ mi}$

The length of one side of the piece of land is 10 mi.

14. A number subtracted from 57
$$57 - x$$
(*Note:* $x - 57$ would be the translation of "57 subtracted from a number.")

15. The sum of 15 and twice a number
$$15 + 2x \quad \text{or} \quad 2x + 15$$

16. The product of ⁻9 and a number
$$^-9x$$

17. Let n represent the unknown number.

The sum of four times a number and 6 is ⁻30.
$$4n + 6 = {}^-30$$

$4n + 6 = {}^-30$
$ {}^-6 \quad {}^-6$
$4n + 0 = {}^-36$
$\dfrac{4n}{4} = \dfrac{^-36}{4}$
$n = {}^-9$

The number is ⁻9.

18. Let n represent the unknown number.

When twice a number is subtracted from 10 the result is 4 plus the number.
$$10 - 2n = 4 + n$$

$10 - 2n = 4 + n$
$ +2n \quad\ +2n$
$10 + 0 = 4 + 3n$
$10 = 4 + 3n$
$^-4 \quad\ ^-4$
$6 = 0 + 3n$
$\dfrac{6}{3} = \dfrac{3n}{3}$
$2 = n$

The number is 2.

86 Chapter 3 Solving Application Problems

19. *Step 1*
Unknown: Amount in Grace's account before writing her rent check

Known: $600 check, $750 deposit, $75 deposit, $309 ending balance

Step 2(a)
Let m represent the amount of money originally in the account.

Step 3
$m - 600 + 750 + 75 = 309$

Step 4
$$\begin{aligned} m + (^-600) + 750 + 75 &= 309 \\ m + 225 &= 309 \\ ^-225 \quad &\quad ^-225 \\ \hline m + 0 &= 84 \\ m &= 84 \end{aligned}$$

Step 5
$84 was originally in Grace's account.

Step 6
$84 - $600 + $750 + $75
$= ^-$516 + $750 + 75
$= $234 + 75
$= 309 (ending balance checks)

20. *Step 1*
Unknown: Number of candles in each box

Known: There were 4 boxes, one candle on each of 25 tables, 23 candles left.

Step 2(a)
Let c represent the number of candles per box.

Step 3
$$\underbrace{4 \cdot c}_{\text{Total number of candles in 4 boxes}} - \underbrace{25}_{\text{number of candles used}} = \underbrace{23}_{\text{number of candles left}}$$

Step 4
$$\begin{aligned} 4c - 25 &= 23 \\ +25 \quad &\quad +25 \\ \hline 4c + 0 &= 48 \\ \frac{4c}{4} &= \frac{48}{4} \\ c &= 12 \end{aligned}$$

Step 5
There were 12 candles in each box.

Step 6
4 boxes with 12 candles per box $= 48$ candles.

25 candles used: $48 - 25 = 23$ left (matches)

21. *Step 1*
Unknowns: How much Reggie earned and how much Donald earned

Known: $1000 prize, Donald should get $300 more than Reggie

Step 2(b)
There are two unknowns. You know the least about how much Reggie received in prize money. Let p represent the prize money earned by Reggie. Then $p + 300$ represents the prize money earned by Donald.

Step 3
$$\underbrace{p}_{\substack{\text{Reggie's} \\ \text{prize} \\ \text{money}}} + \underbrace{p + 300}_{\substack{\text{Donald's} \\ \text{prize} \\ \text{money}}} = \underbrace{1000}_{\substack{\text{Total} \\ \text{prize} \\ \text{money}}}$$

Step 4
$$\begin{aligned} p + p + 300 &= 1000 \\ 2p + 300 &= 1000 \\ ^-300 \quad &\quad ^-300 \\ \hline 2p + 0 &= 700 \\ \frac{2p}{2} &= \frac{700}{2} \\ p &= 350 \end{aligned}$$

Step 5
Reggie earns $350 and Donald earns $350 + $300 = $650.

Step 6
$650 is $300 more than $350 and the sum of $650 and $350 is $1000.

22. *Step 1*
Unknowns: length and width of photograph

Known: Perimeter is 84 cm, the photo is twice as long as it is wide

Step 2(b)
There are two unknowns. You know the least about the width. Let w represent the width. Then $2w$ represents the length.

Step 3
$2l + 2w = P$

Step 4
$$\begin{aligned} 2l + 2w &= 84 \text{ cm} \quad P = 84 \text{ cm} \\ 2 \cdot 2w + 2w &= 84 \text{ cm} \quad \text{length } l = 2w \\ 4w + 2w &= 84 \text{ cm} \\ 6w &= 84 \text{ cm} \\ \frac{6w}{6} &= \frac{84}{6} \text{ cm} \\ w &= 14 \text{ cm} \end{aligned}$$

Step 5
The width is 14 cm and the length is 28 cm.

Step 6
$2 \cdot 14 \text{ cm} + 2 \cdot 28 \text{ cm} = P$
$28 \text{ cm} + 56 \text{ cm} = P$
$84 \text{ cm} = P$ (matches)

23. **[3.1] (a)** Kit #2 has 36 feet of fencing to make a square dog pen.

$P = 4s$ Replace P with 36 feet.
$36 \text{ ft} = 4s$
$\dfrac{36 \text{ ft}}{4} = \dfrac{4s}{4}$
$9 \text{ ft} = s$

The length of one side of the dog pen is 9 ft.

[3.2] (b) The area of the square dog pen
$A = s^2$
$A = s \cdot s$
$A = 9 \text{ ft} \cdot 9 \text{ ft}$
$A = 81 \text{ ft}^2$

The area of the dog pen is 81 ft^2.

24. **[3.2]** Rectangles will vary. To create your rectangles, choose a length and width that add up to 10, which is one-half of the perimeter of 20. Two possibilities are:

```
      7 ft                    6 ft
   ┌────────┐             ┌────────┐
3 ft│        │3 ft     4 ft│        │4 ft
   └────────┘             └────────┘
      7 ft                    6 ft
```

$A = l \cdot w$ $A = l \cdot w$
$A = 7 \text{ ft} \cdot 3 \text{ ft}$ $A = 6 \text{ ft} \cdot 4 \text{ ft}$
$A = 21 \text{ ft}^2$ $A = 24 \text{ ft}^2$

25. **[3.3]** *Step 1*
Unknown: Amount of fencing Timotha put around her garden

Known: Kit #1 contains 20 ft, Kit #2 contains 36 ft, 41 ft of fencing available for dog pen

Step 2(a)
Let f represent the number of feet of fencing for the garden.

Step 3

Fencing in Kit #2	−	Fencing around garden	+	Fencing in Kit #1	=	Fencing for dog pen
↓		↓		↓		↓
36	−	f	+	20	=	41

Step 4
$36 - f + 20 = 41$
$56 - f = 41$
$\underline{^-56 ^-56}$
$^-f = {}^-15$
$\dfrac{^-1f}{^-1} = \dfrac{^-15}{^-1}$
$f = 15$

Step 5
15 ft of fencing went around the garden.

Step 6
$15 \text{ ft} + 41 \text{ ft} = 56 \text{ ft}$ matches the total fencing from Kits #1 and #2 [20 + 36].

26. **[3.4]** *Step 1*
Unknowns: The width and the length of the dog pen

Known: 36 ft of fencing was used, the length was 2 ft more than the width

Step 2(b)
There are two unknowns. You know the least about the width. Let w represent the width. Then $w + 2$ represents the length.

```
          w + 2
       ┌────────┐
     w │        │ w
       └────────┘
          w + 2
```

Step 3
$2l + 2w = P$
$2 \cdot (w + 2) + 2w = 36$

Step 4
$2(w + 2) + 2w = 36$
$2w + 4 + 2w = 36$
$4w + 4 = 36$
$\underline{ ^-4 ^-4}$
$4w = 32$
$\dfrac{4w}{4} = \dfrac{32}{4}$
$w = 8$

Step 5
The width is 8 ft and the length is $8 + 2 = 10$ ft.

Step 6
10 ft is 2 ft longer than 8 ft.

$P = 2 \cdot 10 \text{ ft} + 2 \cdot 8 \text{ ft}$
$P = 20 \text{ ft} + 16 \text{ ft}$
$P = 36 \text{ ft}$ (matches)

Chapter 3 Test

1. $P = 59 \text{ m} + 72 \text{ m} + 59 \text{ m} + 72 \text{ m}$
   $P = 262 \text{ m}$

88 Chapter 3 Solving Application Problems

2. $P = 8$ in. $+ 4$ in. $+ 4$ in. $+ 4$ in. $+ 8$ in. $+$
   4 in. $+ 4$ in. $+ 4$ in.
   $P = 40$ in.

3. $P = 4s$
   $P = 4 \cdot 3$ miles
   $P = 12$ miles

4. $P = 2l + 2w$
   $P = 2 \cdot 4$ ft $+ 2 \cdot 2$ ft
   $P = 8$ ft $+ 4$ ft
   $P = 12$ ft

5. $P = 45$ cm $+ 50$ cm $+ 15$ cm
   $P = 110$ cm

6. $A = l \cdot w$
   $A = 27$ mm $\cdot 18$ mm
   $A = 486$ mm^2

7. *Note:* The base is perpendicular to the height.
   $A = b \cdot h$
   $A = 14$ cm $\cdot 10$ cm
   $A = 140$ cm^2

8. $A = l \cdot w$
   $A = 68$ mi $\cdot 55$ mi
   $A = 3740$ mi^2

9. $A = s^2$
   $A = s \cdot s$
   $A = 6$ m $\cdot 6$ m
   $A = 36$ m^2

10. $P = 4s$
    12 ft $= 4s$
    $\dfrac{12 \text{ ft}}{4} = \dfrac{4s}{4}$
    3 ft $= s$

    The length of one side of the table is 3 ft.

11. $P = 2l + 2w$
    34 ft $= 2l + 2(6$ ft$)$
    34 ft $= 2l + 12$ ft
    $\phantom{34 \text{ ft}} {}^-12$ ft $\phantom{={}} {}^-12$ ft
    $\overline{22 \text{ ft} = 2l + 0}$
    $\dfrac{22 \text{ ft}}{2} = \dfrac{2l}{2}$
    11 ft $= l$

    The length of the plot is 11 ft.

12. $A = bh$
    65 in.$^2 = 13$ in. $\cdot h$
    $\dfrac{65 \text{ in.} \cdot \text{in.}}{13 \text{ in.}} = \dfrac{13 \text{ in.} \cdot h}{13 \text{ in.}}$
    5 in. $= h$

    The parallelogram's height is 5 inches.

13. $A = l \cdot w$
    12 cm$^2 = 4$ cm $\cdot w$
    $\dfrac{12 \text{ cm} \cdot \text{cm}}{4 \text{ cm}} = \dfrac{4 \text{ cm} \cdot w}{4 \text{ cm}}$
    3 cm $= w$

    The width of the postage stamp is 3 cm.

14. $A = s^2$
    $A = s \cdot s$
    16 ft$^2 = s \cdot s$ Replace A with 16 ft^2 and
    $\phantom{16 \text{ ft}^2 = s \cdot s}$ solve by inspection.
    16 ft$^2 = 4$ ft $\cdot 4$ ft

    Each side of the bulletin board is 4 feet.

15. Answers will vary; one possibility follows. Linear units like ft are used to measure length, width, height, and perimeter. Area is measured in square units like ft^2 (squares that measure 1 ft on each side).

16. Let n represent the unknown number.

If 40	is added to	four times a number	the result is	zero.
↓	⏟	⏟	⏟	↓
40	+	$4n$	=	0

    $40 + 4n = 0$
    $ {}^-40 \phantom{={}} {}^-40$
    $0 + 4n = {}^-40$
    $\dfrac{4n}{4} = \dfrac{{}^-40}{4}$
    $n = {}^-10$

    The number is $^-10$.

17. Let n represent the unknown number.

When 7 times a number is decreased by 23	the result is	the number plus 7.
⏟	⏟	⏟
$7n - 23$	=	$n + 7$

    $7n - 23 = n + 7$
    $ {}^-n \phantom{={}} {}^-n$
    $6n - 23 = 7$
    $ {+ 23} \phantom{={}} {+ 23}$
    $6n + 0 = 30$
    $6n = 30$
    $\dfrac{6n}{6} = \dfrac{30}{6}$
    $n = 5$

    The number is 5.

18. *Step 1*
Unknown: Amount of money used by son

Known: $43 originally, some spent by son on groceries, $16 put into wallet, and $44 at the end

Step 2(a)
Let m represent the money spent on groceries.

Step 3
$43 - m + 16 = 44$

Step 4
$$\begin{aligned} 43 - m + 16 &= 44 \\ 59 - m &= 44 \\ -59 & \quad -59 \\ \hline 0 - m &= {}^-15 \\ \frac{{}^-1m}{{}^-1} &= \frac{{}^-15}{{}^-1} \\ m &= 15 \end{aligned}$$

Step 5
Her son spent $15.

Step 6
$43 - $15 + $16 = 44 (matches)

19. *Step 1*
Unknown: Daughter's age

Known: Ray is 39 years old and his age is 4 years more than five times his daughter's age.

Step 2(a)
Let d represent the daughter's age.

Step 3
$$\underbrace{\text{Ray's age}}_{39} \underbrace{\text{is}}_{=} \underbrace{\text{4 years more than five times his daughter's age.}}_{4 + 5d}$$

Step 4
$$\begin{aligned} 39 &= 4 + 5d \\ {}^-4 & \quad {}^-4 \\ \hline 35 &= 0 + 5d \\ \frac{35}{5} &= \frac{5d}{5} \\ 7 &= d \end{aligned}$$

Step 5
The daughter is 7 years old.

Step 6
Four years more than five times 7 years is $4 + 5 \cdot 7 = 4 + 35 = 39$ yrs (matches).

20. *Step 1*
Unknowns: The length of each piece

Known: Total board length is 118 cm, one piece is 4 cm longer than the other

Step 2(b)
There are two unknowns. Let p represent the length of the shorter piece. Then $p + 4$ represents the length of the longer piece.

Step 3
$$\underbrace{\text{long piece}}_{p+4} + \underbrace{\text{short piece}}_{p} = \underbrace{\text{total length}}_{118}$$

Step 4
$$\begin{aligned} p + 4 + p &= 118 \\ 2p + 4 &= 118 \\ {}^-4 & \quad {}^-4 \\ \hline 2p + 0 &= 114 \\ \frac{2p}{2} &= \frac{114}{2} \\ p &= 57 \end{aligned}$$

Step 5
One piece is 57 cm and the other is $57 + 4 = 61$ cm.

Step 6
61 cm is 4 cm longer than 57 cm and the sum of 61 cm and 57 cm is 118 cm.

21. *Step 1*
Unknowns: Length and width

Known: Perimeter is 420 ft, length is four times as long as the width

Step 2(b)
Let w represent the width.
Then $4w$ represents the length.

Step 3
$P = 2l + 2w$
$420 \text{ ft} = 2 \cdot 4w + 2 \cdot w$

Step 4
$$\begin{aligned} 420 &= 2 \cdot 4w + 2 \cdot w \\ 420 &= 8w + 2w \\ 420 &= 10w \\ \frac{420}{10} &= \frac{10w}{10} \\ 42 &= w \end{aligned}$$

Step 5
The width is 42 ft and the length is $4(42) = 168$ ft.

Step 6
168 ft is four times 42 ft.

$P = 2 \cdot 168 \text{ ft} + 2 \cdot 42 \text{ ft}$
$P = 336 \text{ ft} + 84 \text{ ft}$
$P = 420 \text{ ft}$ (matches)

22. *Step 1*
Unknowns: How many hours each person worked

Known: 19 hours worked by both, Tim worked 3 hours less than Marcella

Step 2(b)
Let h represent the hours worked by Marcella. Then $h - 3$ represents the hours worked by Tim.

Step 3

Tim's hours	+	Marcella's hours	=	Total hours
$h - 3$	+	h	=	19

Step 4
$$h - 3 + h = 19$$
$$2h - 3 = 19$$
$$\underline{+3 \qquad +3}$$
$$2h + 0 = 22$$
$$\frac{2h}{2} = \frac{22}{2}$$
$$h = 11$$

Step 5
Marcella worked 11 hours and Tim worked $11 - 3 = 8$ hours.

Step 6
8 hours is 3 hours less than 11 hours and the sum of 8 and 11 is 19.

Cumulative Review Exercises (Chapters 1–3)

1. 4,000,206,300 in words: four billion, two hundred six thousand, three hundred

2. Seventy million, five thousand, four hundred eighty-nine written using digits is 70,005,489.

3. $^-7$ lies to the *left* of $^-1$ on the number line. Write $^-7 < {}^-1$.

0 lies to the *right* of $^-5$ on the number line. Write $0 > {}^-5$.

4. (a) $1(97) = 97$; multiplication property of 1

(b) $^-10 + 0 = {}^-10$; addition property of 0

(c) $(3 \cdot {}^-7) \cdot 6 = 3 \cdot ({}^-7 \cdot 6)$; associative property of multiplication

5. (a) Underline the tens place. 37$\underline{9}$5
The next digit is 5 or more. Add one to 9. Write 0 and carry one to the hundred place. Change 5 to 0. **3800**

(b) Underline the ten-thousands place. 4$\underline{9}$3,662
The next digit is 4 or less. Leave 9 as 9 and change 3, 6, 6, and 2 to zeros. **490,000**

6. $^-12 - 12$
$= {}^-12 + ({}^-12)$
$= {}^-24$

7. $^-3({}^-9) = 27$

8. $|7| - |{}^-10|$
$= 7 - 10$
$= 7 + ({}^-10)$
$= {}^-3$

9. $^-40 \div 2 \cdot 5$ Divide.
$= {}^-20 \cdot 5$ Multiply.
$= {}^-100$

10. $3 - 8 + 10$
$= 3 + ({}^-8) + 10$
$= {}^-5 + 10$
$= 5$

11. $\dfrac{0}{^-6} = 0$
Zero divided by any nonzero number is 0.

12. $^-8 + 5(2 - 3)$ Change to addition.
$= {}^-8 + 5(2 + {}^-3)$ Parentheses
$= {}^-8 + 5({}^-1)$ Multiply.
$= {}^-8 + {}^-5$ Add.
$= {}^-13$

13. $({}^-3)^2 + 4^2$ Exponents
$= {}^-3 \cdot {}^-3 + 4 \cdot 4$ Multiply.
$= 9 + 16$ Add.
$= 25$

14. $4 - 3({}^-6 \div 3) + 7(0 - 6)$
$= 4 - 3 \cdot ({}^-2) + 7 \cdot ({}^-6)$
$= 4 - ({}^-6) + ({}^-42)$
$= 4 + ({}^+6) + ({}^-42)$
$= 10 + ({}^-42)$
$= {}^-32$

15. $\dfrac{4 - 2^3 + 5^2 - 3(^-2)}{^-1(3) - 6(^-2) - 9}$

    Numerator:
    $4 - 2^3 + 5^2 - 3(^-2)$
    $= 4 - (2 \cdot 2 \cdot 2) + (5 \cdot 5) - 3 \cdot (^-2)$
    $= 4 - 8 + 25 - (^-6)$
    $= 4 + (^-8) + 25 + (^+6)$
    $= {}^-4 + 25 + 6$
    $= 21 + 6$
    $= 27$

    Denominator:
    $^-1(3) - 6(^-2) - 9$
    $= {}^-3 - (^-12) - 9$
    $= {}^-3 + (^+12) + (^-9)$
    $= 9 + (^-9)$
    $= 0$

    Last step is division: $\frac{27}{0}$ is undefined.

16. $10w^2xy^4$ can be written as
    $$10 \cdot w \cdot w \cdot x \cdot y \cdot y \cdot y \cdot y.$$

17. $^-6cd^3$

    $= {}^-6 \cdot c \cdot d \cdot d \cdot d$ Replace c with 5 and d with $^-2$.
    $= {}^-6 \cdot 5 \cdot {}^-2 \cdot {}^-2 \cdot {}^-2$
    $= {}^-30 \cdot {}^-2 \cdot {}^-2 \cdot {}^-2$
    $= 60 \cdot {}^-2 \cdot {}^-2$
    $= {}^-120 \cdot {}^-2$
    $= 240$

18. $^-4k + k + 5k$
    $= {}^-4k + 1k + 5k$
    $= {}^-3k + 5k$
    $= 2k$

19. $m^2 + 2m + 2m^2 = \underline{1m^2} + \underline{2m^2} + 2m = 3m^2 + 2m$

20. $xy^3 - xy^3 = 0$

    Anything minus itself equals zero.

21. $5(^-4a) = (5 \cdot {}^-4)a = {}^-20a$

22. $^-8 + x + 5 - 2x^2 - x$
    $= {}^-8 + 1x + 5 + (^-2x^2) + (^-1x)$
    $= {}^-8 + 5 + 1x + (^-1x) + (^-2x^2)$
    $= {}^-3 + 0x + (^-2x^2)$
    $= {}^-3 + (^-2x^2)$ or $^-2x^2 - 3$

23. $^-3(4n + 3) + 10$
    $= {}^-3 \cdot 4n + {}^-3 \cdot 3 + 10$
    $= {}^-12n + (^-9) + 10$
    $= {}^-12n + 1$

24. $6 - 20 = 2x - 9x$
    $6 + (^-20) = 2x + (^-9x)$
    $^-14 = {}^-7x$
    $\dfrac{^-14}{^-7} = \dfrac{^-7x}{^-7}$
    $2 = x$

    The solution is 2.

    Check: $6 - 20 = 2(2) - 9(2)$
    $6 - 20 = 4 - 18$
    $6 + (^-20) = 4 + (^-18)$
    $^-14 = {}^-14$ Balances

25. $^-5y = y + 6$
    $^-5y = 1y + 6$
    $\dfrac{^-1y}{} \quad \dfrac{^-1y}{}$
    $^-6y = 0 + 6$
    $\dfrac{^-6y}{^-6} = \dfrac{6}{^-6}$
    $y = {}^-1$

    The solution is $^-1$.

    Check: $^-5(^-1) = {}^-1 + 6$
    $5 = 5$ Balances

26. $3b - 9 = 19 - 4b$
    $\underline{+ 4b} \quad\quad \underline{+ 4b}$
    $7b - 9 = 19 + 0$
    $7b - 9 = 19$
    $\underline{+ 9} \quad\quad \underline{+ 9}$
    $7b + 0 = 28$
    $\dfrac{7b}{7} = \dfrac{28}{7}$
    $b = 4$

    The solution is 4.

27. $^-16 - h + 2 = h - 10$
    $^-16 + (^-h) + 2 = h - 10$
    $^-14 + (^-h) = h - 10$
    $\underline{+ h} \quad\quad\quad \underline{+ h}$
    $^-14 + 0 = 2h - 10$
    $^-14 = 2h - 10$
    $\underline{+ 10} \quad\quad \underline{+ 10}$
    $^-4 = 2h + 0$
    $\dfrac{^-4}{2} = \dfrac{2h}{2}$
    $^-2 = h$

    The solution is $^-2$.

28. $$-5(2x+4) = 3x - 20$$
$$-10x + (-20) = 3x - 20$$
$$\underline{+20 +20}$$
$$-10x + 0 = 3x + 0$$
$$-10x = 3x$$
$$\underline{+10x +10x}$$
$$0 = 13x$$
$$\frac{0}{13} = \frac{13x}{13}$$
$$0 = x$$

The solution is 0.

29. $$6 + 4(a+8) = -8a + 5 + a$$
$$6 + 4a + 32 = -8a + 5 + 1a$$
$$38 + 4a = -7a + 5$$
$$\underline{+7a +7a}$$
$$38 + 11a = 0 + 5$$
$$38 + 11a = 5$$
$$\underline{-38 -38}$$
$$0 + 11a = -33$$
$$\frac{11a}{11} = \frac{-33}{11}$$
$$a = -3$$

The solution is -3.

30. $P = 12$ in. $+ 18$ in. $+ 12$ in. $+ 18$ in.
$P = 60$ in.

$A = b \cdot h$
$A = 18$ in. $\cdot 11$ in.
$A = 198$ in.^{2}

31. $P = 4s \qquad A = s^2$
$P = 4 \cdot 15$ m $\qquad A = s \cdot s$
$P = 60$ m $\qquad A = 15$ m $\cdot 15$ m
$\qquad\qquad\qquad A = 225$ m^2

32. $P = 2l + 2w \qquad A = l \cdot w$
$P = 2 \cdot 8$ ft $+ 2 \cdot 4$ ft $\qquad A = 8$ ft $\cdot 4$ ft
$P = 16$ ft $+ 8$ ft $\qquad A = 32$ ft^2
$P = 24$ ft

33. Let n represent the unknown number.

$\underbrace{-50 \text{ is added to five}}_{5n + {}^{-}50} \underbrace{\text{the}}_{=} \underbrace{\text{zero}}_{0}$
times a number
result is

$$5n + {}^{-}50 = 0$$
$$\underline{+50 +50}$$
$$0 + 5n = 50$$
$$\frac{5n}{5} = \frac{50}{5}$$
$$n = 10$$

The number is 10.

34. Let n represent the unknown number.

$\underbrace{\text{Three times a number}}_{10 - 3n} \underbrace{\text{the}}_{=} \underbrace{\text{two times}}_{2n}$
$\text{is subtracted from 10}$
result is
the number.

$$10 - 3n = 2n$$
$$\underline{+3n +3n}$$
$$10 + 0 = 5n$$
$$\frac{10}{5} = \frac{5n}{5}$$
$$2 = n$$

The number is 2.

35. *Step 1*
Unknown: Number of people originally in line

Known: 3 left, 6 got in line, 2 left, 5 still in line

Step 2(a)
Let p represent the number of people in line originally.

Step 3
$p - 3 + 6 - 2 = 5$

Step 4
$$p + (-3) + 6 + (-2) = 5$$
$$p + 1 = 5$$
$$\underline{-1 -1}$$
$$p + 0 = 4$$
$$p = 4$$

Step 5
There were 4 people in line originally.

Step 6
4 in line
3 left $\qquad 4 - 3 = 1$
6 got in line $\quad 1 + 6 = 7$
2 left $\qquad 7 - 2 = 5$
5 left $\qquad$ 5 left (matches)

36. *Step 1*
Unknown: Amount paid by each player

Known: 12 players, $2200 total expenses, account overdrawn by $40

Step 2(a)
Let m be the amount paid by each player.

Step 3
$\underbrace{\text{amount}}_{12 \cdot m} - \underbrace{\text{total}}_{2200} = \underbrace{\text{amount in the}}_{-40}$
$\text{collected} \quad\quad \text{expenses} \quad\quad \text{team bank account}$

Step 4
$$12m - 2200 = {}^-40$$
$$\underline{+2200 = +2200}$$
$$12m + 0 = 2160$$
$$\frac{12m}{12} = \frac{2160}{12}$$
$$m = 180$$

Step 5
Each player paid $180.

Step 6
$12 \cdot \$180 = \$2160$ collected
$\$2160 - \$2200 = {}^-\$40$ (matches)

37. *Step 1*
Unknowns: Number of students in small group and in large group

Known: 192 students, one group was three times the size of the other group

Step 2(b)
There are 2 unknowns. You know the least about the small group. Let g be the size of the small group. Then $3g$ is the size of the large group.

Step 3
$$\underbrace{\text{small}}_{g} \; + \; \underbrace{\text{large}}_{3g} \; = \; \underbrace{\text{total number}}_{192} \text{ of students}$$

Step 4
$$g + 3g = 192$$
$$4g = 192$$
$$g = 48$$

Step 5
There were 48 students in the small group and $3(48) = 144$ students in the large group.

Step 6
144 is 3 times 48 and the sum of 48 and 144 is 192.

38. *Step 1*
Unknowns: length and width

Known: Perimeter is 92 ft, length is 14 ft longer than the width

Step 2(b)
You know the least about the width.
Let w represent the width.
Then $w + 14$ represents the length.

Step 3
$$P = 2l + 2w$$
$$92 = 2 \cdot (w + 14) + 2 \cdot w$$

Step 4
$$92 = 2(w + 14) + 2w$$
$$92 = 2w + 28 + 2w$$
$$92 = 4w + 28$$
$$\underline{{}^-28 \qquad\qquad {}^-28}$$
$$\frac{64}{4} = \frac{4w}{4}$$
$$16 = w$$

Step 5
The width is 16 ft and the length is $16 + 14 = 30$ ft.

Step 6
30 ft is 14 ft longer than 16 ft.
$$P = 2 \cdot 30 \text{ ft} + 2 \cdot 16 \text{ ft}$$
$$P = 60 \text{ ft} + 32 \text{ ft}$$
$$P = 92 \text{ ft (matches)}$$

CHAPTER 4 RATIONAL NUMBERS: POSITIVE AND NEGATIVE FRACTIONS

4.1 Introduction to Signed Fractions

4.1 Margin Exercises

1. **(a)** The 3 shaded parts are represented by the fraction $\frac{3}{5}$; the 2 unshaded parts by $\frac{2}{5}$.

   (b) The 1 shaded part is represented by the fraction $\frac{1}{6}$; the unshaded parts by $\frac{5}{6}$.

   (c) The 7 shaded parts are represented by the fraction $\frac{7}{8}$; the unshaded part by $\frac{1}{8}$.

2. **(a)** An area equal to 8 of the $\frac{1}{7}$ parts is shaded, so $\frac{8}{7}$ is shaded.

   (b) An area equal to 7 of the $\frac{1}{4}$ parts is shaded, so $\frac{7}{4}$ is shaded.

3. **(a)** $\frac{2}{3}$ ← 2 is the *numerator*
   ← 3 is the *denominator*

   (b) $\frac{1}{4}$ ← 1 is the *numerator*
   ← 4 is the *denominator*

   (c) $\frac{8}{5}$ ← 8 is the *numerator*
   ← 5 is the *denominator*

   (d) $\frac{5}{2}$ ← 5 is the *numerator*
   ← 2 is the *denominator*

   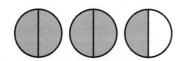

4. **(a)** Proper fractions have a numerator that is *smaller* than the denominator.

   $\frac{3}{4}, \frac{5}{7}, \frac{1}{2}$

   (b) Improper fractions have a numerator that is *equal to or greater* than the denominator.

   $\frac{8}{7}, \frac{6}{6}, \frac{2}{1}$

5. **(a)** $\frac{2}{4}$ is positive. Divide the space between 0 and 1 into 4 equal parts. Start at zero and count 2 parts to the right.

   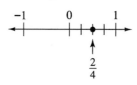

   (b) $\frac{1}{2}$ is positive. Divide the space between 0 and 1 into 2 equal parts. Start at zero and count 1 part to the right.

   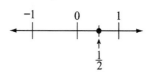

   (c) $-\frac{2}{3}$ is negative. Divide the space between 0 and -1 into 3 equal parts. Start at zero and count 2 parts to the left.

   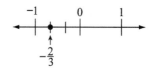

6. **(a)** $\left|-\frac{3}{4}\right| = \frac{3}{4}$ because the distance from 0 to $-\frac{3}{4}$ on the number line is $\frac{3}{4}$.

   (b) $\left|\frac{5}{8}\right| = \frac{5}{8}$ because the distance from 0 to $\frac{5}{8}$ on the number line is $\frac{5}{8}$.

   (c) $|0| = 0$ because the distance from 0 to 0 on the number line is 0.

7. **(a)** $\frac{2}{5} = \frac{?}{20}$

   Multiply 5 by 4 to get the denominator of 20. So multiply both the numerator and denominator by 4.

   $$\frac{2}{5} = \frac{2 \cdot 4}{5 \cdot 4} = \frac{8}{20}$$

   (b) $-\frac{21}{28} = \frac{?}{4}$

   Divide 28 by 7 to get the denominator of 4. So divide both the numerator and denominator by 7.

   $$-\frac{21}{28} = -\frac{21 \div 7}{28 \div 7} = -\frac{3}{4}$$

8. **(a)** $\frac{10}{10} = 1$. Think of $\frac{10}{10}$ as $10 \div 10$.

   (b) $-\frac{3}{1} = -3$. Think of $-\frac{3}{1}$ as $-3 \div 1$.

   (c) $\frac{8}{2} = 4$. Think of $\frac{8}{2}$ as $8 \div 2$.

   (d) $-\frac{25}{5} = -5$. Think of $-\frac{25}{5}$ as $-25 \div 5$.

96 Chapter 4 Rational Numbers: Positive and Negative Fractions

4.1 Section Exercises

1. The figure has 8 equal parts. The 5 shaded parts are represented by the fraction $\frac{5}{8}$; the unshaded parts by $\frac{3}{8}$.

3. The figure has 3 equal parts. The 2 shaded parts are represented by the fraction $\frac{2}{3}$; the unshaded part by $\frac{1}{3}$.

5. An area equal to 3 of the $\frac{1}{2}$ parts is shaded: $\frac{3}{2}$

   An area equal to 1 of the $\frac{1}{2}$ parts is unshaded: $\frac{1}{2}$

7. An area equal to 11 of the $\frac{1}{6}$ parts is shaded: $\frac{11}{6}$

   An area equal to 1 of the $\frac{1}{6}$ parts is unshaded: $\frac{1}{6}$

9. Two of the 11 coins are dimes: $\frac{2}{11}$

   Three of the 11 coins are pennies: $\frac{3}{11}$

   Four of the 11 coins are nickels: $\frac{4}{11}$

11. $71 - 58 = 13$

    $\frac{13}{71}$ of the computers are *not* laptops.

    $\frac{58}{71}$ of the computers are laptops.

13. (a) 6 of 20 women would like flowers delivered at work: $\frac{6}{20}$

    (b) $13 + 6 = 19$ of 20 women would like flowers delivered either at home or at work: $\frac{19}{20}$

15. $\dfrac{3}{4}\ \ \begin{array}{l}\leftarrow 3\text{ is the Numerator}\\ \leftarrow 4\text{ is the Denominator}\end{array}$

    There are 4 equal parts in the whole.

17. $\dfrac{12}{7}\ \ \begin{array}{l}\leftarrow 12\text{ is the Numerator}\\ \leftarrow 7\text{ is the Denominator}\end{array}$

    There are 7 equal parts in the whole.

19. Proper fractions have a numerator that is *smaller* than the denominator.

    $\frac{1}{3}, \frac{5}{8}, \frac{7}{16}$

    Improper fractions have a numerator that is *equal to or greater* than the denominator.

    $\frac{8}{5}, \frac{6}{6}, \frac{12}{2}$

21. To graph $\frac{1}{4}$ and $-\frac{1}{4}$ on the number line, divide the space between 0 and 1 into 4 equal parts. Start at zero and count 1 part to the right. This spot represents $\frac{1}{4}$. Now divide the space between -1 and 0 into 4 equal parts. Start at zero and count 1 part to the left. This spot represents $-\frac{1}{4}$.

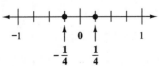

23. Graph $-\frac{3}{5}$ and $\frac{3}{5}$ on the number line.

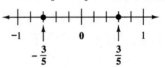

25. Graph $\frac{7}{8}$ and $-\frac{7}{8}$ on the number line.

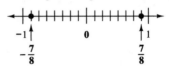

27. "lost" $\frac{3}{4}$ pound: $-\frac{3}{4}$ pound

29. $\frac{3}{10}$ mile long: $+\frac{3}{10}$ mile or $\frac{3}{10}$ mile

31. $\left|-\frac{2}{5}\right| = \frac{2}{5}$ because the distance from 0 to $-\frac{2}{5}$ on the number line is $\frac{2}{5}$.

33. $|0| = 0$ because the distance from 0 to 0 on the number line is 0.

35. (a) $\dfrac{1}{2} = \dfrac{1 \cdot 12}{2 \cdot 12} = \dfrac{12}{24}$

    (b) $\dfrac{1}{3} = \dfrac{1 \cdot 8}{3 \cdot 8} = \dfrac{8}{24}$

    (c) $\dfrac{2}{3} = \dfrac{2 \cdot 8}{3 \cdot 8} = \dfrac{16}{24}$

    (d) $\dfrac{1}{4} = \dfrac{1 \cdot 6}{4 \cdot 6} = \dfrac{6}{24}$

    (e) $\dfrac{3}{4} = \dfrac{3 \cdot 6}{4 \cdot 6} = \dfrac{18}{24}$

    (f) $\dfrac{1}{6} = \dfrac{1 \cdot 4}{6 \cdot 4} = \dfrac{4}{24}$

    (g) $\dfrac{5}{6} = \dfrac{5 \cdot 4}{6 \cdot 4} = \dfrac{20}{24}$

    (h) $\dfrac{1}{8} = \dfrac{1 \cdot 3}{8 \cdot 3} = \dfrac{3}{24}$

    (i) $\dfrac{3}{8} = \dfrac{3 \cdot 3}{8 \cdot 3} = \dfrac{9}{24}$

    (j) $\dfrac{5}{8} = \dfrac{5 \cdot 3}{8 \cdot 3} = \dfrac{15}{24}$

37. (a) $-\dfrac{2}{6} = -\dfrac{2 \div 2}{6 \div 2} = -\dfrac{1}{3}$

    (b) $-\dfrac{4}{6} = -\dfrac{4 \div 2}{6 \div 2} = -\dfrac{2}{3}$

    (c) $-\dfrac{12}{18} = -\dfrac{12 \div 6}{18 \div 6} = -\dfrac{2}{3}$

    (d) $-\dfrac{6}{18} = -\dfrac{6 \div 6}{18 \div 6} = -\dfrac{1}{3}$

    (e) $-\dfrac{200}{300} = -\dfrac{200 \div 100}{300 \div 100} = -\dfrac{2}{3}$

(f) Some possibilities are: $-\frac{4}{12} = -\frac{1}{3}$; $-\frac{8}{24} = -\frac{1}{3}$; $-\frac{20}{30} = -\frac{2}{3}$; $-\frac{24}{36} = -\frac{2}{3}$.

39. (a) $\dfrac{3}{8} = \dfrac{3 \cdot 489}{8 \cdot 489} = \dfrac{1467}{3912}$

(b) Divide 3912 by 8 to get 489; multiply 3 by 489 to get 1467.

40. (a) $\dfrac{7}{9} = \dfrac{7 \cdot 608}{9 \cdot 608} = \dfrac{4256}{5472}$

(b) Divide 5472 by 9 to get 608; multiply 7 by 608 to get 4256.

41. (a) $-\dfrac{697}{3485} = -\dfrac{697 \div 697}{3485 \div 697} = -\dfrac{1}{5}$

(b) Divide 3485 by 2, by 3, and by 5 to see that dividing by 5 gives 697. Or divide 3485 by 697 to get 5.

42. (a) $-\dfrac{817}{4902} = -\dfrac{817 \div 817}{4902 \div 817} = -\dfrac{1}{6}$

(b) Divide 4902 by 4, by 6, and by 8 to see that dividing by 6 gives 817. Or divide 4902 by 817 to get 6.

43. You cannot do it if you want the numerator to be a whole number, because 5 does not divide into 18 evenly. You could use multiples of 5 as the denominator, such as 10, 15, 20, etc.

45. $\dfrac{10}{1} = 10 \div 1 = 10$

47. $-\dfrac{16}{16} = -16 \div 16 = -1$

49. $-\dfrac{18}{3} = -18 \div 3 = -6$

51. $\dfrac{24}{8} = 24 \div 8 = 3$

53. $\dfrac{14}{7} = 14 \div 7 = 2$

55. $-\dfrac{90}{10} = -90 \div 10 = -9$

57. $\dfrac{150}{150} = 150 \div 150 = 1$

59. $-\dfrac{32}{4} = -32 \div 4 = -8$

61.

$\frac{3}{5}$ is shaded; $\frac{2}{5}$ is unshaded.

4.2 Writing Fractions in Lowest Terms

63.

$\frac{3}{8}$ is shaded; $\frac{5}{8}$ is unshaded.

65. One possibility is shown.

○ ■ □ □ □ □ □ ▲ ▲ △

67. One possibility is shown. (! !) ! ! ! , . . . ? ? ?

4.2 Writing Fractions in Lowest Terms

4.2 Margin Exercises

1. (a) $\frac{2}{3}$ is in lowest terms since the numerator and denominator have no common factor other than 1.

(b) $-\frac{8}{10}$ is not in lowest terms. A common factor of 8 and 10 is 2.

(c) $-\frac{9}{11}$ is in lowest terms.

(d) $\frac{15}{20}$ is not in lowest terms. A common factor of 15 and 20 is 5.

2. (a) $\dfrac{5}{10} = \dfrac{5 \div 5}{10 \div 5} = \dfrac{1}{2}$

(b) $\dfrac{9}{12} = \dfrac{9 \div 3}{12 \div 3} = \dfrac{3}{4}$

(c) $-\dfrac{24}{30} = -\dfrac{24 \div 6}{30 \div 6} = -\dfrac{4}{5}$

(d) $\dfrac{15}{40} = \dfrac{15 \div 5}{40 \div 5} = \dfrac{3}{8}$

(e) $-\dfrac{50}{90} = -\dfrac{50 \div 10}{90 \div 10} = -\dfrac{5}{9}$

3. Prime: 2, 3, 7, 13, 19, 29
Composite: 4 because 4 can be divided by 2.
9 because 9 can be divided by 3.
25 because 25 can be divided by 5.

1 is neither prime nor composite.

4. (a) $8 \div 2 = 4$ Divide 8 by 2.
$4 \div 2 = 2$ Divide 4 by 2.
$2 \div 2 = 1$ Divide 2 by 2. Quotient is 1.

The prime factorization of 8 is $2 \cdot 2 \cdot 2$.

(b) $42 \div 2 = 21$ Divide 42 by 2.
$21 \div 3 = 7$ 21 is not divisible by 2; use 3.
$7 \div 7 = 1$ 7 is not divisible by 2, 3, or 5; use 7. Quotient is 1.

The prime factorization of 42 is $2 \cdot 3 \cdot 7$.

98   Chapter 4 Rational Numbers: Positive and Negative Fractions

(c) $90 \div 2 = 45$ Divide 90 by 2.
$45 \div 3 = 15$ 45 is not divisible by 2; use 3.
$15 \div 3 = 5$ Divide 15 by 3.
$5 \div 5 = 1$ 5 is not divisible by 2 or 3; use 5. Quotient is 1.

The prime factorization of 90 is $2 \cdot 3 \cdot 3 \cdot 5$.

(d) $100 \div 2 = 50$ Divide 100 by 2.
$50 \div 2 = 25$ Divide 50 by 2.
$25 \div 5 = 5$ 25 is not divisible by 2 or 3; use 5.
$5 \div 5 = 1$ Divide 5 by 5. Quotient is 1.

The prime factorization of 100 is $2 \cdot 2 \cdot 5 \cdot 5$.

(e) $81 \div 3 = 27$ 81 is not divisible by 2; use 3.
$27 \div 3 = 9$ Divide 27 by 3.
$9 \div 3 = 3$ Divide 9 by 3.
$3 \div 3 = 1$ Divide 3 by 3. Quotient is 1.

The prime factorization of 81 is $3 \cdot 3 \cdot 3 \cdot 3$.

5. (a)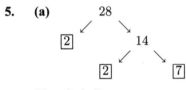

$28 = 2 \cdot 2 \cdot 7$

(b)
$$35$$
$$5 \quad 7$$
$35 = 5 \cdot 7$

(c)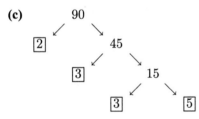

$90 = 2 \cdot 3 \cdot 3 \cdot 5$

6. (a) $\dfrac{16}{48} = \dfrac{\cancel{2}\cdot\cancel{2}\cdot\cancel{2}\cdot\cancel{2}}{\cancel{2}\cdot\cancel{2}\cdot\cancel{2}\cdot\cancel{2}\cdot 3}$
$= \dfrac{1}{3}$ ← Multiply $1 \cdot 1 \cdot 1 \cdot 1$
    ← Multiply $1 \cdot 1 \cdot 1 \cdot 1 \cdot 3$

(b) $\dfrac{28}{60} = \dfrac{\cancel{2}\cdot\cancel{2}\cdot 7}{\cancel{2}\cdot\cancel{2}\cdot 3\cdot 5}$
$= \dfrac{7}{15}$ ← Multiply $1 \cdot 1 \cdot 7$
    ← Multiply $1 \cdot 1 \cdot 3 \cdot 5$

(c) $\dfrac{74}{111} = \dfrac{2\cdot\cancel{37}}{3\cdot\cancel{37}}$
$= \dfrac{2}{3}$ ← Multiply $2 \cdot 1$
    ← Multiply $3 \cdot 1$

(d) $\dfrac{124}{340} = \dfrac{\cancel{2}\cdot\cancel{2}\cdot 31}{\cancel{2}\cdot\cancel{2}\cdot 5\cdot 17}$
$= \dfrac{31}{85}$ ← Multiply $1 \cdot 1 \cdot 31$
    ← Multiply $1 \cdot 1 \cdot 5 \cdot 17$

7. (a) $\dfrac{5c}{15} = \dfrac{\cancel{5}\cdot c}{3\cdot\cancel{5}} = \dfrac{1c}{3}$, or $\dfrac{c}{3}$

(b) $\dfrac{10x^2}{8x^2} = \dfrac{\cancel{2}\cdot 5\cdot\cancel{x}\cdot\cancel{x}}{\cancel{2}\cdot 2\cdot 2\cdot\cancel{x}\cdot\cancel{x}} = \dfrac{5}{4}$

(c) $\dfrac{9a^3}{11b^3} = \dfrac{9\cdot a\cdot a\cdot a}{11\cdot b\cdot b\cdot b}$

There are no common factors, so the fraction is already in lowest terms.

(d) $\dfrac{6m^2n}{9n^2} = \dfrac{2\cdot\cancel{3}\cdot m\cdot m\cdot\cancel{n}}{\cancel{3}\cdot 3\cdot\cancel{n}\cdot n} = \dfrac{2m^2}{3n}$

4.2 Section Exercises

1. (a) $-\dfrac{3}{10}$ is in lowest terms since the numerator and denominator have no common factor other than 1.

(b) $\dfrac{10}{15}$ is not in lowest terms. A common factor of 10 and 15 is 5.

(c) $\dfrac{9}{16}$ is in lowest terms.

(d) $-\dfrac{4}{21}$ is in lowest terms.

(e) $\dfrac{6}{9}$ is not in lowest terms. A common factor of 6 and 9 is 3.

(f) $-\dfrac{7}{28}$ is not in lowest terms. A common factor of 7 and 28 is 7.

3. (a) $\dfrac{10}{15} = \dfrac{10 \div 5}{15 \div 5} = \dfrac{2}{3}$

(b) $\dfrac{6}{9} = \dfrac{6 \div 3}{9 \div 3} = \dfrac{2}{3}$

(c) $-\dfrac{7}{28} = -\dfrac{7 \div 7}{28 \div 7} = -\dfrac{1}{4}$

(d) $-\dfrac{25}{50} = -\dfrac{25 \div 25}{50 \div 25} = -\dfrac{1}{2}$

(e) $\dfrac{16}{18} = \dfrac{16 \div 2}{18 \div 2} = \dfrac{8}{9}$

4.2 Writing Fractions in Lowest Terms

(f) $-\dfrac{8}{20} = -\dfrac{8 \div 4}{20 \div 4} = -\dfrac{2}{5}$

5. Prime: 2, 5, 11
   Composite: 9 since 9 is divisible by 3
   8 since 8 is divisible by 2 and 4
   10 since 10 is divisible by 2 and 5
   21 since 21 is divisible by 3 and 7

   1 is neither prime nor composite.

7. $6 \div 2 = 3$ Divide 6 by 2.
   $3 \div 3 = 1$ 3 is not divisible by 2; use 3.
   Quotient is 1.

   The prime factorization of 6 is $2 \cdot 3$.

9. $20 \div 2 = 10$ Divide 20 by 2.
   $10 \div 2 = 5$ Divide 10 by 2.
   $5 \div 5 = 1$ 5 is not divisible by 2 or 3; use 5.
   Quotient is 1.

   The prime factorization of 20 is $2 \cdot 2 \cdot 5$.

11.

   The prime factorization of 25 is $5 \cdot 5$.

13.

   The prime factorization of 36 is $2 \cdot 2 \cdot 3 \cdot 3$.

15. (a)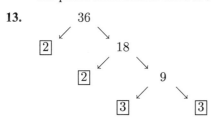

   The prime factorization of 44 is $2 \cdot 2 \cdot 11$.

   (b)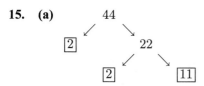

   The prime factorization of 88 is $2 \cdot 2 \cdot 2 \cdot 11$.

17. (a)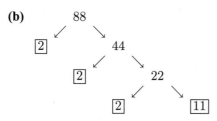

   The prime factorization of 75 is $3 \cdot 5 \cdot 5$.

   (b)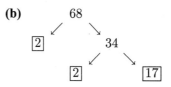

   The prime factorization of 68 is $2 \cdot 2 \cdot 17$.

   (c)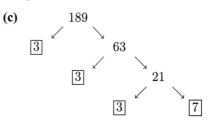

   The prime factorization of 189 is $3 \cdot 3 \cdot 3 \cdot 7$.

19. $\dfrac{8}{16} = \dfrac{\cancel{2} \cdot \cancel{2} \cdot \cancel{2}}{\cancel{2} \cdot \cancel{2} \cdot \cancel{2} \cdot 2} = \dfrac{1}{2}$

21. $\dfrac{32}{48} = \dfrac{\cancel{2} \cdot \cancel{2} \cdot \cancel{2} \cdot \cancel{2} \cdot 2}{\cancel{2} \cdot \cancel{2} \cdot \cancel{2} \cdot \cancel{2} \cdot 3} = \dfrac{2}{3}$

23. $\dfrac{14}{21} = \dfrac{2 \cdot \cancel{7}}{3 \cdot \cancel{7}} = \dfrac{2}{3}$

25. $\dfrac{36}{42} = \dfrac{\cancel{2} \cdot 2 \cdot \cancel{3} \cdot 3}{\cancel{2} \cdot \cancel{3} \cdot 7} = \dfrac{6}{7}$

27. $\dfrac{50}{63} = \dfrac{2 \cdot 5 \cdot 5}{3 \cdot 3 \cdot 7}$

   The fraction is already in lowest terms.

29. $\dfrac{27}{45} = \dfrac{\cancel{3} \cdot \cancel{3} \cdot 3}{\cancel{3} \cdot \cancel{3} \cdot 5} = \dfrac{3}{5}$

31. $\dfrac{12}{18} = \dfrac{\cancel{2} \cdot 2 \cdot \cancel{3}}{\cancel{2} \cdot \cancel{3} \cdot 3} = \dfrac{2}{3}$

33. $\dfrac{35}{40} = \dfrac{\cancel{5} \cdot 7}{2 \cdot 2 \cdot 2 \cdot \cancel{5}} = \dfrac{7}{8}$

35. $\dfrac{90}{180} = \dfrac{\cancel{2} \cdot \cancel{3} \cdot \cancel{3} \cdot \cancel{5}}{\cancel{2} \cdot 2 \cdot \cancel{3} \cdot \cancel{3} \cdot \cancel{5}} = \dfrac{1}{2}$

37. $\dfrac{210}{315} = \dfrac{2 \cdot \cancel{3} \cdot \cancel{5} \cdot \cancel{7}}{3 \cdot \cancel{3} \cdot \cancel{5} \cdot \cancel{7}} = \dfrac{2}{3}$

39. $\dfrac{429}{495} = \dfrac{\cancel{3} \cdot \cancel{11} \cdot 13}{\cancel{3} \cdot 3 \cdot 5 \cdot \cancel{11}} = \dfrac{13}{15}$

41. 60 minutes in an hour $\Rightarrow$ 60 parts in the whole

(a) $\dfrac{15}{60} = \dfrac{\cancel{3}\cdot\cancel{5}}{2\cdot 2\cdot\cancel{3}\cdot\cancel{5}} = \dfrac{1}{4}$ of an hour

(b) $\dfrac{30}{60} = \dfrac{\cancel{2}\cdot\cancel{3}\cdot\cancel{5}}{2\cdot\cancel{2}\cdot\cancel{3}\cdot\cancel{5}} = \dfrac{1}{2}$ of an hour

(c) $\dfrac{6}{60} = \dfrac{\cancel{2}\cdot\cancel{3}}{\cancel{2}\cdot 2\cdot\cancel{3}\cdot 5} = \dfrac{1}{10}$ of an hour

(d) $\dfrac{60}{60} = 1$ hour

43. (a) $\dfrac{800}{2400} = \dfrac{8\cdot\cancel{100}}{24\cdot\cancel{100}} = \dfrac{8}{24} = \dfrac{\cancel{8}}{3\cdot\cancel{8}} = \dfrac{1}{3}$

(b) $\dfrac{400}{2400} = \dfrac{4\cdot\cancel{100}}{24\cdot\cancel{100}} = \dfrac{4}{24} = \dfrac{\cancel{4}}{\cancel{4}\cdot 6} = \dfrac{1}{6}$

(c) $\$2400 - \$800 - \$400 = \$1200$

$\dfrac{1200}{2400} = \dfrac{12\cdot\cancel{100}}{24\cdot\cancel{100}} = \dfrac{12}{24} = \dfrac{\cancel{12}}{2\cdot\cancel{12}} = \dfrac{1}{2}$

45. (a) $\dfrac{10}{48} = \dfrac{\cancel{2}\cdot 5}{\cancel{2}\cdot 2\cdot 2\cdot 2\cdot 3} = \dfrac{5}{24}$

(b) $\dfrac{32}{48} = \dfrac{\cancel{2}\cdot\cancel{2}\cdot\cancel{2}\cdot\cancel{2}\cdot 2}{\cancel{2}\cdot\cancel{2}\cdot\cancel{2}\cdot\cancel{2}\cdot 3} = \dfrac{2}{3}$

(c) $\dfrac{6}{48} = \dfrac{\cancel{2}\cdot\cancel{3}}{\cancel{2}\cdot 2\cdot 2\cdot 2\cdot\cancel{3}} = \dfrac{1}{8}$

47. (a) The result of dividing 3 by 3 is 1, so 1 should be written above and below all the slashes. The numerator is $1\cdot 1$, so the correct answer is $\frac{1}{4}$.

$\dfrac{9}{36} = \dfrac{\cancel{3}\cdot\cancel{3}}{2\cdot 2\cdot\cancel{3}\cdot\cancel{3}} = \dfrac{1}{4}$

(b) You must divide numerator and denominator by the *same* number. The fraction is already in lowest terms because 9 and 16 have no common factor besides 1.

49. $\dfrac{16c}{40} = \dfrac{\cancel{2}\cdot\cancel{2}\cdot\cancel{2}\cdot 2\cdot c}{\cancel{2}\cdot\cancel{2}\cdot\cancel{2}\cdot 5} = \dfrac{2c}{5}$

51. $\dfrac{20x}{35x} = \dfrac{2\cdot 2\cdot\cancel{5}\cdot\cancel{x}}{\cancel{5}\cdot 7\cdot\cancel{x}} = \dfrac{4}{7}$

53. $\dfrac{18r^2}{15rs} = \dfrac{2\cdot\cancel{3}\cdot 3\cdot\cancel{r}\cdot r}{\cancel{3}\cdot 5\cdot\cancel{r}\cdot s} = \dfrac{6r}{5s}$

55. $\dfrac{6m}{42mn^2} = \dfrac{\cancel{2}\cdot\cancel{3}\cdot\cancel{m}}{\cancel{2}\cdot\cancel{3}\cdot 7\cdot\cancel{m}\cdot n\cdot n} = \dfrac{1}{7n^2}$

57. $\dfrac{9x^2}{16y^2} = \dfrac{3\cdot 3\cdot x\cdot x}{2\cdot 2\cdot 2\cdot 2\cdot y\cdot y}$

There are no common factors, so the fraction is already in lowest terms.

59. $\dfrac{7xz}{9xyz} = \dfrac{7\cdot\cancel{x}\cdot\cancel{z}}{3\cdot 3\cdot\cancel{x}\cdot y\cdot\cancel{z}} = \dfrac{7}{9y}$

61. $\dfrac{21k^3}{6k^2} = \dfrac{\cancel{3}\cdot 7\cdot\cancel{k}\cdot\cancel{k}\cdot k}{2\cdot\cancel{3}\cdot\cancel{k}\cdot\cancel{k}} = \dfrac{7k}{2}$

63. $\dfrac{13a^2bc^3}{39a^2bc^3} = \dfrac{\cancel{13}\cdot\cancel{a}\cdot\cancel{a}\cdot\cancel{b}\cdot\cancel{c}\cdot\cancel{c}\cdot\cancel{c}}{3\cdot\cancel{13}\cdot\cancel{a}\cdot\cancel{a}\cdot\cancel{b}\cdot\cancel{c}\cdot\cancel{c}\cdot\cancel{c}} = \dfrac{1}{3}$

65. $\dfrac{14c^2d}{14cd^2} = \dfrac{\cancel{2}\cdot\cancel{7}\cdot\cancel{c}\cdot c\cdot\cancel{d}}{\cancel{2}\cdot\cancel{7}\cdot\cancel{c}\cdot\cancel{d}\cdot d} = \dfrac{c}{d}$

67. $\dfrac{210ab^3c}{35b^2c^2} = \dfrac{2\cdot 3\cdot\cancel{5}\cdot\cancel{7}\cdot a\cdot\cancel{b}\cdot\cancel{b}\cdot b\cdot\cancel{c}}{\cancel{5}\cdot\cancel{7}\cdot\cancel{b}\cdot\cancel{b}\cdot\cancel{c}\cdot c} = \dfrac{6ab}{c}$

69. $\dfrac{25m^3rt^2}{36n^2s^3w^2} = \dfrac{5\cdot 5\cdot m\cdot m\cdot m\cdot r\cdot t\cdot t}{2\cdot 2\cdot 3\cdot 3\cdot n\cdot n\cdot s\cdot s\cdot s\cdot w\cdot w}$

There are no common factors, so the fraction is already in lowest terms.

71. $\dfrac{33e^2fg^3}{11efg} = \dfrac{3\cdot\cancel{11}\cdot\cancel{e}\cdot e\cdot\cancel{f}\cdot\cancel{g}\cdot g\cdot g}{\cancel{11}\cdot\cancel{e}\cdot\cancel{f}\cdot\cancel{g}} = 3eg^2$

4.3 Multiplying and Dividing Signed Fractions

4.3 Margin Exercises

1. (a) $-\dfrac{3}{4}\cdot\dfrac{1}{2} = -\dfrac{3\cdot 1}{4\cdot 2} = -\dfrac{3}{8}$

(b) $\left(-\dfrac{2}{5}\right)\left(-\dfrac{2}{3}\right) = \dfrac{2\cdot 2}{5\cdot 3} = \dfrac{4}{15}$

(c) $\dfrac{3}{4}\left(\dfrac{3}{8}\right) = \dfrac{3\cdot 3}{4\cdot 8} = \dfrac{9}{32}$

4.3 Multiplying and Dividing Signed Fractions

2. (a) $\dfrac{15}{28}\left(-\dfrac{6}{5}\right) = -\dfrac{3 \cdot \cancel{5} \cdot \cancel{2} \cdot 3}{\cancel{2} \cdot 2 \cdot 7 \cdot \cancel{5}} = -\dfrac{9}{14}$

   (b) $\dfrac{12}{7} \cdot \dfrac{7}{24} = \dfrac{\cancel{2} \cdot \cancel{2} \cdot \cancel{3} \cdot \cancel{7}}{\cancel{7} \cdot \cancel{2} \cdot \cancel{2} \cdot 2 \cdot \cancel{3}} = \dfrac{1}{2}$

   (c) $\left(-\dfrac{11}{18}\right)\left(-\dfrac{9}{20}\right) = \dfrac{11 \cdot \cancel{3} \cdot \cancel{3}}{2 \cdot \cancel{3} \cdot \cancel{3} \cdot 2 \cdot 2 \cdot 5} = \dfrac{11}{40}$

3. (a) $\dfrac{3}{4}$ of $36 = \dfrac{3}{4} \cdot \dfrac{36}{1} = \dfrac{3 \cdot \cancel{2} \cdot \cancel{2} \cdot 3 \cdot 3}{\cancel{2} \cdot \cancel{2} \cdot 1}$
   $= \dfrac{27}{1} = 27$

   (b) $-10 \cdot \dfrac{2}{5} = -\dfrac{10}{1} \cdot \dfrac{2}{5} = -\dfrac{2 \cdot \cancel{5} \cdot 2}{1 \cdot \cancel{5}}$
   $= -\dfrac{4}{1} = -4$

   (c) $\left(-\dfrac{7}{8}\right)(-24) = \left(-\dfrac{7}{8}\right)\left(-\dfrac{24}{1}\right)$
   $= \dfrac{7 \cdot \cancel{2} \cdot \cancel{2} \cdot \cancel{2} \cdot 3}{\cancel{2} \cdot \cancel{2} \cdot \cancel{2} \cdot 1}$
   $= \dfrac{21}{1} = 21$

4. (a) $\dfrac{2c}{5} \cdot \dfrac{c}{4} = \dfrac{\cancel{2} \cdot c \cdot c}{5 \cdot \cancel{2} \cdot 2} = \dfrac{c^2}{10}$

   (b) $\left(\dfrac{m}{6}\right)\left(\dfrac{9}{m^2}\right) = \dfrac{\cancel{m} \cdot \cancel{3} \cdot 3}{2 \cdot \cancel{3} \cdot \cancel{m} \cdot m} = \dfrac{3}{2m}$

   (c) $\left(\dfrac{w^2}{y}\right)\left(\dfrac{x^2 y}{w}\right) = \dfrac{\cancel{w} \cdot w \cdot x \cdot x \cdot \cancel{y}}{\cancel{y} \cdot \cancel{w}} = \dfrac{wx^2}{1}$
   $= wx^2$

5. (a) $-\dfrac{3}{4} \div \dfrac{5}{8} = -\dfrac{3}{4} \cdot \dfrac{8}{5} = -\dfrac{3 \cdot 2 \cdot \cancel{4}}{\cancel{4} \cdot 5} = -\dfrac{6}{5}$

   (b) $0 \div \left(-\dfrac{7}{12}\right) = 0 \cdot \left(-\dfrac{12}{7}\right) = 0$

   (c) $\dfrac{5}{6} \div 10 = \dfrac{5}{6} \cdot \dfrac{1}{10} = \dfrac{\cancel{5} \cdot 1}{6 \cdot 2 \cdot \cancel{5}} = \dfrac{1}{12}$

   (d) $-9 \div \left(-\dfrac{9}{16}\right) = -\dfrac{9}{1} \div \left(-\dfrac{9}{16}\right)$
   $= -\dfrac{9}{1} \cdot \left(-\dfrac{16}{9}\right)$
   $= \dfrac{\cancel{9} \cdot 16}{1 \cdot \cancel{9}} = \dfrac{16}{1} = 16$

   (e) $\dfrac{2}{5} \div 0$ is undefined.

   This division can't be written as multiplication because 0 does not have a reciprocal.

6. (a) $\dfrac{c^2 d^2}{4} \div \dfrac{c^2 d}{4} = \dfrac{c^2 d^2}{4} \cdot \dfrac{4}{c^2 d}$
   $= \dfrac{\cancel{c} \cdot \cancel{c} \cdot \cancel{d} \cdot d \cdot \cancel{4}}{\cancel{4} \cdot \cancel{c} \cdot \cancel{c} \cdot \cancel{d}}$
   $= \dfrac{d}{1} = d$

   (b) $\dfrac{20}{7h} \div \dfrac{5h}{7} = \dfrac{20}{7h} \cdot \dfrac{7}{5h} = \dfrac{4 \cdot \cancel{5} \cdot \cancel{7}}{\cancel{7} \cdot h \cdot \cancel{5} \cdot h} = \dfrac{4}{h^2}$

   (c) $\dfrac{n}{8} \div mn = \dfrac{n}{8} \div \dfrac{mn}{1} = \dfrac{n}{8} \cdot \dfrac{1}{mn} = \dfrac{\cancel{n} \cdot 1}{8 \cdot m \cdot \cancel{n}}$
   $= \dfrac{1}{8m}$

7. (a) Rewording the information might help. 18 quarts will be used up by repeatedly filling a $\frac{2}{3}$ quart spray bottle.

   $18 \div \dfrac{2}{3} = \dfrac{18}{1} \cdot \dfrac{3}{2} = \dfrac{\cancel{2} \cdot 9 \cdot 3}{1 \cdot \cancel{2}}$
   $= \dfrac{27}{1} = 27$

   The spray bottle can be filled 27 times.

   (b) $\frac{5}{8}$ *of* her highest annual salary suggests multiplication.

   $\dfrac{5}{8} \cdot 64{,}000 = \dfrac{5}{8} \cdot \dfrac{64{,}000}{1}$
   $= \dfrac{5 \cdot \cancel{8} \cdot 8000}{\cancel{8} \cdot 1}$
   $= \dfrac{40{,}000}{1} = 40{,}000$

   The officer will receive $40,000.

4.3 Section Exercises

1. $-\dfrac{3}{8} \cdot \dfrac{1}{2} = -\dfrac{3 \cdot 1}{8 \cdot 2} = -\dfrac{3}{16}$

Chapter 4 Rational Numbers: Positive and Negative Fractions

3. $\left(-\dfrac{3}{8}\right)\left(-\dfrac{12}{5}\right) = \dfrac{3 \cdot 3 \cdot \overset{1}{\cancel{4}}}{2 \cdot \underset{1}{\cancel{4}} \cdot 5} = \dfrac{9}{10}$

5. $\dfrac{21}{30}\left(\dfrac{5}{7}\right) = \dfrac{\overset{1}{\cancel{3}} \cdot \overset{1}{\cancel{7}} \cdot \overset{1}{\cancel{5}}}{2 \cdot \underset{1}{\cancel{3}} \cdot \underset{1}{\cancel{5}} \cdot \underset{1}{\cancel{7}}} = \dfrac{1}{2}$

7. $10\left(-\dfrac{3}{5}\right) = \dfrac{10}{1}\left(-\dfrac{3}{5}\right) = -\dfrac{2 \cdot \overset{1}{\cancel{5}} \cdot 3}{1 \cdot \underset{1}{\cancel{5}}} = -\dfrac{6}{1} = -6$

9. $\dfrac{4}{9}$ of $81 = \dfrac{4}{9} \cdot \dfrac{81}{1} = \dfrac{4 \cdot \overset{1}{\cancel{9}} \cdot 9}{\underset{1}{\cancel{9}} \cdot 1} = \dfrac{36}{1} = 36$

11. $\left(\dfrac{3x}{4}\right)\left(\dfrac{5}{xy}\right) = \dfrac{3 \cdot \overset{1}{\cancel{x}} \cdot 5}{4 \cdot \underset{1}{\cancel{x}} \cdot y} = \dfrac{15}{4y}$

13. $\dfrac{1}{6} \div \dfrac{1}{3} = \dfrac{1}{6} \cdot \dfrac{3}{1} = \dfrac{1 \cdot \overset{1}{\cancel{3}}}{2 \cdot \underset{1}{\cancel{3}} \cdot 1} = \dfrac{1}{2}$

15. $-\dfrac{3}{4} \div \left(-\dfrac{5}{8}\right) = -\dfrac{3}{4} \cdot \left(-\dfrac{8}{5}\right) = \dfrac{3 \cdot 2 \cdot \overset{1}{\cancel{4}}}{\underset{1}{\cancel{4}} \cdot 5} = \dfrac{6}{5}$

17. $6 \div \left(-\dfrac{2}{3}\right) = \dfrac{6}{1} \cdot \left(-\dfrac{3}{2}\right) = -\dfrac{\overset{1}{\cancel{2}} \cdot 3 \cdot 3}{1 \cdot \underset{1}{\cancel{2}}}$

    $= -\dfrac{9}{1} = -9$

19. $-\dfrac{2}{3} \div 4 = -\dfrac{2}{3} \div \dfrac{4}{1} = -\dfrac{2}{3} \cdot \dfrac{1}{4} = -\dfrac{\overset{1}{\cancel{2}} \cdot 1}{3 \cdot \underset{1}{\cancel{2}} \cdot 2} = -\dfrac{1}{6}$

21. $\dfrac{11c}{5d} \div 3c = \dfrac{11c}{5d} \div \dfrac{3c}{1} = \dfrac{11c}{5d} \cdot \dfrac{1}{3c}$

    $= \dfrac{11 \cdot \overset{1}{\cancel{c}} \cdot 1}{5 \cdot d \cdot 3 \cdot \underset{1}{\cancel{c}}} = \dfrac{11}{15d}$

23. $\dfrac{ab^2}{c} \div \dfrac{ab}{c} = \dfrac{ab^2}{c} \cdot \dfrac{c}{ab} = \dfrac{\overset{1}{\cancel{a}} \cdot \overset{1}{\cancel{b}} \cdot b \cdot \overset{1}{\cancel{c}}}{\underset{1}{\cancel{c}} \cdot \underset{1}{\cancel{a}} \cdot \underset{1}{\cancel{b}}} = \dfrac{b}{1} = b$

25. (a) Forgot to write 1s in numerator when dividing out common factors. Answer is $\dfrac{1}{6}$.

    (b) Used reciprocal of $\dfrac{2}{3}$ in multiplication, but the reciprocal is used only in division. Correct answer is $8 \cdot \dfrac{2}{3} = \dfrac{8 \cdot 2}{3} = \dfrac{16}{3}$.

27. (a) Forgot to use reciprocal of $\dfrac{4}{1}$; correct answer is

    $\dfrac{2}{3} \cdot \dfrac{1}{4} = \dfrac{\overset{1}{\cancel{2}} \cdot 1}{3 \cdot 2 \cdot \underset{1}{\cancel{2}}} = \dfrac{1}{6}$.

   (b) Used reciprocal of $\dfrac{5}{6}$ instead of reciprocal of $\dfrac{10}{9}$; correct answer is

    $\dfrac{5}{6} \cdot \dfrac{9}{10} = \dfrac{\overset{1}{\cancel{5}} \cdot \overset{1}{\cancel{3}} \cdot 3}{2 \cdot \underset{1}{\cancel{3}} \cdot 2 \cdot \underset{1}{\cancel{5}}} = \dfrac{3}{4}$.

29. Rewrite division as multiplication. Leave the first number (dividend) the same. Change the second number (divisor) to its reciprocal by "flipping" it. Then multiply.

31. $\dfrac{4}{5} \div 3 = \dfrac{4}{5} \div \dfrac{3}{1} = \dfrac{4}{5} \cdot \dfrac{1}{3} = \dfrac{4 \cdot 1}{5 \cdot 3} = \dfrac{4}{15}$

33. $-\dfrac{3}{8}\left(\dfrac{3}{4}\right) = -\dfrac{3 \cdot 3}{8 \cdot 4} = -\dfrac{9}{32}$

35. $\dfrac{3}{5}$ of $35 = \dfrac{3}{5} \cdot 35 = \dfrac{3}{5} \cdot \dfrac{35}{1} = \dfrac{3 \cdot \overset{1}{\cancel{5}} \cdot 7}{\underset{1}{\cancel{5}} \cdot 1} = \dfrac{21}{1} = 21$

37. $-9 \div \left(-\dfrac{3}{5}\right) = -\dfrac{9}{1} \cdot \left(-\dfrac{5}{3}\right) = \dfrac{\overset{1}{\cancel{3}} \cdot 3 \cdot 5}{1 \cdot \underset{1}{\cancel{3}}}$

    $= \dfrac{15}{1} = 15$

39. $\dfrac{12}{7} \div 0$ is undefined.

41. $\left(\dfrac{11}{2}\right)\left(-\dfrac{5}{6}\right) = -\dfrac{11 \cdot 5}{2 \cdot 6} = -\dfrac{55}{12}$

43. $\dfrac{4}{7}$ of $14b = \dfrac{4}{7} \cdot 14b = \dfrac{4}{7} \cdot \dfrac{14b}{1} = \dfrac{4 \cdot 2 \cdot \overset{1}{\cancel{7}} \cdot b}{\underset{1}{\cancel{7}} \cdot 1}$

    $= \dfrac{8b}{1} = 8b$

45. $\dfrac{12}{5} \div 4d = \dfrac{12}{5} \div \dfrac{4d}{1} = \dfrac{12}{5} \cdot \dfrac{1}{4d}$

    $= \dfrac{3 \cdot \overset{1}{\cancel{4}} \cdot 1}{5 \cdot \underset{1}{\cancel{4}} \cdot d} = \dfrac{3}{5d}$

47. $\dfrac{x^2}{y} \div \dfrac{w}{2y} = \dfrac{x^2}{y} \cdot \dfrac{2y}{w} = \dfrac{x^2 \cdot 2 \cdot \overset{1}{\cancel{y}}}{\underset{1}{\cancel{y}} \cdot w} = \dfrac{2x^2}{w}$

49. The top of a table is a rectangle.

    $A = l \cdot w$

    $A = \dfrac{4}{5}$ yd $\cdot \dfrac{3}{8}$ yd

    $A = \dfrac{\overset{1}{\cancel{4}} \cdot 3}{5 \cdot 2 \cdot \underset{1}{\cancel{4}}}$ yd^2

    $A = \dfrac{3}{10}$ yd^2 or $\dfrac{3}{10}$ square yard

51. Splitting 10 ounces into equal size parts indicates division. Each part will contain $\frac{1}{8}$ ounce.

$$10 \div \frac{1}{8} = \frac{10}{1} \cdot \frac{8}{1} = \frac{80}{1} = 80$$

80 eyedrop dispensers can be filled.

53. Todd must earn $\frac{3}{4}$ of the cost:

$$\frac{3}{4} \cdot \$12{,}400 = \frac{3}{4} \cdot \frac{12{,}400}{1} = \frac{3 \cdot \overset{1}{\cancel{4}} \cdot 3100}{\underset{1}{\cancel{4}} \cdot 1}$$
$$= \frac{9300}{1} = \$9300$$

Todd must borrow the rest:

$$\$12{,}400 - \$9300 = \$3100$$

Thus, he must earn $9300 and borrow $3100.

55. The total number of cords divided by the number of cords per trip will give us the number of trips.

$$6 \div \frac{2}{3} = \frac{6}{1} \div \frac{2}{3} = \frac{6}{1} \cdot \frac{3}{2} = \frac{\overset{1}{\cancel{2}} \cdot 3 \cdot 3}{1 \cdot \underset{1}{\cancel{2}}} = \frac{9}{1} = 9$$

9 trips are needed to deliver 6 cords.

57. $234 \cdot \frac{1}{3} = \frac{234}{1} \cdot \frac{1}{3} = \frac{78 \cdot \overset{1}{\cancel{3}} \cdot 1}{1 \cdot \underset{1}{\cancel{3}}} = 78$

About 78 infield players are in the Hall of Fame.

59. The weight of the adult divided by the weight of the hatchling is

$$400 \div \frac{1}{8} = \frac{400}{1} \cdot \frac{8}{1} = \frac{400 \cdot 8}{1 \cdot 1} = 3200.$$

The adult weighs 3200 times the hatchling.

61. **(a)** Add the monthly income.

$$\$4575 + \$4312 + \cdots + \$6458 = \$58{,}000$$

(b) Rent is $\frac{1}{5}$ of the income.

$$\frac{1}{5} \cdot 58{,}000 = \frac{1 \cdot \overset{1}{\cancel{5}} \cdot 11{,}600}{\underset{1}{\cancel{5}}} = 11{,}600$$

The family's rent is $11,600.

63. The circle graph shows that $\frac{5}{16}$ of their total income is spent on food and $\frac{1}{8}$ is spent on clothing.

From Exercise 61(a), their total income is $58,000.

Food Expense:

$$\frac{5}{16} \cdot 58{,}000 = \frac{5}{16} \cdot \frac{58{,}000}{1} = \frac{5 \cdot \overset{1}{\cancel{16}} \cdot 3625}{\underset{1}{\cancel{16}} \cdot 1} = 18{,}125$$

$18,125 is spent on food.

Clothing Expense:

$$\frac{1}{8} \cdot 58{,}000 = \frac{1}{8} \cdot \frac{58{,}000}{1} = \frac{1 \cdot \overset{1}{\cancel{8}} \cdot 7250}{\underset{1}{\cancel{8}} \cdot 1} = 7250$$

$7250 is spent on clothing.

$18{,}125 + \$7250 = \$25{,}375$ was spent on food and clothing.

65. Horses: $175 \cdot \frac{1}{25} = \frac{175}{1} \cdot \frac{1}{25} = \frac{7 \cdot \overset{1}{\cancel{25}} \cdot 1}{1 \cdot \underset{1}{\cancel{25}}} = 7$

7 million U.S. pets are horses.

67. Dogs: $175 \cdot \frac{2}{5} = \frac{175}{1} \cdot \frac{2}{5} = \frac{\overset{1}{\cancel{5}} \cdot 35 \cdot 2}{1 \cdot \underset{1}{\cancel{5}}} = 70$

Cats: $175 \cdot \frac{12}{25} = \frac{175 \cdot 12}{1 \cdot 25} = \frac{7 \cdot \overset{1}{\cancel{25}} \cdot 12}{\underset{1}{\cancel{25}}} = 84$

$70 + 84 = 154$ million U.S. pets are dogs and cats.

4.4 Adding and Subtracting Signed Fractions

4.4 Margin Exercises

1. **(a)** $\frac{1}{6} + \frac{5}{6} = \frac{1+5}{6} = \frac{6}{6} = 1$

(b) $-\frac{11}{12} + \frac{5}{12} = \frac{-11+5}{12}$
$= \frac{-6}{12}$ or $-\frac{6}{12}$ Reduce to lowest terms
$= -\frac{\overset{1}{\cancel{6}}}{2 \cdot \underset{1}{\cancel{6}}} = -\frac{1}{2}$

(c) $-\frac{2}{9} - \frac{3}{9} = \frac{-2-3}{9} = \frac{-2+(-3)}{9}$
$= \frac{-5}{9}$ or $-\frac{5}{9}$

104 Chapter 4 Rational Numbers: Positive and Negative Fractions

(d) $\dfrac{8}{ab} + \dfrac{3}{ab} = \dfrac{8+3}{ab} = \dfrac{11}{ab}$

2. (a) The LCD for $\frac{3}{5}$ and $\frac{3}{10}$ is 10, since the larger denominator, 10, is divisible by the smaller denominator, 5.

   (b) The LCD of $\frac{1}{2}$ and $\frac{2}{5}$ is 10.

   Since the larger denominator, 5, is not divisible by the smaller denominator, 2, check multiples of 5: 5, 10, 15, 20, etc. 10 is the smallest one divisible by 2.

   (c) The LCD of $\frac{3}{4}$ and $\frac{1}{6}$ is 12.

   Since the larger denominator, 6, is not divisible by the smaller denominator, 4, check multiples of 6: 6, 12, 18, 24, 30, etc. 12 is the smallest one divisible by 4.

   (d) The LCD of $\frac{5}{6}$ and $\frac{7}{18}$ is 18, since the larger denominator, 18, is divisible by the smaller denominator, 6.

3. (a) $\frac{1}{10}$ and $\frac{13}{14}$

   $\left.\begin{array}{l} 10 = 2 \cdot 5 \\ 14 = 2 \cdot 7 \end{array}\right\}$ LCD $= 2 \cdot 5 \cdot 7 = 70$

   (b) $\frac{5}{12}$ and $\frac{17}{20}$

   $\left.\begin{array}{l} 12 = 2 \cdot 2 \cdot 3 \\ 20 = 2 \cdot 2 \cdot 5 \end{array}\right\}$ LCD $= 2 \cdot 2 \cdot 3 \cdot 5 = 60$

   (c) $\frac{7}{15}$ and $\frac{7}{9}$

   $\left.\begin{array}{l} 15 = 3 \cdot 5 \\ 9 = 3 \cdot 3 \end{array}\right\}$ LCD $= 3 \cdot 3 \cdot 5 = 45$

4. (a) $\dfrac{2}{3} + \dfrac{1}{6}$

   Step 1
   The LCD is 6, the larger denominator.

   Step 2
   $\frac{2}{3} = \frac{2 \cdot 2}{3 \cdot 2} = \frac{4}{6}$ and $\frac{1}{6}$ already has the LCD.

   Step 3
   $\dfrac{2}{3} + \dfrac{1}{6} = \dfrac{4}{6} + \dfrac{1}{6} = \dfrac{4+1}{6} = \dfrac{5}{6}$

   Step 4
   $\frac{5}{6}$ is in lowest terms.

   (b) $\dfrac{1}{12} - \dfrac{5}{6}$

   Step 1
   The LCD is 12, the larger denominator.

   Step 2
   $\frac{1}{12}$ already has the LCD and $\frac{5}{6} = \frac{5 \cdot 2}{6 \cdot 2} = \frac{10}{12}$.

   Step 3
   $\dfrac{1}{12} - \dfrac{5}{6} = \dfrac{1}{12} - \dfrac{10}{12} = \dfrac{1-10}{12} = \dfrac{1+(-10)}{12}$
   $= \dfrac{-9}{12}$ or $-\dfrac{9}{12}$

   Step 4
   $-\dfrac{9}{12} = -\dfrac{\overset{1}{\cancel{3}} \cdot 3}{\cancel{3} \cdot 4} = -\dfrac{3}{4}$

   (c) $3 - \dfrac{4}{5}$

   Step 1
   Think of 3 as $\frac{3}{1}$. The LCD for $\frac{3}{1}$ and $\frac{4}{5}$ is 5, the larger denominator.

   Step 2
   $\frac{3}{1} = \frac{3 \cdot 5}{1 \cdot 5} = \frac{15}{5}$ and $\frac{4}{5}$ already has the LCD.

   Step 3
   $3 - \dfrac{4}{5} = \dfrac{15}{5} - \dfrac{4}{5} = \dfrac{15-4}{5} = \dfrac{11}{5}$

   Step 4
   $\frac{11}{5}$ is in lowest terms.

   (d) $-\dfrac{5}{12} + \dfrac{9}{16}$

   Step 1
   Use prime factorization to find the LCD.

   $\left.\begin{array}{l} 12 = 2 \cdot 2 \cdot 3 \\ 16 = 2 \cdot 2 \cdot 2 \cdot 2 \end{array}\right\}$ LCD $= 2 \cdot 2 \cdot 2 \cdot 2 \cdot 3 = 48$

   Step 2
   $-\dfrac{5}{12} = -\dfrac{5 \cdot 4}{12 \cdot 4} = -\dfrac{20}{48}$ and $\dfrac{9}{16} = \dfrac{9 \cdot 3}{16 \cdot 3} = \dfrac{27}{48}$

   Step 3
   $-\dfrac{5}{12} + \dfrac{9}{16} = -\dfrac{20}{48} + \dfrac{27}{48}$
   $= \dfrac{-20+27}{48} = \dfrac{7}{48}$

   Step 4
   $\frac{7}{48}$ is in lowest terms.

5. (a) $\dfrac{5}{6} - \dfrac{h}{2}$

   Step 1
   The LCD is 6, the larger denominator.

   Step 2
   $\frac{5}{6}$ already has the LCD and $\dfrac{h}{2} = \dfrac{h \cdot 3}{2 \cdot 3} = \dfrac{3h}{6}$.

   Step 3
   $\dfrac{5}{6} - \dfrac{h}{2} = \dfrac{5}{6} - \dfrac{3h}{6}$
   $= \dfrac{5 - 3h}{6}$ Lowest terms

4.4 Adding and Subtracting Signed Fractions

(b) $\dfrac{7}{t} + \dfrac{3}{5}$

Step 1
The LCD is $5 \cdot t$ or $5t$.

Step 2
$\dfrac{7}{t} = \dfrac{7 \cdot 5}{t \cdot 5} = \dfrac{35}{5t}$ and $\dfrac{3}{5} = \dfrac{3 \cdot t}{5 \cdot t} = \dfrac{3t}{5t}$

Step 3
$\dfrac{7}{t} + \dfrac{3}{5} = \dfrac{35}{5t} + \dfrac{3t}{5t}$
$= \dfrac{35 + 3t}{5t}$ Lowest terms

(c) $\dfrac{4}{x} - \dfrac{8}{3}$

Step 1
The LCD is $3 \cdot x$ or $3x$.

Step 2
$\dfrac{4}{x} = \dfrac{4 \cdot 3}{x \cdot 3} = \dfrac{12}{3x}$ and $\dfrac{8}{3} = \dfrac{8 \cdot x}{3 \cdot x} = \dfrac{8x}{3x}$

Step 3
$\dfrac{4}{x} - \dfrac{8}{3} = \dfrac{12}{3x} - \dfrac{8x}{3x}$
$= \dfrac{12 - 8x}{3x}$ Lowest terms

4.4 Section Exercises

1. $\dfrac{3}{4} + \dfrac{1}{8}$

Step 1
The LCD is 8.

Step 2
$\dfrac{3}{4} = \dfrac{3 \cdot 2}{4 \cdot 2} = \dfrac{6}{8}$, $\dfrac{1}{8}$ has the LCD.

Step 3
$\dfrac{3}{4} + \dfrac{1}{8} = \dfrac{6}{8} + \dfrac{1}{8} = \dfrac{6+1}{8} = \dfrac{7}{8}$

Step 4
$\dfrac{7}{8}$ is already in lowest terms.

3. $-\dfrac{1}{14} + \left(-\dfrac{3}{7}\right)$

Step 1
The LCD is 14.

Step 2
$-\dfrac{1}{14}$ has the LCD.

$-\dfrac{3}{7} = -\dfrac{3 \cdot 2}{7 \cdot 2} = -\dfrac{6}{14}$.

Step 3
$-\dfrac{1}{14} + \left(-\dfrac{3}{7}\right) = -\dfrac{1}{14} + \left(-\dfrac{6}{14}\right)$
$= \dfrac{-1 + (-6)}{14}$
$= -\dfrac{7}{14}$

Step 4
$-\dfrac{7}{14} = -\dfrac{\overset{1}{\cancel{7}}}{2 \cdot \underset{1}{\cancel{7}}} = -\dfrac{1}{2}$

5. $\dfrac{2}{3} - \dfrac{1}{6}$

Step 1
The LCD is 6.

Step 2
$\dfrac{2}{3} = \dfrac{2 \cdot 2}{3 \cdot 2} = \dfrac{4}{6}$, $\dfrac{1}{6}$ has the LCD.

Step 3
$\dfrac{2}{3} - \dfrac{1}{6} = \dfrac{4}{6} - \dfrac{1}{6} = \dfrac{4-1}{6} = \dfrac{3}{6}$

Step 4
$\dfrac{3}{6} = \dfrac{\overset{1}{\cancel{3}}}{2 \cdot \underset{1}{\cancel{3}}} = \dfrac{1}{2}$

7. $\dfrac{3}{8} - \dfrac{3}{5}$

Step 1
The LCD is $8 \cdot 5 = 40$.

Step 2
$\dfrac{3}{8} = \dfrac{3 \cdot 5}{8 \cdot 5} = \dfrac{15}{40}$, $\dfrac{3}{5} = \dfrac{3 \cdot 8}{5 \cdot 8} = \dfrac{24}{40}$

Step 3
$\dfrac{3}{8} - \dfrac{3}{5} = \dfrac{15}{40} - \dfrac{24}{40} = \dfrac{15 - 24}{40} = -\dfrac{9}{40}$

Step 4
$-\dfrac{9}{40}$ is already in lowest terms.

9. $-\dfrac{5}{8} + \dfrac{1}{12}$

Step 1
Use prime factorization to find the LCD.

$\left.\begin{array}{l} 8 = 2 \cdot 2 \cdot 2 \\ 12 = 2 \cdot 2 \cdot 3 \end{array}\right\}$ LCD $= 2 \cdot 2 \cdot 2 \cdot 3 = 24$

Step 2
$-\dfrac{5}{8} = -\dfrac{5 \cdot 3}{8 \cdot 3} = -\dfrac{15}{24}$, $\dfrac{1}{12} = \dfrac{1 \cdot 2}{12 \cdot 2} = \dfrac{2}{24}$

Step 3
$$-\frac{5}{8} + \frac{1}{12} = -\frac{15}{24} + \frac{2}{24} = \frac{-15+2}{24}$$
$$= \frac{-13}{24} \text{ or } -\frac{13}{24}$$

Step 4
$-\frac{13}{24}$ is already in lowest terms.

11. $-\frac{7}{20} - \frac{5}{20}$

Step 1
The LCD is 20.

Step 2
Each fraction has the LCD.

Step 3
$$-\frac{7}{20} - \frac{5}{20} = \frac{-7-5}{20} = -\frac{12}{20}$$

Step 4
$$-\frac{12}{20} = -\frac{\cancel{4} \cdot 3}{\cancel{4} \cdot 5} = -\frac{3}{5}$$

13. $0 - \frac{7}{18} = 0 + \left(-\frac{7}{18}\right) = -\frac{7}{18}$

Addition property of zero

15. $2 - \frac{6}{7} = \frac{2}{1} - \frac{6}{7}$

Step 1
The LCD is 7.

Step 2
$2 = \frac{2 \cdot 7}{7} = \frac{14}{7}$, $\frac{6}{7}$ has the LCD.

Step 3
$$2 - \frac{6}{7} = \frac{14}{7} - \frac{6}{7} = \frac{14-6}{7} = \frac{8}{7}$$

Step 4
$\frac{8}{7}$ is already in lowest terms.

17. $-\frac{1}{2} + \frac{3}{24}$

Step 1
The LCD is 24, the larger denominator.

Step 2
$-\frac{1}{2} = -\frac{1 \cdot 12}{2 \cdot 12} = -\frac{12}{24}$, $\frac{3}{24}$ has the LCD.

Step 3
$$-\frac{1}{2} + \frac{3}{24} = -\frac{12}{24} + \frac{3}{24} = \frac{-12+3}{24}$$
$$= \frac{-9}{24} \text{ or } -\frac{9}{24}$$

Step 4
$$-\frac{9}{24} = -\frac{\cancel{3} \cdot 3}{\cancel{3} \cdot 8} = -\frac{3}{8}$$

19. $\frac{1}{5} + \frac{c}{3}$

Step 1
The LCD is $5 \cdot 3 = 15$.

Step 2
$\frac{1}{5} = \frac{1 \cdot 3}{5 \cdot 3} = \frac{3}{15}$, $\frac{c}{3} = \frac{c \cdot 5}{3 \cdot 5} = \frac{5c}{15}$

Step 3
$$\frac{1}{5} + \frac{c}{3} = \frac{3}{15} + \frac{5c}{15} = \frac{3+5c}{15}$$

Step 4
$\frac{3+5c}{15}$ is already in lowest terms.

21. $\frac{5}{m} - \frac{1}{2}$

Step 1
The LCD is $2 \cdot m$ or $2m$.

Step 2
$\frac{5}{m} = \frac{5 \cdot 2}{m \cdot 2} = \frac{10}{2m}$, $\frac{1}{2} = \frac{1 \cdot m}{2 \cdot m} = \frac{1m}{2m}$

Step 3
$$\frac{5}{m} - \frac{1}{2} = \frac{10}{2m} - \frac{1m}{2m}$$
$$= \frac{10-1m}{2m}, \text{ or } \frac{10-m}{2m}$$

Step 4
$\frac{10-m}{2m}$ is already in lowest terms.

23. $\frac{3}{b^2} + \frac{5}{b^2} = \frac{3+5}{b^2} = \frac{8}{b^2}$, which is in lowest terms.

25. $\frac{c}{7} + \frac{3}{b}$

Step 1
The LCD is $7 \cdot b$ or $7b$.

Step 2
$\frac{c}{7} = \frac{c \cdot b}{7 \cdot b} = \frac{bc}{7b}$, $\frac{3}{b} = \frac{3 \cdot 7}{b \cdot 7} = \frac{21}{7b}$

Step 3
$$\frac{c}{7} + \frac{3}{b} = \frac{bc}{7b} + \frac{21}{7b} = \frac{bc+21}{7b}$$

Step 4
$\frac{bc+21}{7b}$ is already in lowest terms.

4.4 Adding and Subtracting Signed Fractions

27. $-\dfrac{4}{c^2} - \dfrac{d}{c}$

Step 1
The LCD is c^2.

Step 2
$\dfrac{4}{c^2}$ has the LCD, $\dfrac{d}{c} = \dfrac{d \cdot c}{c \cdot c} = \dfrac{cd}{c^2}$.

Step 3
$-\dfrac{4}{c^2} - \dfrac{d}{c} = -\dfrac{4}{c^2} - \dfrac{cd}{c^2} = \dfrac{-4 - cd}{c^2}$

Step 4
$\dfrac{-4 - cd}{c^2}$ is already in lowest terms.

29. $-\dfrac{11}{42} - \dfrac{11}{70}$

Step 1
Use prime factorization to find the LCD.

$\left.\begin{array}{l} 42 = 2 \cdot 3 \cdot 7 \\ 70 = 2 \cdot 5 \cdot 7 \end{array}\right\}$ LCD $= 2 \cdot 3 \cdot 5 \cdot 7 = 210$

Step 2
$-\dfrac{11}{42} = -\dfrac{11 \cdot 5}{42 \cdot 5} = -\dfrac{55}{210}, \quad \dfrac{11}{70} = \dfrac{11 \cdot 3}{70 \cdot 3} = \dfrac{33}{210}$

Step 3
$-\dfrac{11}{42} - \dfrac{11}{70} = -\dfrac{55}{210} - \dfrac{33}{210} = \dfrac{-55 - 33}{210}$
$= \dfrac{-55 + (-33)}{210}$
$= \dfrac{-88}{210}$ or $-\dfrac{88}{210}$

Step 4
$-\dfrac{88}{210} = -\dfrac{2 \cdot 44}{2 \cdot 105} = -\dfrac{44}{105}$

31. You cannot add or subtract until all the fractional pieces are the same size.

33. (a) You cannot add fractions with unlike denominators; use 20 as the LCD.

$\dfrac{3}{4} + \dfrac{2}{5} = \dfrac{15}{20} + \dfrac{8}{20} = \dfrac{23}{20}$

(b) When rewriting fractions with 18 as the denominator, you must multiply denominator and numerator by the same number.

$\dfrac{5}{6} - \dfrac{4}{9} = \dfrac{15}{18} - \dfrac{8}{18} = \dfrac{7}{18}$

35. (a) $-\dfrac{2}{3} + \dfrac{3}{4} = -\dfrac{8}{12} + \dfrac{9}{12} = \dfrac{-8 + 9}{12} = \dfrac{1}{12}$

$\dfrac{3}{4} + \left(-\dfrac{2}{3}\right) = \dfrac{9}{12} + \left(-\dfrac{8}{12}\right) = \dfrac{9 + (-8)}{12} = \dfrac{1}{12}$

Both sums are $\dfrac{1}{12}$ because addition is commutative.

(b) $\dfrac{5}{6} - \dfrac{1}{2} = \dfrac{5}{6} - \dfrac{3}{6} = \dfrac{5 - 3}{6} = \dfrac{2}{6} = \dfrac{\overset{1}{\cancel{2}}}{\cancel{2} \cdot 3} = \dfrac{1}{3}$

$\dfrac{1}{2} - \dfrac{5}{6} = \dfrac{3}{6} - \dfrac{5}{6} = \dfrac{3 - 5}{6} = \dfrac{-2}{6} = -\dfrac{2}{6}$
$= \dfrac{-\overset{1}{\cancel{2}}}{\cancel{2} \cdot 3} = -\dfrac{1}{3}$

The answers are different because subtraction is not commutative.

(c) $\left(-\dfrac{2}{3}\right)\left(\dfrac{9}{10}\right) = -\dfrac{\overset{1}{\cancel{2}} \cdot \overset{3}{\cancel{9}} \cdot 3}{\cancel{3} \cdot \cancel{2} \cdot 5} = -\dfrac{3}{5}$

$\left(\dfrac{9}{10}\right)\left(-\dfrac{2}{3}\right) = -\dfrac{\overset{1}{\cancel{3}} \cdot 3 \cdot \overset{1}{\cancel{2}}}{\cancel{2} \cdot 5 \cdot \cancel{3}} = -\dfrac{3}{5}$

Multiplication is commutative.

(d) $\dfrac{2}{5} \div \dfrac{1}{15} = \dfrac{2}{5} \cdot \dfrac{15}{1} = \dfrac{2 \cdot 3 \cdot \overset{1}{\cancel{5}}}{\cancel{5} \cdot 1} = 6$

$\dfrac{1}{15} \div \dfrac{2}{5} = \dfrac{1}{15} \cdot \dfrac{5}{2} = \dfrac{1 \cdot \overset{1}{\cancel{5}}}{3 \cdot \cancel{5} \cdot 2} = \dfrac{1}{6}$

Division is *not* commutative.

36. (a) $-\dfrac{7}{12} + \dfrac{7}{12} = \dfrac{-7 + 7}{12} = \dfrac{0}{12} = 0$

$\dfrac{3}{5} + \left(-\dfrac{3}{5}\right) = \dfrac{3 + (-3)}{5} = \dfrac{0}{5} = 0$

The sum of a number and its opposite is 0.

(b) $-\dfrac{13}{16} \div \left(-\dfrac{13}{16}\right) = -\dfrac{\overset{1}{\cancel{13}}}{\cancel{16}} \cdot \left(-\dfrac{\overset{1}{\cancel{16}}}{\cancel{13}}\right) = 1$

$\dfrac{1}{8} \div \dfrac{1}{8} = \dfrac{1}{\cancel{8}} \cdot \dfrac{\overset{1}{\cancel{8}}}{1} = 1$

When a nonzero number is divided by itself, the quotient is 1.

(c) $\dfrac{5}{6} \cdot 1 = \dfrac{5}{6}$

$1\left(-\dfrac{17}{20}\right) = -\dfrac{17}{20}$

Multiplying by 1 leaves a number unchanged.

(d) $\left(-\dfrac{4}{5}\right)\left(-\dfrac{5}{4}\right) = \left(-\dfrac{\overset{1}{\cancel{4}}}{\cancel{5}}\right)\left(-\dfrac{\overset{1}{\cancel{5}}}{\cancel{4}}\right) = 1$

$7 \cdot \dfrac{1}{7} = \dfrac{\overset{1}{\cancel{7}}}{1} \cdot \dfrac{1}{\cancel{7}} = 1$

A number times its reciprocal is 1.

108 Chapter 4 Rational Numbers: Positive and Negative Fractions

37. $\frac{1}{5} + \frac{1}{3} + \frac{1}{4} = \frac{12}{60} + \frac{20}{60} + \frac{15}{60} = \frac{12+20+15}{60} = \frac{47}{60}$

The total length of the bolt is $\frac{47}{60}$ inches.

39. $\frac{1}{3} + \frac{3}{8} + \frac{1}{4} = \frac{8}{24} + \frac{9}{24} + \frac{6}{24} = \frac{8+9+6}{24} = \frac{23}{24}$

$\frac{23}{24}$ cubic yard of material was ordered.

41. Fraction of an acre planted:

$\frac{5}{12} + \frac{11}{12} = \frac{16}{12}$

Lost $\frac{7}{12}$ acre due to fire: $-\frac{7}{12}$

Remaining seedlings:

$\frac{16}{12} + \left(-\frac{7}{12}\right) = \frac{16+(-7)}{12} = \frac{9}{12} = \frac{3}{4}$

$\frac{3}{4}$ acre of seedlings remained.

43. $\frac{2}{5} + \frac{3}{50} = \frac{20}{50} + \frac{3}{50} = \frac{20+3}{50} = \frac{23}{50}$

$\frac{23}{50}$ of workers are self-taught or learned from friends or family.

45. $\frac{2}{5} - \frac{6}{25} = \frac{10}{25} - \frac{6}{25} = \frac{10-6}{25} = \frac{4}{25}$

$\frac{4}{25}$ of workers is the difference.

47. $\frac{1}{8} + \frac{1}{6} = \frac{3}{24} + \frac{4}{24} = \frac{3+4}{24} = \frac{7}{24}$

$\frac{7}{24} \cdot 24 = \frac{7}{24} \cdot \frac{24}{1} = \frac{7 \cdot \cancel{24}}{\cancel{24}} = 7$

$\frac{7}{24}$ of the day (or 7 hours) was spent in class and study.

49. $\frac{7}{24} - \frac{1}{6} = \frac{7}{24} - \frac{4}{24} = \frac{3}{24} = \frac{\cancel{3}^1}{\cancel{3} \cdot 8} = \frac{1}{8}$

$\frac{1}{8}$ of the day more was spent sleeping than studying.

51. rightmost size minus leftmost size:

$\frac{1}{2} - \frac{3}{16} = \frac{8}{16} - \frac{3}{16} = \frac{8-3}{16} = \frac{5}{16}$

The rightmost driver fits a nut that is $\frac{5}{16}$ inch larger than the nut for the leftmost driver.

53. If the total perimeter is $\frac{7}{8}$ mile, use subtraction to find the length of the fourth side.

Length $= \frac{7}{8} - \frac{1}{4} - \frac{1}{6} - \frac{3}{8}$

$= \frac{7}{8} + \left(-\frac{1}{4}\right) + \left(-\frac{1}{6}\right) + \left(-\frac{3}{8}\right)$

$= \frac{21}{24} + \left(-\frac{6}{24}\right) + \left(-\frac{4}{24}\right) + \left(-\frac{9}{24}\right)$

$= \frac{21+(-6)+(-4)+(-9)}{24}$

$= \frac{2}{24} = \frac{1}{12}$ mile

The fourth side is $\frac{1}{12}$ mile long.

4.5 Problem Solving: Mixed Numbers and Estimating

4.5 Margin Exercises

1. **(a)** $\frac{5}{3}$: 5 sections shaded from "whole" parts divided into 3 sections.

   (b)

   (c) $\frac{9}{4}$: 9 sections shaded from "whole" parts divided into 4 sections.

   (d)

2. **(a)** $3\frac{2}{3}$

   Step 1 $3 \cdot 3 = 9; 9 + 2 = 11$

   Step 2 $3\frac{2}{3} = \frac{11}{3}$

   (b) $4\frac{7}{10}$

   Step 1 $10 \cdot 4 = 40; 40 + 7 = 47$

   Step 2 $4\frac{7}{10} = \frac{47}{10}$

   (c) $5\frac{3}{4}$

   Step 1 $4 \cdot 5 = 20; 20 + 3 = 23$

   Step 2 $5\frac{3}{4} = \frac{23}{4}$

   (d) $8\frac{5}{6}$

   Step 1 $6 \cdot 8 = 48; 48 + 5 = 53$

   Step 2 $8\frac{5}{6} = \frac{53}{6}$

4.5 Problem Solving: Mixed Numbers and Estimating

3. **(a)** Divide 5 by 2.

$$2\overline{)5} \quad \text{so} \quad \frac{5}{2} = 2\frac{1}{2}$$
$$\underline{4}$$
$$1$$

(b) Divide 14 by 4.

$$4\overline{)14} \quad \text{so} \quad \frac{14}{4} = 3\frac{2}{4} = 3\frac{1}{2}$$
$$\underline{12}$$
$$2$$

or reduce $\frac{14}{4}$ to lowest terms first.

$$\frac{14}{4} = \frac{\cancel{2}\cdot 7}{\cancel{2}\cdot 2} = \frac{7}{2}$$

$$2\overline{)7} \quad \text{so} \quad \frac{7}{2} = 3\frac{1}{2}$$
$$\underline{6}$$
$$1$$

(c) Divide 33 by 5.

$$5\overline{)33} \quad \text{so} \quad \frac{33}{5} = 6\frac{3}{5}$$
$$\underline{30}$$
$$3$$

(d) Reduce first. $\dfrac{58}{10} = \dfrac{\cancel{2}\cdot 29}{\cancel{2}\cdot 5} = \dfrac{29}{5}$

Divide 29 by 5.

$$5\overline{)29} \quad \text{so} \quad \frac{29}{5} = 5\frac{4}{5}$$
$$\underline{25}$$
$$4$$

4. **(a)** $2\frac{3}{4}$ rounds up to 3 since the numerator, 3, is more than half of the denominator, 4.

(b) $6\frac{3}{8}$ rounds to 6 since the numerator, 3, is less than half of the denominator, 8.

(c) $4\frac{2}{3}$ rounds up to 5.

2 is more than half of 3.

(d) $1\frac{7}{10}$ rounds up to 2.

7 is more than half of 10.

(e) $3\frac{1}{2}$ rounds up to 4.

1 is half of 2.

(f) $5\frac{4}{9}$ rounds to 5.

4 is less than half of 9.

5. **(a)** $2\frac{1}{4}$ rounds to 2. $7\frac{1}{3}$ rounds to 7.

Estimate: $2 \cdot 7 = 14$

Exact: $2\dfrac{1}{4} \cdot 7\dfrac{1}{3} = \dfrac{9}{4} \cdot \dfrac{22}{3} = \dfrac{\cancel{3}\cdot 3 \cdot \cancel{2} \cdot 11}{\cancel{2}\cdot 2 \cdot \cancel{3}}$

$$= \dfrac{33}{2} = 16\dfrac{1}{2}$$

(b) $4\frac{1}{2}$ rounds to 5. $1\frac{2}{3}$ rounds to 2.

Estimate: $(5)(2) = 10$

Exact: $\left(4\dfrac{1}{2}\right)\left(1\dfrac{2}{3}\right) = \dfrac{9}{2} \cdot \dfrac{5}{3} = \dfrac{3\cdot \cancel{3} \cdot 5}{2\cdot \cancel{3}}$

$$= \dfrac{15}{2} = 7\dfrac{1}{2}$$

(c) $3\frac{3}{5}$ rounds to 4. $4\frac{4}{9}$ rounds to 4.

Estimate: $4 \cdot 4 = 16$

Exact: $3\dfrac{3}{5} \cdot 4\dfrac{4}{9} = \dfrac{18}{5} \cdot \dfrac{40}{9} = \dfrac{2\cdot \cancel{9} \cdot \cancel{5} \cdot 8}{\cancel{5}\cdot \cancel{9}}$

$$= \dfrac{16}{1} = 16$$

(d) $3\frac{1}{5}$ rounds to 3. $5\frac{3}{8}$ rounds to 5.

Estimate: $(3)(5) = 15$

Exact: $\left(3\dfrac{1}{5}\right)\left(5\dfrac{3}{8}\right) = \dfrac{16}{5} \cdot \dfrac{43}{8} = \dfrac{2\cdot \cancel{8} \cdot 43}{5\cdot \cancel{8}}$

$$= \dfrac{86}{5} = 17\dfrac{1}{5}$$

6. **(a)** $6\frac{1}{4}$ rounds to 6. $3\frac{1}{3}$ rounds to 3.

Estimate: $6 \div 3 = 2$

Exact: $6\dfrac{1}{4} \div 3\dfrac{1}{3} = \dfrac{25}{4} \div \dfrac{10}{3} = \dfrac{25}{4} \cdot \dfrac{3}{10}$

$$= \dfrac{5\cdot \cancel{5}\cdot 3}{4\cdot 2\cdot \cancel{5}} = \dfrac{15}{8} = 1\dfrac{7}{8}$$

(b) $3\frac{3}{8}$ rounds to 3. $2\frac{4}{7}$ rounds to 3.

Estimate: $3 \div 3 = 1$

Exact: $3\dfrac{3}{8} \div 2\dfrac{4}{7} = \dfrac{27}{8} \div \dfrac{18}{7} = \dfrac{27}{8} \cdot \dfrac{7}{18}$

$$= \dfrac{3\cdot \cancel{9}\cdot 7}{8\cdot 2\cdot \cancel{9}} = \dfrac{21}{16} = 1\dfrac{5}{16}$$

(c) $5\frac{1}{3}$ rounds to 5.

Estimate: $8 \div 5 = \frac{8}{5} = 1\frac{3}{5}$

Exact: $8 \div 5\frac{1}{3} = \frac{8}{1} \div \frac{16}{3} = \frac{8}{1} \cdot \frac{3}{16}$

$= \frac{\overset{1}{\cancel{8}} \cdot 3}{1 \cdot 2 \cdot \underset{1}{\cancel{8}}} = \frac{3}{2} = 1\frac{1}{2}$

(d) $4\frac{1}{2}$ rounds to 5.

Estimate: $5 \div 6 = \frac{5}{6}$

Exact: $4\frac{1}{2} \div 6 = \frac{9}{2} \div \frac{6}{1} = \frac{9}{2} \cdot \frac{1}{6}$

$= \frac{3 \cdot \overset{1}{\cancel{3}} \cdot 1}{2 \cdot 2 \cdot \underset{1}{\cancel{3}}} = \frac{3}{4}$

7. (a) $5\frac{1}{3}$ rounds to 5. $2\frac{5}{6}$ rounds to 3.

Estimate: $5 - 3 = 2$

Exact: $5\frac{1}{3} - 2\frac{5}{6} = \frac{16}{3} - \frac{17}{6} = \frac{32}{6} - \frac{17}{6}$

$= \frac{15}{6} = 2\frac{3}{6} = 2\frac{1}{2}$

(b) $\frac{3}{4}$ rounds to 1. $3\frac{1}{8}$ rounds to 3.

Estimate: $1 + 3 = 4$

Exact: $\frac{3}{4} + 3\frac{1}{8} = \frac{3}{4} + \frac{25}{8} = \frac{6}{8} + \frac{25}{8}$

$= \frac{31}{8} = 3\frac{7}{8}$

(c) $3\frac{4}{5}$ rounds to 4.

Estimate: $6 - 4 = 2$

Exact: $6 - 3\frac{4}{5} = \frac{6}{1} - \frac{19}{5} = \frac{30}{5} - \frac{19}{5}$

$= \frac{11}{5} = 2\frac{1}{5}$

8. (a) $3\frac{5}{8}$ inches rounds to 4 inches.

$2\frac{1}{4}$ inches rounds to 2 inches.

Richard's son grew about 4 inches last year and 2 inches this year. How much has his height increased over the two years? Using rounded numbers makes it easier to see that you need to add.

Estimate: $4 + 2 = 6$ inches

Exact: $3\frac{5}{8} + 2\frac{1}{4} = \frac{29}{8} + \frac{9}{4} = \frac{29}{8} + \frac{18}{8}$

$= \frac{47}{8} = 5\frac{7}{8}$ inches

(b) $2\frac{1}{2}$ packages rounds to 3 packages.

$5\frac{1}{2}$ ounces rounds to 6 ounces.

Ernestine used about 3 packages of chips and each package contained about 6 ounces. Using rounded numbers makes it easier to see that you need to *multiply*.

Estimate: $3 \cdot 6 = 18$ ounces

Exact: $2\frac{1}{2} \cdot 5\frac{1}{2} = \frac{5}{2} \cdot \frac{11}{2} = \frac{55}{4}$

$= 13\frac{3}{4}$ ounces

4.5 Section Exercises

1. Graph $2\frac{1}{3}$ and $-2\frac{1}{3}$.

3. Graph $\frac{3}{2} = 1\frac{1}{2}$ and $-\frac{3}{2} = -1\frac{1}{2}$.

5. $4\frac{1}{2}$

Step 1: $2 \cdot 4 = 8;\ 8 + 1 = 9$

Step 2: $4\frac{1}{2} = \frac{9}{2}$

7. $-1\frac{3}{5}$

Step 1: $5 \cdot 1 = 5;\ 5 + 3 = 8$

Step 2: $-1\frac{3}{5} = -\frac{8}{5}$

9. $2\frac{3}{8}$

Step 1: $8 \cdot 2 = 16;\ 16 + 3 = 19$

Step 2: $2\frac{3}{8} = \frac{19}{8}$

11. $-5\frac{7}{10}$

Step 1: $10 \cdot 5 = 50;\ 50 + 7 = 57$

Step 2: $-5\frac{7}{10} = -\frac{57}{10}$

13. $10\frac{11}{15}$

Step 1: $15 \cdot 10 = 150;\ 150 + 11 = 161$

Step 2: $10\frac{11}{15} = \frac{161}{15}$

15. Divide 13 by 3.

$\begin{array}{r}4\\3\overline{)13}\\\underline{12}\\1\end{array}$ so $\frac{13}{3} = 4\frac{1}{3}$

17. Divide 10 by 4.

$$4\overline{)10} \quad \begin{array}{c} 2 \\ \underline{8} \\ 2 \end{array} \quad \text{so} \quad -\frac{10}{4} = -2\frac{2}{4} = -2\frac{1}{2}$$

19. Divide 22 by 6.

$$6\overline{)22} \quad \begin{array}{c} 3 \\ \underline{18} \\ 4 \end{array} \quad \text{so} \quad \frac{22}{6} = 3\frac{4}{6} = 3\frac{2}{3}$$

21. Divide 51 by 9.

$$9\overline{)51} \quad \begin{array}{c} 5 \\ \underline{45} \\ 6 \end{array} \quad \text{so} \quad -\frac{51}{9} = -5\frac{6}{9} = -5\frac{2}{3}$$

23. Divide 188 by 16.

$$16\overline{)188} \quad \begin{array}{c} 11 \\ \underline{16} \\ 28 \\ \underline{16} \\ 12 \end{array} \quad \text{so} \quad \frac{188}{16} = 11\frac{12}{16} = 11\frac{3}{4}$$

25. $2\frac{1}{4}$ rounds to 2. $3\frac{1}{2}$ rounds to 4.

    Estimate: $2 \cdot 4 = 8$

    Exact: $2\frac{1}{4} \cdot 3\frac{1}{2} = \frac{9}{4} \cdot \frac{7}{2} = \frac{63}{8} = 7\frac{7}{8}$

27. $3\frac{1}{4}$ rounds to 3. $2\frac{5}{8}$ rounds to 3.

    Estimate: $3 \div 3 = 1$

    Exact: $3\frac{1}{4} \div 2\frac{5}{8} = \frac{13}{4} \div \frac{21}{8} = \frac{13}{4} \cdot \frac{8}{21}$

    $= \frac{13 \cdot 2 \cdot \overset{1}{\cancel{4}}}{\underset{1}{\cancel{4}} \cdot 21} = \frac{26}{21} = 1\frac{5}{21}$

29. $3\frac{2}{3}$ rounds to 4. $1\frac{5}{6}$ rounds to 2.

    Estimate: $4 + 2 = 6$

    Exact: $3\frac{2}{3} + 1\frac{5}{6} = \frac{11}{3} + \frac{11}{6} = \frac{22}{6} + \frac{11}{6}$

    $= \frac{22 + 11}{6} = \frac{33}{6}$

    $= 5\frac{3}{6} = 5\frac{1}{2}$

31. $4\frac{1}{4}$ rounds to 4. $\frac{7}{12}$ rounds to 1.

    Estimate: $4 - 1 = 3$

    Exact: $4\frac{1}{4} - \frac{7}{12} = \frac{17}{4} - \frac{7}{12} = \frac{51}{12} - \frac{7}{12}$

    $= \frac{51 - 7}{12} = \frac{44}{12} = \frac{11}{3} = 3\frac{2}{3}$

33. $5\frac{2}{3}$ rounds to 6.

    Estimate: $6 \div 6 = 1$

    Exact: $5\frac{2}{3} \div 6 = \frac{17}{3} \div \frac{6}{1} = \frac{17}{3} \cdot \frac{1}{6} = \frac{17}{18}$

35. $1\frac{4}{5}$ rounds to 2.

    Estimate: $8 - 2 = 6$

    Exact: $8 - 1\frac{4}{5} = \frac{8}{1} - \frac{9}{5} = \frac{40}{5} - \frac{9}{5}$

    $= \frac{40 - 9}{5} = \frac{31}{5} = 6\frac{1}{5}$

37. The figure is a square.

    Perimeter $= 4s = 4 \cdot 1\frac{3}{4} = \frac{4}{1} \cdot \frac{7}{4} = \frac{\overset{1}{\cancel{4}} \cdot 7}{1 \cdot \underset{1}{\cancel{4}}}$

    $= \frac{7}{1} = 7$ inches

    Area $= s \cdot s = 1\frac{3}{4} \cdot 1\frac{3}{4} = \frac{7}{4} \cdot \frac{7}{4} = \frac{49}{16}$

    $= 3\frac{1}{16}$ square inches

39. The figure is a rectangle.

    $P = 2l + 2w = 2 \cdot 6\frac{1}{2} + 2 \cdot 3\frac{1}{4}$

    $= \frac{2}{1} \cdot \frac{13}{2} + \frac{2}{1} \cdot \frac{13}{4} = \frac{26}{2} + \frac{13}{2} = \frac{39}{2} = 19\frac{1}{2}$

    $A = lw = 6\frac{1}{2} \cdot 3\frac{1}{4} = \frac{13}{2} \cdot \frac{13}{4} = \frac{169}{8} = 21\frac{1}{8}$

41. $12\frac{1}{2}$ ft rounds to 13 ft.

    $8\frac{2}{3}$ ft rounds to 9 ft.

    "In all" implies addition.

    Estimate: $13 + 9 = 22$ ft

    Exact: $12\frac{1}{2} + 8\frac{2}{3} = \frac{25}{2} + \frac{26}{3} = \frac{75}{6} + \frac{52}{6}$

    $= \frac{127}{6} = 21\frac{1}{6}$ ft of trim

    He has a total of $21\frac{1}{6}$ ft of oak trim.

112 Chapter 4 Rational Numbers: Positive and Negative Fractions

43. $1\frac{3}{4}$ ounces/gallon rounds to 2 ounces/gallon.

$5\frac{1}{2}$ gallons rounds to 6 gallons.

Estimate: $2 \cdot 6 = 12$ ounces

Exact: $1\frac{3}{4} \cdot 5\frac{1}{2} = \frac{7}{4} \cdot \frac{11}{2}$

$= \frac{77}{8} = 9\frac{5}{8}$ ounces

$9\frac{5}{8}$ ounces of chemical should be mixed with $5\frac{1}{2}$ gallons of water.

45. $1\frac{7}{10}$ miles rounds to 2 miles.

Amount left to be picked up implies subtraction.

Estimate: $4 - 2 = 2$ miles

Exact: $4 - 1\frac{7}{10} = \frac{4}{1} - \frac{17}{10}$

$= \frac{40}{10} - \frac{17}{10}$

$= \frac{23}{10} = 2\frac{3}{10}$

$2\frac{3}{10}$ miles of highway remain to be picked up by the Boy Scout troop.

47. $3\frac{3}{4}$ yd rounds to 4 yd.

Estimate: $4 \cdot 5 = 20$ yd

Exact: $3\frac{3}{4} \cdot 5 = \frac{15}{4} \cdot \frac{5}{1} = \frac{75}{4} = 18\frac{3}{4}$ yd

$18\frac{3}{4}$ yards of material are needed to make dresses for five bridesmaids.

49. The distance from Devils Kitchen to the beach parking is the *difference* in the given distances.

$2\frac{3}{8} - 1\frac{3}{4} = \frac{19}{8} - \frac{7}{4} = \frac{19}{8} - \frac{14}{8}$

$= \frac{19 - 14}{8} = \frac{5}{8}$

The distance is $\frac{5}{8}$ mile.

51. *To the picnic parking:* $2\frac{1}{4}$ miles

Back to Face Rock: $2\frac{1}{4} - \frac{1}{4} = 2$ miles

Face Rock to the beach parking:

$2\frac{3}{8} - \frac{1}{4} = \frac{19}{8} - \frac{1}{4} = \frac{19}{8} - \frac{2}{8}$

$= \frac{19 - 2}{8} = \frac{17}{8} = 2\frac{1}{8}$

Add the three distances to find the sum.

$2\frac{1}{4} + 2 + 2\frac{1}{8} = 6 + \frac{1}{4} + \frac{1}{8}$

$= 6 + \frac{2}{8} + \frac{1}{8} = 6\frac{3}{8}$

You will have traveled $6\frac{3}{8}$ miles.

53. $29\frac{1}{2} - 6\frac{1}{4} - 1\frac{7}{8} = \frac{59}{2} - \frac{25}{4} - \frac{15}{8}$

$= \frac{236}{8} - \frac{50}{8} - \frac{15}{8}$

$= \frac{236 - 50 - 15}{8}$

$= \frac{171}{8} = 21\frac{3}{8}$ in.

The length of the arrow shaft is $21\frac{3}{8}$ inches.

55. $23\frac{3}{4}$ in. rounds to 24 in.

$34\frac{1}{2}$ in. rounds to 35 in.

Estimate: $24 + 35 + 24 + 35 = 118$ in.

Exact: $23\frac{3}{4} + 34\frac{1}{2} + 23\frac{3}{4} + 34\frac{1}{2}$

$= 116.5$ (by calculator)

The length of lead stripping needed is $116\frac{1}{2}$ in.

57. $10\frac{3}{8}$ pounds rounds to 10 pounds.

25,730 pounds of steel are available and each anchor requires about 10 pounds. Using estimated numbers it is easier to see that you need *division*.

Estimate: $25{,}730 \div 10 = 2573$ anchors

Exact: $25{,}730 \div 10\frac{3}{8} = 2480$ anchors

2480 anchors can be manufactured from 25,730 pounds of steel.

59. Round the finished lengths, $21\frac{7}{8}, 22\frac{5}{8},$ and $23\frac{1}{2}$, to 22, 23, and 24. Round the $\frac{3}{4}$ inch seam allowance to 1 inch. There are $4 + 5 + 3 = 12$ total bands.

Estimate:
$(4 \cdot 22) + (5 \cdot 23) + (3 \cdot 24) + (12 \cdot 1) = 287$ in.

Exact:

$\left(4 \cdot 21\frac{7}{8}\right) + \left(5 \cdot 22\frac{5}{8}\right) + \left(3 \cdot 23\frac{1}{2}\right) + \left(12 \cdot \frac{3}{4}\right)$

$= 280.125$ in.

$280\frac{1}{8}$ inches of fabric strip are needed to make the bands including the seam allowance.

Summary Exercises on Fractions

1. **(a)** 3 of 8 equally sized portions are shaded: $\frac{3}{8}$

5 of 8 equally sized portions are unshaded: $\frac{5}{8}$

(b) 4 of 5 equally sized portions are shaded: $\frac{4}{5}$

1 of 5 equally sized portions are unshaded: $\frac{1}{5}$

3. **(a)** $30 \div 5 = 6$

$-\frac{4}{5} = -\frac{4 \cdot 6}{5 \cdot 6} = -\frac{24}{30}$

4.6 Exponents, Order of Operations, and Complex Fractions

(b) $14 \div 7 = 2$

$\dfrac{2}{7} = \dfrac{2 \cdot 2}{7 \cdot 2} = \dfrac{4}{14}$

5. **(a)**

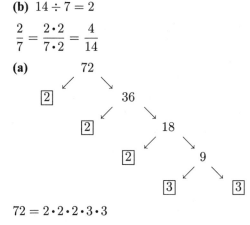

$72 = 2 \cdot 2 \cdot 2 \cdot 3 \cdot 3$

(b)

$105 = 3 \cdot 5 \cdot 7$

7. $\left(-\dfrac{3}{4}\right)\left(-\dfrac{2}{3}\right) = \dfrac{\overset{1}{\cancel{3}} \cdot \overset{1}{\cancel{2}}}{\underset{1}{\cancel{2}} \cdot 2 \cdot \underset{1}{\cancel{3}}} = \dfrac{1}{2}$

9. $\dfrac{7}{16} + \dfrac{5}{8} = \dfrac{7}{16} + \dfrac{10}{16} = \dfrac{7+10}{16} = \dfrac{17}{16}$

11. $\dfrac{2}{3} - \dfrac{4}{5} = \dfrac{2 \cdot 5}{3 \cdot 5} - \dfrac{4 \cdot 3}{5 \cdot 3} = \dfrac{10}{15} - \dfrac{12}{15} = \dfrac{10-12}{15} = -\dfrac{2}{15}$

13. $-21 \div \left(-\dfrac{3}{8}\right) = \dfrac{21}{1} \cdot \left(\dfrac{8}{3}\right) = \dfrac{\overset{1}{\cancel{3}} \cdot 7 \cdot 8}{\underset{1}{\cancel{3}}} = 56$

15. $-\dfrac{35}{45} \div \dfrac{10}{15} = -\dfrac{35}{45} \cdot \dfrac{15}{10} = -\dfrac{\overset{1}{\cancel{5}} \cdot 7 \cdot \overset{1}{\cancel{3}} \cdot \overset{1}{\cancel{5}}}{\underset{1}{\cancel{5}} \cdot 3 \cdot \underset{1}{\cancel{3}} \cdot 2 \cdot \underset{1}{\cancel{5}}} = -\dfrac{7}{6}$

17. $\dfrac{7}{12} + \dfrac{5}{6} + \dfrac{2}{3} = \dfrac{7}{12} + \dfrac{10}{12} + \dfrac{8}{12} = \dfrac{7+10+8}{12} = \dfrac{25}{12}$

19. $4\dfrac{3}{4}$ rounds to 5. $2\dfrac{5}{6}$ rounds to 3.

    Estimate: $5 + 3 = 8$

    Exact: $4\dfrac{3}{4} + 2\dfrac{5}{6} = \dfrac{19}{4} + \dfrac{17}{6} = \dfrac{57}{12} + \dfrac{34}{12}$
    $= \dfrac{91}{12} = 7\dfrac{7}{12}$

21. $2\dfrac{7}{10}$ rounds to 3.

    Estimate: $6 - 3 = 3$

    Exact: $6 - 2\dfrac{7}{10} = \dfrac{60}{10} - \dfrac{27}{10} = \dfrac{60-27}{10}$
    $= \dfrac{33}{10} = 3\dfrac{3}{10}$

23. $4\dfrac{2}{3}$ rounds to 5. $1\dfrac{1}{6}$ rounds to 1.

    Estimate: $5 \div 1 = 5$

    Exact: $4\dfrac{2}{3} \div 1\dfrac{1}{6} = \dfrac{14}{3} \div \dfrac{7}{6} = \dfrac{14}{3} \cdot \dfrac{6}{7}$
    $= \dfrac{2 \cdot \overset{1}{\cancel{7}} \cdot 2 \cdot \overset{1}{\cancel{3}}}{\underset{1}{\cancel{3}} \cdot \underset{1}{\cancel{7}}} = \dfrac{4}{1} = 4$

25. **(a)** $\dfrac{1}{4} + \dfrac{11}{16} + 1\dfrac{1}{8} = \dfrac{4}{16} + \dfrac{11}{16} + \dfrac{18}{16} = \dfrac{33}{16} = 2\dfrac{1}{16}$

    The total length is $2\dfrac{1}{16}$ inches.

    (b) $2\dfrac{1}{16} - 1\dfrac{3}{4} = \dfrac{33}{16} - \dfrac{7}{4} = \dfrac{33}{16} - \dfrac{28}{16}$
    $= \dfrac{33-28}{16} = \dfrac{5}{16}$

    $\dfrac{5}{16}$ in. will stick out the back of the board.

27. $9 \div \dfrac{3}{4} = \dfrac{9}{1} \cdot \dfrac{4}{3} = \dfrac{3 \cdot \overset{1}{\cancel{3}} \cdot 4}{1 \cdot \underset{1}{\cancel{3}}} = 12$

    You can make 12 batches of cookies.

29. Not sure: $\dfrac{3}{20} \cdot 1500 = \dfrac{3 \cdot 75 \cdot \overset{1}{\cancel{20}}}{\underset{1}{\cancel{20}}} = 225$ adults

    Real: $\dfrac{9}{20} \cdot 1500 = \dfrac{9 \cdot 75 \cdot \overset{1}{\cancel{20}}}{\underset{1}{\cancel{20}}} = 675$ adults

    Imaginary: $\dfrac{2}{5} \cdot 1500 = \dfrac{2 \cdot 300 \cdot \overset{1}{\cancel{5}}}{\underset{1}{\cancel{5}}} = 600$ adults

31. $15\dfrac{1}{3} \div \dfrac{2}{3} = \dfrac{46}{3} \cdot \dfrac{3}{2} = \dfrac{\overset{1}{\cancel{2}} \cdot 23 \cdot \overset{1}{\cancel{3}}}{\underset{1}{\cancel{3}} \cdot \underset{1}{\cancel{2}}} = 23$

    23 bottles can be filled.

4.6 Exponents, Order of Operations, and Complex Fractions

4.6 Margin Exercises

1. **(a)** $\left(-\dfrac{3}{5}\right)^2 = \left(-\dfrac{3}{5}\right)\left(-\dfrac{3}{5}\right) = \dfrac{3 \cdot 3}{5 \cdot 5} = \dfrac{9}{25}$

   (b) $\left(\dfrac{1}{3}\right)^4 = \dfrac{1}{3} \cdot \dfrac{1}{3} \cdot \dfrac{1}{3} \cdot \dfrac{1}{3} = \dfrac{1 \cdot 1 \cdot 1 \cdot 1}{3 \cdot 3 \cdot 3 \cdot 3} = \dfrac{1}{81}$

   (c) $\left(-\dfrac{2}{3}\right)^3\left(\dfrac{1}{2}\right)^2$
   $= \left(-\dfrac{2}{3}\right)\left(-\dfrac{2}{3}\right)\left(-\dfrac{2}{3}\right)\left(\dfrac{1}{2}\right)\left(\dfrac{1}{2}\right)$
   $= -\dfrac{2 \cdot \overset{1}{\cancel{2}} \cdot \overset{1}{\cancel{2}} \cdot 1 \cdot 1}{3 \cdot 3 \cdot 3 \cdot \underset{1}{\cancel{2}} \cdot \underset{1}{\cancel{2}}} = -\dfrac{2}{27}$

Exact: $4\dfrac{2}{3} \div 1\dfrac{1}{6} = \dfrac{14}{3} \div \dfrac{7}{6} = \dfrac{14}{3} \cdot \dfrac{6}{7}$
$= \dfrac{2 \cdot \overset{1}{\cancel{7}} \cdot 2 \cdot \overset{1}{\cancel{3}}}{\underset{1}{\cancel{3}} \cdot \underset{1}{\cancel{7}}} = \dfrac{4}{1} = 4$

114 Chapter 4 Rational Numbers: Positive and Negative Fractions

(d) $\left(-\dfrac{1}{2}\right)^2\left(\dfrac{1}{4}\right)^2 = \left(-\dfrac{1}{2}\right)\left(-\dfrac{1}{2}\right)\left(\dfrac{1}{4}\right)\left(\dfrac{1}{4}\right)$
$= \dfrac{1\cdot1\cdot1\cdot1}{2\cdot2\cdot4\cdot4} = \dfrac{1}{64}$

2. (a) $\dfrac{1}{3} - \dfrac{5}{9}\left(\dfrac{3}{4}\right)$ Multiply.

$= \dfrac{1}{3} - \dfrac{5\cdot\cancel{3}^1}{\cancel{9}_3\cdot 4}$

$= \dfrac{1}{3} - \dfrac{5}{12}$ The LCD is 12.

$= \dfrac{4}{12} - \dfrac{5}{12}$ Subtract.

$= \dfrac{4-5}{12}$

$= \dfrac{-1}{12}$ or $-\dfrac{1}{12}$

(b) $-\dfrac{3}{4} + \left(-\dfrac{1}{2}\right)^2 \div \dfrac{2}{3}$ Exponent first

$= -\dfrac{3}{4} + \dfrac{1}{4} \div \dfrac{2}{3}$ Change division.

$= -\dfrac{3}{4} + \dfrac{1}{4} \cdot \dfrac{3}{2}$ Multiply.

$= -\dfrac{3}{4} + \dfrac{3}{8}$ The LCD is 8.

$= -\dfrac{6}{8} + \dfrac{3}{8}$ Add.

$= \dfrac{-6+3}{8}$

$= \dfrac{-3}{8}$ or $-\dfrac{3}{8}$

(c) $\dfrac{12}{5} - \dfrac{1}{6}\left(3 - \dfrac{3}{5}\right)$ Parentheses first

$= \dfrac{12}{5} - \dfrac{1}{6}\left(\dfrac{15}{5} - \dfrac{3}{5}\right)$ Subtract.

$= \dfrac{12}{5} - \dfrac{1}{6}\left(\dfrac{12}{5}\right)$ Multiply.

$= \dfrac{12}{5} - \dfrac{1\cdot\cancel{12}^2}{\cancel{6}\cdot 5}$

$= \dfrac{12}{5} - \dfrac{2}{5}$ Subtract.

$= \dfrac{12-2}{5}$

$= \dfrac{10}{5}$

$= 2$

3. (a) $\dfrac{-\frac{3}{5}}{\frac{9}{10}} = -\dfrac{3}{5} \div \dfrac{9}{10} = -\dfrac{3}{5} \cdot \dfrac{10}{9}$

$= -\dfrac{\cancel{3}\cdot 2 \cdot \cancel{5}}{\cancel{5}\cdot\cancel{9}\cdot 3} = -\dfrac{2}{3}$

(b) $\dfrac{6}{\frac{3}{4}} = 6 \div \dfrac{3}{4} = \dfrac{6}{1}\cdot\dfrac{4}{3} = \dfrac{2\cdot\cancel{3}\cdot 4}{1\cdot\cancel{3}}$

$= \dfrac{8}{1} = 8$

(c) $\dfrac{-\frac{15}{16}}{-5} = -\dfrac{15}{16} \div -\dfrac{5}{1} = -\dfrac{15}{16}\cdot -\dfrac{1}{5}$

$= \dfrac{3\cdot\cancel{5}\cdot 1}{16\cdot\cancel{5}} = \dfrac{3}{16}$

(d) $\dfrac{-3}{\left(-\frac{3}{4}\right)^2} = -3 \div \left(-\dfrac{3}{4}\right)^2 = -\dfrac{3}{1} \div \dfrac{9}{16}$

$= -\dfrac{3}{1}\cdot\dfrac{16}{9} = -\dfrac{\cancel{3}\cdot 16}{1\cdot\cancel{9}_3}$

$= -\dfrac{16}{3}$, or $-5\dfrac{1}{3}$

4.6 Section Exercises

1. $\left(-\dfrac{3}{4}\right)^2 = \left(-\dfrac{3}{4}\right)\left(-\dfrac{3}{4}\right) = \dfrac{3\cdot 3}{4\cdot 4} = \dfrac{9}{16}$

3. $\left(\dfrac{2}{5}\right)^3 = \dfrac{2}{5}\cdot\dfrac{2}{5}\cdot\dfrac{2}{5}$

$= \dfrac{4}{25}\cdot\dfrac{2}{5}$

$= \dfrac{8}{125}$

5. $\left(-\dfrac{1}{3}\right)^3 = \left(-\dfrac{1}{3}\right)\left(-\dfrac{1}{3}\right)\left(-\dfrac{1}{3}\right)$

$= \dfrac{1}{9}\left(-\dfrac{1}{3}\right)$

$= -\dfrac{1}{27}$

7. $\left(\dfrac{1}{2}\right)^5 = \dfrac{1}{2}\cdot\dfrac{1}{2}\cdot\dfrac{1}{2}\cdot\dfrac{1}{2}\cdot\dfrac{1}{2}$

$= \dfrac{1}{4}\cdot\dfrac{1}{2}\cdot\dfrac{1}{2}\cdot\dfrac{1}{2}$

$= \dfrac{1}{8}\cdot\dfrac{1}{2}\cdot\dfrac{1}{2}$

$= \dfrac{1}{16}\cdot\dfrac{1}{2}$

$= \dfrac{1}{32}$

9. $\left(\dfrac{7}{10}\right)^2 = \dfrac{7}{10}\cdot\dfrac{7}{10} = \dfrac{49}{100}$

11. $\left(-\dfrac{6}{5}\right)^2 = \left(-\dfrac{6}{5}\right)\left(-\dfrac{6}{5}\right)$

$= \dfrac{36}{25}$, or $1\dfrac{11}{25}$

4.6 Exponents, Order of Operations, and Complex Fractions

13. $\dfrac{15}{16}\left(\dfrac{4}{5}\right)^3$ Exponent first

$= \dfrac{15}{16}\left(\dfrac{4}{5} \cdot \dfrac{4}{5} \cdot \dfrac{4}{5}\right)$

$= \dfrac{15}{16}\left(\dfrac{64}{125}\right)$ Multiply.

$= \dfrac{3 \cdot \cancel{5} \cdot \cancel{16} \cdot 4}{\cancel{16} \cdot \cancel{5} \cdot 25}$ Reduce.

$= \dfrac{12}{25}$

15. $\left(\dfrac{1}{3}\right)^4 \left(\dfrac{9}{10}\right)^2 = \left(\dfrac{1}{3} \cdot \dfrac{1}{3} \cdot \dfrac{1}{3} \cdot \dfrac{1}{3}\right)\left(\dfrac{9}{10} \cdot \dfrac{9}{10}\right)$

$= \dfrac{1 \cdot 1 \cdot 1 \cdot 1 \cdot \cancel{3} \cdot \cancel{3} \cdot \cancel{3} \cdot \cancel{3}}{\cancel{3} \cdot \cancel{3} \cdot \cancel{3} \cdot \cancel{3} \cdot 10 \cdot 10}$

$= \dfrac{1}{100}$

17. $\left(-\dfrac{3}{2}\right)^3 \left(-\dfrac{2}{3}\right)^2$

$= \left(-\dfrac{3}{2}\right)\left(-\dfrac{3}{2}\right)\left(-\dfrac{3}{2}\right)\left(-\dfrac{2}{3}\right)\left(-\dfrac{2}{3}\right)$

$= -\dfrac{\cancel{3} \cdot \cancel{3} \cdot 3 \cdot \cancel{2} \cdot \cancel{2}}{\cancel{2} \cdot \cancel{2} \cdot 2 \cdot \cancel{3} \cdot \cancel{3}}$

$= -\dfrac{3}{2}, \text{ or } -1\dfrac{1}{2}$

19. **(a)** $\left(-\dfrac{1}{2}\right)^2 = \left(-\dfrac{1}{2}\right)\left(-\dfrac{1}{2}\right)$

$= \dfrac{1}{4}$

$\left(-\dfrac{1}{2}\right)^3 = \left(-\dfrac{1}{2}\right)^2\left(-\dfrac{1}{2}\right)$

$= \left(\dfrac{1}{4}\right)\left(-\dfrac{1}{2}\right) = -\dfrac{1}{8}$

$\left(-\dfrac{1}{2}\right)^4 = \left(-\dfrac{1}{2}\right)^3\left(-\dfrac{1}{2}\right)$

$= \left(-\dfrac{1}{8}\right)\left(-\dfrac{1}{2}\right) = \dfrac{1}{16}$

$\left(-\dfrac{1}{2}\right)^5 = \left(-\dfrac{1}{2}\right)^4\left(-\dfrac{1}{2}\right)$

$= \left(\dfrac{1}{16}\right)\left(-\dfrac{1}{2}\right) = -\dfrac{1}{32}$

$\left(-\dfrac{1}{2}\right)^6 = \left(-\dfrac{1}{2}\right)^5\left(-\dfrac{1}{2}\right)$

$= \left(-\dfrac{1}{32}\right)\left(-\dfrac{1}{2}\right) = \dfrac{1}{64}$

$\left(-\dfrac{1}{2}\right)^7 = \left(-\dfrac{1}{2}\right)^6\left(-\dfrac{1}{2}\right)$

$= \left(\dfrac{1}{64}\right)\left(-\dfrac{1}{2}\right) = -\dfrac{1}{128}$

$\left(-\dfrac{1}{2}\right)^8 = \left(-\dfrac{1}{2}\right)^7\left(-\dfrac{1}{2}\right)$

$= \left(-\dfrac{1}{128}\right)\left(-\dfrac{1}{2}\right) = \dfrac{1}{256}$

$\left(-\dfrac{1}{2}\right)^9 = \left(-\dfrac{1}{2}\right)^8\left(-\dfrac{1}{2}\right)$

$= \left(\dfrac{1}{256}\right)\left(-\dfrac{1}{2}\right) = -\dfrac{1}{512}$

(b) When a negative number is raised to an even power, the answer is positive. When a negative number is raised to an odd power, the answer is negative.

20. **(a)** Ask yourself, "What number, times itself, is 4?" This is the numerator. Then ask, "What number, times itself, is 9?" This is the denominator. The number under the ketchup is either $\dfrac{2}{3}$ or $-\dfrac{2}{3}$.

(b) The number under the ketchup is $-\dfrac{1}{3}$ because $\left(-\dfrac{1}{3}\right)\left(-\dfrac{1}{3}\right)\left(-\dfrac{1}{3}\right) = -\dfrac{1}{27}$.

(c) Either $\dfrac{1}{2}$ or $-\dfrac{1}{2}$.

(d) No real number works, because both $\left(\dfrac{3}{4}\right)^2$ and $\left(-\dfrac{3}{4}\right)^2$ give a *positive* result.

(e) Either $\dfrac{1}{3}$ or $-\dfrac{1}{3}$ inside one set of parentheses and $\dfrac{1}{2}$ or $-\dfrac{1}{2}$ inside the other.

21. $\dfrac{1}{5} - 6\left(\dfrac{7}{10}\right)$ Rewrite 6 as $\dfrac{6}{1}$.

$= \dfrac{1}{5} - \dfrac{6}{1}\left(\dfrac{7}{10}\right)$ Multiply.

$= \dfrac{1}{5} - \dfrac{42}{10}$ LCD is 10.

$= \dfrac{2}{10} - \dfrac{42}{10}$ Subtract.

$= \dfrac{2 - 42}{10}$

$= \dfrac{-40}{10}$ Reduce.

$= -4$

23. $\left(\dfrac{4}{3} \div \dfrac{8}{3}\right) + \left(-\dfrac{3}{4} \cdot \dfrac{1}{4}\right)$

$= \left(\dfrac{4}{3} \cdot \dfrac{3}{8}\right) + \left(-\dfrac{3}{4} \cdot \dfrac{1}{4}\right)$ Change division.

$= \left(\dfrac{\cancel{4}\cdot\cancel{3}}{\cancel{3}\cdot 2\cdot\cancel{4}}\right) + \left(-\dfrac{3}{16}\right)$ Parentheses first

$= \dfrac{1}{2} - \dfrac{3}{16}$ LCD is 16.

$= \dfrac{8}{16} - \dfrac{3}{16}$ Subtract.

$= \dfrac{8-3}{16} = \dfrac{5}{16}$

25. $-\dfrac{3}{10} \div \dfrac{3}{5}\left(-\dfrac{2}{3}\right)$ Change division.

$= -\dfrac{3}{10} \cdot \dfrac{5}{3}\left(-\dfrac{2}{3}\right)$

$= -\dfrac{\cancel{3}\cdot\cancel{5}}{2\cdot\cancel{5}\cdot\cancel{3}}\left(-\dfrac{2}{3}\right)$

$= -\dfrac{1}{2}\left(-\dfrac{2}{3}\right)$

$= \dfrac{1\cdot\cancel{2}}{\cancel{2}\cdot 3}$

$= \dfrac{1}{3}$

27. $\dfrac{8}{3}\left(\dfrac{1}{4} - \dfrac{1}{2}\right)^2$ LCD is 4.

$= \dfrac{8}{3}\left(\dfrac{1}{4} - \dfrac{2}{4}\right)^2$ Subtract.

$= \dfrac{8}{3}\left(-\dfrac{1}{4}\right)^2$ Exponent

$= \dfrac{8}{3}\left(-\dfrac{1}{4}\right)\left(-\dfrac{1}{4}\right)$

$= \dfrac{8}{3}\left(\dfrac{1}{16}\right)$

$= \dfrac{\cancel{8}\cdot 1}{3\cdot 2\cdot\cancel{8}}$

$= \dfrac{1}{6}$

29. $-\dfrac{3}{8} + \dfrac{2}{3}\left(-\dfrac{2}{3} + \dfrac{1}{6}\right)$ LCD inside parentheses is 6.

$= -\dfrac{3}{8} + \dfrac{2}{3}\left(-\dfrac{4}{6} + \dfrac{1}{6}\right)$ Add.

$= -\dfrac{3}{8} + \dfrac{2}{3}\left(\dfrac{-4+1}{6}\right)$

$= -\dfrac{3}{8} + \dfrac{2}{3}\left(\dfrac{-3}{6}\right)$ Reduce.

$= -\dfrac{3}{8} + \dfrac{2}{3}\left(-\dfrac{1}{2}\right)$ Multiply.

$= -\dfrac{3}{8} + \left(-\dfrac{1}{3}\right)$ LCD is 24.

$= -\dfrac{9}{24} + \left(-\dfrac{8}{24}\right)$ Add.

$= \dfrac{-9+(-8)}{24}$

$= \dfrac{-17}{24}$ or $-\dfrac{17}{24}$

31. $2\left(\dfrac{1}{3}\right)^3 - \dfrac{2}{9}$ Exponent

$= 2\left(\dfrac{1}{3}\right)\left(\dfrac{1}{3}\right)\left(\dfrac{1}{3}\right) - \dfrac{2}{9}$

$= 2\left(\dfrac{1}{9}\right)\left(\dfrac{1}{3}\right) - \dfrac{2}{9}$

$= 2\left(\dfrac{1}{27}\right) - \dfrac{2}{9}$

$= \dfrac{2}{27} - \dfrac{2}{9}$ LCD is 27.

$= \dfrac{2}{27} - \dfrac{6}{27}$ Subtract.

$= \dfrac{2-6}{27} = -\dfrac{4}{27}$

4.6 Exponents, Order of Operations, and Complex Fractions

33. $\left(-\dfrac{2}{3}\right)^3\left(\dfrac{1}{8}-\dfrac{1}{2}\right)-\dfrac{2}{3}\left(\dfrac{1}{8}\right)$ LCD inside paren. is 8.

$=\left(-\dfrac{2}{3}\right)^3\left(\dfrac{1}{8}-\dfrac{4}{8}\right)-\dfrac{2}{3}\left(\dfrac{1}{8}\right)$ Subtract inside paren.

$=\left(-\dfrac{2}{3}\right)^3\left(-\dfrac{3}{8}\right)-\dfrac{2}{3}\left(\dfrac{1}{8}\right)$ Exponent

$=\left(-\dfrac{2}{3}\right)\left(-\dfrac{2}{3}\right)\left(-\dfrac{2}{3}\right)\left(-\dfrac{3}{8}\right)-\dfrac{2}{3}\left(\dfrac{1}{8}\right)$

$=\left(-\dfrac{8}{27}\right)\left(-\dfrac{3}{8}\right)-\dfrac{2}{3}\left(\dfrac{1}{8}\right)$ Multiply.

$=\dfrac{\overset{1}{\cancel{8}}\cdot\overset{1}{\cancel{3}}}{\underset{1}{\cancel{3}}\cdot 9\cdot\underset{1}{\cancel{8}}}-\dfrac{\overset{1}{\cancel{2}}\cdot 1}{3\cdot\underset{1}{\cancel{2}}\cdot 4}$

$=\dfrac{1}{9}-\dfrac{1}{12}$ LCD is 36.

$=\dfrac{4}{36}-\dfrac{3}{36}$ Subtract.

$=\dfrac{4-3}{36}$

$=\dfrac{1}{36}$

35. $A = s^2 = \left(\dfrac{3}{8}\right)^2$

$=\left(\dfrac{3}{8}\right)\left(\dfrac{3}{8}\right)$

$=\dfrac{3\cdot 3}{8\cdot 8}$

$=\dfrac{9}{64}$ in.^{2}

37. $P = 2l + 2w$

$= 2\left(\dfrac{7}{10}\right) + 2\left(\dfrac{1}{4}\right)$ Rewrite as $\dfrac{2}{1}$.

$= \dfrac{2}{1}\left(\dfrac{7}{10}\right) + \dfrac{2}{1}\left(\dfrac{1}{4}\right)$ Multiply.

$= \dfrac{\overset{1}{\cancel{2}}\cdot 7}{1\cdot\underset{1}{\cancel{2}}\cdot 5} + \dfrac{\overset{1}{\cancel{2}}\cdot 1}{1\cdot\underset{1}{\cancel{2}}\cdot 2}$ Reduce.

$= \dfrac{7}{5} + \dfrac{1}{2}$ LCD is 10.

$= \dfrac{14}{10} + \dfrac{5}{10}$ Add.

$= \dfrac{14+5}{10}$

$= \dfrac{19}{10}$, or $1\dfrac{9}{10}$ miles

39. $\dfrac{-\frac{7}{9}}{-\frac{7}{36}} = -\dfrac{7}{9} \div \left(-\dfrac{7}{36}\right)$

$= -\dfrac{7}{9}\cdot\left(-\dfrac{36}{7}\right)$

$= \dfrac{\overset{1}{\cancel{7}}\cdot\overset{1}{\cancel{9}}\cdot 4}{\underset{1}{\cancel{9}}\cdot\underset{1}{\cancel{7}}}$

$= \dfrac{4}{1} = 4$

41. $\dfrac{-15}{\frac{6}{5}} = -15 \div \dfrac{6}{5}$

$= -\dfrac{15}{1} \div \dfrac{6}{5}$

$= -\dfrac{15}{1} \cdot \dfrac{5}{6}$

$= -\dfrac{\overset{1}{\cancel{3}}\cdot 5\cdot 5}{1\cdot 2\cdot\underset{1}{\cancel{3}}} = -\dfrac{25}{2}$, or $-12\dfrac{1}{2}$

43. $\dfrac{\frac{4}{7}}{8} = \dfrac{4}{7} \div 8$

$= \dfrac{4}{7} \div \dfrac{8}{1}$

$= \dfrac{4}{7} \cdot \dfrac{1}{8}$

$= \dfrac{\overset{1}{\cancel{4}}\cdot 1}{7\cdot\underset{2}{\cancel{8}}}$

$= \dfrac{1}{14}$

45. $\dfrac{-\frac{2}{3}}{-2\frac{2}{5}} = \dfrac{-\frac{2}{3}}{-\frac{12}{5}}$ Write as division.

$= \dfrac{2}{3} \div \dfrac{12}{5}$ Change division.

$= \dfrac{2}{3} \cdot \dfrac{5}{12}$

$= \dfrac{\overset{1}{\cancel{2}}\cdot 5}{3\cdot\underset{6}{\cancel{12}}}$

$= \dfrac{5}{18}$

47. $\dfrac{-4\frac{1}{2}}{(\frac{3}{4})^2} = \dfrac{-4\frac{1}{2}}{\frac{9}{16}}$ Exponent first

$= \dfrac{-\frac{9}{2}}{\frac{9}{16}}$ Write as division.

$= -\dfrac{9}{2} \div \dfrac{9}{16}$ Change division.

$= -\dfrac{9}{2} \cdot \dfrac{16}{9}$

$= -\dfrac{\cancel{9} \cdot \cancel{2} \cdot 8}{\cancel{2} \cdot \cancel{9}}$

$= -\dfrac{8}{1} = -8$

49. $\dfrac{(\frac{2}{5})^2}{(-\frac{4}{3})^2}$ Exponents first

$= \dfrac{\frac{4}{25}}{\frac{16}{9}}$ Write as division.

$= \dfrac{4}{25} \div \dfrac{16}{9}$ Change division.

$= \dfrac{4}{25} \cdot \dfrac{9}{16}$

$= \dfrac{\cancel{4} \cdot 9}{25 \cdot \cancel{4} \cdot 4} = \dfrac{9}{100}$

4.7 Problem Solving: Equations Containing Fractions

4.7 Margin Exercises

1. (a) $\dfrac{1}{6}m = 3$ Multiply both sides by $\frac{6}{1}$, the reciprocal of $\frac{1}{6}$.

$\dfrac{\cancel{6}}{1}\left(\dfrac{1}{\cancel{6}}m\right) = \dfrac{6}{1}(3)$

$m = 18$

Check: $\dfrac{1}{6}(18) = 3$

$\dfrac{18}{6} = 3$

$3 = 3$ Balances

The solution is 18.

(b) $\dfrac{3}{2}a = -9$ Multiply both sides by $\frac{2}{3}$, the reciprocal of $\frac{3}{2}$.

$\dfrac{\cancel{2}}{\cancel{3}}\left(\dfrac{\cancel{3}}{\cancel{2}}a\right) = \dfrac{2}{3}\left(-\dfrac{9}{1}\right)$

$a = -\dfrac{18}{3}$

$a = -6$

Check: $\dfrac{3}{2}(-6) = -9$

$\dfrac{-18}{2} = -9$

$-9 = -9$ Balances

The solution is -6.

(c) $\dfrac{3}{14} = -\dfrac{2}{7}x$ Multiply both sides by $-\frac{7}{2}$.

$-\dfrac{7}{2}\left(\dfrac{3}{14}\right) = -\dfrac{\cancel{7}}{\cancel{2}}\left(-\dfrac{\cancel{2}}{\cancel{7}}x\right)$

$-\dfrac{\cancel{7} \cdot 3}{2 \cdot 2 \cdot \cancel{7}} = x$

$-\dfrac{3}{4} = x$

Check: $\dfrac{3}{14} = -\dfrac{2}{7}\left(-\dfrac{3}{4}\right)$

$\dfrac{3}{14} = \dfrac{\cancel{2} \cdot 3}{7 \cdot \cancel{2} \cdot 2}$

$\dfrac{3}{14} = \dfrac{3}{14}$ Balances

The solution is $-\dfrac{3}{4}$.

2. (a) $18 = \dfrac{4}{5}x + 2$ Add -2 to both sides.

$\underline{ -2 -2 }$

$16 = \dfrac{4}{5}x + 0$

$\dfrac{5}{4}(16) = \dfrac{\cancel{5}}{\cancel{4}}\left(\dfrac{\cancel{4}}{\cancel{5}}x\right)$ Multiply both sides by $\frac{5}{4}$, the reciprocal of $\frac{4}{5}$.

$\dfrac{5 \cdot \cancel{4} \cdot 4}{\cancel{4}} = x$

$20 = x$

Check: $18 = \dfrac{4}{5}(20) + 2$

$18 = \dfrac{4 \cdot \cancel{5} \cdot 4}{\cancel{5}} + 2$

$18 = 16 + 2$

$18 = 18$ Balances

The solution is 20.

4.7 Problem Solving: Equations Containing Fractions

(b) $\frac{1}{4}h - 5 = 1$ Add 5 to both sides.

$$\frac{5}{\frac{1}{4}h - 0 = 6} \quad \frac{5}{}$$

$\frac{\cancel{4}}{1}\left(\frac{1}{\cancel{4}}h\right) = \frac{4}{1}(6)$ Multiply both sides by $\frac{4}{1}$, the reciprocal of $\frac{1}{4}$.

$h = 24$

Check: $\frac{1}{4}(24) - 5 = 1$

$6 - 5 = 1$

$1 = 1$ Balances

The solution is 24.

(c) $\frac{4}{3}r + 4 = -8$ Add -4 to both sides.

$$\frac{-4 \quad\quad -4}{\frac{4}{3}r + 0 = -12}$$

$\frac{\cancel{3}}{\cancel{4}}\left(\frac{\cancel{4}}{\cancel{3}}r\right) = \frac{3}{4}\left(-\frac{12}{1}\right)$ Multiply both sides by $\frac{3}{4}$.

$r = -\frac{3 \cdot 3 \cdot \cancel{4}}{\cancel{4}}$

$r = -9$

Check: $\frac{4}{3}\left(-\frac{9}{1}\right) + 4 = -8$

$-\frac{36}{3} + 4 = -8$

$-12 + 4 = -8$

$-8 = -8$ Balances

The solution is -9.

3. Let a represent the woman's age.

$100 + \frac{a}{2} =$ Systolic Blood Pressure

$100 + \frac{a}{2} = 111$ Add -100 to both sides.

$$\frac{-100 \quad\quad -100}{\frac{a}{2} = 11}$$ $\frac{a}{2}$ is equivalent to $\frac{1}{2}a$.

$\frac{\cancel{2}}{1}\left(\frac{1}{\cancel{2}}a\right) = \frac{2}{1}(11)$ Multiply both sides by $\frac{2}{1}$, the reciprocal of $\frac{1}{2}$.

$a = 22$

The woman is 22 years old.

Check: $100 + \frac{22}{2} = 111$

$100 + 11 = 111$

$111 = 111$ Balances

4.7 Section Exercises

1. $\frac{1}{3}a = 10$ Multiply both sides by $\frac{3}{1}$, the reciprocal of $\frac{1}{3}$.

$\frac{3}{1}\left(\frac{1}{3}a\right) = \frac{3}{1}(10)$

$a = \frac{3}{1} \cdot \frac{10}{1}$

$a = 30$

Check: $\frac{1}{3}(30) = 10$

$\frac{30}{3} = 10$

$10 = 10$ Balances

The solution is 30.

3. $-20 = \frac{5}{6}b$ Multiply both sides by $\frac{6}{5}$, the reciprocal of $\frac{5}{6}$.

$\frac{6}{5}(-20) = \frac{\cancel{6}}{\cancel{5}}\left(\frac{\cancel{5}}{\cancel{6}}b\right)$

$-\frac{6 \cdot 4 \cdot \cancel{5}}{\cancel{5}} = b$

$-24 = b$

Check: $-20 = \frac{5}{6}(-24)$

$-20 = -20$ Balances

The solution is -24.

5. $-\frac{7}{2}c = -21$ Multiply both sides by $-\frac{2}{7}$, the reciprocal of $-\frac{7}{2}$.

$-\frac{\cancel{2}}{\cancel{7}}\left(-\frac{\cancel{7}}{\cancel{2}}c\right) = -\frac{2}{\cancel{7}}\left(-\frac{\cancel{21}}{1}\right)$

$c = \frac{6}{1}$

$c = 6$

Check: $-\frac{7}{2}(6) = -21$

$-\frac{42}{2} = -21$

$-21 = -21$ Balances

The solution is 6.

120 Chapter 4 Rational Numbers: Positive and Negative Fractions

7. $\dfrac{9}{16} = \dfrac{3}{4}m$ Multiply both sides by $\dfrac{4}{3}$, the reciprocal of $\dfrac{3}{4}$.

$$\dfrac{\cancel{4}^1}{\cancel{3}_1} \cdot \dfrac{\cancel{9}^3}{\cancel{16}_4} = \dfrac{\cancel{4}^1}{\cancel{3}_1}\left(\dfrac{\cancel{3}^1}{\cancel{4}_1}m\right)$$

$$\dfrac{3}{4} = m$$

Check: $\dfrac{9}{16} = \dfrac{3}{4}\left(\dfrac{3}{4}\right)$

$\dfrac{9}{16} = \dfrac{9}{16}$ Balances

The solution is $\dfrac{3}{4}$.

9. $\dfrac{3}{10} = -\dfrac{1}{4}d$ Multiply both sides by $-\dfrac{4}{1}$, the reciprocal of $-\dfrac{1}{4}$.

$$-\dfrac{4}{1}\left(\dfrac{3}{10}\right) = -\dfrac{\cancel{4}^1}{1}\left(\dfrac{-1}{\cancel{4}_1}d\right)$$

$$-\dfrac{12}{10} = d$$

$$-\dfrac{6}{5} = d$$

Check: $\dfrac{3}{10} = -\dfrac{1}{4}\left(-\dfrac{6}{5}\right)$

$$\dfrac{3}{10} = \dfrac{1 \cdot \cancel{2} \cdot 3}{\cancel{2} \cdot 2 \cdot 5}$$

$$\dfrac{3}{10} = \dfrac{3}{10}$$ Balances

The solution is $-\dfrac{6}{5}$.

11. $\dfrac{1}{6}n + 7 = 9$ Add -7 to both sides.

$\quad\quad\quad -7\quad\quad -7$

$\dfrac{1}{6}n = 2$ Multiply both sides by $\dfrac{6}{1}$, the reciprocal of $\dfrac{1}{6}$.

$$\dfrac{\cancel{6}^1}{1} \cdot \dfrac{1}{\cancel{6}_1}n = \dfrac{6}{1} \cdot 2$$

$$n = 12$$

Check: $\dfrac{1}{6}(12) + 7 = 9$

$2 + 7 = 9$

$9 = 9$ Balances

The solution is 12.

13. $-10 = \dfrac{5}{3}r + 5$ Add -5 to both sides.

$\quad -5 \quad\quad\quad -5$

$-15 = \dfrac{5}{3}r + 0$ Multiply both sides by $\dfrac{3}{5}$.

$$\dfrac{3}{5}\left(-\dfrac{15}{1}\right) = \dfrac{\cancel{3}^1}{\cancel{5}_1}\left(\dfrac{\cancel{5}^1}{\cancel{3}_1}r\right)$$

$$-\dfrac{3 \cdot 3 \cdot \cancel{5}^1}{\cancel{5}_1} = r$$

$$-9 = r$$

Check: $-10 = \dfrac{5}{3}(-9) + 5$

$-10 = -\dfrac{45}{3} + 5$

$-10 = -15 + 5$

$-10 = -10$ Balances

The solution is -9.

15. $\dfrac{3}{8}x - 9 = 0$ Add 9 to both sides.

$\quad\quad\quad 9 \quad\quad 9$

$\dfrac{3}{8}x = 9$ Multiply both sides by $\dfrac{8}{3}$.

$$\dfrac{\cancel{8}^1}{\cancel{3}_1} \cdot \dfrac{\cancel{3}^1}{\cancel{8}_1}x = \dfrac{8}{\cancel{3}_1} \cdot \dfrac{\cancel{9}^3}{1}$$

$$x = 24$$

Check: $\dfrac{3}{8}(24) - 9 = 0$

$9 - 9 = 0$

$0 = 0$ Balances

The solution is 24.

17. $7 - 2 = \dfrac{1}{5}y - 4$ Add the opposite.

$7 + (-2) = \dfrac{1}{5}y + (-4)$ Combine like terms.

$5 = \dfrac{1}{5}y + (-4)$ Add 4 to both sides.

$\quad 4 \quad\quad\quad\quad\quad 4$

$9 = \dfrac{1}{5}y + 0$ Multiply both sides by 5.

$$\dfrac{5}{1}(9) = \dfrac{\cancel{5}^1}{1}\left(\dfrac{1}{\cancel{5}_1}y\right)$$

$$45 = y$$

The solution is 45.

4.7 Problem Solving: Equations Containing Fractions

19. $4 + \frac{2}{3}n = -10 + 2$ Combine like terms.

$\frac{2}{3}n = -8$ Add -4 to both sides.

$\underline{-4 -4 }$

$\frac{2}{3}n = -12$ Multiply both sides by $\frac{3}{2}$.

$\frac{\cancel{3}}{\cancel{2}}\left(\frac{\cancel{2}}{\cancel{3}}n\right) = \frac{3}{2}(-12)$

$ n = -\frac{36}{2}$

$ n = -18$

The solution is -18.

21. $3x + \frac{1}{2} = \frac{3}{4}$

$\underline{-\frac{1}{2} -\frac{1}{2}}$

$3x + 0 = \frac{3}{4} + \left(-\frac{1}{2}\right)$

$3x = \frac{3}{4} + \left(-\frac{2}{4}\right)$

$3x = \frac{3 + (-2)}{4}$

$3x = \frac{1}{4}$

$\frac{1}{\cancel{3}}(\cancel{3}x) = \frac{1}{3}\left(\frac{1}{4}\right)$

$ x = \frac{1}{12}$

The solution is $\frac{1}{12}$.

23. $\frac{3}{10} = -4b - \frac{1}{5}$

$\underline{\frac{1}{5} \frac{1}{5}}$

$\frac{3}{10} + \frac{1}{5} = -4b$

$\frac{3}{10} + \frac{2}{10} = -4b$

$\frac{5}{10} = -4b$

$-\frac{1}{4} \cdot \frac{5}{10} = -\frac{1}{4}(-4b)$

$-\frac{5}{40} = b$

$-\frac{\cancel{5}}{\cancel{5} \cdot 8} = b$

$-\frac{1}{8} = b$

The solution is $-\frac{1}{8}$.

25. (a) $\frac{1}{6}x + 1 = -2$ Let $x = 18$.

$\frac{1}{6}\left(\frac{18}{1}\right) + 1 = -2$

$\frac{18}{6} + 1 = -2$

$3 + 1 = -2$

$ 4 \neq -2$

No, it does not balance. $x = 18$ is not the solution, so we'll solve the equation.

$\frac{1}{6}x + 1 = -2$

$\underline{-1 -1}$

$\frac{1}{6}x + 0 = -3$

$\frac{1}{6}x = -3$

$\frac{\cancel{6}}{1}\left(\frac{1}{\cancel{6}}x\right) = \frac{6}{1}(-3)$

$ x = -18$

The correct solution is -18.

(b) $-\frac{3}{2} = \frac{9}{4}k$ Let $k = -\frac{2}{3}$.

$-\frac{3}{2} = \frac{9}{4}\left(-\frac{2}{3}\right)$

$-\frac{3}{2} = -\frac{18}{12}$

$-\frac{3}{2} = -\frac{6 \cdot 3}{6 \cdot 2}$

$-\frac{3}{2} = -\frac{3}{2}$ Balances

Yes, $k = -\frac{2}{3}$ is the correct solution because the equation balances.

27. Some possibilities are:

$\frac{1}{2}x = 4;\ -\frac{1}{4}a = -2;\ \frac{3}{4}b = 6.$

29. Let a be the man's age.

$100 + \frac{a}{2} = $ Systolic Blood Pressure

$100 + \frac{a}{2} = 109$

$\underline{-100 -100}$

$0 + \frac{a}{2} = 9$

$\frac{a}{2} = 9$

$\frac{1}{2}a = 9$

$\frac{\cancel{2}}{1}\left(\frac{1}{\cancel{2}}a\right) = \frac{2}{1}(9)$

$ a = 18$

The man is 18 years old.

31. Let a be the woman's age.

$$100 + \frac{a}{2} = \text{Systolic Blood Pressure}$$

$$100 + \frac{a}{2} = 122$$

$$\underline{-100 \qquad\qquad -100}$$

$$0 + \frac{a}{2} = 22$$

$$\frac{\cancel{2}^{1}}{\cancel{1}_{1}}\left(\frac{1}{\cancel{2}_{1}}a\right) = 2(22)$$

$$a = 44$$

The woman is 44 years old.

33. Let p be the penny size.

$$\frac{p}{4} + \frac{1}{2} = 3$$

$$\underline{-\frac{1}{2} \qquad -\frac{1}{2}}$$

$$\frac{p}{4} = 3 - \frac{1}{2}$$

$$\frac{p}{4} = \frac{6}{2} - \frac{1}{2}$$

$$\frac{p}{4} = \frac{5}{2}$$

$$\frac{\cancel{4}^{1}}{1} \cdot \frac{p}{\cancel{4}_{1}} = \frac{\cancel{4}^{2}}{1} \cdot \frac{5}{\cancel{2}_{1}}$$

$$p = 10$$

The penny size is 10.

35. Let p be the penny size.

$$\frac{p}{4} + \frac{1}{2} = \frac{5}{2}$$

$$\underline{-\frac{1}{2} \qquad -\frac{1}{2}}$$

$$\frac{p}{4} = 2$$

$$\frac{\cancel{4}^{1}}{1} \cdot \frac{p}{\cancel{4}_{1}} = 4 \cdot 2$$

$$p = 8$$

The penny size is 8.

37. Let h be the man's height.

$$\frac{11}{2}h - 220 = \text{Recommended weight}$$

$$\frac{11}{2}h - 220 = 209$$

$$\underline{\qquad\quad 220 \qquad\quad 220}$$

$$\frac{11}{2}h + 0 = 429$$

$$\frac{\cancel{2}^{1}}{\cancel{11}_{1}}\left(\frac{\cancel{11}^{1}}{\cancel{2}_{1}}h\right) = \frac{2}{11}(429)$$

$$h = \frac{2 \cdot \cancel{11} \cdot \overset{1}{39}}{\cancel{11}_{1}}$$

$$h = 78$$

The man is 78 inches tall.

39. Let h be the woman's height.

$$\frac{11}{2}h - 220 = \text{Recommended weight}$$

$$\frac{11}{2}h - 220 = 132$$

$$\underline{\qquad\quad 220 \qquad\quad 220}$$

$$\frac{11}{2}h = 352$$

$$\frac{\cancel{2}^{1}}{\cancel{11}_{1}}\left(\frac{\cancel{11}^{1}}{\cancel{2}_{1}}h\right) = \frac{2}{11}(352)$$

$$h = \frac{2 \cdot \cancel{11} \cdot \overset{1}{32}}{\cancel{11}_{1}}$$

$$h = 64$$

The woman is 64 inches tall.

4.8 Geometry Applications: Area and Volume

4.8 Margin Exercises

1. (a) $A = \frac{1}{2}bh$

$$A = \frac{1}{2} \cdot 26 \text{ m} \cdot 20 \text{ m}$$

$$A = \frac{1}{2} \cdot \frac{26 \text{ m}}{1} \cdot \frac{20 \text{ m}}{1}$$

$$A = \frac{1 \cdot \overset{1}{\cancel{2}} \cdot 13 \text{ m} \cdot 20 \text{ m}}{\cancel{2}_{1} \cdot 1 \cdot 1}$$

$$A = 260 \text{ m}^2$$

4.8 Geometry Applications: Area and Volume

(b) $A = \dfrac{1}{2}bh$

$A = \dfrac{1}{2} \cdot 6 \text{ yd} \cdot 5 \text{ yd}$

$A = \dfrac{1}{2} \cdot \dfrac{6 \text{ yd}}{1} \cdot \dfrac{5 \text{ yd}}{1}$

$A = \dfrac{1 \cdot \cancel{2} \cdot 3 \text{ yd} \cdot 5 \text{ yd}}{\underset{1}{\cancel{2}} \cdot 1 \cdot 1}$

$A = 15 \text{ yd}^2$

(c) $A = \dfrac{1}{2}bh$

$A = \dfrac{1}{2} \cdot 9\dfrac{1}{2} \text{ ft} \cdot 7 \text{ ft}$

$A = \dfrac{1}{2} \cdot \dfrac{19}{2} \text{ ft} \cdot \dfrac{7}{1} \text{ ft}$

$A = \dfrac{1 \cdot 19 \text{ ft} \cdot 7 \text{ ft}}{2 \cdot 2 \cdot 1}$

$A = \dfrac{133}{4}$, or $33\dfrac{1}{4} \text{ ft}^2$

2. The entire figure is a square.

$A = s^2 = (25 \text{ m})^2$
$= 25 \text{ m} \cdot 25 \text{ m}$
$= 625 \text{ m}^2$

The unshaded parts are triangles of equal size.

$A = \dfrac{1}{2}bh = \dfrac{1}{2} \cdot 25 \text{ m} \cdot 10 \text{ m}$

$= \dfrac{1}{2} \cdot \dfrac{25 \text{ m}}{1} \cdot \dfrac{10 \text{ m}}{1}$

$= \dfrac{1 \cdot 25 \text{ m} \cdot \cancel{2} \cdot 5 \text{ m}}{\underset{1}{\cancel{2}} \cdot 1 \cdot 1}$

$= 125 \text{ m}^2$

So for both triangles the area is

$2(125 \text{ m}^2) = 250 \text{ m}^2$

Subtract to find the area of the shaded part.

entire area	−	unshaded parts	=	shaded part
↓		↓		↓
$A = 625 \text{ m}^2$	−	250 m^2	=	375 m^2

3. (a) $V = l \cdot w \cdot h$
$V = 8 \text{ m} \cdot 3 \text{ m} \cdot 3 \text{ m}$
$V = 72 \text{ m}^3$

(b) $V = l \cdot w \cdot h$

$V = 6\dfrac{1}{4} \text{ ft} \cdot 3\dfrac{1}{2} \text{ ft} \cdot 2 \text{ ft}$

$V = \dfrac{25 \text{ ft}}{4} \cdot \dfrac{7 \text{ ft}}{2} \cdot \dfrac{2 \text{ ft}}{1}$

$V = \dfrac{25 \text{ ft} \cdot 7 \text{ ft} \cdot \cancel{2} \text{ ft}}{4 \cdot \underset{1}{\cancel{2}} \cdot 1}$

$V = \dfrac{175}{4}$, or $43\dfrac{3}{4} \text{ ft}^3$

4. The base is square and measures 10 ft by 10 ft.

$B = s^2 = (10 \text{ ft})^2$
$= 10 \text{ ft} \cdot 10 \text{ ft}$
$= 100 \text{ ft}^2$

$V = \dfrac{1}{3}Bh = \dfrac{1}{3} \cdot 100 \text{ ft}^2 \cdot 6 \text{ ft}$

$= \dfrac{1}{3} \cdot \dfrac{100 \text{ ft}^2}{1} \cdot \dfrac{6 \text{ ft}}{1}$

$= \dfrac{1 \cdot 100 \text{ ft}^2 \cdot 2 \cdot \cancel{3} \text{ ft}}{\underset{1}{\cancel{3}} \cdot 1 \cdot 1}$

$= 200 \text{ ft}^3$

4.8 Section Exercises

1. $P = 58 \text{ m} + 72 \text{ m} + 72 \text{ m} = 202 \text{ m}$

$A = \dfrac{1}{2}bh$

$A = \dfrac{1}{2}(58 \text{ m})(66 \text{ m})$

$A = \dfrac{1}{2} \cdot \dfrac{58 \text{ m}}{1} \cdot \dfrac{66 \text{ m}}{1}$

$A = \dfrac{1 \cdot \cancel{2} \cdot 29 \text{ m} \cdot 66 \text{ m}}{\underset{1}{\cancel{2}} \cdot 1 \cdot 1}$

$A = 1914 \text{ m}^2$

3. $P = 2\dfrac{1}{4} \text{ ft} + 1\dfrac{1}{2} \text{ ft} + 1\dfrac{1}{4} \text{ ft}$

$P = \dfrac{9}{4} \text{ ft} + \dfrac{3}{2} \text{ ft} + \dfrac{5}{4} \text{ ft}$

$P = \dfrac{9}{4} \text{ ft} + \dfrac{6}{4} \text{ ft} + \dfrac{5}{4} \text{ ft}$

$P = \dfrac{9 + 6 + 5}{4} \text{ ft}$

$P = \dfrac{20}{4} \text{ ft} = 5 \text{ ft}$

continued

$A = \frac{1}{2}bh$

$A = \frac{1}{2}\left(2\frac{1}{4} \text{ ft}\right)\left(\frac{3}{4} \text{ ft}\right)$

$A = \frac{1}{2}\left(\frac{9}{4} \text{ ft}\right)\left(\frac{3}{4} \text{ ft}\right)$

$A = \frac{1 \cdot 9 \cdot 3}{2 \cdot 4 \cdot 4} \text{ ft}^2 = \frac{27}{32} \text{ ft}^2$

5. $P = 9 \text{ yd} + 7 \text{ yd} + 10\frac{1}{4} \text{ yd}$

$P = 16 \text{ yd} + 10\frac{1}{4} \text{ yd}$

$P = 26\frac{1}{4} \text{ yd}$

$A = \frac{1}{2}bh$

$A = \frac{1}{2}\left(10\frac{1}{4} \text{ yd}\right)(6 \text{ yd})$

$A = \frac{1}{2}\left(\frac{41}{4} \text{ yd}\right)\left(\frac{6}{1} \text{ yd}\right)$

$A = \frac{1 \cdot 41 \cdot \overset{1}{\cancel{2}} \cdot 3}{\underset{1}{\cancel{2}} \cdot 4 \cdot 1} \text{ yd}^2$

$A = \frac{123}{4}, \text{ or } 30\frac{3}{4} \text{ yd}^2$

7. $P = 12\frac{3}{5} \text{ yd} + 7\frac{2}{3} \text{ yd} + 10 \text{ yd}$

$P = \frac{63}{5} \text{ yd} + \frac{23}{3} \text{ yd} + 10 \text{ yd}$

$P = \frac{189}{15} \text{ yd} + \frac{115}{15} \text{ yd} + \frac{150}{15} \text{ yd}$

$P = \frac{454}{15}, \text{ or } 30\frac{4}{15} \text{ yd}$

$A = \frac{1}{2}bh$

$A = \frac{1}{2}(10 \text{ yd})\left(7\frac{2}{3} \text{ yd}\right)$

$A = (5 \text{ yd})\left(7\frac{2}{3} \text{ yd}\right)$

$A = \left(\frac{5}{1} \text{ yd}\right)\left(\frac{23}{3} \text{ yd}\right)$

$A = \frac{115}{3}, \text{ or } 38\frac{1}{3} \text{ yd}^2$

9. Area of rectangle − Area of triangle
= Area of shaded portion

$lw - \frac{1}{2}bh = (52 \text{ m})(37 \text{ m}) - \frac{1}{2}(52 \text{ m})(8 \text{ m})$

$= 1924 \text{ m}^2 - 208 \text{ m}^2$

$= 1716 \text{ m}^2$

The shaded area is 1716 m².

11. $A = \frac{1}{2}bh$

$A = \frac{1}{2}\left(3\frac{1}{2} \text{ ft}\right)\left(4\frac{1}{2} \text{ ft}\right)$

$A = \frac{1}{2}\left(\frac{7}{2} \text{ ft}\right)\left(\frac{9}{2} \text{ ft}\right)$

$A = \frac{63}{8}, \text{ or } 7\frac{7}{8} \text{ ft}^2$

$7\frac{7}{8}$ ft² of material is needed.

13. Amount of curbing to go "around" the space implies perimeter.

$P = 33 \text{ m} + 55 \text{ m} + 44 \text{ m}$

$P = 132 \text{ m}$

132 m of curbing will be needed.

Amount of sod to "cover" the space implies area.

The figure is a triangle.

$A = \frac{1}{2}bh$

$A = \frac{1}{2} \cdot 44 \text{ m} \cdot 33 \text{ m}$

$A = \frac{1}{2} \cdot \frac{44 \text{ m}}{1} \cdot \frac{33 \text{ m}}{1}$

$A = \frac{1 \cdot \overset{1}{\cancel{2}} \cdot 22 \text{ m} \cdot 33 \text{ m}}{\underset{1}{\cancel{2}} \cdot 1 \cdot 1}$

$A = 726 \text{ m}^2$

726 m² of sod will be needed.

15. The figure is a rectangular solid.

$V = lwh$

$V = (12 \text{ cm})(4 \text{ cm})(11 \text{ cm})$

$V = 528 \text{ cm}^3$

17. The figure is a rectangular solid or cube.

$V = lwh$

$V = 2\frac{1}{2} \text{ in.} \cdot 2\frac{1}{2} \text{ in.} \cdot 2\frac{1}{2} \text{ in.}$

$V = \frac{5}{2} \text{ in.} \cdot \frac{5}{2} \text{ in.} \cdot \frac{5}{2} \text{ in.}$

$V = \frac{125}{8}, \text{ or } 15\frac{5}{8} \text{ in.}^3$

19. The figure is a pyramid.

First find the area of the rectangular base.

$B = l \cdot w$

$= 15 \text{ cm} \cdot 8 \text{ cm}$

$= 120 \text{ cm}^2$

Now find the volume of the pyramid.

$$V = \frac{B \cdot h}{3}$$
$$= \frac{120 \text{ cm}^2 \cdot 20 \text{ cm}}{3}$$
$$= 800 \text{ cm}^3$$

21. First find the area of the square base.

    $B = l \cdot w$
    $= 8 \text{ ft} \cdot 8 \text{ ft}$
    $= 64 \text{ ft}^2$

    Now find the volume of the pyramid.

    $$V = \frac{B \cdot h}{3}$$
    $$= \frac{64 \text{ ft}^2 \cdot 5 \text{ ft}}{3}$$
    $$= \frac{320}{3}, \text{ or } 106\frac{2}{3} \text{ ft}^3$$

23. The figure is a rectangular box.

    $V = lwh$
    $= 8 \text{ in.} \cdot 3 \text{ in.} \cdot \frac{3}{4} \text{ in.}$
    $= \frac{8 \text{ in.}}{1} \cdot \frac{3 \text{ in.}}{1} \cdot \frac{3 \text{ in.}}{4}$
    $= \frac{2 \cdot \cancel{4} \text{ in.} \cdot 3 \text{ in.} \cdot 3 \text{ in.}}{1 \cdot 1 \cdot \cancel{4}}$
    $= 18 \text{ in.}^3$

25. Use the formula for the volume of a pyramid.

    First find the area of the square base.

    $B = s \cdot s$
    $= 145 \text{ m} \cdot 145 \text{ m}$
    $= 21{,}025 \text{ m}^2$

    Now find the volume of the pyramid.

    $$V = \frac{B \cdot h}{3}$$
    $$= \frac{21{,}025 \text{ m}^2 \cdot 93 \text{ m}}{3}$$
    $$= 651{,}775 \text{ m}^3$$

    The volume of the ancient stone pyramid is $651{,}775 \text{ m}^3$.

Chapter 4 Review Exercises

1. 2 of the 5 figures are squares: $\frac{2}{5}$

    1 of the 5 figures is a circle: $\frac{1}{5}$

2. Shaded: $\frac{3}{10}$ Unshaded: $\frac{7}{10}$

3. Graph $-\frac{1}{2}$ and $1\frac{1}{2}$ on the number line.

4. (a) $-\frac{20}{5} = -\frac{20 \div 5}{5 \div 5} = -\frac{4}{1} = -4$

    (b) $\frac{8}{1} = 8 \div 1 = 8$

    (c) $-\frac{3}{3} = -3 \div 3 = -1$

5. $\frac{28}{32} = \frac{\cancel{2} \cdot \cancel{2} \cdot 7}{\cancel{2} \cdot \cancel{2} \cdot 2 \cdot 2 \cdot 2} = \frac{7}{8}$

6. $\frac{54}{90} = \frac{\cancel{2} \cdot \cancel{3} \cdot \cancel{3} \cdot 3}{\cancel{2} \cdot \cancel{3} \cdot \cancel{3} \cdot 5} = \frac{3}{5}$

7. $\frac{16}{25} = \frac{2 \cdot 2 \cdot 2 \cdot 2}{5 \cdot 5} = \frac{16}{25}$

    The fraction is already in lowest terms.

8. $\frac{15x^2}{40x} = \frac{3 \cdot \cancel{5} \cdot \cancel{x} \cdot x}{2 \cdot 2 \cdot 2 \cdot \cancel{5} \cdot \cancel{x}} = \frac{3x}{8}$

9. $\frac{7a^3}{35a^3 b} = \frac{\cancel{7} \cdot \cancel{a} \cdot \cancel{a} \cdot \cancel{a}}{5 \cdot \cancel{7} \cdot \cancel{a} \cdot \cancel{a} \cdot \cancel{a} \cdot b} = \frac{1}{5b}$

10. $\frac{12mn^2}{21m^3 n} = \frac{2 \cdot 2 \cdot \cancel{3} \cdot \cancel{m} \cdot \cancel{n} \cdot n}{\cancel{3} \cdot 7 \cdot \cancel{m} \cdot m \cdot m \cdot \cancel{n}} = \frac{4n}{7m^2}$

11. $-\frac{3}{8} \div (-6) = -\frac{3}{8} \div \frac{-6}{1} = -\frac{3}{8} \cdot \frac{1}{-6}$
    $= -\frac{3}{8}\left(-\frac{1}{6}\right)$
    $= \frac{\cancel{3} \cdot 1}{8 \cdot 2 \cdot \cancel{3}} = \frac{1}{16}$

12. $\frac{2}{5}$ of (-30)

    $\frac{2}{5} \cdot (-30) = \frac{2}{5} \cdot -\frac{30}{1} = -\frac{2 \cdot 30}{5 \cdot 1}$
    $= -\frac{2 \cdot \cancel{5} \cdot 6}{\cancel{5} \cdot 1} = -\frac{12}{1} = -12$

13. $\frac{4}{9}\left(\frac{2}{3}\right) = \frac{4 \cdot 2}{9 \cdot 3} = \frac{8}{27}$

14. $\left(\frac{7}{3x^3}\right)\left(\frac{x^2}{14}\right) = \frac{\cancel{7} \cdot \cancel{x} \cdot \cancel{x}}{3 \cdot \cancel{x} \cdot \cancel{x} \cdot x \cdot 2 \cdot \cancel{7}} = \frac{1}{6x}$

15. $\dfrac{ab}{5} \div \dfrac{b}{10a} = \dfrac{ab}{5} \cdot \dfrac{10a}{b} = \dfrac{a \cdot \cancel{b} \cdot 2 \cdot \cancel{5} \cdot a}{\cancel{5} \cdot \cancel{b}} = 2a^2$

16. $\dfrac{18}{7} \div 3k = \dfrac{18}{7} \div \dfrac{3k}{1}$

$= \dfrac{18}{7} \cdot \dfrac{1}{3k}$

$= \dfrac{6 \cdot \cancel{3} \cdot 1}{7 \cdot \cancel{3} \cdot k}$

$= \dfrac{6}{7k}$

17. $-\dfrac{5}{12} + \dfrac{5}{8}$

Step 1
The LCD is 24.

Step 2
$-\dfrac{5}{12} = -\dfrac{5 \cdot 2}{12 \cdot 2} = -\dfrac{10}{24}$
$\dfrac{5}{8} = \dfrac{5 \cdot 3}{8 \cdot 3} = \dfrac{15}{24}$

Step 3
$-\dfrac{5}{12} + \dfrac{5}{8} = -\dfrac{10}{24} + \dfrac{15}{24}$

$= \dfrac{-10 + 15}{24} = \dfrac{5}{24}$

Step 4
$\dfrac{5}{24}$ is in lowest terms.

18. $\dfrac{2}{3} - \dfrac{4}{5}$

Step 1
The LCD is 15.

Step 2
$\dfrac{2}{3} = \dfrac{2 \cdot 5}{3 \cdot 5} = \dfrac{10}{15}$
$\dfrac{4}{5} = \dfrac{4 \cdot 3}{5 \cdot 3} = \dfrac{12}{15}$

Step 3
$\dfrac{2}{3} - \dfrac{4}{5} = \dfrac{10}{15} - \dfrac{12}{15}$

$= \dfrac{10 - 12}{15} = -\dfrac{2}{15}$

Step 4
$-\dfrac{2}{15}$ is in lowest terms.

19. $4 - \dfrac{5}{6} = \dfrac{4}{1} - \dfrac{5}{6}$

Step 1
The LCD is 6.

Step 2
$\dfrac{4}{1} = \dfrac{4 \cdot 6}{1 \cdot 6} = \dfrac{24}{6}$

Step 3
$\dfrac{4}{1} - \dfrac{5}{6} = \dfrac{24}{6} - \dfrac{5}{6} = \dfrac{24 - 5}{6} = \dfrac{19}{6}$, or $3\dfrac{1}{6}$

Step 4
$\dfrac{19}{6}$ is in lowest terms.

20. $\dfrac{7}{9} + \dfrac{13}{18}$

Step 1
The LCD is 18.

Step 2
$\dfrac{7}{9} = \dfrac{7 \cdot 2}{9 \cdot 2} = \dfrac{14}{18}$

Step 3
$\dfrac{7}{9} + \dfrac{13}{18} = \dfrac{14}{18} + \dfrac{13}{18} = \dfrac{14 + 13}{18} = \dfrac{27}{18}$

Step 4
$\dfrac{27}{18} = \dfrac{3 \cdot \cancel{9}}{2 \cdot \cancel{9}} = \dfrac{3}{2}$, or $1\dfrac{1}{2}$

21. $\dfrac{n}{5} + \dfrac{3}{4}$

Step 1
The LCD is 20.

Step 2
$\dfrac{n}{5} = \dfrac{n \cdot 4}{5 \cdot 4} = \dfrac{4n}{20}, \dfrac{3}{4} = \dfrac{3 \cdot 5}{4 \cdot 5} = \dfrac{15}{20}$

Step 3
$\dfrac{n}{5} + \dfrac{3}{4} = \dfrac{4n}{20} + \dfrac{15}{20} = \dfrac{4n + 15}{20}$

Step 4
$\dfrac{4n + 15}{20}$ is in lowest terms.

22. $\dfrac{3}{10} - \dfrac{7}{y}$

Step 1
The LCD is $10 \cdot y$ or $10y$.

Step 2
$\dfrac{3}{10} = \dfrac{3 \cdot y}{10 \cdot y} = \dfrac{3y}{10y}, \dfrac{7}{y} = \dfrac{7 \cdot 10}{y \cdot 10} = \dfrac{70}{10y}$

Step 3
$\dfrac{3}{10} - \dfrac{7}{y} = \dfrac{3y}{10y} - \dfrac{70}{10y} = \dfrac{3y - 70}{10y}$

Step 4
$\dfrac{3y - 70}{10y}$ is in lowest terms.

23. $2\frac{1}{4}$ rounds to 2. $1\frac{5}{8}$ rounds to 2.

Estimate: $2 \div 2 = 1$

Exact: $2\frac{1}{4} \div 1\frac{5}{8} = \frac{9}{4} \div \frac{13}{8} = \frac{9}{4} \cdot \frac{8}{13}$

$$= \frac{9 \cdot 2 \cdot \cancel{4}}{\cancel{4} \cdot 13} = \frac{18}{13}, \text{ or } 1\frac{5}{13}$$

24. $7\frac{1}{3}$ rounds to 7. $4\frac{5}{6}$ rounds to 5.

Estimate: $7 - 5 = 2$

Exact: $7\frac{1}{3} - 4\frac{5}{6} = \frac{22}{3} - \frac{29}{6} = \frac{44}{6} - \frac{29}{6}$

$$= \frac{15}{6} = \frac{\cancel{3} \cdot 5}{\cancel{3} \cdot 2} = \frac{5}{2}, \text{ or } 2\frac{1}{2}$$

25. $1\frac{3}{4}$ rounds to 2. $2\frac{3}{10}$ rounds to 2.

Estimate: $2 + 2 = 4$

Exact: $1\frac{3}{4} + 2\frac{3}{10} = \frac{7}{4} + \frac{23}{10} = \frac{35}{20} + \frac{46}{20}$

$$= \frac{81}{20}, \text{ or } 4\frac{1}{20}$$

26. $\left(-\frac{3}{4}\right)^3 = -\frac{3}{4}\left(-\frac{3}{4}\right)\left(-\frac{3}{4}\right)$

$$= \frac{9}{16}\left(-\frac{3}{4}\right)$$

$$= -\frac{27}{64}$$

27. $\left(\frac{2}{3}\right)^2\left(-\frac{1}{2}\right)^4$

$$= \frac{2}{3} \cdot \frac{2}{3}\left(-\frac{1}{2}\right)\left(-\frac{1}{2}\right)\left(-\frac{1}{2}\right)\left(-\frac{1}{2}\right)$$

$$= \frac{\cancel{2} \cdot \cancel{2}}{3 \cdot 3 \cdot \cancel{2} \cdot \cancel{2} \cdot 2 \cdot 2} = \frac{1}{36}$$

28. $\frac{2}{5} + \frac{3}{10}(-4)$

$$= \frac{2}{5} + \frac{3}{10} \cdot \left(\frac{-4}{1}\right)$$

$$= \frac{2}{5} + \left(-\frac{12}{10}\right)$$

$$= \frac{2}{5} + \left(-\frac{6}{5}\right)$$

$$= \frac{2 + (-6)}{5} = \frac{-4}{5} \text{ or } -\frac{4}{5}$$

29. $-\frac{5}{8} \div \left(-\frac{1}{2}\right)\left(\frac{14}{15}\right)$

$= -\frac{5}{8} \div \left(-\frac{1}{2}\right) \cdot \left(\frac{14}{15}\right)$ Change division to multiplication.

$= -\frac{5}{8}\left(-\frac{2}{1}\right)\left(\frac{14}{15}\right)$ Multiply.

$= \frac{\cancel{5} \cdot \cancel{2} \cdot \cancel{2} \cdot 7}{\cancel{2} \cdot \cancel{2} \cdot 2 \cdot 3 \cdot \cancel{5}}$

$= \frac{7}{6}, \text{ or } 1\frac{1}{6}$

30. $\frac{\frac{5}{8}}{\frac{1}{16}} = \frac{5}{8} \div \frac{1}{16} = \frac{5}{8} \cdot \frac{16}{1}$

$$= \frac{5 \cdot 2 \cdot \cancel{8}}{\cancel{8} \cdot 1}$$

$$= \frac{10}{1} = 10$$

31. $\frac{\frac{8}{9}}{-6} = \frac{8}{9} \div (-6) = \frac{8}{9} \div \left(-\frac{6}{1}\right)$

$$= \frac{8}{9}\left(-\frac{1}{6}\right)$$

$$= -\frac{\cancel{2} \cdot 4 \cdot 1}{9 \cdot \cancel{2} \cdot 3} = -\frac{4}{27}$$

32. $\qquad -12 = -\frac{3}{5}w$

$-\frac{5}{3}\left(-\frac{12}{1}\right) = -\frac{5}{3}\left(-\frac{3}{5}w\right)$

$-\frac{5}{\cancel{3}}\left(-\frac{\cancel{3} \cdot 4}{1}\right) = -\frac{\cancel{5}}{\cancel{3}}\left(-\frac{\cancel{3}}{\cancel{5}}\right)w$

$\qquad 20 = w$

The solution is 20.

33. $\quad 18 + \frac{6}{5}r = \quad 0 \qquad$ Add -18 to both sides.

$\quad \underline{-18 \qquad\qquad -18}$

$\qquad 0 + \frac{6}{5}r = -18 \qquad$ Multiply both sides by $\frac{5}{6}$.

$\quad \frac{5}{6} \cdot \left(\frac{6}{5}r\right) = \frac{5}{6}(-18)$

$\qquad\qquad r = -15$

The solution is -15.

128   Chapter 4   Rational Numbers: Positive and Negative Fractions

34. $3x - \dfrac{2}{3} = \dfrac{5}{6}$
     $ + \dfrac{2}{3} + \dfrac{2}{3}$
     $\overline{}$
     $3x + 0 = \dfrac{5}{6} + \dfrac{2}{3}$ $\dfrac{5}{6} + \dfrac{2}{3} = \dfrac{5}{6} + \dfrac{4}{6} = \dfrac{9}{6}$
     $3x = \dfrac{9}{6}$ $\dfrac{9}{6} = \dfrac{\cancel{3} \cdot 3}{\cancel{3} \cdot 2} = \dfrac{3}{2}$
     $\dfrac{1}{3}(3x) = \dfrac{1}{3}\left(\dfrac{3}{2}\right)$
     $x = \dfrac{1}{2}$

     The solution is $\dfrac{1}{2}$.

35. $A = \dfrac{1}{2}bh$
     $A = \dfrac{1}{2} \cdot 8 \text{ ft} \cdot 3\dfrac{1}{2} \text{ ft}$
     $A = \dfrac{1}{2} \cdot \dfrac{8}{1} \text{ ft} \cdot \dfrac{7}{2} \text{ ft}$
     $A = \dfrac{1 \cdot \cancel{2} \cdot \cancel{2} \cdot 2 \cdot 7}{\cancel{2} \cdot 1 \cdot \cancel{2}} \text{ ft}^2$
     $A = 14 \text{ ft}^2$

36. The figure is a rectangular solid.
     $V = l \cdot w \cdot h$
     $V = 4 \text{ in.} \cdot 3\dfrac{1}{4} \text{ in.} \cdot 2\dfrac{1}{2} \text{ in.}$
     $V = \dfrac{4}{1} \text{ in.} \cdot \dfrac{13}{4} \text{ in.} \cdot \dfrac{5}{2} \text{ in.}$
     $V = \dfrac{\cancel{4} \cdot 13 \cdot 5}{1 \cdot \cancel{4} \cdot 2} \text{ in.}^3$
     $V = \dfrac{65}{2}, \text{ or } 32\dfrac{1}{2} \text{ in.}^3$

37. The figure is a pyramid.
     First find the area of the rectangular base.
     $B = l \cdot w$
     $B = 8 \text{ yd} \cdot 5 \text{ yd}$
     $B = 40 \text{ yd}^2$

     Now find the volume of the pyramid.
     $V = \dfrac{1}{3}B \cdot h$
     $V = \dfrac{1}{3} \cdot \dfrac{40 \text{ yd}^2}{1} \cdot \dfrac{7 \text{ yd}}{1}$
     $V = \dfrac{280}{3}, \text{ or } 93\dfrac{1}{3} \text{ yd}^3$

38. [4.3] The $2\dfrac{1}{2}$ pounds of meat will be used to make 10 servings:

     "$2\dfrac{1}{2}$ pounds divided or separated into 10 servings"

     $2\dfrac{1}{2} \div 10 = \dfrac{5}{2} \div \dfrac{10}{1} = \dfrac{5}{2} \cdot \dfrac{1}{10} = \dfrac{\cancel{5} \cdot 1}{2 \cdot 2 \cdot \cancel{5}}$
     $= \dfrac{1}{4}$

     $\dfrac{1}{4}$ pound will be used in each serving.

     For 30 servings:

     $30\left(\dfrac{1}{4}\right) = \dfrac{30}{1} \cdot \dfrac{1}{4} = \dfrac{\cancel{2} \cdot 15 \cdot 1}{1 \cdot \cancel{2} \cdot 2}$
     $= \dfrac{15}{2} = 7\dfrac{1}{2}$

     You would need $7\dfrac{1}{2}$ pounds of meat.

39. [4.4] To find out how much longer she worked on Monday than on Friday, use subtraction.
     $4\dfrac{1}{2} - 3\dfrac{2}{3} = \dfrac{9}{2} - \dfrac{11}{3}$ LCD is 6.
     $= \dfrac{27}{6} - \dfrac{22}{6}$
     $= \dfrac{5}{6}$ hour

     She worked $\dfrac{5}{6}$ hour longer on Monday than on Friday.

     To find how many hours she worked in all, use addition.
     $4\dfrac{1}{2} + 2\dfrac{3}{4} + 3\dfrac{2}{3}$
     $= \dfrac{9}{2} + \dfrac{11}{4} + \dfrac{11}{3}$ LCD is 12.
     $= \dfrac{54}{12} + \dfrac{33}{12} + \dfrac{44}{12}$
     $= \dfrac{54 + 33 + 44}{12} = \dfrac{131}{12} = 10\dfrac{11}{12}$ hours

     She worked a total of $10\dfrac{11}{12}$ hours.

40. [4.3] 60 total children is the "whole."

     $\dfrac{1}{5}$ of the "whole" are preschoolers:

     $\dfrac{1}{5} \cdot 60 = \dfrac{1}{5} \cdot \dfrac{60}{1} = \dfrac{60}{5} = 12$ preschoolers

     $\dfrac{2}{3}$ of the "whole" are toddlers:

     $\dfrac{2}{3} \cdot 60 = \dfrac{2}{3} \cdot \dfrac{60}{1} = \dfrac{120}{3} = 40$ toddlers

     "The rest" are infants; use subtraction:

     $60 - 12 - 40 = 8$ infants

41. [4.5] $P = 2l + 2w$

$$P = \frac{2}{1} \cdot \frac{3}{4} \text{ mi} + \frac{2}{1} \cdot \frac{3}{10} \text{ mi}$$

$$P = \frac{3}{2} \text{ mi} + \frac{3}{5} \text{ mi}$$

$$P = \frac{15}{10} \text{ mi} + \frac{6}{10} \text{ mi}$$

$$P = \frac{21}{10}, \text{ or } 2\frac{1}{10} \text{ miles}$$

$A = l \cdot w$

$A = \frac{3}{4} \text{ mi} \cdot \frac{3}{10} \text{ mi}$

$A = \frac{9}{40} \text{ mi}^2$

Chapter 4 Test

1. Shaded: $\frac{5}{6}$ Unshaded: $\frac{1}{6}$

2. Graph $-\frac{2}{3}$ and $2\frac{1}{3}$ on the number line.

[Number line showing $-\frac{2}{3}$ and $2\frac{1}{3}$ marked between -3 and 3]

3. $\dfrac{21}{84} = \dfrac{\overset{1}{\cancel{3}} \cdot \overset{1}{\cancel{7}}}{2 \cdot 2 \cdot \underset{1}{\cancel{3}} \cdot \underset{1}{\cancel{7}}} = \dfrac{1}{4}$

4. $\dfrac{25}{54} = \dfrac{5 \cdot 5}{2 \cdot 3 \cdot 3 \cdot 3}$ is already in lowest terms.

5. $\dfrac{6a^2b}{9b^2} = \dfrac{2 \cdot \overset{1}{\cancel{3}} \cdot a \cdot a \cdot \overset{1}{\cancel{b}}}{\underset{1}{\cancel{3}} \cdot 3 \cdot \underset{1}{\cancel{b}} \cdot b} = \dfrac{2a^2}{3b}$

6. $\dfrac{1}{6} + \dfrac{7}{10}$ LCD is 30

$= \dfrac{5}{30} + \dfrac{21}{30} = \dfrac{5+21}{30} = \dfrac{26}{30} = \dfrac{13}{15}$

7. $-\dfrac{3}{4} \div \dfrac{3}{8} = -\dfrac{3}{4} \cdot \dfrac{8}{3} = -\dfrac{\overset{1}{\cancel{3}} \cdot 2 \cdot \overset{1}{\cancel{4}}}{\underset{1}{\cancel{4}} \cdot \underset{1}{\cancel{3}}}$

$= -\dfrac{2}{1} = -2$

8. $\dfrac{5}{8} - \dfrac{4}{5} = \dfrac{25}{40} - \dfrac{32}{40} = \dfrac{25-32}{40}$

$= \dfrac{-7}{40}$ or $-\dfrac{7}{40}$

9. $(-20)\left(-\dfrac{7}{10}\right) = -\dfrac{20}{1}\left(-\dfrac{7}{10}\right) = \dfrac{2 \cdot \overset{1}{\cancel{10}} \cdot 7}{1 \cdot \underset{1}{\cancel{10}}}$

$= \dfrac{14}{1} = 14$

10. $\dfrac{\frac{4}{9}}{-6} = \dfrac{4}{9} \div (-6) = \dfrac{4}{9} \div \left(-\dfrac{6}{1}\right)$

$= \dfrac{4}{9}\left(-\dfrac{1}{6}\right) = -\dfrac{\overset{1}{\cancel{2}} \cdot 2 \cdot 1}{9 \cdot \underset{1}{\cancel{2}} \cdot 3}$

$= -\dfrac{2}{27}$

11. $4 - \dfrac{7}{8} = \dfrac{4}{1} - \dfrac{7}{8} = \dfrac{32}{8} - \dfrac{7}{8} = \dfrac{32-7}{8}$

$= \dfrac{25}{8}, \text{ or } 3\dfrac{1}{8}$

12. $-\dfrac{2}{9} + \dfrac{2}{3} = -\dfrac{2}{9} + \dfrac{6}{9} = \dfrac{-2+6}{9} = \dfrac{4}{9}$

13. $\dfrac{21}{24}\left(\dfrac{9}{14}\right) = \dfrac{\overset{1}{\cancel{3}} \cdot \overset{1}{\cancel{7}} \cdot 3 \cdot 3}{\underset{1}{\cancel{3}} \cdot 8 \cdot 2 \cdot \underset{1}{\cancel{7}}} = \dfrac{9}{16}$

14. $\dfrac{12x}{7y} \div 3x = \dfrac{12x}{7y} \div \dfrac{3x}{1} = \dfrac{12x}{7y} \cdot \dfrac{1}{3x}$

$= \dfrac{\overset{1}{\cancel{3}} \cdot 4 \cdot \overset{1}{\cancel{x}} \cdot 1}{7 \cdot y \cdot \underset{1}{\cancel{3}} \cdot \underset{1}{\cancel{x}}} = \dfrac{4}{7y}$

15. $\dfrac{6}{n} - \dfrac{1}{4} = \dfrac{24}{4n} - \dfrac{n}{4n} = \dfrac{24-n}{4n}$

16. $\dfrac{2}{3} + \dfrac{a}{5} = \dfrac{10}{15} + \dfrac{3a}{15} = \dfrac{10+3a}{15}$

17. $\left(\dfrac{5}{9b^2}\right)\left(\dfrac{b}{10}\right) = \dfrac{\overset{1}{\cancel{5}} \cdot \overset{1}{\cancel{b}}}{3 \cdot 3 \cdot \underset{1}{\cancel{b}} \cdot b \cdot 2 \cdot \underset{1}{\cancel{5}}} = \dfrac{1}{18b}$

18. $\left(-\dfrac{1}{2}\right)^3\left(\dfrac{2}{3}\right)^2 = -\dfrac{1}{2}\left(-\dfrac{1}{\underset{1}{\cancel{2}}}\right)\left(-\dfrac{1}{\underset{1}{\cancel{2}}}\right) \cdot \dfrac{\overset{1}{\cancel{2}}}{3} \cdot \dfrac{\overset{1}{\cancel{2}}}{3}$

$= -\dfrac{1}{18}$

19. $\frac{1}{6} + 4\left(\frac{2}{5} - \frac{7}{10}\right)$ Parentheses

$= \frac{1}{6} + 4\left(\frac{4}{10} - \frac{7}{10}\right)$

$= \frac{1}{6} + 4\left(\frac{-3}{10}\right)$

$= \frac{1}{6} + \frac{4}{1} \cdot \frac{-3}{10}$

Note that $\frac{4}{1} \cdot \frac{-3}{10} = \frac{\overset{1}{\cancel{2}} \cdot 2 \cdot (-3)}{1 \cdot \underset{1}{\cancel{2}} \cdot 5} = \frac{-6}{5}$.

$= \frac{1}{6} + \frac{-6}{5}$

$= \frac{5}{30} + \frac{-36}{30}$

$= \frac{5 + (-36)}{30}$

$= -\frac{31}{30}$, or $-1\frac{1}{30}$

20. $4\frac{4}{5}$ rounds to 5. $1\frac{1}{8}$ rounds to 1.

Estimate: $5 \div 1 = 5$

Exact: $4\frac{4}{5} \div 1\frac{1}{8} = \frac{24}{5} \div \frac{9}{8} = \frac{24}{5} \cdot \frac{8}{9}$

$= \frac{\overset{1}{\cancel{3}} \cdot 8 \cdot 8}{5 \cdot \underset{1}{\cancel{3}} \cdot 3}$

$= \frac{64}{15}$, or $4\frac{4}{15}$

21. $3\frac{2}{5}$ rounds to 3. $1\frac{9}{10}$ rounds to 2.

Estimate: $3 - 2 = 1$

Exact: $3\frac{2}{5} - 1\frac{9}{10} = \frac{17}{5} - \frac{19}{10} = \frac{34}{10} - \frac{19}{10}$

$= \frac{15}{10} = \frac{3}{2}$, or $1\frac{1}{2}$

22. $7 = \frac{1}{5}d$

$\frac{5}{1}(7) = \frac{5}{1}\left(\frac{1}{5}d\right)$

$35 = d$

The solution is 35.

23. $-\frac{3}{10}t = \frac{9}{14}$

$-\frac{10}{3}\left(-\frac{3}{10}t\right) = -\frac{10}{3}\left(\frac{9}{14}\right)$

$t = -\frac{\overset{1}{\cancel{2}} \cdot 5 \cdot \overset{1}{\cancel{3}} \cdot 3}{\underset{1}{\cancel{3}} \cdot \underset{1}{\cancel{2}} \cdot 7}$

$t = -\frac{15}{7}$, or $-2\frac{1}{7}$

The solution is $-\frac{15}{7}$.

24. $0 = \frac{1}{4}b - 2$ Add 2 to both sides.

$\underline{+2 \qquad +2}$

$2 = \frac{1}{4}b + 0$

$\frac{4}{1}(2) = \frac{4}{1}\left(\frac{1}{4}b\right)$

$8 = b$

The solution is 8.

25. $\frac{4}{3}x + 7 = -13$ Add -7 to both sides.

$\underline{-7 \qquad -7}$

$\frac{4}{3}x + 0 = -20$

$\frac{3}{4}\left(\frac{4}{3}x\right) = \frac{3}{4}(-20)$

$x = -15$

The solution is -15.

26. $A = \frac{1}{2}bh$

$A = \frac{1}{2} \cdot 13 \text{ m} \cdot 8 \text{ m}$

$A = \frac{1}{2} \cdot \frac{13 \text{ m}}{1} \cdot \frac{8 \text{ m}}{1}$

$A = \frac{13 \cdot \overset{1}{\cancel{2}} \cdot 4}{\underset{1}{\cancel{2}}} \text{ m}^2$

$A = 52 \text{ m}^2$

27. $A = \frac{1}{2}bh$

$A = \frac{1}{2} \cdot 9 \text{ yd} \cdot 13 \text{ yd}$

$A = \frac{1}{2} \cdot \frac{9 \text{ yd}}{1} \cdot \frac{13 \text{ yd}}{1}$

$A = \frac{117}{2}$, or $58\frac{1}{2}$ yd^2

28. The figure is a rectangular solid.

$V = l \cdot w \cdot h$

$V = 30 \text{ m} \cdot 18 \text{ m} \cdot 12 \text{ m}$

$V = 6480 \text{ m}^3$

29. The figure is a pyramid.

The base is a rectangle.
The area of the base is:

$B = l \cdot w$

$B = 4 \text{ yd} \cdot 3 \text{ yd}$

$B = 12 \text{ yd}^2$

$V = \frac{1}{3}Bh$

$V = \frac{1}{3} \cdot 12 \text{ yd}^2 \cdot 4 \text{ yd}$

$V = \frac{48}{3} \text{ yd}^3$

$V = 16 \text{ yd}^3$

30. To find her total training hours, use addition.

$4\frac{5}{6} + 6\frac{2}{3} + 3\frac{1}{4}$

$= \frac{29}{6} + \frac{20}{3} + \frac{13}{4}$ LCD is 12

$= \frac{58}{12} + \frac{80}{12} + \frac{39}{12}$

$= \frac{177}{12} = \frac{59}{4}$, or $14\frac{3}{4}$

She spent $14\frac{3}{4}$ hours training.

Use subtraction to find how many more hours she trained on Tuesday than on Monday.

$6\frac{2}{3} - 4\frac{5}{6}$

$= \frac{20}{3} - \frac{29}{6}$ LCD is 6

$= \frac{40}{6} - \frac{29}{6}$

$= \frac{11}{6}$, or $1\frac{5}{6}$

She spent $1\frac{5}{6}$ hours more on Tuesday than on Monday.

31. $8\frac{3}{4}$ ounces can be synthesized in how many days if $2\frac{1}{2}$ ounces can be synthesized in a day?

Estimation will help. 9 ounces is needed. 3 ounces can be synthesized per day.

$9 \div 3 = 3$ days

Exact: $8\frac{3}{4} \div 2\frac{1}{2}$

$= \frac{35}{4} \div \frac{5}{2}$

$= \frac{35}{4} \cdot \frac{2}{5}$

$= \frac{\overset{1}{\cancel{5}} \cdot 7 \cdot \overset{1}{\cancel{2}}}{\underset{1}{\cancel{2}} \cdot 2 \cdot \underset{1}{\cancel{5}}}$

$= \frac{7}{2}$, or $3\frac{1}{2}$ days

32. $\frac{7}{8}$ of the 8448 students work:

$\frac{7}{8} \cdot 8448 = \frac{7 \cdot \overset{1}{\cancel{8}} \cdot 1056}{\underset{1}{\cancel{8}}} = 7392$

7392 students work.

Cumulative Review Exercises (Chapters 1–4)

1. In words, 505,008,238 is five hundred five million, eight thousand, two hundred thirty-eight.

2. Thirty-five billion, six hundred million, nine hundred sixteen can be written as 35,600,000,916

3. **(a)** 60,719

Underline the hundreds place: 60,<u>7</u>19
The next digit is 4 or less. Leave the 7, change 1 and 9 to 0. **60,700**

(b) 99,505

Underline the thousands place: 9<u>9</u>,505
The next digit is 5 or more, so add one to 9.
Write 0 and carry 1 to the ten-thousands place.
Write 0 and carry 1 to the hundred-thousands place. Change 5 and 5 to 0. **100,000**

(c) 3206

Underline the tens place: 32<u>0</u>6
The next digit is 5 or more, so add one to 0.
Change 6 to 0. **3210**

4. **(a)** $-3 \cdot 6 = 6 \cdot (-3)$

Commutative property of multiplication

(b) $(7 + 18) + 2 = 7 + (18 + 2)$

Associative property of addition

(c) $5(-10 + 7) = 5 \cdot (-10) + 5 \cdot 7$

Distributive property

5. $9(-6) = -54$

6. $-10 - 10$
$= -10 + (-10)$
$= -20$

7. $\frac{-14}{0}$ is undefined.

8. $6 + 3(2 - 7)^2$ Parentheses first
$= 6 + 3[2 + (-7)]^2$
$= 6 + 3(-5)^2$ Exponents next
$= 6 + 3 \cdot 25$ Multiply.
$= 6 + 75$ Add.
$= 81$

9. $|-8| + |2|$
$= 8 + 2$
$= 10$

10. $-8 + 24 \div 2$ Division first
$= -8 + 12$
$= 4$

11. $(-4)^2 - 2^5$ Exponents
$= -4(-4) - 2 \cdot 2 \cdot 2 \cdot 2 \cdot 2$
$= 16 - 32$
$= -16$

12. $\dfrac{-45 \div (-5)(3)}{-5 - 4(0 - 8)}$

Numerator:
$-45 \div (-5)(3) = 9(3) = 27$

Denominator:
$-5 - 4(0 - 8)$
$= -5 - 4(-8)$
$= -5 - (-32)$
$= -5 + 32$
$= 27$

Last step is division: $\dfrac{27}{27} = 1$

13. $-3 \div \dfrac{3}{8} = -\dfrac{3}{1} \div \dfrac{3}{8} = -\dfrac{3}{1} \cdot \dfrac{8}{3} = -\dfrac{\cancel{3} \cdot 8}{1 \cdot \cancel{3}}$
$= -\dfrac{8}{1} = -8$

14. $-\dfrac{5}{6}(-42) = -\dfrac{5}{6}\left(-\dfrac{42}{1}\right) = \dfrac{5 \cdot \cancel{6} \cdot 7}{\cancel{6} \cdot 1}$
$= \dfrac{35}{1} = 35$

15. $\left(\dfrac{5a^2}{12}\right)\left(\dfrac{18}{a}\right) = \dfrac{5 \cdot a \cdot \cancel{a} \cdot 3 \cdot \cancel{6}}{2 \cdot \cancel{6} \cdot \cancel{a}} = \dfrac{15a}{2}$

16. $\dfrac{7}{x^2} \div \dfrac{7y^2}{3x} = \dfrac{7}{x^2} \cdot \dfrac{3x}{7y^2} = \dfrac{\cancel{7} \cdot 3 \cdot \cancel{x}}{x \cdot \cancel{x} \cdot \cancel{7} \cdot y^2}$
$= \dfrac{3}{xy^2}$

17. $\dfrac{3}{10} - \dfrac{5}{6} = \dfrac{9}{30} - \dfrac{25}{30} = \dfrac{9 - 25}{30}$
$= \dfrac{-16}{30} = \dfrac{-16 \div 2}{30 \div 2}$
$= \dfrac{-8}{15}$ or $-\dfrac{8}{15}$

18. $-\dfrac{3}{8} + \dfrac{11}{16} = -\dfrac{6}{16} + \dfrac{11}{16}$
$= \dfrac{-6 + 11}{16} = \dfrac{5}{16}$

19. $\dfrac{2}{3} - \dfrac{b}{7} = \dfrac{14}{21} - \dfrac{3b}{21} = \dfrac{14 - 3b}{21}$

20. $\dfrac{8}{5} + \dfrac{3}{n} = \dfrac{8n}{5n} + \dfrac{15}{5n} = \dfrac{8n + 15}{5n}$

21. $3\dfrac{1}{4} \div 2\dfrac{1}{4} = \dfrac{13}{4} \div \dfrac{9}{4} = \dfrac{13}{4} \cdot \dfrac{4}{9}$
$= \dfrac{13 \cdot \cancel{4}}{\cancel{4} \cdot 9} = \dfrac{13}{9}$, or $1\dfrac{4}{9}$

22. $2\dfrac{2}{5} - 1\dfrac{3}{4} = \dfrac{12}{5} - \dfrac{7}{4} = \dfrac{48}{20} - \dfrac{35}{20}$
$= \dfrac{48 - 35}{20} = \dfrac{13}{20}$

23. $\left(\dfrac{1}{2}\right)^3 (-2)^3$
$= \dfrac{1}{2} \cdot \dfrac{1}{2} \cdot \dfrac{1}{2}(-2)(-2)(-2)$
$= \dfrac{1}{8}(-8)$
$= \dfrac{1}{8}\left(\dfrac{-8}{1}\right)$
$= \dfrac{-8}{8} = -1$

24. $\dfrac{\frac{7}{12}}{-\frac{14}{15}} = \dfrac{7}{12} \div \left(-\dfrac{14}{15}\right) = \dfrac{7}{12} \cdot \left(-\dfrac{15}{14}\right)$
$= -\dfrac{\cancel{7} \cdot \cancel{3} \cdot 5}{\cancel{3} \cdot 4 \cdot 2 \cdot \cancel{7}}$
$= -\dfrac{5}{8}$

25. $-5a + 2a = 12 - 15$
$-5a + 2a = 12 + (-15)$
$-3a = -3$ Divide both sides by -3.
$\dfrac{-3a}{-3} = \dfrac{-3}{-3}$
$a = 1$

The solution is 1.

26. $y = -7y - 40$
$1y = -7y + (-40)$
$\underline{+7y \quad\quad +7y\quad\quad\quad}$
$8y = 0 + (-40)$
$\dfrac{8y}{8} = \dfrac{-40}{8}$
$y = -5$

The solution is -5.

27.
$$-10 = 6 + \frac{4}{9}k$$
$$\underline{-6 \qquad -6\phantom{+\frac{4}{9}k}}$$
$$-16 = 0 + \frac{4}{9}k$$
$$\frac{9}{4}(-16) = \frac{9}{4}\left(\frac{4}{9}k\right)$$
$$-36 = k$$

The solution is -36.

28.
$$20 + 5x = -2x - 8$$
$$\underline{+2x \qquad +2x}$$
$$20 + 7x = 0 - 8$$
$$20 + 7x = -8$$
$$\underline{-20 \qquad\qquad -20}$$
$$0 + 7x = -28$$
$$\frac{7x}{7} = \frac{-28}{7}$$
$$x = -4$$

The solution is -4.

29.
$$3(-2m + 5) = -4m + 9 + m$$
$$-6m + 15 = -3m + 9$$
$$\underline{+6m \qquad\quad +6m}$$
$$0 + 15 = 3m + 9$$
$$15 = 3m + 9$$
$$\underline{-9 \qquad\quad -9}$$
$$6 = 3m + 0$$
$$\frac{6}{3} = \frac{3m}{3}$$
$$2 = m$$

The solution is 2.

30. The figure is a square.
$$P = 4s$$
$$P = 4 \cdot 4\frac{1}{2} \text{ in.}$$
$$P = \frac{4}{1} \cdot \frac{9}{2} \text{ in.}$$
$$P = 18 \text{ in.}$$
$$A = s^2$$
$$A = s \cdot s$$
$$A = 4\frac{1}{2} \text{ in.} \cdot 4\frac{1}{2} \text{ in.}$$
$$A = \frac{9}{2} \text{ in.} \cdot \frac{9}{2} \text{ in.}$$
$$A = \frac{81}{4}, \text{ or } 20\frac{1}{4} \text{ in.}^2$$

31. The figure is a triangle.
$$P = 6\frac{1}{4} \text{ yd} + 6\frac{1}{4} \text{ yd} + 7\frac{1}{2} \text{ yd}$$
$$P = \frac{25}{4} \text{ yd} + \frac{25}{4} \text{ yd} + \frac{15}{2} \text{ yd}$$
$$P = \frac{25}{4} \text{ yd} + \frac{25}{4} \text{ yd} + \frac{30}{4} \text{ yd}$$
$$P = \frac{25 + 25 + 30}{4} \text{ yd}$$
$$P = \frac{80}{4} \text{ yd} = 20 \text{ yd}$$

$$A = \frac{1}{2}bh$$
$$A = \frac{1}{2} \cdot 7\frac{1}{2} \text{ yd} \cdot 5 \text{ yd}$$
$$A = \frac{1}{2} \cdot \frac{15}{2} \text{ yd} \cdot \frac{5}{1} \text{ yd}$$
$$A = \frac{75}{4}, \text{ or } 18\frac{3}{4} \text{ yd}^2$$

32. The figure is a parallelogram.
$$P = 15 \text{ mm} + 24 \text{ mm} + 15 \text{ mm} + 24 \text{ mm}$$
$$P = 78 \text{ mm}$$

$$A = bh$$
$$A = 24 \text{ mm} \cdot 12 \text{ mm}$$
$$A = 288 \text{ mm}^2$$

33. *Step 1*
Unknown: Weight of each bag

Known: Each bag weighs the same, 16 pounds of seed used, 25 pounds of seed used, 79 pounds of seed left

Step 2
Let b be the original weight of each bag.

Step 3

Original weight of 3 bags	–	pounds of seed used	=	pounds of seed remaining
$3b$	–	$(16 + 25)$	=	79

Step 4
$$3b - (16 + 25) = 79$$
$$3b - 41 = 79$$
$$\underline{+41 \quad +41}$$
$$3b + 0 = 120$$
$$\frac{3b}{3} = \frac{120}{3}$$
$$b = 40$$

Step 5
Each bag originally weighed 40 pounds.

Step 6
First bag: $40 - 16 = 24$ pounds
Second bag: $40 - 25 = 15$ pounds
Third bag: 40 pounds
The sum is $24 + 15 + 40 = 79$ pounds, which is the amount of seed left.

34. *Step 1*
Unknowns: Length and width

Known: Perimeter is 102 yd, lot is twice as long as it is wide.

Step 2
Let w represent the width.
Then $2w$ represents the length.

Step 3
$P = 2l + 2w$

Step 4
$102 \text{ yd} = 2(2w) + 2(w)$
$102 \text{ yd} = 4w + 2w$
$102 \text{ yd} = 6w$
$\dfrac{102 \text{ yd}}{6} = \dfrac{6w}{6}$
$17 \text{ yd} = w$

Step 5
The width is 17 yd and the length is $2(17) = 34$ yd.

Step 6
34 yd is twice as long as 17 yd.

$P = 2l + 2w$
$P = 2 \cdot 34 \text{ yd} + 2 \cdot 17 \text{ yd}$
$P = 68 \text{ yd} + 34 \text{ yd}$
$P = 102 \text{ yd}$

CHAPTER 5 RATIONAL NUMBERS: POSITIVE AND NEGATIVE DECIMALS

5.1 Reading and Writing Decimal Numbers

5.1 Margin Exercises

1. (a) The figure has 10 equal parts; 1 part is shaded.

    $\frac{1}{10}$; 0.1; one tenth

   (b) The figure has 10 equal parts; 3 parts are shaded.

    $\frac{3}{10}$; 0.3; three tenths

   (c) The figure has 10 equal parts; 9 parts are shaded.

    $\frac{9}{10}$; 0.9; nine tenths

2. (a) The figure has 10 equal parts; 3 parts are shaded.

    $\frac{3}{10}$; 0.3; three tenths

   (b) The figure has 100 equal parts; 41 parts are shaded.

    $\frac{41}{100}$; 0.41; forty-one hundredths

3. (a) $-0.7 = -\frac{7}{10}$

   (b) $0.2 = \frac{2}{10}$

   (c) $-0.03 = -\frac{3}{100}$

   (d) $0.69 = \frac{69}{100}$

   (e) $0.047 = \frac{47}{1000}$

   (f) $-0.351 = -\frac{351}{1000}$

4. (a) hundreds tens ones . tenths hundredths
       9   7   1 . 5   4

   (b) ones . tenths
       0   . 4

   (c) ones . tenths hundredths
       5   . 6   0

   (d) ones . tenths hundredths thousandths ten-thousandths
       0   . 0   8   3   5

5. (a) 0.6 is six tenths.

   (b) 0.46 is forty-six hundredths.

   (c) 0.05 is five hundredths.

   (d) 0.409 is four hundred nine thousandths.

   (e) 0.0003 is three ten-thousandths.

   (f) 0.2703 is two thousand seven hundred three ten-thousandths.

   (g) 0.088 is eighty-eight thousandths.

6. (a) 3.8 is three and eight tenths.

   (b) 15.001 is fifteen and one thousandth.

   (c) 0.0073 is seventy-three ten-thousandths.

   (d) 764.309 is seven hundred sixty-four and three hundred nine thousandths.

7. (a) $0.7 = \frac{7}{10}$

   (b) $12.21 = 12\frac{21}{100}$

   (c) $0.101 = \frac{101}{1000}$

   (d) $0.007 = \frac{7}{1000}$

   (e) $1.3717 = 1\frac{3717}{10,000}$

8. (a) $0.5 = \frac{5}{10} = \frac{5 \div 5}{10 \div 5} = \frac{1}{2}$

   (b) $12.6 = 12\frac{6}{10} = 12\frac{6 \div 2}{10 \div 2} = 12\frac{3}{5}$

   (c) $0.85 = \frac{85}{100} = \frac{85 \div 5}{100 \div 5} = \frac{17}{20}$

   (d) $3.05 = 3\frac{5}{100} = 3\frac{5 \div 5}{100 \div 5} = 3\frac{1}{20}$

   (e) $0.225 = \frac{225}{1000} = \frac{225 \div 25}{1000 \div 25} = \frac{9}{40}$

   (f) $420.0802 = 420\frac{802}{10,000} = 420\frac{802 \div 2}{10,000 \div 2} = 420\frac{401}{5000}$

5.1 Section Exercises

1.

tens	ones		tenths		
7	0	.	4	8	9

3.

			hundredths	thousandths	ten-thousandths
0	.	2	5	1	8

5.

			tenths		thousandths	ten-thousandths	
9	3	.	0	1	4	7	2

7.

hundreds	tens			tenths		
3	1	4	.	6	5	8

9.

hundreds		ones			hundredths		
1	4	9	.	0	8	3	2

11.

thousands						hundredths	thousandths	
6	2	8	5	.	7	1	2	5

13. 0 ones, 5 hundredths, 1 ten, 4 hundreds, 2 tenths:

410.25

15. 3 thousandths, 4 hundredths, 6 ones, 2 ten-thousandths, 5 tenths:
6.5432

17. 4 hundredths, 4 hundreds, 0 tens, 0 tenths, 5 thousandths, 5 thousands, 6 ones:
5406.045

19. $0.7 = \frac{7}{10}$

21. $13.4 = 13\frac{4}{10} = 13\frac{4 \div 2}{10 \div 2} = 13\frac{2}{5}$

23. $0.35 = \frac{35}{100} = \frac{35 \div 5}{100 \div 5} = \frac{7}{20}$

25. $0.66 = \frac{66}{100} = \frac{66 \div 2}{100 \div 2} = \frac{33}{50}$

27. $10.17 = 10\frac{17}{100}$

29. $0.06 = \frac{6}{100} = \frac{6 \div 2}{100 \div 2} = \frac{3}{50}$

31. $0.205 = \frac{205}{1000} = \frac{205 \div 5}{1000 \div 5} = \frac{41}{200}$

33. $5.002 = 5\frac{2}{1000} = 5\frac{2 \div 2}{1000 \div 2} = 5\frac{1}{500}$

35. $0.686 = \frac{686}{1000} = \frac{686 \div 2}{1000 \div 2} = \frac{343}{500}$

37. 0.5 is five tenths.

39. 0.78 is seventy-eight hundredths.

41. 0.105 is one hundred five thousandths.

43. 12.04 is twelve and four hundredths.

45. 1.075 is one and seventy-five thousandths.

47. Six and seven tenths:

$6\frac{7}{10} = 6.7$

49. Thirty-two hundredths:

$\frac{32}{100} = 0.32$

51. Four hundred twenty and eight thousandths:

$420\frac{8}{1000} = 420.008$

53. Seven hundred three ten-thousandths:

$\frac{703}{10{,}000} = 0.0703$

55. Seventy-five and thirty thousandths:

$75\frac{30}{1000} = 75.030$

57. Anne should not say "and" because that denotes a decimal point.

59. 8-pound test line has a diameter of 0.010 inch. 0.010 inch is read ten thousandths of an inch.

$0.010 = \frac{10}{1000} = \frac{10 \div 10}{1000 \div 10} = \frac{1}{100}$ inch

61. $\frac{13}{1000}$ inch $= 0.013$ inch

A diameter of 0.013 inch corresponds to a test strength of 12 pounds.

63. Six tenths is 0.6, so the correct part number is 3-C.

65. One and six thousandths is 1.006, which is part number 4-A.

67. The size of part number 4-E is 1.602 centimeters, which in words is one and six hundred two thousandths centimeters.

69. Use the Whole Number Place Value Chart on text page 2 to find the names after hundred thousands. Then change "*s*" to "*ths*" in those names. So millions becomes million*ths* when used on the right side of the decimal point, and so on for ten-millionths, hundred-millionths, and billionths.

70. The first place to the left of the decimal point is ones. "Oneths" would mean a fraction with a denominator of 1, which would equal 1 or more. Anything that is 1 or more is to the *left* of the decimal point.

71. The eighth place value to the right of the decimal point is hundred-millionths, so 0.72436955 is seventy-two million four hundred thirty-six thousand nine hundred fifty-five hundred-millionths.

72. The ninth place value to the right of the decimal point is billionths, so 0.000678554 is six hundred seventy-eight thousand five hundred fifty-four billionths.

73. 8006.500001 is eight thousand six and five hundred thousand one millionths.

74. 20,060.000505 is twenty thousand sixty and five hundred five millionths.

75. Three hundred two thousand forty ten-millionths:

0.0302040

76. Nine billion, eight hundred seventy-six million, five hundred forty-three thousand, two hundred ten and one hundred million two hundred thousand three hundred billionths:

9,876,543,210.100200300

5.2 Rounding Decimal Numbers

5.2 Margin Exercises

1. Round to the nearest thousandth.

(a) Draw a cut-off line: 0.334|92
The first digit cut is 9, which is 5 or more, so round up.

$$\begin{array}{r} 0.334 \\ + 0.001 \\ \hline 0.335 \end{array}$$

Answer: ≈ 0.335

(b) Draw a cut-off line: 8.008|51
The first digit cut is 5, which is 5 or more, so round up.

$$\begin{array}{r} 8.008 \\ + 0.001 \\ \hline 8.009 \end{array}$$

Answer: ≈ 8.009

(c) Draw a cut-off line: 265.420|68
The first digit cut is 6, which is 5 or more, so round up.

$$\begin{array}{r} 265.420 \\ + 0.001 \\ \hline 265.421 \end{array}$$

Answer: ≈ 265.421

(d) Draw a cut-off line: 10.701|80
The first digit cut is 8, which is 5 or more, so round up.

$$\begin{array}{r} 10.701 \\ + 0.001 \\ \hline 10.702 \end{array}$$

Answer: ≈ 10.702

2. **(a)** 0.8988 to the nearest hundredth

Draw a cut-off line: 0.89|88
The first digit cut is 8, which is 5 or more, so round up.

$$\begin{array}{r} 0.89 \\ + 0.01 \\ \hline 0.90 \end{array}$$

Answer: ≈ 0.90

(b) 5.8903 to the nearest hundredth

Draw a cut-off line: 5.89|03
The first digit cut is 0, which is 4 or less. The part you keep stays the same.

Answer: ≈ 5.89

(c) 11.0299 to the nearest thousandth

Draw a cut-off line: 11.029|9
The first digit cut is 9, which is 5 or more, so round up.

$$\begin{array}{r} 11.029 \\ + 0.001 \\ \hline 11.030 \end{array}$$

Answer: ≈ 11.030

(d) 0.545 to the nearest tenth

Draw a cut-off line: 0.5|45
The first digit cut is 4, which is 4 or less. The part you keep stays the same.

Answer: ≈ 0.5

3. Round to the nearest cent (hundredths).

(a) Draw a cut-off line: $14.59|5
The first digit cut is 5, which is 5 or more, so round up.

$$\begin{array}{r} \$14.59 \\ + 0.01 \\ \hline \$14.60 \end{array}$$

Answer: ≈ $14.60

(b) Draw a cut-off line: $578.06|63
The first digit cut is 6, which is 5 or more, so round up.

$$\begin{array}{r} \$578.06 \\ + 0.01 \\ \hline \$578.07 \end{array}$$

Answer: ≈ $578.07

(c) Draw a cut-off line: $0.84|9
The first digit cut is 9, which is 5 or more, so round up.

$$\begin{array}{r} \$0.84 \\ + 0.01 \\ \hline \$0.85 \end{array}$$

Answer: ≈ $0.85

(d) Draw a cut-off line: $0.05|48
The first digit cut is 4, which is 4 or less. The part you keep stays the same.

Answer: ≈ $0.05

4. Round to the nearest dollar.

(a) Draw a cut-off line: $29.|10
The first digit cut is 1, which is 4 or less. The part you keep stays the same.

Answer: ≈ $29

(b) Draw a cut-off line: $136.|49
The first digit cut is 4, which is 4 or less. The part you keep stays the same.

Answer: ≈ $136

(c) Draw a cut-off line: $990.|91
The first digit cut is 9, which is 5 or more, so round up by adding $1.

$$\begin{array}{r} \$990 \\ + 1 \\ \hline \$991 \end{array}$$

Answer: ≈ $991

(d) Draw a cut-off line: $5999.|88
The first digit cut is 8, which is 5 or more, so round up by adding $1.

$$\begin{array}{r} \$5999 \\ + 1 \\ \hline \$6000 \end{array}$$

Answer: ≈ $6000

(e) Draw a cut-off line: $49.|60
The first digit cut is 6, which is 5 or more, so round up by adding $1.

$$\begin{array}{r} \$49 \\ + 1 \\ \hline \$50 \end{array}$$

Answer: ≈ $50

(f) Draw a cut-off line: $0.|55
The first digit cut is 5, which is 5 or more, so round up by adding $1.

$$\begin{array}{r} \$0 \\ + 1 \\ \hline \$1 \end{array}$$

Answer: ≈ $1

(g) Draw a cut-off line: $1.|08
The first digit cut is 0, which is 4 or less. The part you keep stays the same.

Answer: ≈ $1

5.2 Section Exercises

1. 16.8974 to the nearest tenth
Draw a cut-off line after the tenths place:
16.8|974
The first digit cut is 9, which is 5 or more, so round up the tenths place.

$$\begin{array}{r} 16.8 \\ + 0.1 \\ \hline 16.9 \end{array}$$

Answer: ≈ 16.9

5.2 Rounding Decimal Numbers   139

3. 0.95647 to the nearest thousandth
    Draw a cut-off line after the thousandths place:
    0.956|47
    The first digit cut is 4, which is 4 or less. The part you keep stays the same.

    Answer: ≈ 0.956

5. 0.799 to the nearest hundredth
    Draw a cut-off line after the hundredths place:
    0.79|9
    The first digit cut is 9, which is 5 or more, so round up the hundredths place.

    $$\begin{array}{r} 0.79 \\ +\,0.01 \\ \hline 0.80 \end{array}$$

    Answer: ≈ 0.80

7. 3.66062 to the nearest thousandth
    Draw a cut-off line after the thousandths place:
    3.660|62
    The first digit cut is 6, which is 5 or more, so round up the thousandths place.

    $$\begin{array}{r} 3.660 \\ +\,0.001 \\ \hline 3.661 \end{array}$$

    Answer: ≈ 3.661

9. 793.988 to the nearest tenth
    Draw a cut-off line after the tenths place:
    793.9|88
    The first digit cut is 8, which is 5 or more, so round up the tenths place.

    $$\begin{array}{r} 793.9 \\ +\,0.1 \\ \hline 794.0 \end{array}$$

    Answer: ≈ 794.0

11. 0.09804 to the nearest ten-thousandth
    Draw a cut-off line after the ten-thousandths place:
    0.0980|4
    The first digit cut is 4, which is 4 or less. The part you keep stays the same.

    Answer: ≈ 0.0980

13. 48.512 to the nearest one
    Draw a cut-off line after the ones place: 48.|512
    The first digit cut is 5, which is 5 or more, so round up the ones place.

    $$\begin{array}{r} 48 \\ +\,1 \\ \hline 49 \end{array}$$

    Answer: ≈ 49

15. 9.0906 to the nearest hundredth
    Draw a cut-off line after the hundredths place:
    9.09|06
    The first digit cut is 0, which is 4 or less. The part you keep stays the same.

    Answer: ≈ 9.09

17. 82.000151 to the nearest ten-thousandth
    Draw a cut-off line after the ten-thousandths place:
    82.0001|51
    The first digit cut is 5, which is 5 or more, so round up the ten-thousandths place.

    $$\begin{array}{r} 82.0001 \\ +\,0.0001 \\ \hline 82.0002 \end{array}$$

    Answer: ≈ 82.0002

19. Round $0.81666 to the nearest cent.
    Draw a cut-off line: $0.81|666
    The first digit cut is 6, which is 5 or more, so round up.

    $$\begin{array}{r} \$0.81 \\ +\,0.01 \\ \hline \$0.82 \end{array}$$

    Nardos pays $0.82.

21. Round $1.2225 to the nearest cent.
    Draw a cut-off line: $1.22|25
    The first digit cut is 2, which is 4 or less. The part you keep stays the same.

    Nardos pays $1.22.

23. Round $0.4983 to the nearest cent.
    Draw a cut-off line: $0.49|83
    The first digit cut is 8, which is 5 or more, so round up.

    $$\begin{array}{r} \$0.49 \\ +\,0.01 \\ \hline \$0.50 \end{array}$$

    Nardos pays $0.50.

25. Round $48,649.60 to the nearest dollar.
    Draw a cut-off line: $48,649.|60
    The first digit cut is 6, which is 5 or more, so round up.

    $$\begin{array}{r} \$48{,}649 \\ +\,1 \\ \hline \$48{,}650 \end{array}$$

    Income from job: $\approx \$48{,}650$

27. Round $310.08 to the nearest dollar.
Draw a cut-off line: $310.|08
The first digit cut is 0, which is 4 or less. The part you keep stays the same.

Union dues: ≈ $310

29. Round $848.91 to the nearest dollar.
Draw a cut-off line: $848.|91
The first digit cut is 9, which is 5 or more, so round up.

$848
+ 1
─────
$849

Donations to charity: ≈ $849

31. $499.98 to the nearest dollar
Draw a cut-off line: $499.|98
The first digit cut is 9, which is 5 or more, so round up.

$499
+ 1
─────
$500

Answer: ≈ $500

33. $0.996 to the nearest cent
Draw a cut-off line: $0.99|6
The first digit cut is 6, which is 5 or more, so round up.

$0.99
+ 0.01
─────
$1.00

Answer: ≈ $1.00

35. $999.73 to the nearest dollar
Draw a cut-off line: $999.|73
The first digit cut is 7, which is 5 or more, so round up.

$999
+ 1
─────
$1000

Answer: ≈ $1000

37. (a) Round 322.16 to the nearest whole number.
Draw a cut-off line: 322.|16
The first digit cut is 1, which is 4 or less. The part you keep stays the same.

The record speed for a motorcycle is about 322 miles per hour.

(b) Round 106.9 to the nearest whole number.
Draw a cut-off line: 106.|9
The first digit cut is 9, which is 5 or more, so round up.

106
+ 1
─────
107

The fastest roller-coaster speed record is about 107 miles per hour.

39. (a) Round 185.981 to the nearest tenth.
Draw a cut-off line: 185.9|81
The first digit cut is 8, which is 5 or more, so round up.

185.9
+ 0.1
─────
186.0

The Indianapolis 500 fastest average winning speed is about 186.0 miles per hour.

(b) Round 763.04 to the nearest tenth.
Draw a cut-off line: 763.0|4
The first digit cut is 4, which is 4 or less. The part you keep stays the same.

The land speed record is about 763.0 miles per hour.

41. If you round $0.499 to the nearest dollar, it will round to $0 (zero dollars) because $0.499 is closer to $0 than to $1.

42. Round $0.499 to the nearest cent to get $0.50. Guideline: Round amounts less than $1.00 to the nearest cent instead of the nearest dollar.

43. If you round $0.0015 to the nearest cent, it will round to $0.00 (zero cents) because $0.0015 is closer to $0.00 than to $0.01.

44. Both $0.5968 and $0.6014 round to $0.60. Rounding to nearest thousandth (tenth of a cent) would allow you to identify $0.597 as less than $0.601.

5.3 Adding and Subtracting Signed Decimal Numbers

5.3 Margin Exercises

1. (a) 2.86 + 7.09

$$\begin{array}{r} \overset{1}{2.86} \\ + 7.09 \\ \hline 9.95 \end{array}$$ *Line up decimal points.*

5.3 Adding and Subtracting Signed Decimal Numbers

(b) $13.761 + 8.325$

$$\begin{array}{r} \overset{11}{13.761} \\ + 8.325 \\ \hline 22.086 \end{array}$$ Line up decimal points.

(c) $0.319 + 56.007 + 8.252$

$$\begin{array}{r} \overset{1}{0}.\overset{1}{3}19 \\ 56.007 \\ + 8.252 \\ \hline 64.578 \end{array}$$ Line up decimal points.

(d) $39.4 + 0.4 + 177.2$

$$\begin{array}{r} \overset{1\,11}{39.4} \\ 0.4 \\ + 177.2 \\ \hline 217.0 \end{array}$$ Line up decimal points.

2. (a) $6.54 + 9.8$

$$\begin{array}{r} \overset{1}{6}.54 \\ + 9.80 \\ \hline 16.34 \end{array}$$ Line up decimal points. Write in one 0.

(b) $0.831 + 222.2 + 10$

$$\begin{array}{r} \overset{1}{0}.831 \\ 222.200 \\ + 10.000 \\ \hline 233.031 \end{array}$$ Line up decimal points. Write in zeros.

(c) $8.64 + 39.115 + 3.0076$

$$\begin{array}{r} \overset{2}{8}.\overset{1}{6}400 \\ 39.1150 \\ + 3.0076 \\ \hline 50.7626 \end{array}$$ Line up decimal points. Write in zeros.

(d) $5 + 429.823 + 0.76$

$$\begin{array}{r} \overset{11}{5}.000 \\ 429.823 \\ + 0.760 \\ \hline 435.583 \end{array}$$ Line up decimal points. Write in zeros.

3. (a) Subtract 22.7 from 72.9.

$$\begin{array}{r} 72.9 \\ - 22.7 \\ \hline 50.2 \end{array} \quad \textbf{Check:} \quad \begin{array}{r} 22.7 \\ + 50.2 \\ \hline 72.9 \end{array}$$

(b) Subtract 6.425 from 11.813.

$$\begin{array}{r} \overset{7\,10\,13}{11.8\cancel{1}\cancel{3}} \\ - 6.425 \\ \hline 5.388 \end{array} \quad \textbf{Check:} \quad \begin{array}{r} 6.425 \\ + 5.388 \\ \hline 11.813 \end{array}$$

(c) $\$20.15 - \$19.67$

$$\begin{array}{r} \overset{1\,9\,10\,15}{\$\cancel{2}\cancel{0}.\cancel{1}\cancel{5}} \\ - 19.67 \\ \hline \$0.48 \end{array} \quad \textbf{Check:} \quad \begin{array}{r} \$19.67 \\ + 0.48 \\ \hline \$20.15 \end{array}$$

4. (a) Subtract 18.651 from 25.3.

$$\begin{array}{r} \overset{1\,14\,12\,9\,10}{\cancel{2}\cancel{5}.\cancel{3}\cancel{0}\cancel{0}} \\ - 18.651 \\ \hline 6.649 \end{array}$$ Write two zeros.

Check: $\begin{array}{r} 18.651 \\ + 6.649 \\ \hline 25.300 \end{array}$

(b) $5.816 - 4.98$

$$\begin{array}{r} \overset{4\,17\,11}{\cancel{5}.\cancel{8}\cancel{1}6} \\ - 4.980 \\ \hline 0.836 \end{array} \quad \textbf{Check:} \quad \begin{array}{r} 4.980 \\ + 0.836 \\ \hline 5.816 \end{array}$$ Write one zero.

(c) 40 less 3.66

$$\begin{array}{r} \overset{3\,9\,9\,10}{\cancel{4}\cancel{0}.\cancel{0}\cancel{0}} \\ - 3.66 \\ \hline 36.34 \end{array} \quad \textbf{Check:} \quad \begin{array}{r} 36.34 \\ + 3.66 \\ \hline 40.00 \end{array}$$

(d) $1 - 0.325$

$$\begin{array}{r} \overset{0\,9\,9\,10}{\cancel{1}.\cancel{0}\cancel{0}\cancel{0}} \\ - 0.325 \\ \hline 0.675 \end{array} \quad \textbf{Check:} \quad \begin{array}{r} 0.325 \\ + 0.675 \\ \hline 1.000 \end{array}$$

5. (a) $13.245 + (-18)$

The addends have unlike signs.

$|13.245| = 13.245; |-18| = 18$

-18 has the larger absolute value, so the sum will be negative. Subtract the absolute values.

$$\begin{array}{r} 18.000 \\ - 13.245 \\ \hline 4.755 \end{array}$$

$13.245 + (-18) = -4.755$

(b) $-0.7 + (-0.33)$

Both addends are negative, so the sum will be negative.

$|-0.7| = 0.7; |-0.33| = 0.33$

$$\begin{array}{r} 0.70 \\ + 0.33 \\ \hline 1.03 \end{array}$$

$-0.7 + (-0.33) = -1.03$

(c) $-6.02 + 100.5$

The addends have unlike signs.

$|-6.02| = 6.02;\ |100.5| = 100.5$

100.5 has the larger absolute value, so the sum will be positive. Subtract the absolute values.

$$\begin{array}{r} 100.50 \\ -\ 6.02 \\ \hline 94.48 \end{array}$$

$-6.02 + 100.5 = 94.48$

6. (a) $-0.37 - (-6)$ Rewrite subtraction to
            ↓   adding the opposite.
       $-0.37 + 6$

6 has the larger absolute value, so the sum will be positive. Subtract the absolute values.

$$\begin{array}{r} 6.00 \\ -\ 0.37 \\ \hline 5.63 \end{array}$$

Answer: 5.63

(b) $5.8 - 10.03$ Rewrite subtraction to
        ↓   adding the opposite.
   $5.8 + (-10.03)$

-10.03 has the larger absolute value, so the sum will be negative. Subtract the absolute values.

$$\begin{array}{r} 10.03 \\ -\ 5.80 \\ \hline 4.23 \end{array}$$

Answer: -4.23

(c) $-312.72 - 65.7$ Change subtraction to
          ↓   adding the opposite.
   $-312.72 + (-65.7)$

Both addends are negative, so the sum will be negative.

$$\begin{array}{r} 312.72 \\ +\ 65.70 \\ \hline 378.42 \end{array}$$

Answer: -378.42

(d) $0.8 - (6 - 7.2)$ Work inside parentheses first. Change subtraction to adding the opposite.

$= 0.8 - (6 + (-7.2))$ -7.2 has the larger absolute value, so the sum will be negative.

$= 0.8 - (-1.2)$ Change subtraction to adding the opposite.

$= 0.8 + 1.2$
$= 2.0$ or   2

7. (a) $2.83 + 5.009 + 76.1$

Estimate: *Exact:*

$$\begin{array}{r} 3 \\ 5 \\ +\ 80 \\ \hline 88 \end{array} \qquad \begin{array}{r} 2.830 \\ 5.009 \\ +\ 76.100 \\ \hline 83.939 \end{array}$$

(b) $19.28 less $1.53

Estimate: *Exact:*

$$\begin{array}{r} \$20 \\ -\ 2 \\ \hline \$18 \end{array} \qquad \begin{array}{r} \$19.28 \\ -\ 1.53 \\ \hline \$17.75 \end{array}$$

(c) $11.365 - 38$

Estimate:

$10 - 40 = 10 + (-40) = -30$

Exact:

$11.365 - 38 = 11.365 + (-38)$

-38 has the larger absolute value and is negative, so the sum is negative.

$$\begin{array}{r} 38.000 \\ -\ 11.365 \\ \hline 26.635 \end{array}$$

Answer: -26.635

(d) $-214.6 + 300.72$

Estimate:

$-200 + 300 = 100$

Exact:

300.72 has the larger absolute value, so the sum is positive.

$$\begin{array}{r} 300.72 \\ -\ 214.60 \\ \hline 86.12 \end{array}$$

5.3 Adding and Subtracting Signed Decimal Numbers

5.3 Section Exercises

1. $\overset{12}{5.69}$ *Line up decimal points.*
    0.24
    $+11.79$
    $\overline{17.72}$

3. 0.38 *Line up decimal points.*
    7.00 *Write in zeros.*
    $+4.60$
    $\overline{11.98}$

5. $14.23 + 8 + 74.63 + 18.715 + 0.286$

    $\overset{211}{14.230}$ *Line up decimal points.*
    8.000 *Write in zeros.*
    74.630
    18.715
    $+0.286$
    $\overline{115.861}$

7. $27.65 + 18.714 + 9.749 + 3.21$

    $\overset{2211}{27.650}$ *Line up decimal points.*
    18.714 *Write in zeros.*
    9.749
    $+3.210$
    $\overline{59.323}$

9. He did not line up the decimal points; 6 should be written as 6.00.

    0.72
    6.00
    $+39.50$
    $\overline{46.22}$

11. $0.3000 = \dfrac{3000}{10{,}000} = \dfrac{3000 \div 1000}{10{,}000 \div 1000} = \dfrac{3}{10} = 0.3$

13. $90.5 - 0.8$

    90.5 *Line up decimal points.*
    -0.8
    $\overline{89.7}$ *Subtract as usual.*

15. 0.4 less 0.291

    0.400 *Line up decimal points.*
    -0.291 *Write in zeros.*
    $\overline{0.109}$ *Subtract as usual.*

17. 6 minus 5.09

    6.00 *Line up decimal points.*
    -5.09 *Write in zeros.*
    $\overline{0.91}$ *Subtract as usual.*

19. Subtract 8.339 from 15

    15.000 *Line up decimal points.*
    -8.339 *Write in zeros.*
    $\overline{6.661}$ *Subtract as usual.*

21. "Subtract 7.45 from 15.32" requires 15.32 to be on top.

    15.32 *Line up decimal points.*
    -7.45
    $\overline{7.87}$ *Subtract as usual.*

23. **(a)** Humerus 14.35 in., radius 10.4 in.

    14.35 *Line up decimal points.*
    $+10.40$ *Write in zero.*
    $\overline{24.75}$

    The combined length is 24.75 inches.

    (b) 14.35
    -10.40
    $\overline{3.95}$

    The difference is 3.95 inches.

25. **(a)** Humerus 14.35 in., ulna 11.1 in., femur 19.88 in., tibia 16.94 in.

    14.35 *Line up decimal points.*
    11.10 *Write in zero.*
    19.88
    $+16.94$
    $\overline{62.27}$

    The sum of the lengths is 62.27 inches.

    (b) 8th rib 9.06 in., 7th rib 9.45 in.

    9.45
    -9.06
    $\overline{0.39}$

    The 8th rib is 0.39 in. shorter than the 7th rib.

27. $24.008 + (-0.995)$

    24.008 has the larger absolute value, so the sum will be positive. Subtract the absolute values.

    24.008
    -0.995
    $\overline{23.013}$

29. $-6.05 + (-39.7)$

    The addends are the same sign, so the sum will be negative.

    6.05
    $+39.70$
    $\overline{45.75}$

    $-6.05 + (-39.7) = -45.75$

31. $0.9 - 7.59 = 0.9 + (-7.59)$

-7.59 has the larger absolute value, so the sum will be negative. Subtract the absolute values.

$$\begin{array}{r} 7.59 \\ -\ 0.90 \\ \hline 6.69 \end{array}$$

$0.9 - 7.59 = -6.69$

33. $-2 - 4.99 = -2 + (-4.99)$

The addends are the same sign, so the sum will be negative.

$$\begin{array}{r} 2.00 \\ +\ 4.99 \\ \hline 6.99 \end{array}$$

$-2 - 4.99 = -6.99$

35. $-5.009 + 0.73$

-5.009 has the larger absolute value, so the sum will be negative. Subtract the absolute values.

$$\begin{array}{r} 5.009 \\ -\ 0.730 \\ \hline 4.279 \end{array}$$

$-5.009 + 0.73 = -4.279$

37. $-1.7035 - (5 - 6.7)$

Work inside parentheses first.

$5 - 6.7 = 5 + (-6.7) = -1.7$

Now the problem becomes:

$= -1.7035 - (-1.7)$
$= -1.7035 + 1.7$
$= -0.0035$

39. $8000 - (8002.63 - 8)$

Work inside parentheses first.

$8002.63 - 8 = 8002.63 + (-8) = 7994.63$

Now the problem becomes:

$8000 - (7994.63)$
$= 8000 + (-7994.63)$
$= 5.37$

41. We can estimate $18 - 11.725$ as $20 - 10 = 10$, so 6.275 is the most reasonable answer.

43. We can estimate $-6.25 + 0.7$ as $-7 + 1 = -6$, so -5.8 is the most reasonable answer.

45. We can estimate $-42.671 - 194.9$ as $-40 + (-200) = -240$, so -237.571 is the most reasonable answer.

47. We can estimate $8.4 - (-50.83)$ as $8 + (+50) = 58$, so 59.23 is the most reasonable answer.

49. Canada: 21.9 million; South Korea: 34 million

Estimate:	Exact:
30	34.0
$-\ 20$	$-\ 21.9$
10	12.1

There are 12.1 million fewer Internet users in Canada than there are in South Korea.

51. Add the number of users from all the countries.

Estimate:	Exact:
200	197.8
100	119.5
90	86.3
50	50.6
30	34.0
20	21.9
$+\ 20$	$+\ 16.9$
510	527.0

There are 527.0 million Internet users in all the countries listed in the table.

53. Add the 3 heights.

Estimate:	Exact:
2	1.83
2	2.16
$+\ 2$	$+\ 2.10$
6 meters	6.09 meters

The NBA stars' combined height is 0.31 meter less than the rhino's height of 6.4 meters.

55. Subtract $9.12 from $20.

Estimate:	Exact:
$20	$20.00
$-\ 9$	$-\ 9.12$
$11	$10.88

He received $10.88 in change.

57. Subtract the price of the regular fishing line, $4.84, from the price of the fluorescent fishing line, $5.14.

Estimate: *Exact:*

$5 $5.14
−5   −4.84
───   ─────
$0 $0.30

The fluorescent fishing line costs $0.30 more than the regular fishing line.

59. Add the price of the four items and the sales tax.

Estimate: *Exact:*

$19 $18.84 *spinning reel*
 2   2.07 *tin split shot*
 2   2.07 *tin split shot*
10   9.96 *tackle box*
+2   +2.31 *sales tax*
───   ──────
$35 $35.25

The total cost of the four items and the sales tax is $35.25.

61. Add the monthly expenses.

$994.00
 190.78
 205.00
  39.95
  19.95
  40.80
  57.32
 186.81
  97.75
+107.00
────────
$1939.36

Olivia's total expenses are $1939.36 per month.

63. Subtract to find the difference.

$190.78
−186.81
───────
$3.97

The difference in the amounts spent for groceries and for the car payment is $3.97.

65. Subtract the two inside measurements from the total length.

 3.00 *Total length*
−0.91 *Leftmost measurement*
─────
 2.09
−0.70 *Middle measurement*
─────
 1.39

$b = 1.39$ cm.

67. Add the given lengths and subtract the sum from the total length.

  2.981   13.905
+ 2.981   − 5.962
───────   ────────
  5.962 feet   7.943 feet

$q = 7.943$ feet

5.4 Multiplying Signed Decimal Numbers

5.4 Margin Exercises

1. (a) $-2.6(0.4)$

2.6 ← 1 *decimal place*
$\times 0.4$ ← 1 *decimal place*
─────
-1.04 ← 2 *decimal places*

The factors have *different* signs, so the product is *negative*.

(b) $(45.2)(0.25)$

45.2 ← 1 *decimal place*
$\times 0.25$ ← 2 *decimal places*
──────
$2\,26\,0$
$9\,04$
──────
$11.30\,0$ ← 3 *decimal places*
or 11.3

The factors have the *same* sign, so the product is *positive*.

(c) 0.104 ← 3 *decimal places*
$\times7$ ← 0 *decimal places*
──────
0.728 ← 3 *decimal places*

(d) $(-3.18)^2$ means $(-3.18)(-3.18)$.

3.18 ← 2 *decimal places*
$\times 3.18$ ← 2 *decimal places*
──────
2544
318
$9\,54$
──────
10.1124 ← 4 *decimal places*

Now count over 4 places and write in the decimal point. The factors have the *same* sign, so the product is *positive*.

2. (a) $0.04(-0.09)$

0.04 ← 2 *decimal places*
$\times 0.09$ ← 2 *decimal places*
──────
-0.0036 ← 4 *decimal places*

Count 4 places. Write in the decimal point and zeros. The factors have *different* signs, so the product is *negative*.

(b) (0.2)(0.008)

$$\begin{array}{r} 0.008 \leftarrow 3 \text{ decimal places} \\ \times\, 0.2 \leftarrow 1 \text{ decimal place} \\ \hline 0.0016 \leftarrow 4 \text{ decimal places} \end{array}$$

Count 4 places. Write in the decimal point and zeros. The factors have the *same* sign, so the product is *positive*.

(c) (−0.063)(−0.04)

$$\begin{array}{r} 0.063 \leftarrow 3 \text{ decimal places} \\ \times\, 0.04 \leftarrow 2 \text{ decimal places} \\ \hline 0.00252 \leftarrow 5 \text{ decimal places} \end{array}$$

Count 5 places. Write in the decimal point and zeros. The factors have the *same* sign, so the product is *positive*.

(d) $(0.003)^2$ means (0.003)(0.003).

$$\begin{array}{r} 0.003 \leftarrow 3 \text{ decimal places} \\ \times\, 0.003 \leftarrow 3 \text{ decimal places} \\ \hline 0.000009 \leftarrow 6 \text{ decimal places} \end{array}$$

Count 6 places. Write in the decimal point and zeros. The factors have the *same* sign, so the product is *positive*.

3. (a) (11.62)(4.01)

Estimate: *Exact:*

$$\begin{array}{r} 10 \\ \times\, 4 \\ \hline 40 \end{array} \qquad \begin{array}{r} 11.62 \leftarrow 2 \text{ decimal places} \\ \times\, 4.01 \leftarrow 2 \text{ decimal places} \\ \hline 1162 \\ 46\,480 \\ \hline 46.5962 \leftarrow 4 \text{ decimal places} \end{array}$$

The factors have the *same* sign, so the product is *positive*.

(b) (−5.986)(−33)

Estimate: *Exact:*

$$\begin{array}{r} 6 \\ \times\, 30 \\ \hline 180 \end{array} \qquad \begin{array}{r} 5.986 \leftarrow 3 \text{ decimal places} \\ \times\, 33 \leftarrow 0 \text{ decimal places} \\ \hline 17\,958 \\ 179\,58 \\ \hline 197.538 \leftarrow 3 \text{ decimal places} \end{array}$$

The factors have the *same* sign, so the product is *positive*.

(c) 8($4.35)

Estimate: *Exact:*

$$\begin{array}{r} 4 \\ \times\, 8 \\ \hline 32 \end{array} \qquad \begin{array}{r} \$4.35 \leftarrow 2 \text{ decimal places} \\ \times\, 8 \leftarrow 0 \text{ decimal places} \\ \hline \$34.80 \leftarrow 2 \text{ decimal places} \end{array}$$

The factors have the *same* sign, so the product is *positive*.

(d) 58.6(−17.4)

Estimate: *Exact:*

$$\begin{array}{r} 60 \\ \times\, 20 \\ \hline -1200 \end{array} \qquad \begin{array}{r} 58.6 \leftarrow 1 \text{ decimal place} \\ \times\, 17.4 \leftarrow 1 \text{ decimal place} \\ \hline 23\,44 \\ 410\,2 \\ 586 \\ \hline -1019.64 \leftarrow 2 \text{ decimal places} \end{array}$$

The factors have *different* signs, so the product is *negative*.

5.4 Section Exercises

1. Multiply the numbers as if they were whole numbers.

$$\begin{array}{r} 0.042 \leftarrow 3 \text{ decimal places} \\ \times\, 3.2 \leftarrow 1 \text{ decimal place} \\ \hline 84 \\ 126 \\ \hline 0.1344 \leftarrow 4 \text{ decimal places} \end{array}$$

Count 4 places. Write in the decimal point and zero. The factors have the *same* sign, so the product is *positive*.

3. −21.5(7.4)

$$\begin{array}{r} 21.5 \leftarrow 1 \text{ decimal place} \\ \times\, 7.4 \leftarrow 1 \text{ decimal place} \\ \hline 8\,60 \\ 150\,5 \\ \hline -159.10 \leftarrow 2 \text{ decimal places} \end{array}$$

The factors have *different* signs, so the product is *negative*.

5. (−23.4)(−0.66)

$$\begin{array}{r} 23.4 \leftarrow 1 \text{ decimal place} \\ \times\, 0.66 \leftarrow 2 \text{ decimal places} \\ \hline 1\,404 \\ 14\,04 \\ \hline 15.444 \leftarrow 3 \text{ decimal places} \end{array}$$

The factors have the *same* sign, so the product is *positive*.

5.4 Multiplying Signed Decimal Numbers

7. Use a calculator.

$$\begin{array}{r} \$51.88 \\ \times 665 \\ \hline \$34{,}500.20 \end{array}$$

9. $72(-0.6) = -43.2$

The factors have *different* signs, so the product is *negative*.

72 has 0 decimal places. $\Big\}\to$ Answer has 1 decimal place.
0.6 has 1 decimal place.

11. $(7.2)(0.06) = 0.432$

The factors have the *same* sign, so the product is *positive*.

7.2 has 1 decimal place. $\Big\}\to$ Answer has 3 decimal places.
0.06 has 2 decimal places.

13. $-0.72(-0.06) = 0.0432$

The factors have the *same* sign, so the product is *positive*.

0.72 has 2 decimal places. $\Big\}\to$ Answer has 4 decimal places.
0.06 has 2 decimal places.

15. $(0.0072)(0.6) = 0.00432$

The factors have the *same* sign, so the product is *positive*.

0.0072 has 4 decimal places. $\Big\}\to$ Answer has 5 decimal places.
0.6 has 1 decimal place.

17. $(0.006)(0.0052)$

$$\begin{array}{rl} 0.0052 & \leftarrow 4 \text{ decimal places} \\ \times\; 0.006 & \leftarrow 3 \text{ decimal places} \\ \hline 0.0000312 & \leftarrow 7 \text{ decimal places} \end{array}$$

Write in 4 zeros to get 7 decimal places.

19. $(-0.003)^2$ means $(-0.003)(-0.003)$.

$$\begin{array}{rl} 0.003 & \leftarrow 3 \text{ decimal places} \\ \times\; 0.003 & \leftarrow 3 \text{ decimal places} \\ \hline 0.000009 & \leftarrow 6 \text{ decimal places} \end{array}$$

Write in 5 zeros to get 6 decimal places. The factors have the *same* sign, so the product is *positive*.

21. $(5.96)(10) = \underline{59.6}$ $\quad (3.2)(10) = \underline{32}$
$(0.476)(10) = \underline{4.76}$ $\quad (80.35)(10) = \underline{803.5}$
$(722.6)(10) = \underline{7226}$ $\quad (0.9)(10) = \underline{9}$

Multiplying by 10, decimal point moves one place to the right; by 100, two places to the right; by 1000, three places to the right.

22. $(59.6)(0.1) = \underline{5.96}$ $\quad (3.2)(0.1) = \underline{0.32}$
$(0.476)(0.1) = \underline{0.0476}$ $\quad (80.35)(0.1) = \underline{8.035}$
$(65)(0.1) = \underline{6.5}$ $\quad (523)(0.1) = \underline{52.3}$

Multiplying by 0.1, decimal point moves one place to the left; by 0.01, two places to the left; by 0.001, three places to the left.

23. Estimate: Exact:

$$\begin{array}{r} 40 \\ \times\; 5 \\ \hline 200 \end{array} \qquad \begin{array}{rl} 39.6 & \leftarrow 1 \text{ decimal place} \\ \times\; 4.8 & \leftarrow 1 \text{ decimal place} \\ \hline 31\;68 \\ 158\;4 \\ \hline 190.08 & \leftarrow 2 \text{ decimal places} \end{array}$$

25. Estimate: Exact:

$$\begin{array}{r} 40 \\ \times\; 40 \\ \hline 1600 \end{array} \qquad \begin{array}{rl} 37.1 & \leftarrow 1 \text{ decimal place} \\ \times\; 42 & \leftarrow 0 \text{ decimal places} \\ \hline 74\;2 \\ 1484 \\ \hline 1558.2 & \leftarrow 1 \text{ decimal place} \end{array}$$

27. Estimate: Exact:

$$\begin{array}{r} 7 \\ \times\; 5 \\ \hline 35 \end{array} \qquad \begin{array}{rl} 6.53 & \leftarrow 2 \text{ decimal places} \\ \times\; 4.6 & \leftarrow 1 \text{ decimal place} \\ \hline 3\;918 \\ 26\;12 \\ \hline 30.038 & \leftarrow 3 \text{ decimal places} \end{array}$$

29. Estimate: Exact:

$$\begin{array}{r} 3 \\ \times\; 7 \\ \hline 21 \end{array} \qquad \begin{array}{rl} 2.809 & \leftarrow 3 \text{ decimal places} \\ \times\; 6.85 & \leftarrow 2 \text{ decimal places} \\ \hline 19.24165 & \leftarrow 5 \text{ decimal places} \end{array}$$

31. An $28.90 car payment is *unreasonable*. A reasonable answer would be $289.00.

33. A height of 60.5 inches (about 5 feet) is *reasonable*.

35. A gallon of milk for $419 is *unreasonable*. A reasonable answer would be $4.19.

37. 0.095 pound for a baby's weight is *unreasonable*. A reasonable answer would be 9.5 pounds.

39. Multiply her pay per hour times the hours she worked.

$$\begin{array}{rl} \$18.73 & \leftarrow 2 \text{ decimal places} \\ \times\; 50.5 & \leftarrow 1 \text{ decimal place} \\ \hline 9\;365 \\ 936\;50 \\ \hline \$945.865 & \leftarrow 3 \text{ decimal places} \end{array}$$

Round $945.865 to the nearest cent. LaTasha made $945.87 (rounded).

41. Multiply the cost of one meter of canvas by the number of meters needed.

$$\begin{array}{r} \$4.09 \\ \times\ 0.6 \\ \hline \$2.454 \end{array}$$

$2.454 rounds to $2.45. Sid will spend $2.45 on the canvas.

43. Multiply the number of gallons that she pumped into her pickup truck by the price per gallon.

$$\begin{array}{r} 20.510 \\ \times\ \ \ 3.979 \\ \hline 81.609290 \end{array}$$

Round 81.609290 to the nearest cent. Michelle paid $81.61 for the gas.

45. Multiply the cost of the home by 0.07.

$$\begin{array}{r} \$289{,}500 \\ \times\ \ \ \ 0.07 \\ \hline \$20{,}265.00 \end{array}$$

Ms. Rolack's fee was $20,265.

47. (a) Area before 1929:

$$\begin{array}{r} 7.4218 \\ \times\ \ \ 3.125 \\ \hline 23.1931250 \end{array}$$

Rounding to the nearest tenth gives us 23.2 in.^{2}.

Area after 1929:

$$\begin{array}{r} 6.14 \\ \times\ \ 2.61 \\ \hline 16.0254 \end{array}$$

Rounding to the nearest tenth gives us 16.0 in.^{2}.

(b) Subtract to find the difference.

$$\begin{array}{r} 23.2 \\ -\ 16.0 \\ \hline 7.2 \end{array}$$

The difference in rounded areas is 7.2 in.^{2}.

49. (a) Multiply the thickness of one bill times the number of bills.

$$\begin{array}{r} 0.0043 \\ \times\ \ \ 100 \\ \hline 0.4300 \end{array}$$

A pile of 100 bills would be 0.43 inch high.

(b)
$$\begin{array}{r} 0.0043 \\ \times\ 1000 \\ \hline 4.3000 \end{array}$$

A pile of 1000 bills would be 4.3 inches high.

51. Multiply the monthly cost of cable times 24 months (two years).

$$\begin{array}{r} \$38.96\ \textit{basic per month} \\ \times\ \ \ \ \ 24 \\ \hline 155\ 84 \\ 779\ 2\ \ \ \\ \hline \$935.04\ \textit{monthly total} \\ +\ 49.00\ \textit{one-time installation} \\ \hline \$984.04\ \textit{two-year total} \end{array}$$

The total cost for basic cable is $984.04.

$$\begin{array}{r} \$89.95\ \textit{deluxe per month} \\ \times\ \ \ \ \ 24 \\ \hline 359\ 80 \\ 1799\ 0\ \ \ \\ \hline \$2158.80\ \textit{monthly total} \\ +\ \ \ 49.00\ \textit{one-time installation} \\ \hline \$2207.80\ \textit{two-year total} \end{array}$$

The total cost for deluxe cable is $2207.80.

53. Multiply the number of sheets by the cost per sheet.

$$\begin{array}{r} 5100 \\ \times\ \$0.015 \\ \hline \$76.500 \end{array}$$

The library will pay $76.50 for the paper.

55. Multiply to find the cost of the rope, then multiply to find the cost of the wire. Add the results to find Barry's total purchases. Subtract the purchases from $15 (three $5 bills is $15).

Cost of rope	Cost of wire
16.5	$1.05
× $0.47	× 3
$7.755	$3.15

The cost of the rope rounds to $7.76.

Purchases	Change
$7.76 *rope*	$15.00
+ 3.15 *wire*	− 10.91
$10.91	$4.09

Barry received $4.09 in change.

57. Find the cost of the 4 long-sleeve, solid color shirts.

$18.95
× 4

$75.80

Then find the cost of the 2 short-sleeve, striped shirts.

$16.75
× 2

$33.50

Add these two amounts and the $2 per shirt charge for the XXL size.

$2 $75.80
× 6   33.50
---   + 12.00
$12   -------
          $121.30

The total cost is $121.30 plus shipping, or $121.30 + $7.95 = $129.25.

59. **(a)** Find the cost for 3 short-sleeved, solid-color shirts.

$14.75
× 3

$44.25

Based on this subtotal, shipping is $5.95. Find the cost of the 3 monograms.

$4.95
× 3

$14.85

Add these amounts, plus $5.00 for a gift box.

$44.25
 5.95
14.85
+ 5.00

$70.05

The total cost is $70.05.

(b) Subtract the cost of the shirts to find the difference.

$70.05
− 44.25

$25.80

The monograms, gift box, and shipping added $25.80 to the cost of the gift.

5.5 Dividing Signed Decimal Numbers

5.5 Margin Exercises

1. **(a)**
$$\begin{array}{r} 23.4 \\ 4\overline{)93.6} \\ \underline{8} \\ 13 \\ \underline{12} \\ 16 \\ \underline{16} \\ 0 \end{array}$$

Check:
23.4
× 4

93.6

(b)
$$\begin{array}{r} 1.134 \\ 6\overline{)6.804} \\ \underline{6} \\ 0\,8 \\ \underline{6} \\ 20 \\ \underline{18} \\ 24 \\ \underline{24} \\ 0 \end{array}$$

Check:
1.134
× 6

6.804

(c) $\dfrac{278.3}{11}$

$$\begin{array}{r} 25.3 \\ 11\overline{)278.3} \\ \underline{22} \\ 58 \\ \underline{55} \\ 33 \\ \underline{33} \\ 0 \end{array}$$

Check:
25.3
× 11
-----
25 3
253

278.3

(d) $-0.51835 \div 5$

Different signs, quotient is negative.

$$\begin{array}{r} -0.10367 \\ 5\overline{)0.51835} \\ \underline{5} \\ 01 \\ \underline{0} \\ 18 \\ \underline{15} \\ 33 \\ \underline{30} \\ 35 \\ \underline{35} \\ 0 \end{array}$$

Check: 0.10367
× 5

0.51835

150 Chapter 5 Rational Numbers: Positive and Negative Decimals

(e) $-213.45 \div (-15)$

Same signs, quotient is positive.

```
      1 4. 2 3        Check:    14.23
15 ) 2 1 3. 4 5                × 15
     1 5                        71 15
     ─                         142 3
     6 3                       ──────
     6 0                       213.45
     ───
       3 4
       3 0
       ───
         4 5
         4 5
         ───
           0
```

2. (a)
```
     1. 2 8              Check:  1.28
5 ) 6. 4 0  ← Write one zero.    × 5
    5                            ─────
    ─                            6.40
    1 4                          or
    1 0                          6.4
    ───
      4 0
      4 0
      ───
        0
```

(b) $30.87 \div (-14)$

Different signs, quotient is negative.

```
      - 2. 2 0 5
14 ) 3 0. 8 7 0  ← Write one zero.
     2 8
     ───
       2 8
       2 8
       ───
         0 7
           0
           ─
           7 0
           7 0
           ───
             0
```

Check:
```
   2.205
 ×   14
 ──────
   8 820
  22 05
  ──────
  30.870  or 30.87
```

(c) $\dfrac{-259.5}{-30}$

Same signs, quotient is positive.

```
         8. 6 5
30 ) 2 5 9. 5 0  ← Write one zero.
     2 4 0
     ─────
       1 9 5
       1 8 0
       ─────
         1 5 0
         1 5 0
         ─────
             0
```

Check:
```
    8.65
  ×  30
  ──────
  259.50  or 259.5
```

(d) $0.3 \div 8$

```
      0. 0 3 7 5
8 ) 0. 3 0 0 0  ← Write three zeros.
    0 0
    ───
      3 0
      2 4
      ───
        6 0
        5 6
        ───
          4 0
          4 0
          ───
            0
```

Check:
```
   0.0375
 ×     8
 ───────
   0.3000  or 0.3
```

3. (a) $13 \overline{) 2\ 6\ 7.\ 0\ 1}$

$267.01 \div 13 \approx 20.539231$

There are no repeating digits visible on the calculator.

20.539|231 rounds to 20.539.

Check: $(20.539)(13) = 267.007 \approx 267.01$

(b) $6 \overline{) 2\ 0.\ 5}$

$20.5 \div 6 \approx 3.416666$

There is a repeating decimal: $3.41\overline{6}$.

3.416|666 rounds to 3.417.

Check: $(3.417)(6) = 20.502 \approx 20.5$

(c) $\dfrac{10.22}{9} = 10.22 \div 9 \approx 1.135555$

There is a repeating decimal: $1.13\overline{5}$.

1.135|555 rounds to 1.136.

Check: $(1.136)(9) = 10.224 \approx 10.22$

5.5 Dividing Signed Decimal Numbers

(d) $16.15 \div 3 \approx 5.383333$

There is a repeating decimal: $5.38\overline{3}$.

$5.383|333$ rounds to 5.383.

Check: $(5.383)(3) = 16.149 \approx 16.15$

(e) $116.3 \div 11 \approx 10.5727272$

The answer has a repeating decimal that starts repeating in the fourth place: $10.5\overline{72}$.

$10.572|72$ rounds to 10.573.

Check: $(10.573)(11) = 116.303 \approx 116.3$

4. (a)
```
            5. 2
0.2∧ )1. 0∧ 4
      1 0
        0 4
          4
          0
```

(b)
```
              3 0. 1 2
0.06∧ )1. 8 0∧ 7 2
       1 8
         0 0
            0
            0 7
              6
              1 2
              1 2
                0
```

(c)
```
              6 4 0 0
0.005∧ )3 2. 0 0 0∧
        3 0
          2 0
          2 0
            0 0
              0
              0 0
                0
                0
```

(d) $-8.1 \div 0.025$

Different signs, quotient is negative.

```
              − 3 2 4
0.025∧ )8. 1 0 0∧
        7 5
          6 0
          5 0
          1 0 0
          1 0 0
              0
```

(e) $\dfrac{7}{1.3}$

```
              5. 3 8 4   ≈ 5.38
1.3∧ )7. 0∧ 0 0 0
      6 5
        5 0
        3 9
        1 1 0
        1 0 4
            6 0
            5 2
              8
```

(f) $-5.3091 \div (-6.2)$

Same signs, quotient is positive.

```
             0. 8 5 6   ≈ 0.86
6.2∧ )5. 3∧ 0 9 1
      4 9 6
        3 4 9
        3 1 0
          3 9 1
          3 7 2
            1 9
```

5. (a) $42.75 \div 3.8 = 1.125$

Estimate: $40 \div 4 = 10$

The answer is not reasonable.

```
              1 1. 2 5
3.8∧ )4 2. 7∧ 5 0
      3 8
        4 7
        3 8
          9 5
          7 6
          1 9 0
          1 9 0
              0
```

The answer should be 11.25

(b) $807.1 \div 1.76 = 458.580$

Estimate: $800 \div 2 = 400$

The answer is reasonable.

(c) $48.63 \div 52 = 93.519$

Estimate: $50 \div 50 = 1$

The answer is not reasonable.

$$\begin{array}{r} 0.9\,3\,5\,1 \approx 0.935 \\ 52\overline{)48.6\,3\,0\,0} \\ \underline{46\;8} \\ 1\;8\;3 \\ \underline{1\;5\;6} \\ 2\;7\;0 \\ \underline{2\;6\;0} \\ 1\;0\;0 \\ \underline{5\;2} \\ 4\;8 \end{array}$$

The answer should be 0.935.

(d) $9.0584 \div 2.68 = 0.338$

Estimate: $9 \div 3 = 3$

The answer is not reasonable.

$$\begin{array}{r} 3.\,3\,8 \\ 2.68_\wedge\overline{)9.\,0\,5_\wedge\,8\,4} \\ \underline{8\;0\;4} \\ 1\;0\;1\;8 \\ \underline{8\;0\;4} \\ 2\;1\;4\;4 \\ \underline{2\;1\;4\;4} \\ 0 \end{array}$$

The answer should be 3.38.

6. (a) $-4.6 - 0.79 + 1.5^2$ *Exponent*
    $= -4.6 - 0.79 + 2.25$ *Add the opposite.*
    $= -4.6 + (-0.79) + 2.25$ *Add.*
    $= -5.39 + 2.25$ *Add.*
    $= -3.14$

(b) $3.64 \div 1.3(3.6)$ *Divide.*
    $= 2.8(3.6)$ *Multiply.*
    $= 10.08$

(c) $0.08 + 0.6(2.99 - 3)$ *Parentheses first*
    $= 0.08 + 0.6(-0.01)$ *Multiply.*
    $= 0.08 + (-0.006)$ *Add.*
    $= 0.074$

(d) $10.85 - 2.3 \cdot (5.2) \div 3.2$ *Multiply.*
    $= 10.85 - 11.96 \div 3.2$ *Divide.*
    $= 10.85 - 3.7375$ *Subtract.*
    $= 7.1125$

5.5 Section Exercises

1. $27.3 \div (-7) = -3.9$

The numbers have different signs, so the quotient is negative. Divide as if both numbers are whole numbers.

$$\begin{array}{r} 3.\,9 \\ 7\overline{)2\,7.\,3} \\ \underline{2\;1} \\ 6\;3 \\ \underline{6\;3} \\ 0 \end{array}$$ *Line up decimal points.*

3. $\dfrac{4.23}{9} = 0.47$

The numbers have the same sign, so the quotient is positive. Divide as if both numbers are whole numbers.

$$\begin{array}{r} 0.\,4\,7 \\ 9\overline{)4.\,2\,3} \\ \underline{3\;6} \\ 6\;3 \\ \underline{6\;3} \\ 0 \end{array}$$ *Line up decimal points.*

5. $-20.01 \div (-0.05) = 400.2$

The numbers have the same sign, so the quotient is positive.

$$\begin{array}{r} 4\;0\,0.\,2 \\ 0.05_\wedge\overline{)2\,0.\,0\,1_\wedge\,0} \\ \underline{2\;0} \\ 0\;0\;1\;0 \\ \underline{1\;0} \\ 0 \end{array}$$ *Move decimal point in divisor and dividend 2 places; write one zero.*

7. $1.5\overline{)54}$

The numbers have the same sign, so the quotient is positive.

$$\begin{array}{r} 3\;6 \\ 1.5_\wedge\overline{)5\,4.\,0_\wedge} \\ \underline{4\;5} \\ 9\;0 \\ \underline{9\;0} \\ 0 \end{array}$$ *Move decimal point in divisor and dividend 1 place; write 0.*

9. Given: $108 \div 18 = 6$

Find: $1.8\overline{)0.108}$ or $0.108 \div 1.8$

The new dividend, 0.108, has the effect of moving the decimal place in the answer *left* three places. The new divisor, 1.8, has the effect of moving the decimal place in the answer *right* one place. So the new answer has the decimal point moved to the left two places.

$0.108 \div 1.8 = 0.06$

5.5 Dividing Signed Decimal Numbers

11. Given: $108 \div 18 = 6$

Find: $0.018 \overline{)108}$ or $108 \div 0.018$

The new divisor, 0.018, has the effect of moving the decimal place in the answer *right* three places.

$108 \div 0.018 = 6000$

13. Given: $108 \div 18 = 6$

Find: $0.18 \overline{)10.8}$ or $10.8 \div 0.18$

The new dividend, 10.8, has the effect of moving the decimal place in the answer *left* one place. The new divisor, 0.18, has the effect of moving the decimal place in the answer *right* two places. So the new answer has the decimal point moved to the right one place.

$10.8 \div 0.18 = 60$

15. Given: $108 \div 18 = 6$

Find: $18 \overline{)0.0108}$ or $0.0108 \div 18$

The new dividend, 0.108, has the effect of moving the decimal place in the answer *left* four places. So the new answer has the decimal point moved to the left four places.

$0.0108 \div 18 = 0.0006$

17. $4.6 \overline{)116.38}$

```
            2 5 . 3    Line up decimal points.
4.6∧ ) 1 1 6 . 3∧ 8    Move decimal point 1
        9 2            place in dividend and
        2 4 3          divisor.
        2 3 0
          1 3 8
          1 3 8
              0
```

19. $\dfrac{-3.1}{-0.006}$

The numbers have the same sign, so the quotient is positive.

```
              5 1 6 . 6 6 6   Line up decimal points.
0.006∧ ) 3 . 1 0 0∧ 0 0 0
        3 0                   Move decimal point in
          1 0                 divisor and dividend 3
            6                 places.
          4 0
          3 6
            4 0               Write 0 in dividend.
            3 6
              4 0             Write 0 in dividend.
              3 6
                4 0           Write 0 in dividend.
                3 6
                  4           Stop and round answer
                              to the nearest hundredth.
```

516.666 rounds to 516.67.

21. $-240.8 \div 9$

```
        2 6 . 7 5 5 5   Line up decimal points.
9 ) 2 4 0 . 8 0 0 0
    1 8
      6 0
      5 4
        6 8
        6 3
          5 0           Write 0 in dividend.
          4 5
            5 0         Write 0 in dividend.
            4 5
              5 0       Write 0 in dividend.
              4 5
                5       Stop and round answer
                        to the nearest thousandth.
```

The numbers have different signs, so the quotient is negative.

Round -26.7555 to -26.756.

23. $0.034 \overline{)342.81}$

Enter on calculator: $342.81 \;\boxed{\div}\; 0.034 \;\boxed{=}$

Round 10,082.64706 to 10,082.647.

25. $3.77 \div 10 = \underline{0.377}$ $\quad 9.1 \div 10 = \underline{0.91}$

$0.886 \div 10 = \underline{0.0886}$ $\quad 30.19 \div 10 = \underline{3.019}$

$406.5 \div 10 = \underline{40.65}$ $\quad 6625.7 \div 10 = \underline{662.57}$

(a) Dividing by 10, decimal point moves one place to the left; by 100, two places to the left; by 1000, three places to the left.

(b) The decimal point moved to the *right* when multiplying by 10, by 100, or by 1000; here it moves to the *left* when dividing by 10, by 100, or by 1000.

26. $40.2 \div 0.1 = \underline{402}$ $7.1 \div 0.1 = \underline{71}$
$0.339 \div 0.1 = \underline{3.39}$ $15.77 \div 0.1 = \underline{157.7}$
$46 \div 0.1 = \underline{460}$ $873 \div 0.1 = \underline{8730}$

(a) Dividing by 0.1, decimal point moves one place to the right; by 0.01, two places to the right; by 0.001, three places to the right.

(b) The decimal point moved to the *left* when multiplying by 0.1, 0.01, or 0.001; here it moves to the *right* when dividing by 0.1, 0.01, or 0.001.

27. $37.8 \div 8 = 47.25$

Estimate: $40 \div 8 = 5$

The answer 47.25 is *unreasonable*.

```
      4. 7 2 5
  8 ) 3 7. 8 0 0
      3 2
      ───
        5 8
        5 6
        ───
          2 0
          1 6
          ───
            4 0
            4 0
            ───
              0
```

The correct answer is 4.725.

29. $54.6 \div 48.1 \approx 1.135$

Estimate: $50 \div 50 = 1$

The answer 1.135 is *reasonable*.

31. $307.02 \div 5.1 = 6.2$

Estimate: $300 \div 5 = 60$

The answer 6.2 is *unreasonable*.

```
          6 0. 2
  5.1∧) 3 0 7. 0∧ 2
        3 0 6
        ─────
            1 0
                0
            ─────
            1 0 2
            1 0 2
            ─────
                0
```

The correct answer is 60.2.

33. $9.3 \div 1.25 = 0.744$

Estimate: $9 \div 1 = 9$

The answer 0.744 is *unreasonable*.

```
             7. 4 4
  1.25∧) 9. 3 0∧ 0 0
         8 7 5
         ─────
           5 5 0
           5 0 0
           ─────
             5 0 0
             5 0 0
             ─────
                 0
```

The correct answer is 7.44.

35. Divide the cost by the number of pairs of tights.

```
         3. 9 9 6
  6 ) 2 3. 9 8 0
      1 8
      ───
        5 9
        5 4
        ───
          5 8
          5 4
          ───
            4 0
            3 6
            ───
              4
```

$3.996 rounds to $4.00.

One pair costs $4.00 (rounded).

37. Divide the balance by the number of months.

```
            6 7. 0 8
  21 ) 1 4 0 8. 6 8
       1 2 6
       ─────
         1 4 8
         1 4 7
         ─────
             1 6 8
             1 6 8
             ─────
                 0
```

Aimee is paying $67.08 per month.

39. Divide the total cost by the number of bricks to find the cost per brick.

```
              0. 3 0
  619 ) 1 8 5. 7 0
        1 8 5 7
        ───────
              0 0
                  0
              ───
                  0
```

One brick costs $0.30.

41. Divide the total earnings by the number of hours.

$$
\begin{array}{r}
11.92 \\
40\overline{)476.80} \\
\underline{40} \\
76 \\
\underline{40} \\
368 \\
\underline{360} \\
80 \\
\underline{80} \\
0
\end{array}
$$

Darren earns $11.92 per hour.

43. Divide the miles driven by the gallons of gas purchased.

$346.2 \div 16.35 \approx 21.17$

She got 21.2 miles per gallon (rounded).

45. First find the sum of lengths for Jackie Joyner-Kersee.

$$
\begin{array}{r}
7.49 \\
7.45 \\
7.40 \\
7.32 \\
+\,7.20 \\
\hline
36.86
\end{array}
$$

Then divide by the number of jumps, namely 5.

$$
\begin{array}{r}
7.372 \\
5\overline{)36.860} \\
\underline{35} \\
18 \\
\underline{15} \\
36 \\
\underline{35} \\
10 \\
\underline{10} \\
0
\end{array}
$$

7.372 rounds to 7.37.

The average length of the long jumps made by Jackie Joyner-Kersee is 7.37 meters (rounded).

47. Subtract to find the difference.

$$
\begin{array}{rl}
7.40 & \textit{fifth longest jump} \\
-\,7.32 & \textit{sixth longest jump} \\
\hline
0.08 & \textit{difference}
\end{array}
$$

The fifth longest jump was 0.08 meter longer than the sixth longest jump.

49. Add the lengths of the jumps of the top three athletes.

$$
\begin{array}{r}
7.52 \\
7.49 \\
+\,7.48 \\
\hline
22.49
\end{array}
$$

The total length jumped by the top three athletes was 22.49 meters.

51. $7.2 - 5.2 + 3.5^2$ *Exponent*
$= 7.2 - 5.2 + 12.25$ *Subtract.*
$= 2 + 12.25$ *Add.*
$= 14.25$

53. $38.6 + 11.6(10.4 - 13.4)$ *Change to addition.*
$= 38.6 + 11.6[10.4 + (-13.4)]$ *Brackets*
$= 38.6 + 11.6(-3)$ *Multiply.*
$= 38.6 + (-34.8)$ *Add.*
$= 3.8$

55. $-8.68 - 4.6(10.4) \div 6.4$ *Multiply.*
$= -8.68 - 47.84 \div 6.4$ *Divide.*
$= -8.68 - 7.475$ *Subtract.*
$= -16.155$

57. $33 - 3.2(0.68 + 9) + (-1.3)^2$ *Parentheses; Exponent*
$= 33 - 3.2(9.68) + 1.69$ *Multiply.*
$= 33 - 30.976 + 1.69$ *Subtract.*
$= 2.024 + 1.69$ *Add.*
$= 3.714$

59. Multiply the price per can by the number of cans.

$$
\begin{array}{r}
\$0.57 \\
\times\,6 \\
\hline
\$3.42
\end{array}
$$

Subtract to find total savings.

$$
\begin{array}{r}
\$3.42 \\
-\,3.25 \\
\hline
\$0.17
\end{array}
$$

There are six cans, so divide by 6 to find the savings per can.

$$
\begin{array}{r}
0.028 \\
6\overline{)0.170} \\
\underline{12} \\
50 \\
\underline{48} \\
2
\end{array}
$$

$0.028 rounds to $0.03.

You will save $0.03 per can (rounded).

61. (a) $\dfrac{38{,}000{,}000}{24} \approx 1{,}583{,}333$ pieces each hour

(b) There are $24 \times 60 = 1440$ minutes in a day.

$\dfrac{38{,}000{,}000}{1440} \approx 26{,}389$ pieces each minute

(c) There are $24 \times 60 \times 60 = 86{,}400$ seconds in a day.

$\dfrac{38{,}000{,}000}{86{,}400} \approx 440$ pieces each second

63. Divide $10,000 by 10¢.

```
          1 0 0, 0 0 0
0.10∧ ⟌ 1 0, 0 0 0. 0 0∧
       1 0
       ─────
       0 0 0 0 0 0
```

The school would need to collect 100,000 box tops.

65. Divide 100,000 (the answer from Exercise 63) by 38.

```
         2 6 3 1. 5
   38 ⟌ 1 0 0, 0 0 0. 0
        7 6
        ───
         2 4 0
         2 2 8
         ─────
           1 2 0
           1 1 4
           ─────
             6 0
             3 8
             ───
             2 2 0
             1 9 0
             ─────
               3 0
```

2631.5 rounds to 2632.

The school needs to collect 2632 box tops (rounded) during each of the 38 weeks.

Summary Exercises on Decimals

1. $0.8 = \dfrac{8}{10} = \dfrac{8 \div 2}{10 \div 2} = \dfrac{4}{5}$

3. $0.35 = \dfrac{35}{100} = \dfrac{35 \div 5}{100 \div 5} = \dfrac{7}{20}$

5. 2.0003 is two and three ten-thousandths.

7. five hundredths:

$\dfrac{5}{100} = 0.05$

9. ten and seven tenths:

$10\dfrac{7}{10} = 10.7$

11. 0.95 to the nearest tenth

Draw a cut-off line: 0.9|5

The first digit cut is 5, which is 5 or more, so round up.

```
  0.9
+ 0.1
─────
  1.0
```

Answer: ≈ 1.0

13. $0.893 to the nearest cent

Draw a cut-off line: $0.89|3

The first digit cut is 3, which is 4 or less. The part you keep stays the same.

Answer: $\approx \$0.89$

15. $99.64 to the nearest dollar

Draw a cut-off line: $99.|64

The first digit cut is 6, which is 5 or more, so round up.

```
 $99
+  1
────
$100
```

Answer: $\approx \$100$

17. $50 - 0.3801$

```
  4 9 9 9 10
  5̸ 0̸.0̸ 0̸ 0̸    Write four zeros.
- 0.3 8 0 1
───────────
 49.6 1 9 9
```

Check: 0.3801
        + 49.6199
        ─────────
          50.0000

19. $\dfrac{-90.18}{-6}$

Same signs, quotient is positive

```
        1 5. 0 3
   6 ⟌ 9 0. 1 8
      6
      ───
      3 0
      3 0
      ───
        0 1
        0
        ───
         1 8
         1 8
         ───
          0
```

21. $1.55 - 3.7 = 1.55 + (-3.7)$

Subtract absolute values.

$$\begin{array}{r} \overset{6\,10}{3.7\cancel{0}} \\ -1.5\,5 \\ \hline 2.1\,5 \end{array} \quad \textit{Check:} \quad \begin{array}{r} 1.55 \\ +2.15 \\ \hline 3.70 \end{array}$$

Since -3.7 has the larger absolute value, the sum is negative.

$1.55 - 3.7 = -2.15$

23. $3.6 + 0.718 + 9 + 5.0829$

$$\begin{array}{r} \overset{1\,11}{3.6000} \quad \textit{Line up decimal points.} \\ 0.7180 \quad \textit{Write in zeros.} \\ 9.0000 \\ +5.0829 \\ \hline 18.4009 \end{array}$$

25. $-8.9 + 4^2 \div (-0.02)$ *Exponent*
  $= -8.9 + 16 \div (-0.02)$ *Divide.*
  $= -8.9 - 800$ *Subtract.*
  $= -808.9$

27. $0.64 \div 16.3 \approx 0.0392638037 \approx 0.04$

29. To find the perimeter, add the lengths of all the sides.

$$\begin{array}{r} 2.000 \\ 1.000 \\ 1.700 \\ 0.860 \\ 2.095 \\ 1.180 \\ +0.900 \\ \hline 9.735 \text{ meters} \end{array}$$

The perimeter is 9.735 meters.

31. Find their total revenue.

$$\begin{array}{r} \overset{47\,4}{\$15.95} \quad \textit{price per blanket} \\ \times 8 \quad \textit{number of blankets} \\ \hline \$127.60 \quad \textit{total revenue} \end{array}$$

Subtract their cost from the revenue.

$$\begin{array}{r} \overset{510}{\$127.\cancel{6}\cancel{0}} \quad \textit{total revenue} \\ -40.3\,2 \quad \textit{cost} \\ \hline \$87.2\,8 \quad \textit{profit} \end{array}$$

They made a profit of $87.28.

33. (a) $A = s^2$
  $= (12 \text{ ft})(12 \text{ ft})$
  $= 144 \text{ ft}^2$

(b) Cost per square foot:

$$\dfrac{\$99}{144 \text{ ft}^2} = \$0.6875 \approx \$0.69 \text{ (rounded)}$$

35. Multiply 0.004 and 80.

$$\begin{array}{r} 0.004 \\ \times 80 \\ \hline 0.320 \end{array}$$

The average weight of food eaten each day by a queen bee is 0.32 ounce.

5.6 Fractions and Decimals

5.6 Margin Exercises

1. (a) $\frac{1}{9}$ is written $9\overline{)1}$.

(b) $\frac{2}{3}$ is written $3\overline{)2}$.

(c) $\frac{5}{4}$ is written $4\overline{)5}$.

(d) $\frac{3}{10}$ is written $10\overline{)3}$.

(e) $\frac{21}{16}$ is written $16\overline{)21}$.

(f) $\frac{1}{50}$ is written $50\overline{)1}$.

2. (a) $\dfrac{1}{4}$

$$\begin{array}{r} 0.2\,5 \\ 4\overline{)1.0\,0} \\ \underline{8} \\ 2\,0 \\ \underline{2\,0} \\ 0 \end{array}$$

$\dfrac{1}{4} = 0.25$

(b) $2\frac{1}{2} = \frac{5}{2}$

$$\begin{array}{r} 2.\,5 \\ 2\overline{)5.\,0} \\ \underline{4} \\ 1\,0 \\ \underline{1\,0} \\ 0 \end{array}$$

$2\frac{1}{2} = 2.5$

(c) $\dfrac{5}{8}$

$$\begin{array}{r} 0.\,6\,2\,5 \\ 8\overline{)5.\,0\,0\,0} \\ \underline{4\,8} \\ 2\,0 \\ \underline{1\,6} \\ 4\,0 \\ \underline{4\,0} \\ 0 \end{array}$$

$\dfrac{5}{8} = 0.625$

(d) $4\tfrac{3}{5}$

$$\begin{array}{r} 0.6 \\ 5\overline{)3.0} \\ \underline{3\ 0} \\ 0 \end{array}$$

$4 + 0.6 = 4.6$
$4\tfrac{3}{5} = 4.6$

(e) $\tfrac{7}{8}$

$$\begin{array}{r} 0.8\ 7\ 5 \\ 8\overline{)7.0\ 0\ 0} \\ \underline{6\ 4} \\ 6\ 0 \\ \underline{5\ 6} \\ 4\ 0 \\ \underline{4\ 0} \\ 0 \end{array}$$

$\tfrac{7}{8} = 0.875$

3. (a) $\tfrac{1}{3}$

$$\begin{array}{r} 0.3\ 3\ 3\ 3 \\ 3\overline{)1.0\ 0\ 0\ 0} \\ \underline{9} \\ 1\ 0 \\ \underline{9} \\ 1\ 0 \\ \underline{9} \\ 1\ 0 \\ \underline{9} \\ 1 \end{array}$$

Rounded to the nearest thousandth, $\tfrac{1}{3} \approx 0.333$.

(b) $2\tfrac{7}{9} = \tfrac{25}{9}$

$$\begin{array}{r} 2.7\ 7\ 7\ 7 \\ 9\overline{)2\ 5.0\ 0\ 0\ 0} \\ \underline{1\ 8} \\ 7\ 0 \\ \underline{6\ 3} \\ 7\ 0 \\ \underline{6\ 3} \\ 7\ 0 \\ \underline{6\ 3} \\ 7\ 0 \\ \underline{6\ 3} \\ 7 \end{array}$$

Rounded to the nearest thousandth, $2\tfrac{7}{9} \approx 2.778$.

(c) $\tfrac{10}{11}$

$$\begin{array}{r} 0.9\ 0\ 9\ 0 \\ 11\overline{)1\ 0.0\ 0\ 0\ 0} \\ \underline{9\ 9} \\ 1\ 0 \\ \underline{0} \\ 1\ 0\ 0 \\ \underline{9\ 9} \\ 1\ 0 \\ \underline{0} \\ 1\ 0 \end{array}$$

Rounded to the nearest thousandth, $\tfrac{10}{11} \approx 0.909$.

(d) $\tfrac{3}{7}$

$$\begin{array}{r} 0.4\ 2\ 8\ 5 \\ 7\overline{)3.0\ 0\ 0\ 0} \\ \underline{2\ 8} \\ 2\ 0 \\ \underline{1\ 4} \\ 6\ 0 \\ \underline{5\ 6} \\ 4\ 0 \\ \underline{3\ 5} \\ 5 \end{array}$$

Rounded to the nearest thousandth, $\tfrac{3}{7} \approx 0.429$.

(e) $3\tfrac{5}{6} = \tfrac{23}{6}$

$$\begin{array}{r} 3.8\ 3\ 3\ 3 \\ 6\overline{)2\ 3.0\ 0\ 0\ 0} \\ \underline{1\ 8} \\ 5\ 0 \\ \underline{4\ 8} \\ 2\ 0 \\ \underline{1\ 8} \\ 2\ 0 \\ \underline{1\ 8} \\ 2\ 0 \\ \underline{1\ 8} \\ 2 \end{array}$$

Rounded to the nearest thousandth, $3\tfrac{5}{6} \approx 3.833$.

4. (a) On the first number line, 0.4375 is to the *left* of 0.5, so use the $\boxed{<}$ symbol: $0.4375 < 0.5$

(b) On the first number line, 0.75 is to the *right* of 0.6875, so use the $\boxed{>}$ symbol: $0.75 > 0.6875$

(c) $0.625 > 0.0625$

(d) $\tfrac{2}{8} = \tfrac{1}{4} = 0.25$

$0.25 < 0.375$, so $\tfrac{2}{8} < 0.375$.

(e) On the second number line, $0.8\overline{3}$ and $\tfrac{5}{6}$ are at the *same point*, so use the $\boxed{=}$ symbol:

$0.8\overline{3} = \tfrac{5}{6}$

5.6 Fractions and Decimals

(f) $\frac{1}{2} < 0.\overline{5}$ ($\frac{1}{2} = 0.5$)

(g) $0.\overline{1} < 0.1\overline{6}$

(h) $\frac{8}{9} = 0.\overline{8}$

(i) $\frac{4}{6} = \frac{2}{3} = 0.\overline{6}$

$0.\overline{7} > 0.\overline{6}$, so $0.\overline{7} > \frac{4}{6}$.

(j) $\frac{1}{4} = 0.25$

5. (a) 0.7   0.703   0.7029
       ↓   ↓   ↓
       0.7000 0.7030   0.7029

From least to greatest:
0.7000, 0.7029, 0.7030
or
0.7, 0.7029, 0.703

(b) 6.39   6.309   6.401   6.4
    ↓   ↓   ↓   ↓
    6.390 6.309   6.401   6.400

From least to greatest:
6.309, 6.390, 6.400, 6.401
or
6.309, 6.39, 6.4, 6.401

(c) 1.085 $1\frac{3}{4}$ 0.9
    ↓   ↓   ↓
    1.085   1.750   0.900

From least to greatest:
0.900, 1.085, 1.750
or
0.9, 1.085, $1\frac{3}{4}$

(d) $\frac{1}{4}, \frac{2}{5}, \frac{3}{7}, 0.428$

To compare, change fractions to decimals.

$\frac{1}{4} = 0.250$ $\frac{2}{5} = 0.400$ $\frac{3}{7} \approx 0.429$

From least to greatest:
0.250, 0.400, 0.428, 0.429
or
$\frac{1}{4}, \frac{2}{5}, 0.428, \frac{3}{7}$

5.6 Section Exercises

1. $\frac{1}{2} = 0.5$ $\begin{array}{r} 0.5 \\ 2\overline{)1.0} \\ \underline{1\ 0} \\ 0 \end{array}$

3. $\frac{3}{4} = 0.75$ $\begin{array}{r} 0.75 \\ 4\overline{)3.00} \\ \underline{2\ 8} \\ 20 \\ \underline{20} \\ 0 \end{array}$

5. $\frac{3}{10} = 0.3$ $\begin{array}{r} 0.3 \\ 10\overline{)3.0} \\ \underline{30} \\ 0 \end{array}$

7. $\frac{9}{10} = 0.9$ $\begin{array}{r} 0.9 \\ 10\overline{)9.0} \\ \underline{90} \\ 0 \end{array}$

9. $\frac{3}{5} = 0.6$ $\begin{array}{r} 0.6 \\ 5\overline{)3.0} \\ \underline{30} \\ 0 \end{array}$

11. $\frac{7}{8} = 0.875$ $\begin{array}{r} 0.875 \\ 8\overline{)7.000} \\ \underline{64} \\ 60 \\ \underline{56} \\ 40 \\ \underline{40} \\ 0 \end{array}$

13. $2\frac{1}{4} = \frac{9}{4}$ $\begin{array}{r} 2.25 \\ 4\overline{)9.00} \\ \underline{8} \\ 10 \\ \underline{8} \\ 20 \\ \underline{20} \\ 0 \end{array}$

15. $14\frac{7}{10} = 14 + \frac{7}{10} = 14 + 0.7 = 14.7$

17. $3\frac{5}{8} = \frac{29}{8} = 3.625$ $\begin{array}{r} 3.625 \\ 8\overline{)29.000} \\ \underline{24} \\ 50 \\ \underline{48} \\ 20 \\ \underline{16} \\ 40 \\ \underline{40} \\ 0 \end{array}$

19. $6\frac{1}{3} = \frac{19}{3} \approx 6.3333333 \approx 6.333$

21. $\frac{5}{6} \approx 0.8333333 \approx 0.833$

23. $1\frac{8}{9} = \frac{17}{9} \approx 1.8888889 \approx 1.889$

25. (a) In $\frac{5}{9}$, the numerator is less than the denominator, so the decimal must be less than 1 (not 1.8).

   (b) $\frac{5}{9}$ means $5 \div 9$ or $9\overline{)5}$, so the correct answer is 0.556 (rounded). This makes sense because both the fraction and decimal are less than 1.

   ```
       0. 5 5 5 5
   9 ) 5. 0 0 0 0
       4 5
       ─────
         5 0
         4 5
         ─────
           5 0
           4 5
           ─────
             5 0
             4 5
             ───
               5
   ```

26. (a) $2.035 = 2\frac{35}{1000} = 2\frac{7}{200}$, not $2\frac{7}{20}$.

   (b) Adding the whole number part gives $2 + 0.35$, which is 2.35, not 2.035. To check, $2.35 = 2\frac{35}{100} = 2\frac{7}{20}$ but $2.035 = 2\frac{35}{1000} = 2\frac{7}{200}$.

27. Just add the whole number part to 0.375. So $1\frac{3}{8} = 1.375$; $3\frac{3}{8} = 3.375$; $295\frac{3}{8} = 295.375$.

28. It works only when the fraction part has a one-digit numerator and a denominator of 10, or a two-digit numerator and a denominator of 100, and so on.

29. $0.4 = \frac{4}{10} = \frac{4 \div 2}{10 \div 2} = \frac{2}{5}$

31. $0.625 = \frac{625}{1000} = \frac{625 \div 125}{1000 \div 125} = \frac{5}{8}$

33. $0.35 = \frac{35}{100} = \frac{35 \div 5}{100 \div 5} = \frac{7}{20}$

35. $\frac{7}{20} = 0.35$

   ```
          0. 3 5
   20 ) 7. 0 0
        6 0
        ─────
          1 0 0
          1 0 0
          ─────
              0
   ```

37. $0.04 = \frac{4}{100} = \frac{4 \div 4}{100 \div 4} = \frac{1}{25}$

39. $0.15 = \frac{15}{100} = \frac{15 \div 5}{100 \div 5} = \frac{3}{20}$

41. $\frac{1}{5} = 0.2$

   ```
         0. 2
   5 ) 1. 0
       1 0
       ───
         0
   ```

43. $0.09 = \frac{9}{100}$

45. Compare the two lengths.
    average length → 20.80 *longer*
    Charlene's baby → 20.08 *shorter*

    ```
      20.80
    − 20.08
    ───────
       0.72
    ```

    Her baby is 0.72 inch *shorter* than the average length.

47. Write two zeros to the right of 0.5 so it has the same number of decimal places as 0.505. Then you can compare the numbers: $0.505 > 0.500$. There was too much calcium in each capsule. Subtract to find the difference.

    ```
      0.505
    − 0.500
    ───────
      0.005
    ```

    There was 0.005 gram too much.

49. Write two zeros so that all the numbers have four decimal places.

    $1.0100 > 1.0020$ *unacceptable*
    $0.9991 > 0.9980$ and $0.9991 < 1.0020$ *acceptable*
    $1.0007 > 0.9980$ and $1.0007 < 1.0020$ *acceptable*
    $0.9900 < 0.9980$ *unacceptable*

    The lengths of 0.9991 cm and 1.0007 cm are acceptable.

51. Compare the two amounts.

    $3\frac{3}{4} \rightarrow 3.75$ *less*
    $3.8 \rightarrow 3.80$ *more*

    ```
      3.80
    − 3.75
    ──────
      0.05
    ```

    3.8 inches is 0.05 inch *more* than Ginny had hoped for.

53. (a) On the first number line, 0.3125 is to the *left* of 0.375, so use the $\boxed{<}$ symbol: $0.3125 < 0.375$

   (b) Write $\frac{6}{8}$ in lowest terms as $\frac{3}{4}$. On the first number line, $\frac{3}{4}$ and 0.75 are at the *same point*, so use the $\boxed{=}$ symbol: $\frac{3}{4} = 0.75$

   (c) On the second number line, $0.\overline{8}$ is to the *right* of $0.8\overline{3}$, so use the $\boxed{>}$ symbol: $0.\overline{8} > 0.8\overline{3}$

   (d) On the second number line, 0.5 is to the *left* of $\frac{5}{9}$, so use the $\boxed{<}$ symbol: $0.5 < \frac{5}{9}$

55. 0.54, 0.5455, 0.5399

$\quad$ 0.54 = 0.5400
$\quad$ 0.5455 = 0.5455 $\quad$ *greatest*
$\quad$ 0.5399 = 0.5399 $\quad$ *least*

From least to greatest: 0.5399, 0.54, 0.5455

57. 5.8, 5.79, 5.0079, 5.804

$\quad$ 5.8 = 5.8000
$\quad$ 5.79 = 5.7900
$\quad$ 5.0079 = 5.0079 $\quad$ *least*
$\quad$ 5.804 = 5.8040 $\quad$ *greatest*

From least to greatest: 5.0079, 5.79, 5.8, 5.804

59. 0.628, 0.62812, 0.609, 0.6009

$\quad$ 0.628 = 0.62800
$\quad$ 0.62812 = 0.62812 $\quad$ *greatest*
$\quad$ 0.609 = 0.60900
$\quad$ 0.6009 = 0.60090 $\quad$ *least*

From least to greatest:
0.6009, 0.609, 0.628, 0.62812

61. 5.8751, 4.876, 2.8902, 3.88

The numbers after the decimal places are irrelevant since the whole number parts are all different.

From least to greatest:
2.8902, 3.88, 4.876, 5.8751

63. 0.043, 0.051, 0.006, $\frac{1}{20}$

$\quad$ 0.043 = 0.043
$\quad$ 0.051 = 0.051 $\quad$ *greatest*
$\quad$ 0.006 = 0.006 $\quad$ *least*
$\quad$ $\frac{1}{20}$ = 0.050

From least to greatest: 0.006, 0.043, $\frac{1}{20}$, 0.051

65. $\frac{3}{8}$, $\frac{2}{5}$, 0.37, 0.4001

$\quad$ $\frac{3}{8}$ = 0.3750
$\quad$ $\frac{2}{5}$ = 0.4000
$\quad$ 0.37 = 0.3700 $\quad$ *least*
$\quad$ 0.4001 = 0.4001 $\quad$ *greatest*

From least to greatest: 0.37, $\frac{3}{8}$, $\frac{2}{5}$, 0.4001

67. (a) Find the greatest of:
0.018, 0.01, 0.008, 0.010

$\quad$ 0.018 = 0.018 $\quad$ *greatest*
$\quad$ 0.01 = 0.010
$\quad$ 0.008 = 0.008 $\quad$ *least*
$\quad$ 0.010 = 0.010

List from least to greatest:

0.008, 0.01 = 0.010, 0.018

The red box, labeled 0.018 in. diameter, has the strongest line.

The green box, labeled 0.008 in. diameter, has the line with the least strength.

(b) Subtract 0.008 from 0.018.

$\quad$ 0.018
$\quad$ − 0.008
$\quad$ ———
$\quad$ 0.010

The difference in diameter is 0.01 inch.

69. $1\frac{7}{16} = \frac{23}{16}$

$$\begin{array}{r} 1.43 \\ 16\overline{)23.00} \\ \underline{16} \\ 70 \\ \underline{64} \\ 60 \\ \underline{48} \\ 12 \end{array}$$

1.43 rounded to the nearest tenth is 1.4.
Length (a) is 1.4 inch (rounded).

71. $\frac{1}{4} = 0.25$

0.25 rounded to the nearest tenth is 0.3.
Length (c) is 0.3 inch (rounded).

73. $\frac{3}{8}$

$$\begin{array}{r} 0.37 \\ 8\overline{)3.00} \\ \underline{24} \\ 60 \\ \underline{56} \\ 4 \end{array}$$

0.37 rounded to the nearest tenth is 0.4.
Length (e) is 0.4 inch (rounded).

5.7 Problem Solving with Statistics: Mean, Median, and Mode

5.7 Margin Exercises

1. $$\text{mean} = \frac{\text{sum of all values}}{\text{number of values}}$$
   $$= \frac{95 + 91 + 81 + 78 + 81 + 90}{6}$$
   $$= \frac{516}{6}$$
   $$= 86$$

2. (a) $25.12 + $42.58 + $76.19 + $32 + $81.11 + $26.41 + $49.76 + $59.32 + $71.18 + $30.09 + $60.50 + $79.84 = $634.10

   $$\text{mean} = \frac{\$634.10}{12} \approx \$52.84$$

   His average monthly cell phone bill was $52.84.

   (b) $749,820 + $765,480 + $643,744 + $824,222 + $485,886 + $668,178 + $702,294 + $525,800 = $5,365,424

   $$\text{mean} = \frac{\$5,365,424}{8} = \$670,678$$

   The average sales for each office supply store were $670,678.

3. | Parking Fee | Frequency | Product |
   |---|---|---|
   | $6 | 2 | $12 |
   | $7 | 6 | $42 |
   | $8 | 3 | $24 |
   | $9 | 4 | $36 |
   | $10 | 6 | $60 |
   | | 21 | $174 |

   $$\text{weighted mean} = \frac{\$174}{21} \approx \$8.29$$

   Her average daily parking cost was about $8.29.

4. | Course | Credits | Grade | Credits · Grade |
   |---|---|---|---|
   | Math | 5 | A (= 4) | 5 · 4 = 20 |
   | English | 3 | C (= 2) | 3 · 2 = 6 |
   | Biology | 4 | B (= 3) | 4 · 3 = 12 |
   | History | 3 | B (= 3) | 3 · 3 = 9 |
   | | 15 | | 47 |

   $$\text{GPA} = \frac{\text{sum of Credits} \cdot \text{Grade}}{\text{total number of credits}}$$
   $$= \frac{47}{15} \approx 3.13 \text{ (rounded)}$$

5. Arrange the 9 numbers in numerical order from least to greatest.

   25, 27, 30, 30, 33, 35, 39, 50, 59

   The middle number is the fifth number, which is 33. The median is 33 customers.

6. Arrange in numerical order.

   121, 121, 126, 178, 189, 195, 200, 261

   The middle values are 178 and 189. The median is the mean of the two middle values.

   $$\text{median} = \frac{178 \text{ ft} + 189 \text{ ft}}{2} = 183.5 \text{ feet}$$

7. (a) <u>28</u>, 21, 16, 22, <u>28</u>, 34, 22, <u>28</u>, 19, 18

   The only number that occurs three times, which is more than any other number, is 28, so the mode is 28 years.

   (b) <u>312</u>, <u>219</u>, 782, <u>312</u>, <u>219</u>, 426, 507, 600

   Because both 219 and 312 occur two times, which is more than any other number, each is a mode. This list is *bimodal*.

   (c) $1706, $1289, $1653, $1892, $1301, $1782, $1450, $1566

   No number occurs more than once. This list has *no mode*.

5.7 Section Exercises

1. $$\text{mean} = \frac{\text{sum of all values}}{\text{number of values}}$$
   $$= \frac{92 + 51 + 59 + 86 + 68 + 73 + 49 + 80}{8}$$
   $$= \frac{558}{8} = 69.75$$

   The mean (average) final exam score was 69.8 (rounded).

3. $$\text{mean} = \frac{\text{sum of all values}}{\text{number of values}}$$
   $$= \frac{(\$31,900 + 32,850 + 34,930 + 39,712 + 38,340 + 60,000)}{6}$$
   $$= \frac{\$237,732}{6} = \$39,622$$

   The mean (average) annual salary was $39,622.

5. mean = $\dfrac{\text{sum of all values}}{\text{number of values}}$

$= \dfrac{(\$75.52 + 36.15 + 58.24 + 21.86 + 47.68 + 106.57 + 82.72 + 52.14 + 28.60 + 72.92)}{10}$

$= \dfrac{\$582.40}{10} = \$58.24$

The mean (average) shoe sales amount was $58.24.

7.

Quiz Score	Frequency	Product
3	4	12
5	2	10
6	5	30
8	5	40
9	2	18
Totals	18	110

weighted mean = $\dfrac{\text{sum of products}}{\text{total number of quizzes}}$

$= \dfrac{110}{18} = 6.\overline{1}$

The mean (average) quiz score was 6.1 (rounded).

9.

Course	Credits	Grade	Credits · Grade
Biology	4	B (= 3)	4·3 = 12
Biology Lab	2	A (= 4)	2·4 = 8
Mathematics	5	C (= 2)	5·2 = 10
Health	1	F (= 0)	1·0 = 0
Psychology	3	B (= 3)	3·3 = 9
Totals	15		39

GPA = $\dfrac{\text{sum of Credits · Grade}}{\text{total number of credits}}$

$= \dfrac{39}{15} = 2.60$

11. **(a)** In Exercise 9, replace 1·0 with 1·3 to get GPA = $\dfrac{42}{15} = 2.80$.

(b) In Exercise 9, replace 5·2 with 5·3 to get GPA = $\dfrac{44}{15} = 2.9\overline{3} = 2.93$ (rounded).

(c) Making both of those changes gives us GPA = $\dfrac{47}{15} = 3.1\overline{3} = 3.13$ (rounded).

13. The numbers are already arranged in numerical order from least to greatest.

9, 12, 14, 15, 23, 24, 28

The list has 7 numbers. The middle number is the 4th number, so the median is 15 messages.

15. Arrange the numbers in numerical order from least to greatest.

328, 420, 483, 549, 592, 715

The list has 6 numbers. The middle numbers are the 3rd and 4th numbers, so the median is

$\dfrac{483 + 549}{2} = 516$ students.

17. Arrange the numbers in numerical order from least to greatest.

34, 40, 40, 47, 48, 49, 51, 56, 95, 96

The list has 10 numbers. The middle numbers are the 5th and 6th numbers, so the median is

$\dfrac{48 + 49}{2} = 48.5$ pounds of shrimp.

19. mean = $\dfrac{\text{sum of all values}}{\text{number of values}}$

$= \dfrac{7650 + 6450 + 1100 + 5225 + 1550 + 2875}{6}$

$= \dfrac{24{,}850}{6} = 4141.\overline{6}$

The mean distance flown without refueling was 4142 miles (rounded).

21. 1100, 1550, 2875, 5225, 6450, 7650

There are 6 distances. The middle numbers are the 3rd and 4th numbers, so the median is

$\dfrac{2875 + 5225}{2} = 4050$ miles.

23. 3, <u>8</u>, 5, 1, 7, 6, <u>8</u>, 4, 5, <u>8</u>

The number 8 occurs three times, which is more often than any other number. Therefore, 8 samples is the mode.

25. <u>74</u>, <u>68</u>, <u>68</u>, <u>68</u>, 75, 75, <u>74</u>, <u>74</u>, 70, 77

Because both 68 and 74 years occur three times, which is more often than any other values, each is a mode. This list is *bimodal*.

27. 5, 9, 17, 3, 2, 8, 19, 1, 4, 20, 10, 6

No number occurs more than once. This list has *no mode*.

164 Chapter 5 Rational Numbers: Positive and Negative Decimals

29. **(i)** Barrow's

$$\text{mean} = \frac{-2 + (-11) + (-13) + (-18) + (-15) + (-2)}{6}$$

$$= \frac{-61}{6} \approx -10°F$$

Fairbank's

$$\text{mean} = \frac{3 + (-7) + (-10) + (-4) + 11 + 31}{6}$$

$$= \frac{24}{6} = 4°F$$

(ii) Find the difference: $4 - (-10) = 14$

Fairbank's mean is 14 degrees warmer than Barrow's mean.

5.8 Geometry Applications: Pythagorean Theorem and Square Roots

5.8 Margin Exercises

1. **(a)** $\sqrt{36} = 6$ because $6 \cdot 6 = 36$.

(b) $\sqrt{25} = 5$ because $5 \cdot 5 = 25$.

(c) $\sqrt{9} = 3$ because $3 \cdot 3 = 9$.

(d) $\sqrt{100} = 10$ because $10 \cdot 10 = 100$.

(e) $\sqrt{121} = 11$ because $11 \cdot 11 = 121$.

2. **(a)** $\sqrt{11}$
Calculator shows 3.31662479; round to 3.317.

(b) $\sqrt{40}$
Calculator shows 6.32455532; round to 6.325.

(c) $\sqrt{56}$
Calculator shows 7.48331477; round to 7.483.

(d) $\sqrt{196}$
Calculator shows 14; $\sqrt{196} = 14$ because $14 \cdot 14 = 196$.

(e) $\sqrt{147}$
Calculator shows 12.12435565; round to 12.124.

3. **(a)** legs: 5 in. and 12 in.

$$\text{hypotenuse} = \sqrt{(\text{leg})^2 + (\text{leg})^2}$$
$$= \sqrt{(5)^2 + (12)^2}$$
$$= \sqrt{25 + 144}$$
$$= \sqrt{169}$$
$$= 13 \text{ in.}$$

(b) hypotenuse: 25 cm, leg: 7 cm

$$\text{leg} = \sqrt{(\text{hypotenuse})^2 - (\text{leg})^2}$$
$$= \sqrt{(25)^2 - (7)^2}$$
$$= \sqrt{625 - 49}$$
$$= \sqrt{576}$$
$$= 24 \text{ cm}$$

(c) legs: 17 m and 13 m

$$\text{hypotenuse} = \sqrt{(\text{leg})^2 + (\text{leg})^2}$$
$$= \sqrt{(17)^2 + (13)^2}$$
$$= \sqrt{289 + 169}$$
$$= \sqrt{458}$$
$$\approx 21.4 \text{ m}$$

(d) hypotenuse: 20 ft, leg: 18 ft

$$\text{leg} = \sqrt{(\text{hypotenuse})^2 - (\text{leg})^2}$$
$$= \sqrt{(20)^2 - (18)^2}$$
$$= \sqrt{400 - 324}$$
$$= \sqrt{76}$$
$$\approx 8.7 \text{ ft}$$

(e) legs: 8 mm and 5 mm

$$\text{hypotenuse} = \sqrt{(\text{leg})^2 + (\text{leg})^2}$$
$$= \sqrt{(8)^2 + (5)^2}$$
$$= \sqrt{64 + 25}$$
$$= \sqrt{89}$$
$$\approx 9.4 \text{ mm}$$

4. **(a)** hypotenuse: 25 ft, leg: 20 ft

$$\text{leg} = \sqrt{(\text{hypotenuse})^2 - (\text{leg})^2}$$
$$= \sqrt{(25)^2 - (20)^2}$$
$$= \sqrt{625 - 400}$$
$$= \sqrt{225}$$
$$= 15$$

The bottom of the ladder is 15 ft from the building.

(b) legs: 11 ft and 8 ft

$$\text{hypotenuse} = \sqrt{(\text{leg})^2 + (\text{leg})^2}$$
$$= \sqrt{(11)^2 + (8)^2}$$
$$= \sqrt{121 + 64}$$
$$= \sqrt{185}$$
$$\approx 13.6 \text{ ft}$$

The ladder is about 13.6 ft long.

(c) hypotenuse: 17 ft, leg: 10 ft

$$\text{leg} = \sqrt{(\text{hypotenuse})^2 - (\text{leg})^2}$$
$$= \sqrt{(17)^2 - (10)^2}$$
$$= \sqrt{289 - 100}$$
$$= \sqrt{189}$$
$$\approx 13.7 \text{ ft}$$

The ladder will reach about 13.7 ft high up on the building.

5.8 Section Exercises

1. $\sqrt{16} = 4$ because $4 \cdot 4 = 16$.

3. $\sqrt{64} = 8$ because $8 \cdot 8 = 64$.

5. $\sqrt{11}$
   Calculator shows 3.31662479; round to 3.317.

7. $\sqrt{5}$
   Calculator shows 2.236067977; rounds to 2.236.

9. $\sqrt{73}$
   Calculator shows 8.544003745; rounds to 8.544.

11. $\sqrt{101}$
    Calculator shows 10.04987562; rounds to 10.050.

13. On a calculator, $\sqrt{361} = 19$.

15. $\sqrt{1000}$
    Calculator shows 31.6227766; rounds to 31.623.

17. 30 is about halfway between 25 and 36, so $\sqrt{30}$ should be about halfway between 5 and 6, or about 5.5. Using a calculator, $\sqrt{30} \approx 5.477$. Similarly, $\sqrt{26}$ should be a little more than $\sqrt{25}$; by calculator, $\sqrt{26} \approx 5.099$. And $\sqrt{35}$ should be a little less than $\sqrt{36}$; by calculator, $\sqrt{35} \approx 5.916$.

19. legs: 15 ft and 36 ft
    $$\text{hypotenuse} = \sqrt{(\text{leg})^2 + (\text{leg})^2}$$
    $$= \sqrt{(15)^2 + (36)^2}$$
    $$= \sqrt{225 + 1296}$$
    $$= \sqrt{1521}$$
    $$= 39 \text{ ft}$$

21. legs: 8 in. and 15 in.
    $$\text{hypotenuse} = \sqrt{(\text{leg})^2 + (\text{leg})^2}$$
    $$= \sqrt{(8)^2 + (15)^2}$$
    $$= \sqrt{64 + 225}$$
    $$= \sqrt{289}$$
    $$= 17 \text{ in.}$$

23. hypotenuse: 20 mm, leg: 16 mm
    $$\text{leg} = \sqrt{(\text{hypotenuse})^2 - (\text{leg})^2}$$
    $$= \sqrt{(20)^2 - (16)^2}$$
    $$= \sqrt{400 - 256}$$
    $$= \sqrt{144}$$
    $$= 12 \text{ mm}$$

25. legs: 8 in. and 3 in.
    $$\text{hypotenuse} = \sqrt{(\text{leg})^2 + (\text{leg})^2}$$
    $$= \sqrt{(8)^2 + (3)^2}$$
    $$= \sqrt{64 + 9}$$
    $$= \sqrt{73}$$
    $$\approx 8.5 \text{ in.}$$

27. legs: 7 yd and 4 yd
    $$\text{hypotenuse} = \sqrt{(\text{leg})^2 + (\text{leg})^2}$$
    $$= \sqrt{(7)^2 + (4)^2}$$
    $$= \sqrt{49 + 16}$$
    $$= \sqrt{65}$$
    $$\approx 8.1 \text{ yd}$$

29. hypotenuse: 22 cm, leg: 17 cm
    $$\text{leg} = \sqrt{(\text{hypotenuse})^2 - (\text{leg})^2}$$
    $$= \sqrt{(22)^2 - (17)^2}$$
    $$= \sqrt{484 - 289}$$
    $$= \sqrt{195}$$
    $$\approx 14.0 \text{ cm}$$

31. legs: 1.3 m and 2.5 m
    $$\text{hypotenuse} = \sqrt{(\text{leg})^2 + (\text{leg})^2}$$
    $$= \sqrt{(1.3)^2 + (2.5)^2}$$
    $$= \sqrt{1.69 + 6.25}$$
    $$= \sqrt{7.94}$$
    $$\approx 2.8 \text{ m}$$

33. hypotenuse: 11.5 cm, leg: 8.2 cm
    $$\text{leg} = \sqrt{(\text{hypotenuse})^2 - (\text{leg})^2}$$
    $$= \sqrt{(11.5)^2 - (8.2)^2}$$
    $$= \sqrt{132.25 - 67.24}$$
    $$= \sqrt{65.01}$$
    $$\approx 8.1 \text{ cm}$$

35. hypotenuse: 21.6 km, leg: 13.2 km

$$\begin{aligned}\text{leg} &= \sqrt{(\text{hypotenuse})^2 - (\text{leg})^2} \\ &= \sqrt{(21.6)^2 - (13.2)^2} \\ &= \sqrt{466.56 - 174.24} \\ &= \sqrt{292.32} \\ &\approx 17.1 \text{ km}\end{aligned}$$

37. legs: 4 ft and 7 ft

$$\begin{aligned}\text{hypotenuse} &= \sqrt{(\text{leg})^2 + (\text{leg})^2} \\ &= \sqrt{(4)^2 + (7)^2} \\ &= \sqrt{16 + 49} \\ &= \sqrt{65} \\ &\approx 8.1 \text{ ft}\end{aligned}$$

The length of the loading ramp is about 8.1 ft.

39. hypotenuse: 1000 m, leg: 800 m

$$\begin{aligned}\text{leg} &= \sqrt{(\text{hypotenuse})^2 - (\text{leg})^2} \\ &= \sqrt{(1000)^2 - (800)^2} \\ &= \sqrt{1{,}000{,}000 - 640{,}000} \\ &= \sqrt{360{,}000} \\ &= 600 \text{ m}\end{aligned}$$

The airplane is 600 meters above the ground.

41. legs: 6.5 ft and 2.5 ft

$$\begin{aligned}\text{hypotenuse} &= \sqrt{(\text{leg})^2 + (\text{leg})^2} \\ &= \sqrt{(6.5)^2 + (2.5)^2} \\ &= \sqrt{42.25 + 6.25} \\ &= \sqrt{48.5} \\ &\approx 6.96 \text{ ft}\end{aligned}$$

The diagonal brace is about 7.0 ft long.

43.

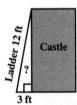

$$\begin{aligned}\text{leg} &= \sqrt{(\text{hypotenuse})^2 - (\text{leg})^2} \\ &= \sqrt{(12)^2 - (3)^2} \\ &= \sqrt{144 - 9} \\ &= \sqrt{135} \\ &\approx 11.6 \text{ ft}\end{aligned}$$

The ladder will reach about 11.6 ft high on the building.

45. The student used the formula for finding the hypotenuse but the unknown side is a leg, so $\text{leg} = \sqrt{(20)^2 - (13)^2}$. Also, the final answer should be m, not m^2. The correct answer is $\sqrt{231} \approx 15.2$ m.

47. legs: 90 ft and 90 ft

$$\begin{aligned}\text{hypotenuse} &= \sqrt{(\text{leg})^2 + (\text{leg})^2} \\ &= \sqrt{(90)^2 + (90)^2} \\ &= \sqrt{8100 + 8100} \\ &= \sqrt{16{,}200} \\ &\approx 127.3 \text{ ft}\end{aligned}$$

The distance from home plate to second base is about 127.3 ft.

48. (a)

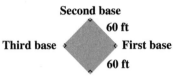

(b) legs: 60 ft and 60 ft

$$\begin{aligned}\text{hypotenuse} &= \sqrt{(\text{leg})^2 + (\text{leg})^2} \\ &= \sqrt{(60)^2 + (60)^2} \\ &= \sqrt{3600 + 3600} \\ &= \sqrt{7200} \\ &\approx 84.9 \text{ ft}\end{aligned}$$

The distance from home plate to second base is about 84.9 ft.

49. The distance from third to first is the same as the distance from home to second because the baseball diamond is a square.

50. (a) Since 80 ft < 84.9 ft, then the side length is less than 60 ft.

(b) $80^2 = 6400$

$\dfrac{6400}{2} = 3200$

$\sqrt{3200} \approx 56.6$

The side length of each side is about 56.6 ft.

5.9 Problem Solving: Equations Containing Decimals

5.9 Margin Exercises

1. (a) $8.1 = h + 9$ To get h by itself, add -9 to both sides.

$$\begin{aligned}8.1 &= h + 9 \\ -9 & -9 \\ \hline -0.9 &= h + 0 \\ -0.9 &= h\end{aligned}$$

Check: $8.1 = h + 9$

$8.1 = -0.9 + 9$

$8.1 = 8.1$ Balances

The solution is -0.9.

5.9 Problem Solving: Equations Containing Decimals

(b) $-0.75 + y = 0$ Add 0.75 to both sides.
$$\begin{array}{r} +0.75 \quad\quad +0.75 \\ \hline 0 + y = 0.75 \\ y = 0.75 \end{array}$$

Check: $-0.75 + y = 0$
$-0.75 + 0.75 = 0$
$0 = 0$ Balances

The solution is 0.75.

(c) $c - 6.8 = -4.8$
$c + (-6.8) = -4.8$
$$\begin{array}{r} +6.8 \quad\quad +6.8 \\ \hline c + 0 = 2 \\ c = 2 \end{array}$$

Check: $c - 6.8 = -4.8$
$2 - 6.8 = -4.8$
$2 + (-6.8) = -4.8$
$-4.8 = -4.8$ Balances

The solution is 2.

2. (a) $-3y = -0.63$ Divide both sides by the coefficient, -3.

$\dfrac{-3y}{-3} = \dfrac{-0.63}{-3}$

$\dfrac{\overset{1}{-\cancel{3}} \cdot y}{\underset{1}{-\cancel{3}}} = 0.21$ On the left, divide out the common factor of -3.

$y = 0.21$

Check: $-3y = -0.63$
$-3(0.21) = -0.63$
$-0.63 = -0.63$ Balances

The solution is 0.21.

(b) $2.25r = -18$ Divide both sides by 2.25.
$\dfrac{2.25r}{2.25} = \dfrac{-18}{2.25}$
$r = -8$

Check: $2.25r = -18$
$2.25(-8) = -18$
$-18 = -18$ Balances

The solution is -8.

(c) $1.7 = 0.5n$
$\dfrac{1.7}{0.5} = \dfrac{0.5n}{0.5}$
$3.4 = n$

Check: $1.7 = 0.5n$
$1.7 = 0.5(3.4)$
$1.7 = 1.7$ Balances

The solution is 3.4.

3. (a) $4 = 0.2c - 2.6$
$4 = 0.2c + (-2.6)$ Add 2.6 to both sides.
$$\begin{array}{r} +2.6 \quad\quad\quad +2.6 \\ \hline 6.6 = 0.2c + 0 \end{array}$$
Divide both sides by 0.2.
$\dfrac{6.6}{0.2} = \dfrac{0.2c}{0.2}$
$33 = c$

Check: $4 = 0.2c - 2.6$
$4 = 0.2(33) - 2.6$
$4 = 6.6 - 2.6$
$4 = 4$ Balances

The solution is 33.

(b) $3.1k - 4 = 0.5k + 13.42$ Add $-0.5k$ to both sides.
$$\begin{array}{r} -0.5k \quad\quad -0.5k \\ \hline 2.6k - 4 = 0 + 13.42 \end{array}$$
$2.6k + (-4) = 13.42$ Add 4 to both sides.
$$\begin{array}{r} +4 \quad\quad +4 \\ \hline 2.6k + 0 = 17.42 \end{array}$$
$2.6k = 17.42$ Divide both sides by 2.6.
$\dfrac{2.6k}{2.6} = \dfrac{17.42}{2.6}$
$k = 6.7$

Check: $3.1k - 4 = 0.5k + 13.42$
$3.1(6.7) - 4 = 0.5(6.7) + 13.42$
$16.77 = 16.77$ Balances

The solution is 6.7.

(c) $-2y + 3 = 3y - 6$ Add $2y$ to both sides.
$$\begin{array}{r} +2y \quad\quad +2y \\ \hline 0 + 3 = 5y - 6 \end{array}$$
$3 = 5y + (-6)$ Add 6 to both sides.
$$\begin{array}{r} +6 \quad\quad +6 \\ \hline 9 = 5y + 0 \end{array}$$
Divide both sides by 5.
$\dfrac{9}{5} = \dfrac{5y}{5}$
$1.8 = y$

Check: $-2y + 3 = 3y - 6$
$-2(1.8) + 3 = 3(1.8) - 6$
$-3.6 + 3 = 5.4 - 6$
$-0.6 = -0.6$ Balances

The solution is 1.8.

Chapter 5 Rational Numbers: Positive and Negative Decimals

4. *Step 1*
It is about the cost of a telephone call.

Unknown: number of minutes the call lasted

Known: Costs are $1.34 connection fee plus $2.69 per minute; total cost was $39.

Step 2
There is only one unknown; so let m be the number of minutes.

Step 3

cost per minute		number of minutes		connection fee		total cost
2.69	·	m	+	1.34	=	39

Step 4
$$2.69m + 1.34 = 39$$
$$ -1.34 -1.34$$
$$2.69m + 0 = 37.66$$
$$\frac{2.69m}{2.69} = \frac{37.66}{2.69}$$
$$m = 14$$

Step 5
The call lasted 14 minutes.

Step 6
$2.69 per minute times 14 minutes = $37.66

$37.66 plus $1.34 connection fee = $39

Maureen was billed $39. (matches)

5.9 Section Exercises

1. $h + 0.63 = 5.1$ To get h by itself, add -0.63 to both sides.
$$ -0.63 -0.63$$
$$h = 4.47$$

Check: $h + 0.63 = 5.1$
$4.47 + 0.63 = 5.1$
$5.1 = 5.1$ Balances

The solution is 4.47.

3. $-20.6 + n = -22$ Add 20.6 to both sides.
$$+20.6 +20.6$$
$$n = -1.4$$

Check: $-20.6 + n = -22$
$-20.6 + (-1.4) = -22$
$-22 = -22$ Balances

The solution is -1.4.

5. $0 = b - 0.008$ Add 0.008 to both sides.
$$+0.008 +0.008$$
$$0.008 = b$$

Check: $0 = b - 0.008$
$0 = 0.008 - 0.008$
$0 = 0$ Balances

The solution is 0.008.

7. $2.03 = 7a$ Divide both sides by 7.
$$\frac{2.03}{7} = \frac{7a}{7}$$
$$0.29 = a$$

Check: $2.03 = 7a$
$2.03 = 7(0.29)$
$2.03 = 2.03$ Balances

The solution is 0.29.

9. $0.8p = -96$ Divide both sides by 0.8.
$$\frac{0.8p}{0.8} = \frac{-96}{0.8}$$
$$p = -120$$

Check: $0.8p = -96$
$0.8(-120) = -96$
$-96 = -96$ Balances

The solution is -120.

11. $-3.3t = -2.31$ Divide both sides by -3.3.
$$\frac{-3.3t}{-3.3} = \frac{-2.31}{-3.3}$$
$$t = 0.7$$

Check: $-3.3t = -2.31$
$-3.3(0.7) = -2.31$
$-2.31 = -2.31$ Balances

The solution is 0.7.

13. $7.5x + 0.15 = -6$ Add -0.15 to both sides.
$$ -0.15 -0.15$$
$$7.5x = -6.15$$ Divide both sides by 7.5.
$$\frac{7.5x}{7.5} = \frac{-6.15}{7.5}$$
$$x = -0.82$$

Check: $7.5x + 0.15 = -6$
$7.5(-0.82) + 0.15 = -6$
$-6.15 + 0.15 = -6$
$-6 = -6$ Balances

The solution is -0.82.

15. $-7.38 = 2.05z - 7.38$ Add 7.38 to both sides.

$$\underline{+7.38 \qquad\qquad +7.38}$$
$$0 = 2.05z \quad \text{Divide both sides by 2.05.}$$
$$\frac{0}{2.05} = \frac{2.05z}{2.05}$$
$$0 = z$$

Check: $-7.38 = 2.05z - 7.38$
$-7.38 = 2.05(0) - 7.38$
$-7.38 = 0 - 7.38$
$-7.38 = -7.38$ Balances

The solution is 0.

17. $3c + 10 = 6c + 8.65$ Add $-3c$ to both sides.

$$\underline{-3c \qquad\qquad -3c}$$
$$10 = 3c + 8.65 \quad \text{Add } -8.65 \text{ to both sides.}$$
$$\underline{-8.65 \qquad -8.65}$$
$$1.35 = 3c \quad \text{Divide both sides by 3.}$$
$$\frac{1.35}{3} = \frac{3c}{3}$$
$$0.45 = c$$

Check: $3c + 10 = 6c + 8.65$
$3(0.45) + 10 = 6(0.45) + 8.65$
$1.35 + 10 = 2.7 + 8.65$
$11.35 = 11.35$ Balances

The solution is 0.45.

19. $0.8w - 0.4 = -6 + w$ Add $-0.8w$ to both sides.

$$\underline{-0.8w \qquad\qquad -0.8w}$$
$$-0.4 = -6 + 0.2w \quad \text{Add 6 to both sides.}$$
$$\underline{+6 \qquad\qquad +6}$$
$$5.6 = 0.2w \quad \text{Divide both sides by 0.2.}$$
$$\frac{5.6}{0.2} = \frac{0.2w}{0.2}$$
$$28 = w$$

Check: $0.8w - 0.4 = -6 + w$
$0.8(28) - 0.4 = -6 + 28$
$22.4 - 0.4 = 22$
$22 = 22$ Balances

The solution is 28.

21. $-10.9 + 0.5p = 0.9p + 5.3$ Add $-0.5p$ to both sides.

$$\underline{-0.5p \qquad\qquad -0.5p}$$
$$-10.9 = 0.4p + 5.3 \quad \text{Add } -5.3 \text{ to both sides.}$$
$$\underline{-5.3 \qquad\qquad -5.3}$$
$$-16.2 = 0.4p \quad \text{Divide both sides by 0.4.}$$
$$\frac{-16.2}{0.4} = \frac{0.4p}{0.4}$$
$$-40.5 = p$$

Check:
$-10.9 + 0.5p = 0.9p + 5.3$
$-10.9 + 0.5(-40.5) = 0.9(-40.5) + 5.3$
$-10.9 - 20.25 = -36.45 + 5.3$
$-31.15 = -31.15$ Balances

The solution is -40.5.

23. *Step 1*
Unknown: the adult dose

Known: child's dose is 0.3 times the adult dose; child's dose is 9 mg

Step 2
Let d be the adult dose.

Step 3
adult dose multiplied by 0.3 is child's dose
$$0.3d = 9$$

Step 4
$0.3d = 9$ Divide both sides by 0.3.
$$\frac{0.3d}{0.3} = \frac{9}{0.3}$$
$$d = 30$$

Step 5
The adult dose is 30 milligrams.

Step 6
$0.3(30) = 9$ (matches)

25. *Step 1*
Unknown: number of days the saw was rented

Known: $65.95 charge per day, $12 sharpening fee, $275.80 total

Step 2
Let d be the number of days.

Step 3

65.95 per day	plus	sharpening fee	equals	total bill
↓	↓	↓	↓	↓
$65.95 \cdot d$	$+$	12	$=$	275.80

Step 4
$$65.95d + 12 = 275.80$$
$$\underline{-12 \quad\ -12}$$
$$65.95d = 263.80$$
$$\frac{65.95d}{65.95} = \frac{263.80}{65.95}$$
$$d = 4$$

Step 5
The saw was rented for 4 days.

Step 6
Charge for 4 days: $4(\$65.95) = \$263.80$
Add sharpening fee: $\$263.80 + \$12 = \$275.80$
This value matches the total charge.

27. $0.7(220 - a) = 140$ Distributive property
$154 - 0.7a = 140$ Add -154 to both sides.
$\underline{-154 \qquad\quad -154}$
$-0.7a = -14$ Divide both sides by -0.7.
$\frac{-0.7a}{-0.7} = \frac{-14}{-0.7}$
$a = 20$

The person is 20 years old.

28. $0.7(220 - a) = 126$ Distributive property
$154 - 0.7a = 126$ Add -154 to both sides.
$\underline{-154 \qquad\quad -154}$
$-0.7a = -28$ Divide both sides by -0.7.
$\frac{-0.7a}{-0.7} = \frac{-28}{-0.7}$
$a = 40$

The person is 40 years old.

29. $0.7(220 - a) = 134$ Distributive property
$154 - 0.7a = 134$ Add -154 to both sides.
$\underline{-154 \qquad\quad -154}$
$-0.7a = -20$ Divide both sides by -0.7.
$\frac{-0.7a}{-0.7} = \frac{-20}{-0.7}$
$a \approx 28.57$, which rounds to 29.

The person is about 29 years old.

30. $0.7(220 - a) = 117$ Distributive property
$154 - 0.7a = 117$ Add -154 to both sides.
$\underline{-154 \qquad\quad -154}$
$-0.7a = -37$ Divide both sides by -0.7.
$\frac{-0.7a}{-0.7} = \frac{-37}{-0.7}$
$a \approx 52.86$, which rounds to 53.

The person is about 53 years old.

5.10 Geometry Applications: Circles, Cylinders, and Surface Area

5.10 Margin Exercises

1. **(a)** diameter: 40 ft
$$r = \frac{d}{2} = \frac{40 \text{ ft}}{2} = 20 \text{ ft}$$

   (b) diameter: 11 cm
$$r = \frac{d}{2} = \frac{11 \text{ cm}}{2} = 5.5 \text{ cm}$$

   (c) radius: 32 yd
$$d = 2 \cdot r = 2 \cdot 32 \text{ yd} = 64 \text{ yd}$$

   (d) radius: 9.5 m
$$d = 2 \cdot r = 2 \cdot 9.5 \text{ m} = 19 \text{ m}$$

2. **(a)** diameter: 150 ft
$$C = \pi \cdot d$$
$$\approx 3.14 \cdot 150 \text{ ft}$$
$$= 471 \text{ ft}$$

   (b) radius: 7 in.
$$C = 2 \cdot \pi \cdot r$$
$$\approx 2 \cdot 3.14 \cdot 7 \text{ in.}$$
$$\approx 44.0 \text{ in.}$$

(c) diameter: 0.9 km

$C = \pi \cdot d$
$\approx 3.14 \cdot 0.9$ km
≈ 2.8 km

(d) radius: 4.6 m

$C = 2 \cdot \pi \cdot r$
$\approx 2 \cdot 3.14 \cdot 4.6$ m
≈ 28.9 m

3. (a) radius: 1 ft

$A = \pi \cdot r \cdot r$
$\approx 3.14 \cdot 1$ ft $\cdot 1$ ft
≈ 3.1 ft^2

(b) diameter: 12 in., so $r = 6$ in.

$A = \pi \cdot r \cdot r$
$\approx 3.14 \cdot 6$ in. $\cdot 6$ in.
≈ 113.0 in.^{2}

(c) radius: 1.8 km

$A = \pi \cdot r \cdot r$
$\approx 3.14 \cdot 1.8$ km $\cdot 1.8$ km
≈ 10.2 km^2

(d) diameter: 8.4 cm, so $r = 4.2$ cm

$A = \pi \cdot r \cdot r$
$\approx 3.14 \cdot 4.2$ cm $\cdot 4.2$ cm
≈ 55.4 cm^2

4. (a) radius: 24 m

Area of circle

$A = \pi \cdot r \cdot r$
$\approx 3.14 \cdot 24$ m $\cdot 24$ m
$= 1808.64$ m^2

Area of semicircle

$\dfrac{1808.64 \text{ m}^2}{2} \approx 904.3$ m^2

(b) diameter: 35.4 ft, so $r = 17.7$ ft

Area of circle

$A = \pi \cdot r \cdot r$
$\approx 3.14 \cdot 17.7$ ft $\cdot 17.7$ ft
$= 983.7306$ ft^2

Area of semicircle

$\dfrac{983.7306 \text{ ft}^2}{2} \approx 491.9$ ft^2

(c) radius: 9.8 m

Area of circle

$A = \pi \cdot r \cdot r$
$\approx 3.14 \cdot 9.8$ m $\cdot 9.8$ m
$= 301.5656$ m^2

Area of semicircle

$\dfrac{301.5656 \text{ m}^2}{2} \approx 150.8$ m^2

5. $C = \pi \cdot d$
$\approx 3.14 \cdot 3$ m
$= 9.42$ m

Cost of binding $= \dfrac{9.42 \text{ m}}{1} \cdot \dfrac{\$4.50}{1 \text{ m}} = \$42.39$

6. diameter: 3 m, so $r = 1\frac{1}{2}$ m

$A = \pi \cdot r \cdot r$
$\approx 3.14 \cdot 1\frac{1}{2}$ m $\cdot 1\frac{1}{2}$ m
$= 7.065$ m^2

Total cost $= \dfrac{7.065 \text{ m}^2}{1} \cdot \dfrac{\$3.89}{\text{m}^2} \approx \$27.48$

7. (a) $r = 4$ ft, $h = 12$ ft

$V = \pi \cdot r \cdot r \cdot h$
$\approx 3.14 \cdot 4$ ft $\cdot 4$ ft $\cdot 12$ ft
≈ 602.9 ft^3

(b) $r = \dfrac{7 \text{ cm}}{2} = 3.5$ cm, $h = 6$ cm

$V = \pi \cdot r \cdot r \cdot h$
$\approx 3.14 \cdot 3.5$ cm $\cdot 3.5$ cm $\cdot 6$ cm
≈ 230.8 cm^3

(c) $r = 14.5$ yd, $h = 3.2$ yd

$V = \pi \cdot r \cdot r \cdot h$
$\approx 3.14 \cdot 14.5$ yd $\cdot 14.5$ yd $\cdot 3.2$ yd
≈ 2112.592 yd^3
≈ 2112.6 yd^3

8. (a) $V = lwh$
$V = (10 \text{ yd})(6 \text{ yd})(9 \text{ yd})$
$V = 540$ yd^3 (*cubic* yd for *volume*)

$SA = 2lw + 2lh + 2wh$
$SA = (2 \cdot 10 \text{ yd} \cdot 6 \text{ yd}) + (2 \cdot 10 \text{ yd} \cdot 9 \text{ yd}) +$
$\quad\quad (2 \cdot 6 \text{ yd} \cdot 9 \text{ yd})$
$SA = 120$ yd^2 $+ 180$ yd^2 $+ 108$ yd^2
$SA = 408$ yd^2 (*square* yd for *area*)

(b) $V = lwh$
$V = (16 \text{ m})(7 \text{ m})(7 \text{ m})$
$V = 784 \text{ m}^3$

$SA = 2lw + 2lh + 2wh$
$SA = (2 \cdot 16 \text{ m} \cdot 7 \text{ m}) + (2 \cdot 16 \text{ m} \cdot 7 \text{ m}) +$
$\quad (2 \cdot 7 \text{ m} \cdot 7 \text{ m})$
$SA = 224 \text{ m}^2 + 224 \text{ m}^2 + 98 \text{ m}^2$
$SA = 546 \text{ m}^2$

9. (a) $V = \pi r^2 h$
$V \approx 3.14 \cdot 5 \text{ cm} \cdot 5 \text{ cm} \cdot 15 \text{ cm}$
$\quad = 1177.5 \text{ cm}^3$ (cubic units for volume)

$SA = 2\pi rh + 2\pi r^2$
$SA \approx (2 \cdot 3.14 \cdot 5 \text{ cm} \cdot 15 \text{ cm}) + (2 \cdot 3.14 \cdot 5 \text{ cm} \cdot 5 \text{ cm})$
$\quad = 471 \text{ cm}^2 + 157 \text{ cm}^2$
$\quad = 628 \text{ cm}^2$ (square units for area)

(b) $r = \dfrac{d}{2} = \dfrac{17 \text{ in.}}{2} = 8.5 \text{ in.}$
$V = \pi r^2 h$
$V \approx 3.14 \cdot 8.5 \text{ in.} \cdot 8.5 \text{ in.} \cdot 8 \text{ in.}$
$\quad = 1814.92 \text{ in.}^3$
$\quad \approx 1814.9 \text{ in.}^3$

$SA = 2\pi rh + 2\pi r^2$
$SA \approx (2 \cdot 3.14 \cdot 8.5 \text{ in.} \cdot 8 \text{ in.}) + (2 \cdot 3.14 \cdot 8.5 \text{ in.} \cdot 8.5 \text{ in.})$
$\quad = 427.04 \text{ in.}^2 + 453.73 \text{ in.}^2$
$\quad = 880.77 \text{ in.}^2$
$\quad \approx 880.8 \text{ in.}^2$

5.10 Section Exercises

1. The radius r is 9 mm, so the diameter d is
$d = 2 \cdot r = 2 \cdot 9 \text{ mm} = 18 \text{ mm}.$

3. The diameter d is 0.7 km, so the radius r is
$r = \dfrac{d}{2} = \dfrac{0.7 \text{ km}}{2} = 0.35 \text{ km}.$

5. The radius r is 11 ft.
$C = 2 \cdot \pi \cdot r$
$\quad \approx 2 \cdot 3.14 \cdot 11 \text{ ft}$
$\quad \approx 69.1 \text{ ft}$
$A = \pi \cdot r \cdot r$
$\quad \approx 3.14 \cdot 11 \text{ ft} \cdot 11 \text{ ft}$
$\quad \approx 379.9 \text{ ft}^2$

7. The diameter d is 2.6 m, so the radius r is
$r = \tfrac{1}{2}d = \tfrac{1}{2}(2.6 \text{ m}) = 1.3 \text{ m}.$
$C = \pi \cdot d$
$\quad \approx 3.14 \cdot 2.6 \text{ m}$
$\quad \approx 8.2 \text{ m}$
$A = \pi \cdot r \cdot r$
$\quad \approx 3.14 \cdot 1.3 \text{ m} \cdot 1.3 \text{ m}$
$\quad \approx 5.3 \text{ m}^2$

9. The diameter d is 15 cm, so the radius r is
$r = \tfrac{1}{2}d = \tfrac{1}{2}(15 \text{ cm}) = 7.5 \text{ cm}.$
$C = \pi \cdot d$
$\quad \approx 3.14 \cdot 15 \text{ cm}$
$\quad = 47.1 \text{ cm}$
$A = \pi \cdot r \cdot r$
$\quad \approx 3.14 \cdot 7.5 \text{ cm} \cdot 7.5 \text{ cm}$
$\quad \approx 176.6 \text{ cm}^2$

11. The diameter d is $7\tfrac{1}{2}$ ft, so the radius r is
$r = \tfrac{1}{2}d = \tfrac{1}{2}(7.5 \text{ ft}) = 3.75 \text{ ft}.$
$C = \pi \cdot d$
$\quad \approx 3.14 \cdot 7.5 \text{ ft}$
$\quad \approx 23.6 \text{ ft}$
$A = \pi \cdot r \cdot r$
$\quad \approx 3.14 \cdot 3.75 \text{ ft} \cdot 3.75 \text{ ft}$
$\quad \approx 44.2 \text{ ft}^2$

13. The diameter d is 8.65 km, so the radius r is
$r = \tfrac{1}{2}d = \tfrac{1}{2}(8.65 \text{ km}) = 4.325 \text{ km}.$
$C = \pi \cdot d$
$\quad \approx 3.14 \cdot 8.65 \text{ km}$
$\quad \approx 27.2 \text{ km}$
$A = \pi \cdot r \cdot r$
$\quad \approx 3.14 \cdot 4.325 \text{ km} \cdot 4.325 \text{ km}$
$\quad \approx 58.7 \text{ km}^2$

15. radius: 7 in.
Area of circle
$A = \pi \cdot r \cdot r$
$\quad \approx 3.14 \cdot 7 \text{ in.} \cdot 7 \text{ in.}$
$\quad = 153.86 \text{ in.}^2$

Area of semicircle
$\dfrac{153.86 \text{ in.}^2}{2} \approx 76.9 \text{ in.}^2$

17. Area of the circle:

$A = \pi \cdot r \cdot r$
$\approx 3.14 \cdot 10 \text{ cm} \cdot 10 \text{ cm}$
$= 314 \text{ cm}^2$

Area of the semicircle:

$\dfrac{314 \text{ cm}^2}{2} = 157 \text{ cm}^2$

Area of the triangle:

$A = \dfrac{1}{2} \cdot b \cdot h$
$= \dfrac{1}{2} \cdot 20 \text{ cm} \cdot 10 \text{ cm}$
$= 100 \text{ cm}^2$

The shaded area is about
$157 \text{ cm}^2 - 100 \text{ cm}^2 = 57 \text{ cm}^2$.

19. $A = \pi \cdot r \cdot r$
$\approx 3.14 \cdot 50 \text{ yd} \cdot 50 \text{ yd}$
$= 7850 \text{ yd}^2$

The watered area is about 7850 yd^2.

21. A point on the tire tread moves the length of the circumference in one complete turn.

$C = \pi \cdot d$
$\approx 3.14 \cdot 29.10 \text{ in.}$
$\approx 91.4 \text{ in.}$

Bonus question:

$\dfrac{1 \text{ revolution}}{91.4 \text{ inches}} \cdot \dfrac{12 \text{ inches}}{1 \text{ foot}} \cdot \dfrac{5280 \text{ feet}}{1 \text{ mile}}$
$\approx 693 \text{ revolutions/mile}$

23.

$A = \pi \cdot r \cdot r$
$\approx 3.14 \cdot 150 \text{ mi} \cdot 150 \text{ mi}$
$= 70{,}650 \text{ mi}^2$

There are about $70{,}650 \text{ mi}^2$ in the broadcast area.

25. **Watch**

Watch: $d = 1 \text{ in.}, r = \dfrac{1}{2} \text{ in.}$

$C = \pi \cdot 1 \text{ in.}$
$\approx 3.14 \cdot 1 \text{ in.}$
$\approx 3.1 \text{ in.}$
$A = \pi \cdot r \cdot r$
$\approx 3.14 \cdot \dfrac{1}{2} \text{ in.} \cdot \dfrac{1}{2} \text{ in.}$
$\approx 0.8 \text{ in.}^2$

Wall Clock

Wall clock: $r = 3 \text{ in.}$

$C = 2 \cdot \pi \cdot 3 \text{ in.}$
$\approx 2 \cdot 3.14 \cdot 3 \text{ in.}$
$\approx 18.8 \text{ in.}$
$A = \pi \cdot r \cdot r$
$\approx 3.14 \cdot 3 \text{ in.} \cdot 3 \text{ in.}$
$\approx 28.3 \text{ in.}^2$

27.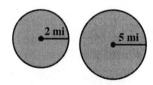

Area of larger circle:

$A = \pi \cdot r \cdot r$
$\approx 3.14 \cdot 5 \text{ mi} \cdot 5 \text{ mi}$
$= 78.5 \text{ mi}^2$

Area of smaller circle:

$A = \pi \cdot r \cdot r$
$\approx 3.14 \cdot 2 \text{ mi} \cdot 2 \text{ mi}$
$\approx 12.6 \text{ mi}^2$

The difference in the area covered is about

$78.5 \text{ mi}^2 - 12.6 \text{ mi}^2 = 65.9 \text{ mi}^2.$

29. (a)
$$C = 144 \text{ cm}$$
$$C = \pi \cdot d$$
$$144 \text{ cm} = \pi \cdot d$$
$$144 \text{ cm} \approx 3.14 \cdot d$$
$$\frac{144 \text{ cm}}{3.14} \approx d$$
$$d \approx 45.9 \text{ cm}$$

The diameter is about 45.9 cm.

(b) Divide the circumference by π (144 cm ÷ 3.14).

31. $C = 2 \cdot \pi \cdot r \approx 2 \cdot 3.14 \cdot 2.33 \text{ ft} \approx 14.6 \text{ ft}$

The rag traveled about 14.6 ft with each revolution of the wheel.

33. Area of large circle $\approx 3.14 \cdot 12 \text{ cm} \cdot 12 \text{ cm}$
$$= 452.16 \text{ cm}^2$$
Area of small circle $\approx 3.14 \cdot 9 \text{ cm} \cdot 9 \text{ cm}$
$$= 254.34 \text{ cm}^2$$
Shaded area $\approx 452.16 \text{ cm}^2 - 254.34 \text{ cm}^2$
$$= 197.8 \text{ cm}^2$$

35. $V = \pi \cdot r^2 \cdot h$
$$\approx 3.14 \cdot 5 \text{ ft} \cdot 5 \text{ ft} \cdot 6 \text{ ft}$$
$$= 471 \text{ ft}^3$$
$SA = 2\pi rh + 2\pi r^2$
$$\approx 2 \cdot 3.14 \cdot 5 \text{ ft} \cdot 6 \text{ ft} + 2 \cdot 3.14 \cdot 5 \text{ ft} \cdot 5 \text{ ft}$$
$$= 345.4 \text{ ft}^2$$

37. $V = lwh$
$$= 16.5 \text{ m} \cdot 9.8 \text{ m} \cdot 10 \text{ m}$$
$$= 1617 \text{ m}^3$$
$SA = 2lw + 2lh + 2wh$
$$= 2 \cdot 16.5 \text{ m} \cdot 9.8 \text{ m} + 2 \cdot 16.5 \text{ m} \cdot 10 \text{ m}$$
$$+ 2 \cdot 9.8 \text{ m} \cdot 10 \text{ m}$$
$$= 849.4 \text{ m}^2$$

39. $r = \dfrac{d}{2} = \dfrac{18 \text{ in.}}{2} = 9 \text{ in.}$
$V = \pi r^2 h$
$$\approx 3.14 \cdot 9 \text{ in.} \cdot 9 \text{ in.} \cdot 3 \text{ in.}$$
$$\approx 763.0 \text{ in.}^3$$
$SA = 2\pi rh + 2\pi r^2$
$$\approx 2 \cdot 3.14 \cdot 9 \text{ in.} \cdot 3 \text{ in.} + 2 \cdot 3.14 \cdot 9 \text{ in.} \cdot 9 \text{ in.}$$
$$\approx 678.2 \text{ in.}^2$$

41. $V = lwh$
$$= 15 \text{ mm} \cdot 10 \text{ mm} \cdot 37 \text{ mm}$$
$$= 5550 \text{ mm}^3$$
$SA = 2lw + 2lh + 2wh$
$$= 2 \cdot 15 \text{ mm} \cdot 10 \text{ mm} + 2 \cdot 15 \text{ mm} \cdot 37 \text{ mm}$$
$$+ 2 \cdot 10 \text{ mm} \cdot 37 \text{ mm}$$
$$= 2150 \text{ mm}^2$$

43. Student should use radius of 3.5 cm instead of diameter of 7 cm in the formula; units for volume are cm^3, not cm^2. Correct answer is $V \approx 192.3 \text{ cm}^3$.

45. Use the formula for the volume of a cylinder.
$$\text{radius} = \frac{5 \text{ ft}}{2} = 2.5 \text{ ft}$$
$V = \pi \cdot r^2 \cdot h$
$$\approx 3.14 \cdot 2.5 \text{ ft} \cdot 2.5 \text{ ft} \cdot 200 \text{ ft}$$
$$= 3925 \text{ ft}^3$$

The volume of the city sewer pipe is about 3925 ft^3.

47. $SA = 2lw + 2lh + 2wh$
$$= 2 \cdot 5.5 \text{ in.} \cdot 2.8 \text{ in.} + 2 \cdot 5.5 \text{ in.} \cdot 8 \text{ in.}$$
$$+ 2 \cdot 2.8 \text{ in.} \cdot 8 \text{ in.}$$
$$= 163.6 \text{ in.}^2$$

The amount of material needed is 163.6 in.^{2}.

Chapter 5 Review Exercises

1. 2 4 3 . 0 5 9 — tenths, hundredths (position of 5 is tenths, 9 is hundredths)

2. 0 . 6 8 1 7 — ones, tenths (0 is ones, 6 is tenths)

3. $5 8 2 4 . 3 9 — hundreds, hundredths (8 is hundreds, 9 is hundredths)

4. 8 9 6 . 5 0 3 — tens, tenths (9 is tens, 5 is tenths)

5.

				tenths		ten-thousandths
2	0	.	7	3	8	6	1

6. $0.5 = \dfrac{5}{10} = \dfrac{1}{2}$

7. $0.75 = \dfrac{75}{100} = \dfrac{75 \div 25}{100 \div 25} = \dfrac{3}{4}$

8. $4.05 = 4\dfrac{5}{100} = 4\dfrac{5 \div 5}{100 \div 5} = 4\dfrac{1}{20}$

9. $0.875 = \dfrac{875}{1000} = \dfrac{875 \div 125}{1000 \div 125} = \dfrac{7}{8}$

10. $0.027 = \dfrac{27}{1000}$

11. $27.8 = 27\dfrac{8}{10} = 27\dfrac{8 \div 2}{10 \div 2} = 27\dfrac{4}{5}$

12. 0.8 is eight tenths.

13. 400.29 is four hundred and twenty-nine hundredths.

14. 12.007 is twelve and seven thousandths.

15. 0.0306 is three hundred six ten-thousandths.

16. eight and three tenths:
$8\dfrac{3}{10} = 8.3$

17. two hundred five thousandths:
$\dfrac{205}{1000} = 0.205$

18. seventy and sixty-six ten-thousandths:
$70\dfrac{66}{10,000} = 70.0066$

19. thirty hundredths:
$\dfrac{30}{100} = 0.30$

20. 275.635 to the nearest tenth:
Draw a cut-off line: 275.6|35
The first digit cut is 3, which is 4 or less. The part you keep stays the same.
Answer: ≈ 275.6

21. 72.789 to the nearest hundredth:
Draw a cut-off line: 72.78|9
The first digit cut is 9, which is 5 or more, so round up.
Answer: ≈ 72.79

22. 0.1604 to the nearest thousandth:
Draw a cut-off line: 0.160|4
The first digit cut is 4, which is 4 or less. The part you keep stays the same.
Answer: ≈ 0.160

23. 0.0905 to the nearest thousandth:
Draw a cut-off line: 0.090|5
The first digit cut is 5, which is 5 or more, so round up.
Answer: ≈ 0.091

24. 0.98 to the nearest tenth:
Draw a cut-off line: 0.9|8
The first digit cut is 8, which is 5 or more, so round up.
Answer: ≈ 1.0

25. $15.8333 to the nearest cent:
Draw a cut-off line: $15.83|33
The first digit cut is 3, which is 4 or less. The part you keep stays the same.
Answer: $\approx \$15.83$

26. $0.698 to the nearest cent:
Draw a cut-off line: $0.69|8
The first digit cut is 8, which is 5 or more, so round up.
Answer: $\approx \$0.70$

27. $17,625.7906 to the nearest cent:
Draw a cut-off line: $17,625.79|06
The first digit cut is 0, which is 4 or less. The part you keep stays the same.
Answer: $\approx \$17,625.79$

28. $350.48 to the nearest dollar:
Draw a cut-off line: $350.|48
The first digit cut is 4, which is 4 or less. The part you keep stays the same.
Answer: $\approx \$350$

29. $129.50 to the nearest dollar:
Draw a cut-off line: $129.|50
The first digit cut is 5, which is 5 or more, so round up.
Answer: $\approx \$130$

30. $99.61 to the nearest dollar:
Draw a cut-off line: $99.|61
The first digit cut is 6, which is 5 or more, so round up.
Answer: $\approx \$100$

31. $29.37 to the nearest dollar:
Draw a cut-off line: $29.|37
The first digit cut is 3, which is 4 or less. The part you keep stays the same.
Answer: $\approx \$29$

32. $0.4 - 6.07$ *Add the opposite.*
      $= 0.4 + (-6.07)$ *Addition of numbers with different signs.*

 $|-6.07| = 6.07, \quad |0.4| = 0.4$

 -6.07 has a larger absolute value, so the sum will be negative. Subtract absolute values.

 6.07
 -0.40
 $\overline{5.67}$

 Answer: -5.67

33. $-20 + 19.97$ *Addition of numbers with different signs.*

 $|-20| = 20, \quad |19.97| = 19.97$

 -20 has the larger absolute value, so the sum will be negative. Subtract absolute values.

 20.00
 -19.97
 $\overline{0.03}$

 Answer: -0.03

34. $-1.35 + 7.229$ *Addition of numbers with different signs.*

 $|-1.35| = 1.35, \quad |7.229| = 7.229$

 7.229 has the larger absolute value, so the sum will be positive. Subtract absolute values.

 $7.\overset{611}{\cancel{2}}\overset{12}{\cancel{2}}9$
 -1.350
 $\overline{5.879}$

 Answer: 5.879

35. $0.005 + (3 - 9.44) = 0.005 + (3 + (-9.44))$
 $ = 0.005 + (-6.44)$
 $ = -6.435$

36. *Estimate:*   *Exact:*
      80 million   81.3 million
 $-$ 50 million $-$ 49.9 million
 $\overline{30\text{ million}}$ $\overline{31.4\text{ million}}$

 31.4 million more people go walking than camping.

37. Add the amounts of the two checks.

 Estimate: *Exact:*
      $200 $215.53
 $+ 40$ $+ 44.67$
 $\overline{\$240}$ $\overline{\$260.20}$

 The total amount of the two checks was $260.20. Now subtract from her balance of $306.

 Estimate: *Exact:*
      $300 $306.00
 $- 240$ $- 260.20$
 $\overline{\$60}$ $\overline{\$45.80}$

 The new balance is $45.80.

38. First total the money that Joey spent.

 Estimate: *Exact:*
      $2 $1.59
      5   5.33
 $+ 20$ $+ 18.94$
 $\overline{\$27}$ $\overline{\$25.86}$

 Then subtract to find the change.

 Estimate: *Exact:*
      $30 $30.00
 $- 27$ $- 25.86$
 $\overline{\$3}$ $\overline{\$4.14}$

 Joey's change was $4.14.

39. Add the kilometers that she raced each day.

 Estimate: *Exact:*
      2   2.30
      4   4.00
 $+ 5$ $+ 5.25$
 $\overline{11\text{ km}}$ $\overline{11.55\text{ km}}$

 Roseanne raced 11.55 kilometers.

40. *Estimate:*   *Exact:*
      6   6.138
 $\times 4$ $\times 3.7$
 $\overline{24}$ $\overline{4\,2966}$
           $18\,414$
           $\overline{22.7106}$

41. *Estimate:*   *Exact:*
      40   42.9
 $\times 3$ $\times 3.3$
 $\overline{120}$ $\overline{12\,87}$
           $128\,7$
           $\overline{141.57}$

42. $(-5.6)(-0.002)$

The signs are the *same*, so the product will be *positive*.

$$\begin{array}{r} 5.6 \\ \times\, 0.002 \\ \hline 0.0112 \end{array} \begin{array}{l} \leftarrow\ 1\ decimal\ place \\ \leftarrow\ 3\ decimal\ places \\ \leftarrow\ 4\ decimal\ places \end{array}$$

43. $(0.071)(-0.005)$

The signs are *different*, so the product will be *negative*.

$$\begin{array}{r} 0.071 \\ \times\, 0.005 \\ \hline -0.000355 \end{array} \begin{array}{l} \leftarrow\ 3\ decimal\ places \\ \leftarrow\ 3\ decimal\ places \\ \leftarrow\ 6\ decimal\ places \end{array}$$

44. $706.2 \div 12 = 58.85$

Estimate: $700 \div 10 = 70$

58.85 is *reasonable*.

45. $26.6 \div 2.8 = 0.95$

Estimate: $30 \div 3 = 10$

0.95 is *not reasonable*.

$$\begin{array}{r} 9.\,5 \\ 2.8_\wedge\overline{\smash{)}2\,6.\,6_\wedge\,0} \\ 2\,5\,2 \\ \hline 1\,4\,\,0 \\ 1\,4\,\,0 \\ \hline 0 \end{array}$$ *Move decimal point 1 place in divisor and dividend; write one zero.*

The correct answer is 9.5.

46.
$$\begin{array}{r} 1\,4.\,4\,6\,6\,6 \\ 3\overline{\smash{)}4\,3.\,4\,0\,0\,0} \\ \underline{3} \\ 1\,3 \\ \underline{1\,2} \\ 1\,4 \\ \underline{1\,2} \\ 2\,0 \\ \underline{1\,8} \\ 2\,0 \\ \underline{1\,8} \\ 2\,0 \\ \underline{1\,8} \\ 2 \end{array}$$
Write one zero.

Write one zero.

Write one zero.

14.4666 rounds to 14.467.

47. $\dfrac{-72}{-0.06} = -72.00 \div (-0.06)$

$\phantom{\dfrac{-72}{-0.06}} = -7200 \div (-6)$ *Move both decimal points two places to the right.*

$\phantom{\dfrac{-72}{-0.06}} = 1200$

48. $-0.00048 \div 0.0012$

$= -4.8 \div 12$ *Move both decimal points four places to the right.*

$= -0.4$

$$\begin{array}{r} 0.\,4 \\ 12\overline{\smash{)}4.\,8} \\ \underline{4\,\,8} \\ 0 \end{array}$$

49. Multiply the hourly wage times the hours worked.

Pay for first 40 hours	Pay rate after 40 hours	Pay for over 40 hours
14.24	14.24	21.36
× 40	× 1.5	× 6.5
569.60	7 120	10 680
	14 24	128 16
	21.360	138.840

Add the two amounts.

$569.60
138.84
$708.44

Adrienne's total earnings were $708 (rounded).

50. Divide the cost of the book by the number of tickets in the book.

$$\begin{array}{r} 2.\,9\,9\,0 \\ 12\overline{\smash{)}3\,5.\,8\,9\,0} \\ \underline{2\,4} \\ 1\,1\,\,8 \\ \underline{1\,0\,\,8} \\ 1\,\,0\,\,9 \\ \underline{1\,\,0\,\,8} \\ 1\,\,0 \end{array}$$

Round to the nearest cent.
Draw a cut-off line: $2.99|0
The first digit cut is 0, which is 4 or less. The part you keep stays the same.

Each ticket costs $2.99 (rounded).

51. Divide the amount of the investment by the price per share.

```
           1  3 3. 3
    3.75∧│5 0 0. 0 0∧ 0
           3 7 5
           ─────
           1 2 5 0
           1 1 2 5
           ───────
             1 2 5 0
             1 1 2 5
             ───────
               1 2 5 0
               1 1 2 5
               ───────
                 1 2 5
```

Round 133.3 down to the nearest whole number. Kenneth could buy 133 shares (rounded).

52. Multiply the price per pound by the amount to be purchased.

```
    $0.99
   × 3.5
   ──────
     495
    2 97
   ──────
   $3.465
```

$3.465 rounds to $3.47.
Ms. Lee will pay $3.47 (rounded).

53. $3.5^2 + 8.7 \cdot (-1.95)$ Exponent
    $= 12.25 + 8.7(-1.95)$ Multiply.
    $= 12.25 + (-16.965)$ Add.
    $= -4.715$

54. $11 - 3.06 \div (3.95 - 0.35)$ Parentheses
    $= 11 - 3.06 \div 3.6$ Divide.
    $= 11 - 0.85$ Subtract.
    $= 10.15$

55. $3\frac{4}{5} = \frac{19}{5} = 3.8$

```
         3. 8
    5│1 9. 0
      1 5
      ───
        4 0
        4 0
        ───
          0
```

56. $\frac{16}{25} = 0.64$

```
         0. 6 4
    25│1 6. 0 0
       1 5 0
       ─────
         1 0 0
         1 0 0
         ─────
             0
```

57. $1\frac{7}{8} = \frac{15}{8} = 1.875$

```
         1. 8 7 5
    8│1 5. 0 0 0
      8
      ─
      7 0
      6 4
      ───
        6 0
        5 6
        ───
          4 0
          4 0
          ───
            0
```

58. $\frac{1}{9}$

```
         0. 1 1 1 1
    9│1. 0 0 0 0
      9
      ─
      1 0
        9
        ─
        1 0
          9
          ─
          1 0
            9
            ─
            1
```

0.1111 rounds to 0.111.
$\frac{1}{9} \approx 0.111$

59. 3.68, 3.806, 3.6008

$3.68 = 3.6800$
$3.806 = 3.8060$ greatest
$3.6008 = 3.6008$ least

From least to greatest:
3.6008, 3.68, 3.806

60. 0.215, 0.22, 0.209, 0.2102

$0.215 = 0.2150$
$0.22 = 0.2200$ greatest
$0.209 = 0.2090$ least
$0.2102 = 0.2102$

From least to greatest:
0.209, 0.2102, 0.215, 0.22

61. $0.17, \frac{3}{20}, \frac{1}{8}, 0.159$

$0.17 = 0.170$ greatest
$\frac{3}{20} = 0.150$
$\frac{1}{8} = 0.125$ least
$0.159 = 0.159$

From least to greatest:
$\frac{1}{8}, \frac{3}{20}, 0.159, 0.17$

62. mean
$$= \frac{\begin{array}{c}(18+12+15+24+9\\+42+54+87+21+3)\end{array}}{10}$$
$$= \frac{285}{10} = 28.5 \text{ digital cameras}$$

To find the median, arrange the data from high to low or low to high.

$$3, 9, 12, 15, 18, 21, 24, 42, 54, 87$$
$$\uparrow \uparrow$$

Since the number of items is even, the median is the average of the middle items.

$$\text{median} = \frac{18+21}{2} = 19.5 \text{ digital cameras}$$

63. mean
$$= \frac{54+28+35+43+17+37+68+75+39}{9}$$
$$= \frac{396}{9} = 44 \text{ claims}$$

To find the median, arrange the data from high to low or low to high.

$$17, 28, 35, 37, 39, 43, 54, 68, 75$$
$$\uparrow$$

Since the number of items is odd, the median is the middle item.

median = 39 claims

64.

Dollar Value	Frequency	Product
$42	3	$126
$47	7	$329
$53	2	$106
$55	3	$165
$59	5	$295
	20	$1021

$$\text{weighted mean} = \frac{\$1021}{20} = \$51.05$$

65. Arrange the data in order:

Store J: $69, $84, $107, $107, $139, $160, $160
There are two modes: $107 and $160 (bimodal);
Store K: $95, $99, $119, $119, $119, $136, $139
mode = $119

66. hypotenuse $= \sqrt{(\text{leg})^2 + (\text{leg})^2}$
$$= \sqrt{(15)^2 + (8)^2}$$
$$= \sqrt{225+64}$$
$$= \sqrt{289}$$
$$= 17 \text{ in.}$$

67. leg $= \sqrt{(\text{hypotenuse})^2 - (\text{leg})^2}$
$$= \sqrt{(25)^2 - (24)^2}$$
$$= \sqrt{625-576}$$
$$= \sqrt{49}$$
$$= 7 \text{ cm}$$

68. leg $= \sqrt{(\text{hypotenuse})^2 - (\text{leg})^2}$
$$= \sqrt{(15)^2 - (11)^2}$$
$$= \sqrt{225-121}$$
$$= \sqrt{104}$$
$$\approx 10.2 \text{ cm}$$

69. hypotenuse $= \sqrt{(\text{leg})^2 + (\text{leg})^2}$
$$= \sqrt{(6)^2 + (4)^2}$$
$$= \sqrt{36+16}$$
$$= \sqrt{52}$$
$$\approx 7.2 \text{ in.}$$

70. hypotenuse $= \sqrt{(\text{leg})^2 + (\text{leg})^2}$
$$= \sqrt{(2.2)^2 + (1.3)^2}$$
$$= \sqrt{4.84+1.69}$$
$$= \sqrt{6.53}$$
$$\approx 2.6 \text{ m}$$

71. leg $= \sqrt{(\text{hypotenuse})^2 - (\text{leg})^2}$
$$= \sqrt{(12)^2 - (8.5)^2}$$
$$= \sqrt{144-72.25}$$
$$= \sqrt{71.75}$$
$$\approx 8.5 \text{ km}$$

72. $-0.1 = b - 0.35$ Add the opposite.
$-0.1 = b + (-0.35)$ Add 0.35 to both sides.
$\underline{+0.35 \qquad\qquad +0.35}$
$0.25 = b + 0$
$0.25 = b$

The solution is 0.25.

73. $-3.8x = 0$ Divide both sides by -3.8.
$$\frac{-3.8x}{-3.8} = \frac{0}{-3.8}$$
$$x = 0$$

The solution is 0.

180 Chapter 5 Rational Numbers: Positive and Negative Decimals

74. $6.8 + 0.4n = 1.6$ Add -6.8 to both sides.

$\underline{-6.8 \qquad\quad -6.8}$

$0 + 0.4n = -5.2$ Divide both sides by 0.4.

$\dfrac{0.4n}{0.4} = \dfrac{-5.2}{0.4}$

$n = -13$

The solution is -13.

75. $-0.375 + 1.75a = 2a$ Add $-1.75a$ to both sides.

$\underline{\qquad -1.75a \quad -1.75a}$

$-0.375 + 0 = 0.25a$

$-0.375 = 0.25a$ Divide both sides by 0.25.

$\dfrac{-0.375}{0.25} = \dfrac{0.25a}{0.25}$

$-1.5 = a$

The solution is -1.5.

76. $0.3y - 5.4 = 2.7 + 0.8y$ Add $-0.3y$ to both sides.

$\underline{-0.3y \qquad\qquad -0.3y}$

$0 - 5.4 = 2.7 + 0.5y$

$-5.4 = 2.7 + 0.5y$ Add -2.7 to both sides.

$\underline{-2.7 \quad -2.7}$

$-8.1 = 0 + 0.5y$ Divide both sides by 0.5.

$\dfrac{-8.1}{0.5} = \dfrac{0.5y}{0.5}$

$-16.2 = y$

The solution is -16.2.

77. $d = 2 \cdot r = 2 \cdot 68.9 \text{ m} = 137.8 \text{ m}$

The diameter of the field is 137.8 m.

78. $r = \dfrac{d}{2} = \dfrac{3 \text{ in.}}{2} = 1\dfrac{1}{2}$ in. or 1.5 in.

The radius of the juice can is $1\frac{1}{2}$ in. or 1.5 in.

79. radius: 1 cm, $d = 2$ cm

$C = \pi \cdot d$

$\approx 3.14 \cdot 2$ cm

≈ 6.3 cm

$A = \pi \cdot r \cdot r$

$\approx 3.14 \cdot 1$ cm $\cdot 1$ cm

≈ 3.1 cm^2

80. radius: 17.4 m, $d = 34.8$ m

$C = \pi \cdot d$

$\approx 3.14 \cdot 34.8$ m

≈ 109.3 m

$A = \pi \cdot r \cdot r$

$\approx 3.14 \cdot 17.4$ m $\cdot 17.4$ m

≈ 950.7 m^2

81. diameter: 12 in., radius: 6 in.

$C = \pi \cdot d$

$\approx 3.14 \cdot 12$ in.

≈ 37.7 in.

$A = \pi \cdot r \cdot r$

$\approx 3.14 \cdot 6$ in. $\cdot 6$ in.

≈ 113.0 in.^{2}

82. The figure is a cylinder.

$V = \pi \cdot r^2 \cdot h$

$\approx 3.14 \cdot 5$ cm $\cdot 5$ cm $\cdot 7$ cm

$= 549.5$ cm^3

$SA = 2\pi rh + 2\pi r^2$

$\approx 2 \cdot 3.14 \cdot 5$ cm $\cdot 7$ cm $+ 2 \cdot 3.14 \cdot 5$ cm $\cdot 5$ cm

≈ 376.8 cm^2

83. The figure is a cylinder.

$V = \pi \cdot r^2 \cdot h$ ($d = 24$ m; $r = 12$ m)

$\approx 3.14 \cdot 12$ m $\cdot 12$ m $\cdot 4$ m

≈ 1808.6 m^3

$SA = 2\pi rh + 2\pi r^2$

$\approx 2 \cdot 3.14 \cdot 12$ m $\cdot 4$ m $+ 2 \cdot 3.14 \cdot 12$ m $\cdot 12$ m

≈ 1205.8 m^2

84. $l = 3.5$ ft, $w = 1.5$ ft, $h = 1.5$ ft

$V = lwh$

$= 3.5$ ft $\cdot 1.5$ ft $\cdot 1.5$ ft

$= 7.875$ ft$^3 \approx 7.9$ ft^3

$SA = 2lw + 2lh + 2wh$

$= 2 \cdot 3.5$ ft $\cdot 1.5$ ft $+ 2 \cdot 3.5$ ft $\cdot 1.5$ ft $+ 2 \cdot 1.5$ ft $\cdot 1.5$ ft

$= 25.5$ ft^2

85. [5.3] $89.19 + 0.075 + 310.6 + 5$

$\overset{111}{89.190}$ *Line up decimal points.*

0.075 *Write in zeros.*

310.600

$\underline{+5.000}$

404.865

86. [5.4] $72.8(-3.5)$

The signs are *different*, so the product will be *negative*.

$$\begin{array}{r} 72.8 \quad \leftarrow 1 \text{ decimal place} \\ \times\, 3.5 \quad \leftarrow 1 \text{ decimal place} \\ \hline 36\,40 \\ 218\,4 \\ \hline 254.80 \quad \leftarrow 2 \text{ decimal places} \end{array}$$

$72.8(-3.5) = -254.80$

87. [5.5] $1648.3 \div 0.46 \approx 3583.2609$

3583.2609 rounds to 3583.261.

88. [5.3]
$$\begin{array}{r} {\scriptstyle 2\,9\,9\,9\,9\,10} \\ \cancel{3}\cancel{0}.\cancel{0}\,\cancel{0}\,\cancel{0}\,\cancel{0} \\ -\,0\,.\,9\,1\,0\,2 \\ \hline 2\,9\,.\,0\,8\,9\,8 \end{array}$$

89. [5.4] $(4.38)(0.007)$

$$\begin{array}{r} 4.38 \quad \leftarrow 2 \text{ decimal places} \\ \times\,0.007 \quad \leftarrow 3 \text{ decimal places} \\ \hline 0.03066 \quad \leftarrow 5 \text{ decimal places} \end{array}$$

90. [5.5]
$$\begin{array}{r} 9.\,4 \\ 0.005_{\wedge}\overline{\smash{)}0.0\,4\,7_{\wedge}0} \quad \text{Move decimal point 3} \\ \underline{4\,5} \quad \text{places in divisor and} \\ 2\,0 \quad \text{dividend; write 0} \\ \underline{2\,0} \\ 0 \end{array}$$

91. [5.3]
$72.105 + 8.2 - 95.37$ Change subt.
$= 72.105 + 8.2 + (-95.37)$ Add.
$= 80.305 + (-95.37)$ Add.
$= -15.065$

92. [5.5] $\dfrac{-81.36}{9}$

The signs are different, so the quotient is negative.

$$\begin{array}{r} 9.04 \\ 9\overline{\smash{)}81.36} \\ \underline{81} \\ 0\,3 \\ \underline{0} \\ 36 \\ \underline{36} \\ 0 \end{array}$$

$\dfrac{-81.36}{9} = -9.04$

93. [5.5]
$(0.6 - 1.22) + 4.8(-3.15)$ Change subt.
$= (0.6 + (-1.22)) + 4.8(-3.15)$ Parentheses
$= (-0.62) + 4.8(-3.15)$ Multiply.
$= -0.62 + (-15.12)$ Add.
$= -15.74$

94. [5.4] $0.455(18)$

$$\begin{array}{r} 0.455 \quad \leftarrow 3 \text{ decimal places} \\ \times\,18 \quad \leftarrow 0 \text{ decimal places} \\ \hline 3\,640 \\ 4\,55 \\ \hline 8.190 \quad \leftarrow 3 \text{ decimal places} \ \text{(or 8.19)} \end{array}$$

95. [5.4] $(-1.6)(-0.58)$

The signs are the same, so the product is positive.

$$\begin{array}{r} 0.58 \quad \leftarrow 2 \text{ decimal places} \\ \times\,1.6 \quad \leftarrow 1 \text{ decimal place} \\ \hline 348 \\ 58 \\ \hline 0.928 \quad \leftarrow 3 \text{ decimal places} \end{array}$$

96. [5.5] $0.218\overline{\smash{)}7.63}$

$7.63 \div 0.218 = 35$

97. [5.3]
$-21.059 - 20.8$ Change subt.
$= -21.059 + (-20.8)$ Add.
$= -41.859$

98. [5.5]
$18.3 - 3^2 \div 0.5$ Exponent
$= 18.3 - 9 \div 0.5$ Divide.
$= 18.3 - 18$ Subtract.
$= 0.3$

99. [5.5] Men's socks are 3 pairs for $8.99. Divide the price by 3 to find the cost of one pair.

$$\begin{array}{r} 2.\,9\,9\,6 \\ 3\overline{\smash{)}8.\,9\,9\,0} \\ \underline{6} \\ 2\,9 \\ \underline{2\,7} \\ 2\,9 \\ \underline{2\,7} \\ 2\,0 \\ \underline{1\,8} \\ 2 \end{array}$$

2.996 rounds to 3.00.

One pair of men's socks costs $3.00 (rounded).

100. [5.5] Children's socks cost 6 pairs for $5.00. Find the cost for one pair.

$$\begin{array}{r} 0.833 \\ 6\overline{)5.000} \\ \underline{48} \\ 20 \\ \underline{18} \\ 20 \\ \underline{18} \\ 2 \end{array}$$

$0.833 rounds to $0.83.

Subtract to find how much more men's socks cost.

$$\begin{array}{r} \$3.00 \\ -\,0.83 \\ \hline \$2.17 \end{array}$$

Men's socks cost $2.17 (rounded) more.

101. [5.4] A dozen pair of socks is $4 \cdot 3 = 12$ pairs of socks. So multiply 4 times $8.99.

$$\begin{array}{r} \$8.99 \\ \times\ 4 \\ \hline \$35.96 \end{array}$$

The cost for a dozen pair of men's socks is $35.96.

102. [5.4] Five pairs of teen jeans cost $19.95 times 5.

$$\begin{array}{r} \$19.95 \\ \times\ 5 \\ \hline \$99.75 \end{array}$$

Four pairs of women's jeans cost $24.99 times 4.

$$\begin{array}{r} \$24.99 \\ \times\ 4 \\ \hline \$99.96 \end{array}$$

Add the two amounts.

$$\begin{array}{r} \$99.75 \\ +\,99.96 \\ \hline \$199.71 \end{array}$$

Akiko would pay $199.71.

103. [5.3] The highest regular price for athletic shoes is $149.50. The cheapest sale price is $71.

Subtract to find the difference.

$$\begin{array}{r} \$149.50 \\ -\,71.00 \\ \hline \$78.50 \end{array}$$

The difference in price is $78.50.

104. [5.9] $4.62 = -6.6y$ Divide both sides by -6.6.

$$\frac{4.62}{-6.6} = \frac{-6.6y}{-6.6}$$
$$-0.7 = y$$

The solution is -0.7.

105. [5.9] $1.05x - 2.5 = 0.8x + 5$ Add $-0.8x$ to both sides.

$$\begin{array}{r} -0.8x \quad -0.8x \\ \hline 0.25x - 2.5 = 0 + 5 \end{array}$$

$0.25x - 2.5 = 5$ Add 2.5 to both sides.

$$\begin{array}{r} +\,2.5 \quad +\,2.5 \\ \hline 0.25x + 0 = 7.5 \end{array}$$

Divide both sides by 0.25.

$$\frac{0.25x}{0.25} = \frac{7.5}{0.25}$$
$$x = 30$$

The solution is 30.

106. [5.10] Rubber strip to go "around" the edge of the table indicates circumference of the table.

$C = \pi \cdot d$
$C = \pi \cdot 5 \text{ ft}$
$C \approx 3.14 \cdot 5 \text{ ft}$
$C \approx 15.7$ ft long rubber strip

Since the diameter is 5, the radius is $\frac{5}{2}$ or 2.5.

$A = \pi \cdot r^2$
$A = \pi \cdot (2.5 \text{ ft})^2$
$A \approx 3.14 \cdot 2.5 \text{ ft} \cdot 2.5 \text{ ft}$
$A \approx 19.6 \text{ ft}^2$

107. [5.7]   mean $= \dfrac{82 + 0 + 78 + 93 + 85}{5}$

$$= \frac{338}{5} = 67.6$$

Order the scores:

$$0, 78, 82, 85, 93$$

The median score for 5 scores is the third score, that is, 82.

108. [5.8]   hypotenuse $= \sqrt{(\text{leg})^2 + (\text{leg})^2}$
$= \sqrt{(16)^2 + (12)^2}$
$= \sqrt{256 + 144}$
$= \sqrt{400}$
$= 20$ miles

She is 20 miles from her starting point.

109. **[5.10]** A "can" is geometrically a cylinder.

$V = \pi \cdot r^2 \cdot h$ ($d = 3$ in., so $r = 1.5$ in.)

$V = \pi \cdot (1.5 \text{ in.})^2 \cdot 7$ in.

$V \approx 3.14 \cdot 1.5$ in. $\cdot 1.5$ in. $\cdot 7$ in.

$V \approx 49.5$ in.^{3}

Chapter 5 Test

1. $18.4 = 18\dfrac{4}{10} = 18\dfrac{4 \div 2}{10 \div 2} = 18\dfrac{2}{5}$

2. $0.075 = \dfrac{75}{1000} = \dfrac{75 \div 25}{1000 \div 25} = \dfrac{3}{40}$

3. 60.007 is sixty and seven thousandths.

4. 0.0208 is two hundred eight ten-thousandths.

5. $7.6 + 82.0128 + 39.59$

Estimate:	*Exact:*	
8	7.6000	*Line up decimals.*
80	82.0128	*Write in zeros.*
+ 40	+ 39.5900	
128	129.2028	

6. $-5.79(1.2)$ The product is negative.

Estimate:	*Exact:*
−6	−5.79
× 1	× 1.2
−6	1 158
	5 79
	−6.948

7. $-79.1 - 3.602$
$= -79.1 + (-3.602)$

Both addends have the same sign, so the sum will be negative.

Estimate:	*Exact:*
−80	−79.100
+ −4	+ −3.602
−84	−82.702

8. $-20.04 \div (-4.8)$
Same signs, quotient is positive

Estimate:
4
$5\overline{)20}$

Exact:
$4.\;1\;7\;5$
$4.8\overline{)2\;0.\;0\wedge 4\;0\;0}$
$1\;9\;2$
$8\;4$
$4\;8$
$3\;6\;0$
$3\;3\;6$
$2\;4\;0$
$2\;4\;0$
0

9. $670 - 0.996$

670.000	*Line up decimals.*
− 0.996	*Write in zeros.*
669.004	

10.
$4\;8\;0.$
$0.15_\wedge\overline{)7\;2.\;0\;0_\wedge}$ *Move decimal point 2 places*
$6\;0$ *in dividend and divisor; write*
$1\;2\;0$ *two zeros.*
$1\;2\;0$
$0\;0$
0
0

11. $(-0.006)(-0.007)$
Same signs, product is positive

0.006	←	*3 decimal places*
× 0.007	←	*3 decimal places*
0.000042	←	*6 decimal places*

12. Divide to find the cost per foot.

$1.\;7\;9$
$6.5_\wedge\overline{)\$1\;1.\;6_\wedge\;4\;0}$ *Move decimal point 1 place*
$6\;5$ *in dividend and divisor; write*
$5\;1\;4$ *one zero.*
$4\;5\;5$
$5\;9\;0$
$5\;8\;5$
5

The chain costs $1.79 per foot (rounded).

13. Compare the two times.

Angela: 3.500 minutes (more time)
Davida: 3.059 minutes (less time)

3.500
-3.059
0.441

Davida won by 0.441 minute.

14. Multiply the price per pound by the amount purchased.

$\$2.89$
$\times\;\;1.85$
1445
$2\;312$
$2\;89$
$\$5.3465$

Round $5.3465 to the nearest cent.
Draw a cut-off line: $5.34|65
The first digit cut is 6, which is 5 or more, so round up.

Mr. Yamamoto paid $5.35 for the cheese.

15. $-5.9 = y + 0.25$ Add -0.25 to both sides.

$$\begin{array}{rcl} \underline{-0.25} & & \underline{-0.25} \\ -6.15 & = & y + 0 \\ -6.15 & = & y \end{array}$$

The solution is -6.15.

16. $-4.2x = 1.47$ Divide both sides by -4.2.

$$\frac{-4.2x}{-4.2} = \frac{1.47}{-4.2}$$
$$x = -0.35$$

The solution is -0.35.

17. $\quad 3a - 22.7 = 10$

$3a + (-22.7) = 10$ Add 22.7 to both sides.

$$\begin{array}{rcl} \underline{+22.7} & & \underline{+22.7} \\ 3a + 0 & = & 32.7 \end{array}$$ Divide both sides by 3.

$$\frac{3a}{3} = \frac{32.7}{3}$$
$$a = 10.9$$

The solution is 10.9.

18. $-0.8n + 1.88 = 2n - 6.1$ Add $0.8n$ to both sides.

$$\begin{array}{rcl} \underline{+0.8n} & & \underline{+0.8n} \\ 0 + 1.88 & = & 2.8n - 6.1 \end{array}$$

$1.88 = 2.8n - 6.1$ Add 6.1 to both sides.

$$\begin{array}{rcl} \underline{+6.1} & & \underline{+6.1} \\ 7.98 & = & 2.8n + 0 \end{array}$$ Divide both sides by 2.8.

$$\frac{7.98}{2.8} = \frac{2.8n}{2.8}$$
$$2.85 = n$$

The solution is 2.85.

19. $0.44, 0.451, \frac{9}{20}, 0.4506$

Change to ten-thousandths, if necessary, and compare.

$0.44 = 0.4400$ least
$0.451 = 0.4510$ greatest
$\frac{9}{20} = 0.4500$
$0.4506 = 0.4506$

From least to greatest: $0.44, \frac{9}{20}, 0.4506, 0.451$

20. $6.3^2 - 5.9 + 3.4(-0.5)$ *Exponent*
$= 39.69 - 5.9 + 3.4(-0.5)$ *Multiply.*
$= 39.69 - 5.9 + (-1.7)$ *Add the opposite.*
$= 39.69 + (-5.9) + (-1.7)$ *Add.*
$= 33.79 + (-1.7)$ *Add.*
$= 32.09$

21. mean
$$= \frac{52 + 61 + 68 + 69 + 73 + 75 + 79 + 84 + 91 + 98}{10}$$
$$= \frac{750}{10} = 75 \text{ books}$$

22. $96°, \underline{104°}, \underline{103°}, \underline{104°}, \underline{103°}, \underline{104°}, 91°, 74°, \underline{103°}$

The mode is the value that occurs most often. The modes are $103°$ and $104°$ (bimodal).

23.

Cost	Frequency	Product
$6	7	$42
$10	3	$30
$11	4	$44
$14	2	$28
$19	3	$57
$24	1	$24
	20	$225

weighted mean $= \dfrac{\$225}{20} = \$11.25$

24. Arrange the prices in numerical order.

$39.75, $46.90, $48, $49.30, $51.80, $54.50, $56.25, $89

Because there is an even number of items, find the middle two numbers (4th and 5th of 8):

median $= \dfrac{\$49.30 + \$51.80}{2} = \dfrac{\$101.10}{2} = \$50.55$

25. hypotenuse $= \sqrt{(\text{leg})^2 + (\text{leg})^2}$
$= \sqrt{(7)^2 + (6)^2}$
$= \sqrt{49 + 36}$
$= \sqrt{85}$
≈ 9.2 cm

26. The unknown side is a leg of the right triangle.

leg $= \sqrt{(\text{hypotenuse})^2 - (\text{leg})^2}$
$= \sqrt{(20)^2 - (11)^2}$
$= \sqrt{400 - 121}$
$= \sqrt{279}$
≈ 16.7 ft (nearest tenth)

Cumulative Review Exercises (Chapters 1–5)   185

27. radius $= \dfrac{d}{2} = \dfrac{25 \text{ in.}}{2} = 12.5$ in. or $12\dfrac{1}{2}$ in.

28. $C = 2 \cdot \pi \cdot r$
    $\approx 2 \cdot 3.14 \cdot 0.9$ km
    ≈ 5.7 km

29. $r = \dfrac{d}{2} = \dfrac{16.2 \text{ cm}}{2} = 8.1$ cm
    $A = \pi \cdot r^2$
    $\approx 3.14 \cdot 8.1 \text{ cm} \cdot 8.1$ cm
    $\approx 206.0 \text{ cm}^2$

30. The figure is a cylinder.
    $V = \pi \cdot r^2 \cdot h$
    $\approx 3.14 \cdot 18 \text{ ft} \cdot 18 \text{ ft} \cdot 5$ ft
    $= 5086.8 \text{ ft}^3$

31. $SA = 2\pi rh + 2\pi r^2$
    $\approx 2 \cdot 3.14 \cdot 18 \text{ ft} \cdot 5 \text{ ft} + 2 \cdot 3.14 \cdot 18 \text{ ft} \cdot 18$ ft
    $\approx 2599.9 \text{ ft}^2$

Cumulative Review Exercises (Chapters 1–5)

1. **(a)** In words, 45.0203 is

    forty-five and two hundred three ten-thousandths.

    (b) In words, 30,000,650,008 is

    thirty billion, six hundred fifty thousand, eight.

2. **(a)** One hundred sixty million, five hundred is written as 160,000,500.

    (b) Seventy-five thousandths is written as 0.075.

3. Underline the hundreds place: 46,$\underline{9}$08

    The next digit is 4 or less. Leave 9 and change 8 to 0.

    46,900

4. Underline the hundredths place: 6.1$\underline{9}$7

    The next digit is 5 or more. Add 1 to 9. Write 0 and carry 1 to the tenths place. Drop all digits to the right of the underlined place.

    6.20

5. Underline the thousandths place: 0.66$\underline{1}$48

    The next digit is 4 or less. Leave 1 and drop all digits to the right of the underlined place.

    0.661

6. Underline the hundreds place: 9$\underline{9}$51

    The next digit is 5 or more. Add 1 to 9. Write 0 and carry 1 to the thousands place. Change 5 and 1 to 0.

    10,000

7. $-5 - 8$ Change subtraction.
   $= -5 + (-8)$ Add.
   $= -13$

8. $-0.003(0.02)$ The product is negative.

    $\begin{array}{r} 0.003 \\ \times\, 0.02 \\ \hline 0.00006 \end{array}$ $\leftarrow$ 3 *decimal places*
                     $\leftarrow$ 2 *decimal places*
                     $\leftarrow$ 5 *decimal places*

    Answer: -0.00006

9. $\dfrac{-7}{0}$ is undefined.

10. $8 + 4(2 - 5)^2$ Parentheses
    $= 8 + 4(-3)^2$ Exponent
    $= 8 + 4(9)$ Multiply.
    $= 8 + 36$ Add.
    $= 44$

11. $-\dfrac{3}{8}(-48) = -\dfrac{3}{8}\left(-\dfrac{48}{1}\right) = \dfrac{3 \cdot 6 \cdot \overset{1}{\cancel{8}}}{\cancel{8} \cdot 1} = \dfrac{18}{1} = 18$

12. $|4| - |-10|$ Absolute values
    $= 4 - 10$ Change subtraction.
    $= 4 + (-10)$ Add.
    $= -6$

13. $0.721 + 55.9$

    $\begin{array}{r} \overset{1}{}0.721 \\ + 55.900 \\ \hline 56.621 \end{array}$ Line up decimals.
               Fill in zeros.

14. $3\dfrac{1}{3} - 1\dfrac{5}{6} = \dfrac{10}{3} - \dfrac{11}{6} = \dfrac{20}{6} - \dfrac{11}{6} = \dfrac{20 - 11}{6}$
    $= \dfrac{9}{6} = \dfrac{9 \div 3}{6 \div 3} = \dfrac{3}{2}$ or $1\dfrac{1}{2}$

15. $\left(\dfrac{3}{b^2}\right)\left(\dfrac{b}{8}\right) = \dfrac{3 \cdot \overset{1}{\cancel{b}}}{\underset{1}{\cancel{b}} \cdot b \cdot 8} = \dfrac{3}{8b}$

16. $12 - 0.853$

    Line up decimal points and fill in zeros for subtracting.

    $\begin{array}{r} 1\overset{1}{\cancel{2}}.\overset{9}{\cancel{0}}\overset{9}{\cancel{0}}\overset{10}{\cancel{0}} \\ -\, 0.853 \\ \hline 11.147 \end{array}$

17. $\dfrac{3}{10} - \dfrac{3}{4} = \dfrac{6}{20} - \dfrac{15}{20} = \dfrac{6 - 15}{20}$
    $= \dfrac{-9}{20}$ or $-\dfrac{9}{20}$

18. $\dfrac{\frac{5}{16}}{-10} = \dfrac{5}{16} \div (-10) = \dfrac{5}{16} \div \left(-\dfrac{10}{1}\right)$

$= \dfrac{5}{16} \cdot \left(-\dfrac{1}{10}\right) = -\dfrac{\overset{1}{\cancel{5}} \cdot 1}{16 \cdot 2 \cdot \underset{1}{\cancel{5}}}$

$= -\dfrac{1}{32}$

19. $-3.75 \div (-2.9)$
$= -37.5 \div (-29)$ Move both decimal points one place to the right.

Same signs, so the quotient is positive.

```
       1. 2 9   ← In order to round to the tenths,
  29 ) 3 7. 5 0    divide out one more place, to
       2 9         the hundredths.
       ———
         8 5
         5 8
         ———
         2 7 0
         2 6 1
         —————
             9
```

Answer: 1.3 (rounded)

20. $\dfrac{x}{2} + \dfrac{3}{5} = \dfrac{x \cdot 5}{2 \cdot 5} + \dfrac{3 \cdot 2}{5 \cdot 2} = \dfrac{5x}{10} + \dfrac{6}{10} = \dfrac{5x + 6}{10}$

21. $5 \div \left(-\dfrac{5}{8}\right) = \dfrac{5}{1} \cdot \left(-\dfrac{8}{5}\right) = -\dfrac{\overset{1}{\cancel{5}} \cdot 8}{1 \cdot \underset{1}{\cancel{5}}}$

$= -\dfrac{8}{1} = -8$

22. $2^5 - 4^3$ Exponents first
$= 32 - 64$ Change subtraction.
$= 32 + (-64)$ Add.
$= -32$

23. $\dfrac{-36 \div (-2)}{-6 - 4(0 - 6)}$

Numerator:

$-36 \div (-2) = \dfrac{36}{2} = \dfrac{18 \cdot \overset{1}{\cancel{2}}}{\underset{1}{\cancel{2}}} = 18$

Denominator:

$-6 - 4(0 - 6) = -6 - 4(-6)$
$ = -6 - (-24)$
$ = -6 + 24$
$ = 18$

Last step is division: $\dfrac{18}{18} = 1$

24. $(0.8)^2 - 3.2 + 4(-0.8)$ Exponent
$= 0.64 - 3.2 + 4(-0.8)$ Multiply.
$= 0.64 - 3.2 + (-3.2)$ Add the opposite.
$= 0.64 + (-3.2) + (-3.2)$ Add.
$= -2.56 + (-3.2)$
$= -5.76$

25. $\dfrac{3}{4} \div \dfrac{3}{10}\left(\dfrac{1}{4} + \dfrac{2}{3}\right)$ Parentheses, LCD is 12

$= \dfrac{3}{4} \div \dfrac{3}{10}\left(\dfrac{3}{12} + \dfrac{8}{12}\right)$ Add.

$= \dfrac{3}{4} \div \dfrac{3}{10} \cdot \dfrac{11}{12}$ Change division.

$= \dfrac{3}{4} \cdot \dfrac{10}{3} \cdot \dfrac{11}{12}$ Multiply.

$= \dfrac{5}{2} \cdot \dfrac{11}{12}$ Multiply.

$= \dfrac{55}{24}$ or $2\dfrac{7}{24}$

26. To find the mean, add the ages and divide by the number of students.

$\text{mean} = \dfrac{\begin{array}{l}(19 + 23 + 24 + 19 + 20 + 29 + 26 + 35 \\ + 20 + 22 + 26 + 23 + 25 + 26 + 20 + 30)\end{array}}{16}$

$= \dfrac{387}{16}$

$= 24.1875$

≈ 24.2 years (nearest tenth)

27. To find the median, first arrange the data from high to low or low to high.

19, 19, 20, 20, 20, 22, 23, 23, 24, 25, 26, 26, 26, 29, 30, 35

Since there is an even number of items (16), the median is defined as the average of the middle two numbers (8th and 9th).

$\dfrac{23 + 24}{2} = 23.5$

The median age is 23.5 years.

28. To find the mode, look for the items which occur most frequently.

This set has 2 modes, 20 and 26 years (each occurs 3 times). We say that the list is bimodal.

29. $3h - 4h = 16 - 12$ Add terms.
$-1h = 4$ Divide both sides by -1.
$\dfrac{-1h}{-1} = \dfrac{4}{-1}$
$h = -4$

The solution is -4.

Cumulative Review Exercises (Chapters 1–5)

30.
$$-2x = x - 15$$
$$-2x = 1x - 15 \quad \text{Add } -1x \text{ to both sides.}$$
$$\underline{-1x \quad\quad -1x}$$
$$-3x = 0 - 15 \quad \text{Divide both sides by } -3.$$
$$\frac{-3x}{-3} = \frac{-15}{-3}$$
$$x = 5$$

The solution is 5.

31.
$$20 = 6r - 45.4 \quad \text{Add 45.4 to both sides.}$$
$$\underline{+45.4 \quad\quad +45.4}$$
$$65.4 = 6r + 0 \quad \text{Divide both sides by 6.}$$
$$\frac{65.4}{6} = \frac{6r}{6}$$
$$10.9 = r$$

The solution is 10.9.

32.
$$-3(y+4) = 7y + 8 \quad \text{Distributive property}$$
$$-3y + (-12) = 7y + 8 \quad \text{Add } 3y \text{ to both sides.}$$
$$\underline{+3y \quad\quad\quad +3y}$$
$$0 + (-12) = 10y + 8$$
$$-12 = 10y + 8 \quad \text{Add } -8 \text{ to both sides.}$$
$$\underline{-8 \quad\quad -8}$$
$$-20 = 10y + 0 \quad \text{Divide both sides by 10.}$$
$$\frac{-20}{10} = \frac{10y}{10}$$
$$-2 = y$$

The solution is -2.

33.
$$-0.8 + 1.4n = 2 + 0.7n \quad \text{Add } -0.7n \text{ to both sides.}$$
$$\underline{-0.7n \quad\quad -0.7n}$$
$$-0.8 + 0.7n = 2 + 0$$
$$-0.8 + 0.7n = 2 \quad \text{Add 0.8 to both sides.}$$
$$\underline{+0.8 \quad\quad\quad +0.8}$$
$$0 + 0.7n = 2.8 \quad \text{Divide both sides by 0.7.}$$
$$\frac{0.7n}{0.7} = \frac{2.8}{0.7}$$
$$n = 4$$

The solution is 4.

34. $P = 10 \text{ ft} + 17 \text{ ft} + 10 \text{ ft} + 17 \text{ ft}$
$P = 54$ ft

$A = bh$
$A = 10 \text{ ft} \cdot 14 \text{ ft}$
$A = 140 \text{ ft}^2$

35. $C = \pi d$
$C \approx 3.14 \cdot 13$ m
$C \approx 40.8$ m (nearest tenth)

$A = \pi \cdot r^2 \quad \text{Note: } r = \frac{d}{2} = \frac{13}{2}$
$A = \pi \cdot r \cdot r$
$A = \pi \cdot \frac{13}{2} \text{ m} \cdot \frac{13}{2} \text{ m}$
$A \approx 3.14 \cdot 6.5 \text{ m} \cdot 6.5 \text{ m}$
$A \approx 132.7 \text{ m}^2$ (nearest tenth)

36. The figure is a right triangle, so use the Pythagorean Theorem.

$\text{leg} = \sqrt{(\text{hypotenuse})^2 - (\text{leg})^2}$
$x = \sqrt{(14)^2 - (6)^2}$
$x = \sqrt{196 - 36}$
$x = \sqrt{160}$
$x \approx 12.6$ km (nearest tenth)

$A = \frac{1}{2}bh$
$A \approx 0.5 \cdot 12.6 \text{ km} \cdot 6 \text{ km}$
$A \approx 37.8 \text{ km}^2$

37. The figure is a cube or rectangular solid.
$V = l \cdot w \cdot h$
$V = 1.5 \text{ in.} \cdot 1.5 \text{ in.} \cdot 1.5 \text{ in.}$
$V = 3.375 \text{ in.}^3 \text{ or } 3.4 \text{ in.}^3$ (rounded)

$SA = 2lw + 2lh + 2wh$
$SA = 2 \cdot 1.5 \text{ in.} \cdot 1.5 \text{ in.} + 2 \cdot 1.5 \text{ in.} \cdot 1.5 \text{ in.}$
$\quad\quad + 2 \cdot 1.5 \text{ in.} \cdot 1.5 \text{ in.}$
$SA = 13.5 \text{ in.}^2$

38. Three $20 bills is worth $60.
$47.96 rounds to $50.
$0.87 rounds to $1.

Estimate: $60 − $50 − $1 = $9
Exact: $60 − $47.96 − $0.87 = $11.17

He has $11.17 left.

39. $2\frac{1}{3}$ yd rounds to 2 yd
$3\frac{7}{8}$ yd rounds to 4 yd

Estimate: 2 yd + 4 yd = 6 yd

Exact: $2\frac{1}{3}$ yd + $3\frac{7}{8}$ yd = $\frac{7}{3}$ yd + $\frac{31}{8}$ yd
$= \frac{56}{24}$ yd + $\frac{93}{24}$ yd
$= \frac{149}{24}$ yd
$= 6\frac{5}{24}$ yd

He bought a total of $6\frac{5}{24}$ yd of fabric.

40. 2.7 pounds rounds to 3 pounds.
$6.18 rounds to $6.

Estimate: $6 ÷ 3 = $2 per pound

Exact: $6.18 ÷ 2.7 = 2.28\overline{8}$
≈ $2.29 per pound

41. $78,000 rounds to $80,000
107 students rounds to 100 students.

Estimate: $80,000 ÷ 100 = $800

Exact: $78,000 ÷ 107
≈ $728.97
≈ $729 (rounded to the nearest dollar)

About $729 could be given to each of the students.

42. *Step 1*
Unknowns: amount each student receives

Knowns: one student receives three times as much as the other; $30,000 total

Step 2
Let x be the amount of the first student's money and $3x$ be the amount of the second student's money.

Step 3

first student's money		second student's money	is	scholarship total
x	+	$3 \cdot x$	=	30,000

Step 4
$x + 3x = 30{,}000$
$4x = 30{,}000$
$\frac{4x}{4} = \frac{30{,}000}{4}$
$x = \$7500$
$3 \cdot x = 3 \cdot \$7500 = \$22{,}500$

Step 5
One student receives $7500 and the other student receives $22,500.

Step 6
$22,500 is three times $7500 and the sum of $7500 and $22,500 is $30,000.

43. *Step 1*
Unknowns: length and width of the rectangle

Knowns: length is 4 in. more than the width; perimeter is 48 in.

Step 2
Let w be the width and $w + 4$ be the length.

Step 3
$P = 2w + 2l$
$48 = 2w + 2(w + 4)$

Step 4
$48 = 2w + 2w + 8$
$48 = 4w + 8$
$\underline{-8 \quad\quad -8}$
$40 = 4w$
$\frac{40}{4} = \frac{4w}{4}$
$10 = w$, so $w + 4 = 14$

Step 5
The width is 10 in. and the length is 14 in.

Step 6
14 in. is 4 in. longer than 10 in. and the perimeter is 2(14 in.) + 2(10 in.) = 48 in.

CHAPTER 6 RATIO, PROPORTION, AND LINE/ANGLE/TRIANGLE RELATIONSHIPS

6.1 Ratios

6.1 Margin Exercises

1. **(a)** $7 spent on fresh fruit to $5 spent on milk
   $$\frac{7}{5}$$

   (b) $5 spent on milk to $14 spent on meat
   $$\frac{5}{14}$$

   (c) $14 spent on meat to $5 spent on milk
   $$\frac{14}{5}$$

2. **(a)** 9 hours to 12 hours
   $$\frac{9}{12} = \frac{9 \div 3}{12 \div 3} = \frac{3}{4}$$

   (b) 100 meters to 50 meters
   $$\frac{100}{50} = \frac{100 \div 50}{50 \div 50} = \frac{2}{1}$$

   (c) width to length or 24 feet to 48 feet
   $$\frac{24}{48} = \frac{24 \div 24}{48 \div 24} = \frac{1}{2}$$

3. **(a)** The increase in price is
   $$\$7.00 - \$5.50 = \$1.50.$$

   The ratio of increase in price to original price is
   $$\frac{1.50}{5.50}.$$

   Rewrite as two whole numbers.
   $$\frac{1.50}{5.50} = \frac{1.50 \times 100}{5.50 \times 100} = \frac{150}{550}$$

   Write in lowest terms.
   $$\frac{150 \div 50}{550 \div 50} = \frac{3}{11}$$

   (b) The decrease in hours is
   $$4.5 \text{ hours } - 3 \text{ hours } = 1.5 \text{ hours}.$$

   The ratio of decrease in hours to the original number of hours is
   $$\frac{1.5}{4.5}.$$

   Rewrite as two whole numbers.
   $$\frac{1.5 \cdot 10}{4.5 \cdot 10} = \frac{15}{45}$$

   Write in lowest terms.
   $$\frac{15 \div 15}{45 \div 15} = \frac{1}{3}$$

4. **(a)** $3\frac{1}{2}$ to 4
   $$\frac{3\frac{1}{2}}{4} = \frac{\frac{7}{2}}{\frac{4}{1}} = \frac{7}{2} \div \frac{4}{1} = \frac{7}{2} \cdot \frac{1}{4} = \frac{7}{8}$$

   (b) $5\frac{5}{8}$ pounds to $3\frac{3}{4}$ pounds
   $$\frac{5\frac{5}{8}}{3\frac{3}{4}} = \frac{\frac{45}{8}}{\frac{15}{4}} = \frac{45}{8} \div \frac{15}{4} = \frac{\cancel{45}^3}{\cancel{8}_2} \cdot \frac{\cancel{4}^1}{\cancel{15}_1} = \frac{3}{2}$$

   (c) $3\frac{1}{3}$ inches to $\frac{5}{6}$ inch
   $$\frac{3\frac{1}{3}}{\frac{5}{6}} = \frac{\frac{10}{3}}{\frac{5}{6}} = \frac{10}{3} \div \frac{5}{6}$$
   $$= \frac{10}{3} \cdot \frac{6}{5} = \frac{\cancel{10}^2 \cdot \cancel{6}^2}{\cancel{3}_1 \cdot \cancel{5}_1} = \frac{4}{1}$$

5. **(a)** 9 inches to 6 feet
   $$6 \text{ feet } = 6 \cdot 12 \text{ inches}$$
   $$= 72 \text{ inches}$$
   $$\frac{9 \text{ inches}}{72 \text{ inches}} = \frac{9}{72} = \frac{9 \div 9}{72 \div 9} = \frac{1}{8}$$

   (b) 2 days to 8 hours
   $$2 \text{ days } = 2 \cdot 24 \text{ hours } = 48 \text{ hours}$$
   $$\frac{48 \text{ hours}}{8 \text{ hours}} = \frac{48}{8} = \frac{48 \div 8}{8 \div 8} = \frac{6}{1}$$

   (c) 7 yards to 14 feet
   $$7 \text{ yards } = 7 \cdot 3 \text{ feet } = 21 \text{ feet}$$
   $$\frac{21 \text{ feet}}{14 \text{ feet}} = \frac{21}{14} = \frac{21 \div 7}{14 \div 7} = \frac{3}{2}$$

   (d) 3 quarts to 3 gallons
   $$3 \text{ gallons } = 3 \cdot 4 = 12 \text{ quarts}$$
   $$\frac{3 \text{ quarts}}{12 \text{ quarts}} = \frac{3}{12} = \frac{3 \div 3}{12 \div 3} = \frac{1}{4}$$

   (e) 25 minutes to 2 hours
   $$2 \text{ hours } = 2 \cdot 60 \text{ minutes}$$
   $$= 120 \text{ minutes}$$
   $$\frac{25 \text{ minutes}}{120 \text{ minutes}} = \frac{25}{120} = \frac{25 \div 5}{120 \div 5} = \frac{5}{24}$$

(f) 4 pounds to 12 ounces

4 pounds = 4 · 16 ounces
= 64 ounces

$$\frac{64 \text{ ounces}}{12 \text{ ounces}} = \frac{64}{12} = \frac{64 \div 4}{12 \div 4} = \frac{16}{3}$$

6.1 Section Exercises

1. 8 days to 9 days: $\frac{8 \text{ days}}{9 \text{ days}} = \frac{8}{9}$

3. $100 to $50

$$\frac{\$100}{\$50} = \frac{100}{50} = \frac{100 \div 50}{50 \div 50} = \frac{2}{1}$$

5. 30 minutes to 90 minutes

$$\frac{30 \text{ minutes}}{90 \text{ minutes}} = \frac{30}{90} = \frac{30 \div 30}{90 \div 30} = \frac{1}{3}$$

7. $4.50 to $3.50

$$\frac{\$4.50}{\$3.50} = \frac{4.50}{3.50} = \frac{4.50 \cdot 10}{3.50 \cdot 10} = \frac{45}{35}$$
$$= \frac{45 \div 5}{35 \div 5} = \frac{9}{7}$$

9. $1\frac{1}{4}$ to $1\frac{1}{2}$

$$\frac{1\frac{1}{4}}{1\frac{1}{2}} = \frac{\frac{5}{4}}{\frac{3}{2}} = \frac{5}{4} \div \frac{3}{2} = \frac{5}{\underset{2}{\cancel{4}}} \cdot \frac{\overset{1}{\cancel{2}}}{3} = \frac{5}{6}$$

11. 4 feet to 30 inches

4 feet = 4 · 12 inches = 48 inches

$$\frac{4 \text{ feet}}{30 \text{ inches}} = \frac{48 \text{ inches}}{30 \text{ inches}} = \frac{48}{30} = \frac{48 \div 6}{30 \div 6} = \frac{8}{5}$$

(Do not write the ratio as $1\frac{3}{5}$.)

13. 5 minutes to 1 hour

1 hour = 60 minutes

$$\frac{5 \text{ minutes}}{1 \text{ hour}} = \frac{5 \text{ minutes}}{60 \text{ minutes}} = \frac{5}{60} = \frac{5 \div 5}{60 \div 5} = \frac{1}{12}$$

15. 5 gallons to 5 quarts

5 gallons = 5 · 4 quarts = 20 quarts

$$\frac{5 \text{ gallons}}{5 \text{ quarts}} = \frac{20 \text{ quarts}}{5 \text{ quarts}} = \frac{20}{5} = \frac{20 \div 5}{5 \div 5} = \frac{4}{1}$$

17. Thanksgiving cards to graduation cards

$$\frac{30 \text{ million}}{60 \text{ million}} = \frac{30}{60} = \frac{30 \div 30}{60 \div 30} = \frac{1}{2}$$

19. Valentine's Day cards to Halloween cards

$$\frac{900 \text{ million}}{25 \text{ million}} = \frac{900}{25} = \frac{900 \div 25}{25 \div 25} = \frac{36}{1}$$

21. Answers will vary. One possibility is stocking cards of various types in the same ratios as those in the table.

23. To get a ratio of $\frac{3}{1}$, we can start with the smallest value in the table and see if there is a value that is three times the smallest. $3 \times 9 = 27$ and 27 million is not in the table. $3 \times 10 = 30$, so there are at least 2 pairs of songs that give a ratio of $\frac{3}{1}$: *White Christmas* to *It's Now or Never* and *White Christmas* to *I Will Always Love You*.
$3 \times 12 = 36$, so there is another pair: *Candle in the Wind* to *I Want to Hold Your Hand*.
$3 \times 17 = 51$, which is greater than the largest value in the table, so we do not need to look for any more pairs.

25. $\frac{\text{longest side}}{\text{shortest side}} = \frac{1.8 \text{ meters}}{0.3 \text{ meters}} = \frac{1.8}{0.3}$

$$= \frac{1.8 \cdot 10}{0.3 \cdot 10} = \frac{18}{3} = \frac{18 \div 3}{3 \div 3} = \frac{6}{1}$$

The ratio of the length of the longest side to the length of the shortest side is $\frac{6}{1}$.

27. $\frac{\text{longest side}}{\text{shortest side}} = \frac{9\frac{1}{2} \text{ inches}}{4\frac{1}{4} \text{ inches}} = \frac{9\frac{1}{2}}{4\frac{1}{4}} = \frac{\frac{19}{2}}{\frac{17}{4}}$

$$= \frac{19}{2} \div \frac{17}{4} = \frac{19}{\underset{1}{\cancel{2}}} \cdot \frac{\overset{2}{\cancel{4}}}{17} = \frac{38}{17}$$

The ratio of the length of the longest side to the length of the shortest side is $\frac{38}{17}$.

29. The increase in price is

$12.50 - $10.00 = $2.50.

$$\frac{\$2.50}{\$10.00} = \frac{2.50}{10.00} = \frac{2.50 \cdot 10}{10.00 \cdot 10} = \frac{25}{100}$$
$$= \frac{25 \div 25}{100 \div 25} = \frac{1}{4}$$

The ratio of the increase in price to the original price is $\frac{1}{4}$.

31. $59\frac{1}{2} \text{ days} \div 7 = \frac{\overset{17}{\cancel{119}}}{2} \cdot \frac{1}{\underset{1}{\cancel{7}}} = \frac{17}{2}$ weeks

$$\frac{\frac{17}{2} \text{ weeks}}{8\frac{3}{4} \text{ weeks}} = \frac{\frac{17}{2}}{\frac{35}{4}} = \frac{17}{2} \div \frac{35}{4} = \frac{17}{\underset{1}{\cancel{2}}} \cdot \frac{\overset{2}{\cancel{4}}}{35} = \frac{34}{35}$$

The ratio of the first movie's filming time to the second movie's filming time is $\frac{34}{35}$.

6.2 Rates

6.2 Margin Exercises

1. **(a)** $6 for 30 packages

$$\frac{\$6 \div 6}{30 \text{ packages} \div 6} = \frac{1 \text{ dollar}}{5 \text{ packages}}$$

6.2 Rates

(b) 500 miles in 10 hours

$$\frac{500 \text{ miles} \div 10}{10 \text{ hours} \div 10} = \frac{50 \text{ miles}}{1 \text{ hour}}$$

(c) 4 teachers for 90 students

$$\frac{4 \text{ teachers} \div 2}{90 \text{ students} \div 2} = \frac{2 \text{ teachers}}{45 \text{ students}}$$

(d) 1270 bushels from 30 acres

$$\frac{1270 \text{ bushels} \div 10}{30 \text{ acres} \div 10} = \frac{127 \text{ bushels}}{3 \text{ acres}}$$

2. **(a)** $4.35 for 3 pounds of cheese

$$\frac{\$4.35 \div 3}{3 \text{ pounds} \div 3} = \frac{\$1.45}{1 \text{ pound}}$$

The unit rate is $1.45/pound.

(b) 304 miles on 9.5 gallons of gas

$$\frac{304 \text{ miles} \div 9.5}{9.5 \text{ gallons} \div 9.5} = \frac{32 \text{ miles}}{1 \text{ gallon}}$$

The unit rate is 32 miles/gallon.

(c) $850 in 5 days

$$\frac{\$850 \div 5}{5 \text{ days} \div 5} = \frac{\$170}{1 \text{ day}}$$

The unit rate is $170/day.

(d) 24-pound turkey for 15 people

$$\frac{24 \text{ pounds} \div 15}{15 \text{ people} \div 15} = \frac{1.6 \text{ pounds}}{1 \text{ person}}$$

The unit rate is 1.6 pounds/person.

3. **(a)**

Size	Cost per Unit
2 quarts	$\frac{\$3.25}{2 \text{ quarts}} = \$1.625$
3 quarts	$\frac{\$4.95}{3 \text{ quarts}} = \$1.65$
4 quarts	$\frac{\$6.48}{4 \text{ quarts}} = \$1.62$ (*)

The lowest cost per quart is $1.62, so the best buy is 4 quarts for $6.48.

(b)

Size	Cost per Unit
6 cans	$\frac{\$1.99}{6 \text{ cans}} \approx \$0.332$
12 cans	$\frac{\$3.49}{12 \text{ cans}} \approx \$0.291$ (*)
24 cans	$\frac{\$7}{24 \text{ cans}} \approx \$0.292$

The lowest price per can is about $0.291, so the best buy is 12 cans of cola for $3.49.

4. **(a)** One battery that lasts twice as long is like getting two batteries.

$$\frac{\$1.19 \div 2}{2 \text{ batteries} \div 2} = \$0.595/\text{battery.} \quad (*)$$

The cost of the four-pack is

$$\frac{\$2.79 \div 4}{4 \text{ batteries} \div 4} = \$0.6975/\text{battery.}$$

The best buy is the single AA-size battery.

(b) **Brand C:**

$3.89 − $0.85 = $3.04

$$\frac{\$3.04 \div 6}{6 \text{ ounces} \div 6} \approx \$0.507/\text{ounce} \quad (*)$$

Brand D:

$1.59 − $0.20 = $1.39

$$\frac{\$1.39 \div 2.5}{2.5 \text{ ounces} \div 2.5} = \$0.556/\text{ounce}$$

Brand C with the 85¢ coupon is the best buy at $0.507 per ounce (rounded).

6.2 Section Exercises

1. 10 cups for 6 people

$$\frac{10 \text{ cups} \div 2}{6 \text{ people} \div 2} = \frac{5 \text{ cups}}{3 \text{ people}}$$

3. 15 feet in 35 seconds

$$\frac{15 \text{ feet} \div 5}{35 \text{ seconds} \div 5} = \frac{3 \text{ feet}}{7 \text{ seconds}}$$

5. 14 people for 28 dresses

$$\frac{14 \text{ people} \div 14}{28 \text{ dresses} \div 14} = \frac{1 \text{ person}}{2 \text{ dresses}}$$

7. 25 letters in 5 minutes

$$\frac{25 \text{ letters} \div 5}{5 \text{ minutes} \div 5} = \frac{5 \text{ letters}}{1 \text{ minute}}$$

9. $63 for 6 visits

$$\frac{\$63 \div 3}{6 \text{ visits} \div 3} = \frac{21 \text{ dollars}}{2 \text{ visits}}$$

11. 72 miles on 4 gallons

$$\frac{72 \text{ miles} \div 4}{4 \text{ gallons} \div 4} = \frac{18 \text{ miles}}{1 \text{ gallon}}$$

13. $60 in 5 hours

$$\frac{\$60 \div 5}{5 \text{ hours} \div 5} = \frac{\$12}{1 \text{ hour}}$$

The unit rate is $12 per hour or $12/hour.

15. 50 eggs from 10 chickens

$$\frac{50 \text{ eggs} \div 10}{10 \text{ chickens} \div 10} = \frac{5 \text{ eggs}}{1 \text{ chicken}}$$

The unit rate is 5 eggs per chicken or 5 eggs/chicken.

17. 7.5 pounds for 6 people

$$\frac{7.5 \text{ pounds} \div 6}{6 \text{ people} \div 6} = \frac{1.25 \text{ pounds}}{1 \text{ person}}$$

The unit rate is 1.25 pounds/person.

19. $413.20 for 4 days

$$\frac{\$413.20 \div 4}{4 \text{ days} \div 4} = \frac{\$103.30}{1 \text{ day}}$$

The unit rate is $103.30/day.

21. Miles traveled:

$27{,}758.2 - 27{,}432.3 = 325.9$

Miles per gallon:

$$\frac{325.9 \text{ miles} \div 15.5}{15.5 \text{ gallons} \div 15.5} \approx 21.03 \approx 21.0$$

23. Miles traveled:

$28{,}396.7 - 28{,}058.1 = 338.6$

Miles per gallon:

$$\frac{338.6 \text{ miles} \div 16.2}{16.2 \text{ gallons} \div 16.2} \approx 20.90 \approx 20.9$$

25.

Size	Cost per Unit	
2 ounces	$\dfrac{\$2.25}{2 \text{ ounces}}$	$= \$1.125$
4 ounces	$\dfrac{\$3.65}{4 \text{ ounces}}$	$= \$0.9125 \; (*)$

The best buy is 4 ounces for $3.65, about $0.913/ounce.

27.

Size	Cost per Unit	
12 ounces	$\dfrac{\$2.49}{12 \text{ ounces}}$	$= \$0.2075$
14 ounces	$\dfrac{\$2.89}{14 \text{ ounces}}$	$\approx \$0.2064 \; (*)$
18 ounces	$\dfrac{\$3.96}{18 \text{ ounces}}$	$= \$0.22$

The best buy is 14 ounces for $2.89, about $0.206/ounce.

29.

Size	Cost per Unit	
12 ounces	$\dfrac{\$1.29}{12 \text{ ounces}}$	$\approx \$0.108$
18 ounces	$\dfrac{\$1.79}{18 \text{ ounces}}$	$\approx \$0.099 \; (*)$
28 ounces	$\dfrac{\$3.39}{28 \text{ ounces}}$	$\approx \$0.121$
40 ounces	$\dfrac{\$4.39}{40 \text{ ounces}}$	$\approx \$0.110$

The best buy is 18 ounces for $1.79, about $0.099/ounce.

31. Answers will vary. For example, you might choose Brand B because you like more chicken, so the cost per chicken chunk may actually be the same or less than Brand A.

33. 10.5 pounds in 6 weeks

$$\frac{10.5 \text{ pounds} \div 6}{6 \text{ weeks} \div 6} = \frac{1.75 \text{ pounds}}{1 \text{ week}}$$

Her rate of loss was 1.75 pounds/week.

35. 7 hours to earn $85.82

$$\frac{\$85.82 \div 7}{7 \text{ hours} \div 7} = \frac{\$12.26}{1 \text{ hour}}$$

His pay rate is $12.26/hour.

37. **(a)** Add the connection fee and five times the cost per minute.

Radiant Penny:
$\$0.39 + 5(\$0.01) = \$0.39 + \$0.05 = \$0.44$

IDT Special:
$\$0.14 + 5(\$0.022) = \$0.14 + \$0.11 = \$0.25$

Access America:
$\$0.00 + 5(\$0.047) = \$0.235$

(b) Divide the total costs by five minutes.

Radiant Penny: $\dfrac{\$0.44 \div 5}{5 \text{ minutes} \div 5} = \$0.088/\text{minute}$

IDT Special: $\dfrac{\$0.25 \div 5}{5 \text{ minutes} \div 5} = \$0.05/\text{minute}$

Access America: $\dfrac{\$0.235 \div 5}{5 \text{ minutes} \div 5} = \$0.047/\text{minute}$

Access America is the best buy.

39. Add the connection fee and fifteen (and the twenty) times the cost per minute.

Radiant Penny:
$0.39 + 15(\$0.01) = \$0.39 + \$0.15 = \$0.54$
$0.39 + 20(\$0.01) = \$0.39 + \$0.20 = \$0.59$

IDT Special:
$0.14 + 15(\$0.022) = \$0.14 + \$0.33 = \$0.47$
$0.14 + 20(\$0.022) = \$0.14 + \$0.44 = \$0.58$

Access America:
$0.00 + 15(\$0.047) = \$0.705$
$0.00 + 20(\$0.047) = \$0.94$

For the 15-minute call, IDT Special is the best buy at $0.47. Alternatively, we could calculate the cost per minute.
For the 20-minute call, IDT Special is the best buy at $0.58.

41. $\dfrac{44 \text{ seconds}}{400 \text{ meters}} = 0.11 \text{ second/meter}$

$\dfrac{400 \text{ meters} \div 44}{44 \text{ seconds} \div 44} = 9.\overline{09} \text{ meters/second}$
$\approx 9.1 \text{ meters/second}$

Michael Johnson's rate was 0.11 second/meter or 9.1 meters/second (rounded).

43. One battery for $1.79 is like getting 3 batteries for $1.79 ÷ 3 ≈ $0.597 per battery.

An eight-pack of AAA batteries for $4.99 is $4.99 ÷ 8 ≈ $0.624 per battery.

The best buy is the one battery package.

45. Brand G: $2.39 − $0.60 coupon = $1.79

$\dfrac{\$1.79}{10 \text{ ounces}} = \$0.179 \text{ per ounce}$

Brand K: $\dfrac{\$3.99}{20.3 \text{ ounces}} \approx \$0.197 \text{ per ounce}$

Brand P: $3.39 − $0.50 coupon = $2.89

$\dfrac{\$2.89}{16.5 \text{ ounces}} \approx \$0.175 \text{ per ounce } (*)$

Brand P with the 50¢ coupon has the lowest cost per ounce and is the best buy.

47. The one-time activation fee is $35 (or $36) for 12 months.

$\dfrac{\$35 \div 12}{12 \text{ months} \div 12} \approx \$2.92/\text{month}$

($36/12 = $3/month)

The activation fee will add $2.92 (rounded) per month for Verizon, T-Mobile, and Nextel; $3 per month for Sprint.

48. Assume there are exactly 52 weeks in a year. Then there are $52 \cdot 5 = 260$ weekdays in a year.

$\dfrac{260 \text{ weekdays} \div 12}{12 \text{ months} \div 12} \approx 21.7 \text{ weekdays/month}$

For Verizon:
$\dfrac{400 \text{ anytime minutes} \div 21.7}{21.7 \text{ weekdays} \div 21.7} \approx 18 \text{ minutes/weekday}$

For T-Mobile:
$\dfrac{600 \text{ anytime minutes} \div 21.7}{21.7 \text{ weekdays} \div 21.7} \approx 28 \text{ minutes/weekday}$

For Nextel and Sprint:
$\dfrac{500 \text{ anytime minutes} \div 21.7}{21.7 \text{ weekdays} \div 21.7} \approx 23 \text{ minutes/weekday}$

49. To find the actual average cost per "anytime minute," add the monthly charge for the activation fee (from Exercise 47) to the monthly charge, then divide by the number of anytime minutes.

For Verizon:
$\dfrac{\$2.92 + \$59.99}{400 \text{ minutes}} = \dfrac{\$62.91}{400 \text{ minutes}} \approx \$0.16/\text{minute}$

For T-Mobile:
$\dfrac{\$2.92 + \$39.99}{600 \text{ minutes}} = \dfrac{\$42.91}{600 \text{ minutes}} \approx \$0.07/\text{min } (*)$

For Nextel:
$\dfrac{\$2.92 + \$45.99}{500 \text{ minutes}} = \dfrac{\$48.91}{500 \text{ minutes}} \approx \$0.10/\text{minute}$

For Sprint:
$\dfrac{\$3 + \$55}{500 \text{ minutes}} = \dfrac{\$58}{500 \text{ minutes}} \approx \$0.12/\text{minute}$

T-Mobile is the best buy at about $0.07 per "anytime minute".

50. The total cost is the sum of the activation fee, two monthly charges, and the termination fee. The number of "anytime minutes" in each case is $2 \times 100 = 200$.

For Verizon:
$\dfrac{\$35 + 2(\$59.99) + \$175}{200 \text{ minutes}} = \dfrac{\$329.98}{200 \text{ minutes}}$
$\approx \$1.65/\text{minute}$

For T-Mobile:
$\dfrac{\$35 + 2(\$39.99) + \$200}{200 \text{ minutes}} = \dfrac{\$314.98}{200 \text{ minutes}}$
$\approx \$1.57/\text{minute}$

For Nextel:
$\dfrac{\$35 + 2(\$45.99) + \$200}{200 \text{ minutes}} = \dfrac{\$326.98}{200 \text{ minutes}}$
$\approx \$1.63/\text{minute}$

For Sprint:
$\dfrac{\$36 + 2(\$55) + \$150}{200 \text{ minutes}} = \dfrac{\$296}{200 \text{ minutes}}$
$= \$1.48/\text{minute}$

6.3 Proportions

6.3 Margin Exercises

1. (a) $7 is to 3 cans as $28 is to 12 cans.
$$\frac{\$7}{3 \text{ cans}} = \frac{\$28}{12 \text{ cans}}$$

   (b) 9 meters is to 16 meters as 18 meters is to 32 meters.
$$\frac{9}{16} = \frac{18}{32} \quad \textit{Common units cancel}$$

   (c) 5 is to 7 as 35 is to 49.
$$\frac{5}{7} = \frac{35}{49}$$

   (d) 10 is to 30 as 60 is to 180.
$$\frac{10}{30} = \frac{60}{180}$$

2. (a) $\frac{6}{12} = \frac{15}{30}$

   $\frac{6 \div 6}{12 \div 6} = \frac{1}{2}$ and $\frac{15 \div 15}{30 \div 15} = \frac{1}{2}$

   Both ratios are equivalent to $\frac{1}{2}$, so the proportion is *true*.

   (b) $\frac{20}{24} = \frac{3}{4}$

   $\frac{20}{24} = \frac{20 \div 4}{24 \div 4} = \frac{5}{6}$ and $\frac{3}{4}$

   Because $\frac{5}{6}$ is *not* equivalent to $\frac{3}{4}$, the proportion is *false*.

   (c) $\frac{25}{40} = \frac{30}{48}$

   $\frac{25 \div 5}{40 \div 5} = \frac{5}{8}$ and $\frac{30 \div 6}{48 \div 6} = \frac{5}{8}$

   Both ratios are equivalent to $\frac{5}{8}$, so the proportion is *true*.

   (d) $\frac{35}{45} = \frac{12}{18}$

   $\frac{35 \div 5}{45 \div 5} = \frac{7}{9}$ and $\frac{12 \div 6}{18 \div 6} = \frac{2}{3}$

   Because $\frac{7}{9}$ is *not* equivalent to $\frac{2}{3}$, the proportion is *false*.

3. (a) $\frac{5}{9} = \frac{10}{18}$

   Cross products: $5 \cdot 18 = 90$; $9 \cdot 10 = 90$

   The cross products are *equal*, so the proportion is *true*.

   (b) $\frac{32}{15} = \frac{16}{8}$

   Cross products: $32 \cdot 8 = 256$; $15 \cdot 16 = 240$

   The cross products are *unequal*, so the proportion is *false*.

   (c) $\frac{10}{17} = \frac{20}{34}$

   Cross products: $10 \cdot 34 = 340$; $17 \cdot 20 = 340$

   The cross products are *equal*, so the proportion is *true*.

   (d) $\frac{2.4}{6} = \frac{5}{12}$

   Cross products: $(2.4)(12) = 28.8$; $6 \cdot 5 = 30$

   The cross products are *unequal*, so the proportion is *false*.

   (e) $\frac{3}{4.25} = \frac{24}{34}$

   Cross products: $3 \cdot 34 = 102$; $(4.25)(24) = 102$

   The cross products are *equal*, so the proportion is *true*.

   (f) $\frac{1\frac{1}{6}}{2\frac{1}{3}} = \frac{4}{8}$

   Cross products:
$$1\frac{1}{6} \cdot 8 = \frac{7}{\cancel{6}_3} \cdot \frac{\cancel{8}^4}{1} = \frac{28}{3}$$
$$2\frac{1}{3} \cdot 4 = \frac{7}{3} \cdot \frac{4}{1} = \frac{28}{3}$$

   The cross products are *equal*, so the proportion is *true*.

4. (a) $\frac{1}{2} = \frac{x}{12}$

   $2 \cdot x = 1 \cdot 12$ *Cross products are equivalent*

   $\frac{\cancel{2} \cdot x}{\cancel{2}_1} = \frac{1 \cdot \cancel{12}^6}{\cancel{2}_1}$

   $x = 6$

   We will check our answers by showing that the cross products are equal.

   Check: $1 \cdot 12 = 12 = 2 \cdot 6$

(b) $\dfrac{6}{10} = \dfrac{15}{x}$

$6 \cdot x = 10 \cdot 15$ *Cross products are equivalent*

$\dfrac{\overset{1}{\cancel{6}} \cdot x}{\underset{1}{\cancel{6}}} = \dfrac{\overset{25}{\cancel{150}}}{\underset{1}{\cancel{6}}}$

$x = 25$

Check: $6 \cdot 25 = 150 = 10 \cdot 15$

(c) $\dfrac{28}{x} = \dfrac{21}{9}$

$x \cdot 21 = 28 \cdot 9$ *Cross products are equivalent*

$\dfrac{x \cdot \overset{1}{\cancel{21}}}{\underset{1}{\cancel{21}}} = \dfrac{\overset{4}{\cancel{28}} \cdot 9}{\underset{3}{\cancel{21}}}$

$x = 12$

Check: $28 \cdot 9 = 252 = 12 \cdot 21$

(d) $\dfrac{x}{8} = \dfrac{3}{5}$

$x \cdot 5 = 8 \cdot 3$ *Cross products are equivalent*

$\dfrac{x \cdot \overset{1}{\cancel{5}}}{\underset{1}{\cancel{5}}} = \dfrac{24}{5}$

$x = \dfrac{24}{5} = 4.8$

Check: $(4.8)(5) = 24 = 8 \cdot 3$

(e) $\dfrac{14}{11} = \dfrac{x}{3}$

$11 \cdot x = 14 \cdot 3$ *Cross products are equivalent*

$\dfrac{\overset{1}{\cancel{11}} \cdot x}{\underset{1}{\cancel{11}}} = \dfrac{14 \cdot 3}{11}$

$x = \dfrac{42}{11} \approx 3.82$ *(rounded to the nearest hundredth)*

Check: $14 \cdot 3 = 42 = 11 \cdot \dfrac{42}{11}$

5. (a) $\dfrac{3\frac{1}{4}}{2} = \dfrac{x}{8}$

$2 \cdot x = 3\dfrac{1}{4} \cdot 8$ *Cross products are equivalent*

$2 \cdot x = \dfrac{13}{\underset{1}{\cancel{4}}} \cdot \dfrac{\overset{2}{\cancel{8}}}{1}$

$\dfrac{\overset{1}{\cancel{2}} \cdot x}{\underset{1}{\cancel{2}}} = \dfrac{\overset{13}{\cancel{26}}}{\underset{1}{\cancel{2}}}$

$x = 13$

Check: $3\tfrac{1}{4} \cdot 8 = 26 = 2 \cdot 13$

(b) $\dfrac{x}{3} = \dfrac{1\frac{2}{3}}{5}$

$5 \cdot x = 3 \cdot 1\dfrac{2}{3}$ *Cross products are equivalent*

$5 \cdot x = \overset{1}{\cancel{3}} \cdot \dfrac{5}{\underset{1}{\cancel{3}}}$

$\dfrac{\overset{1}{\cancel{5}} \cdot x}{\underset{1}{\cancel{5}}} = \dfrac{1 \cdot \overset{1}{\cancel{5}}}{\underset{1}{\cancel{5}}}$

$x = 1$

Check: $1 \cdot 5 = 5 = 3 \cdot 1\tfrac{2}{3}$

(c) $\dfrac{0.06}{x} = \dfrac{0.3}{0.4}$

$0.3 \cdot x = 0.06 \cdot 0.4$ *Cross products are equivalent*

$\dfrac{\overset{1}{\cancel{0.3}} \cdot x}{\underset{1}{\cancel{0.3}}} = \dfrac{\overset{0.08}{\cancel{0.024}}}{\underset{1}{\cancel{0.3}}}$

$x = 0.08$

Check: $(0.06)(0.4) = 0.024 = (0.08)(0.3)$

(d) $\dfrac{2.2}{5} = \dfrac{13}{x}$

$2.2 \cdot x = 5 \cdot 13$ *Cross products are equivalent*

$\dfrac{\overset{1}{\cancel{2.2}} \cdot x}{\underset{1}{\cancel{2.2}}} = \dfrac{65}{2.2}$

$x = \dfrac{65}{2.2} \approx 29.55$ *(rounded to the nearest hundredth)*

Check: $2.2\left(\dfrac{65}{2.2}\right) = 65 = 5 \cdot 13$

(e) $\dfrac{x}{6} = \dfrac{0.5}{1.2}$

$1.2 \cdot x = 6(0.5)$ *Cross products are equivalent*

$\dfrac{\overset{1}{\cancel{1.2}} \cdot x}{\underset{1}{\cancel{1.2}}} = \dfrac{3}{1.2} = \dfrac{30}{12} = \dfrac{5}{2}$

$x = 2.5$

Check: $(2.5)(1.2) = 3 = 6(0.5)$

(f) $\dfrac{0}{2} = \dfrac{x}{7.092}$

Since $\dfrac{0}{2} = 0$, x *must* equal 0. The denominator 7.092 could be any number (except 0).

6.3 Section Exercises

1. $9 is to 12 cans as $18 is to 24 cans.

$\dfrac{\$9}{12 \text{ cans}} = \dfrac{\$18}{24 \text{ cans}}$

3. 200 adults is to 450 children as 4 adults is to 9 children.

$$\frac{200 \text{ adults}}{450 \text{ children}} = \frac{4 \text{ adults}}{9 \text{ children}}$$

5. 120 feet is to 150 feet as 8 feet is to 10 feet.

$$\frac{120}{150} = \frac{8}{10} \quad \text{The common units (feet) cancel.}$$

7. 2.2 hours is to 3.3 hours as 3.2 hours is to 4.8 hours.

$$\frac{2.2}{3.3} = \frac{3.2}{4.8} \quad \text{The common units (hours) cancel.}$$

9. $\frac{6}{10} = \frac{3}{5}$

$\frac{6 \div 2}{10 \div 2} = \frac{3}{5}$ and $\frac{3}{5}$

Both ratios are equivalent to $\frac{3}{5}$, so the proportion is *true*.

11. $\frac{5}{8} = \frac{25}{40}$

$\frac{5}{8}$ and $\frac{25 \div 5}{40 \div 5} = \frac{5}{8}$

Both ratios are equivalent to $\frac{5}{8}$, so the proportion is *true*.

13. $\frac{150}{200} = \frac{200}{300}$

$\frac{150 \div 50}{200 \div 50} = \frac{3}{4}$ and $\frac{200 \div 100}{300 \div 100} = \frac{2}{3}$

Because $\frac{3}{4}$ is *not* equivalent to $\frac{2}{3}$, the proportion is *false*.

15. $\frac{2}{9} = \frac{6}{27}$

Cross products: $2 \cdot 27 = 54$; $9 \cdot 6 = 54$

The cross products are *equal*, so the proportion is *true*.

17. $\frac{20}{28} = \frac{12}{16}$

Cross products: $20 \cdot 16 = 320$; $28 \cdot 12 = 336$

The cross products are *unequal*, so the proportion is *false*.

19. $\frac{110}{18} = \frac{160}{27}$

Cross products: $110 \cdot 27 = 2970$; $18 \cdot 160 = 2880$

The cross products are *unequal*, so the proportion is *false*.

21. $\frac{3.5}{4} = \frac{7}{8}$

Cross products: $(3.5)(8) = 28$; $4 \cdot 7 = 28$

The cross products are *equal*, so the proportion is *true*.

23. $\frac{18}{15} = \frac{2\frac{5}{6}}{2\frac{1}{2}}$

Cross products: $18 \cdot 2\frac{1}{2} = \frac{\overset{9}{\cancel{18}}}{1} \cdot \frac{5}{\underset{1}{\cancel{2}}} = 45$

$15 \cdot 2\frac{5}{6} = \frac{\overset{5}{\cancel{15}}}{1} \cdot \frac{17}{\underset{2}{\cancel{6}}} = \frac{85}{2} = 42\frac{1}{2}$

The cross products are *unequal*, so the proportion is *false*.

25. $\frac{6}{3\frac{2}{3}} = \frac{18}{11}$

Cross products: $6 \cdot 11 = 66$;

$3\frac{2}{3} \cdot 18 = \frac{11}{\underset{1}{\cancel{3}}} \cdot \frac{\overset{6}{\cancel{18}}}{1} = 66$

The cross products are *equal*, so the proportion is *true*.

27. (Answers may vary: proportion may also be set up with "at bats" in both numerators and "hits" in both denominators. Cross product will be the same.)

$$\frac{\textbf{Joe Mauer}}{50 \text{ at bats}} \frac{17 \text{ hits}}{} = \frac{\textbf{Freddy Sanchez}}{450 \text{ at bats}} \frac{153 \text{ hits}}{}$$

Cross products:
$17 \cdot 450 = 7650$; $50 \cdot 153 = 7650$

The cross products are *equal*, so the proportion is *true*. Paul is correct; the two men hit equally well.

29. $\frac{1}{3} = \frac{x}{12}$

$3 \cdot x = 1 \cdot 12$ *Cross products are equivalent*

$\frac{\overset{1}{\cancel{3}} \cdot x}{\underset{1}{\cancel{3}}} = \frac{12}{3}$

$x = 4$

We will check our answers by showing that the cross products are equal.

Check: $1 \cdot 12 = 12 = 3 \cdot 4$

6.3 Proportions

31. $\dfrac{15}{10} = \dfrac{3}{x}$ OR $\dfrac{3}{2} = \dfrac{3}{x}$

Since the numerators are equal, the denominators must be equal, so $x = 2$.

33. $\dfrac{x}{11} = \dfrac{32}{4}$ OR $\dfrac{x}{11} = \dfrac{8}{1}$

$1 \cdot x = 11 \cdot 8$ *Cross products are equivalent*

$x = 88$

Check: $88 \cdot 4 = 352 = 11 \cdot 32$

35. $\dfrac{42}{x} = \dfrac{18}{39}$

$x \cdot 18 = 42 \cdot 39$ *Cross products are equivalent*

$\dfrac{x \cdot \cancel{18}^{1}}{\cancel{18}_{1}} = \dfrac{\cancel{42}^{7} \cdot \cancel{39}^{13}}{\cancel{18}_{\cancel{6}_{1}}}$

$x = 91$

Check: $42 \cdot 39 = 1638 = 91 \cdot 18$

37. $\dfrac{x}{25} = \dfrac{4}{20}$ OR $\dfrac{x}{25} = \dfrac{1}{5}$

$5 \cdot x = 25 \cdot 1$ *Cross products are equivalent*

$\dfrac{\cancel{5}^{1} \cdot x}{\cancel{5}_{1}} = \dfrac{\cancel{25}^{5}}{\cancel{5}_{1}}$

$x = 5$

Check: $5 \cdot 20 = 100 = 25 \cdot 4$

39. $\dfrac{8}{x} = \dfrac{24}{30}$ OR $\dfrac{8}{x} = \dfrac{4}{5}$

$4 \cdot x = 8 \cdot 5$ *Cross products are equivalent*

$\dfrac{\cancel{4}^{1} \cdot x}{\cancel{4}_{1}} = \dfrac{\cancel{8}^{2} \cdot 5}{\cancel{4}_{1}}$

$x = 10$

Check: $8 \cdot 30 = 240 = 10 \cdot 24$

41. $\dfrac{99}{55} = \dfrac{44}{x}$ OR $\dfrac{9}{5} = \dfrac{44}{x}$

$9 \cdot x = 5 \cdot 44$ *Cross products are equivalent*

$\dfrac{\cancel{9}^{1} \cdot x}{\cancel{9}_{1}} = \dfrac{220}{9}$

$x = \dfrac{220}{9}$

≈ 24.44

Check: $99 \cdot \dfrac{220}{9} = 2420 = 55 \cdot 44$

43. $\dfrac{0.7}{9.8} = \dfrac{3.6}{x}$

$0.7 \cdot x = 9.8 \, (3.6)$ *Cross products are equivalent*

$\dfrac{\cancel{0.7}^{1} \cdot x}{\cancel{0.7}_{1}} = \dfrac{35.28}{0.7}$

$x = 50.4$

Check: $0.7 (50.4) = 35.28 = 9.8(3.6)$

45. $\dfrac{250}{24.8} = \dfrac{x}{1.75}$

$24.8 \cdot x = 250(1.75)$ *Cross products are equivalent*

$\dfrac{\cancel{24.8}^{1} \cdot x}{\cancel{24.8}_{1}} = \dfrac{437.5}{24.8}$

≈ 17.64

Check: $250(1.75) = 437.5 = 24.8 \cdot \dfrac{437.5}{24.8}$

47. $\dfrac{15}{1\frac{2}{3}} = \dfrac{9}{x}$

$15 \cdot x = 1\dfrac{2}{3} \cdot 9$ *Cross products are equivalent*

$15 \cdot x = \dfrac{5}{\cancel{3}_{1}} \cdot \dfrac{\cancel{9}^{3}}{1}$

$\dfrac{\cancel{15}^{1} \cdot x}{\cancel{15}_{1}} = \dfrac{15}{15}$

$x = 1$

49. $\dfrac{2\frac{1}{3}}{1\frac{1}{2}} = \dfrac{x}{2\frac{1}{4}}$

$1\dfrac{1}{2} \cdot x = 2\dfrac{1}{3} \cdot 2\dfrac{1}{4}$

$\dfrac{3}{2} \cdot x = \dfrac{7}{\cancel{3}_{1}} \cdot \dfrac{\cancel{9}^{3}}{4}$

$\dfrac{3}{2} \cdot x = \dfrac{21}{4}$

$\dfrac{\frac{3}{2} \cdot x}{\frac{3}{2}} = \dfrac{\frac{21}{4}}{\frac{3}{2}}$

$x = \dfrac{21}{4} \div \dfrac{3}{2} = \dfrac{\cancel{21}^{7}}{\cancel{4}_{2}} \cdot \dfrac{\cancel{2}^{1}}{\cancel{3}_{1}} = \dfrac{7}{2} = 3\dfrac{1}{2}$

51. Change $\frac{1}{2}$ to a decimal by dividing: $1 \div 2 = 0.5$

$$\frac{\frac{1}{2}}{x} = \frac{2}{0.8}$$

$$\frac{0.5}{x} = \frac{2}{0.8}$$

$$x \cdot 2 = 0.5(0.8)$$

$$x \cdot 2 = 0.4$$

$$\frac{x \cdot \cancel{2}^1}{\cancel{2}_1} = \frac{0.4}{2}$$

$$x = 0.2$$

Now change 0.8 to a fraction and write it in lowest terms:

$$0.8 = \frac{8 \div 2}{10 \div 2} = \frac{4}{5}$$

$$\frac{\frac{1}{2}}{x} = \frac{2}{\frac{4}{5}}$$

$$x \cdot 2 = \frac{1}{\cancel{2}} \cdot \frac{\cancel{4}^2}{5}$$

$$x \cdot 2 = \frac{2}{5}$$

$$\frac{x \cdot \cancel{2}^1}{\cancel{2}_1} = \frac{\frac{2}{5}}{2}$$

$$x = \frac{2}{5} \div 2 = \frac{\cancel{2}^1}{5} \cdot \frac{1}{\cancel{2}_1} = \frac{1}{5}$$

53. $\dfrac{x}{\frac{3}{50}} = \dfrac{0.15}{1\frac{4}{5}}$

Change to decimals.

$$\frac{x}{0.06} = \frac{0.15}{1.8}$$

$$x \cdot 1.8 = 0.15(0.06)$$

$$\frac{x \cdot 1.8}{1.8} = \frac{0.009}{1.8}$$

$$x = 0.005$$

Change to fractions.

$$\frac{x}{\frac{3}{50}} = \frac{\frac{3}{20}}{1\frac{4}{5}}$$

$$1\frac{4}{5} \cdot x = \frac{3}{50} \cdot \frac{3}{20}$$

$$\frac{1\frac{4}{5} \cdot x}{1\frac{4}{5}} = \frac{\frac{9}{1000}}{1\frac{4}{5}}$$

$$x = \frac{9}{1000} \div 1\frac{4}{5} = \frac{\cancel{9}^1}{\cancel{1000}_{200}} \cdot \frac{\cancel{5}^1}{\cancel{9}_1} = \frac{1}{200}$$

55. $\dfrac{10}{4} = \dfrac{5}{3}$

Find the cross products.

$10 \cdot 3 = 30$

$4 \cdot 5 = 20$

The cross products are *unequal*, so the proportion is *not* true.

Replace 10 with x.

$$\frac{x}{4} = \frac{5}{3}$$

$$3 \cdot x = 4 \cdot 5$$

$$x = \frac{20}{3} = 6\frac{2}{3}, \text{ so}$$

$\dfrac{6\frac{2}{3}}{4} = \dfrac{5}{3}$ is a true proportion.

Replace 4 with x.

$$\frac{10}{x} = \frac{5}{3}$$

$$5 \cdot x = 10 \cdot 3$$

$$x = \frac{30}{5} = 6, \text{ so}$$

$\dfrac{10}{6} = \dfrac{5}{3}$ is a true proportion.

Replace 5 with x.

$$\frac{10}{4} = \frac{x}{3}$$

$$4 \cdot x = 10 \cdot 3$$

$$x = \frac{30}{4} = \frac{15}{2} = 7.5, \text{ so}$$

$\dfrac{10}{4} = \dfrac{7.5}{3}$ is a true proportion.

Replace 3 with x.

$$\frac{10}{4} = \frac{5}{x}$$

$$10 \cdot x = 4 \cdot 5$$

$$x = \frac{20}{10} = 2, \text{ so}$$

$\dfrac{10}{4} = \dfrac{5}{2}$ is a true proportion.

56. $\dfrac{6}{8} = \dfrac{24}{30}$

Find the cross products.

$6 \cdot 30 = 180$

$8 \cdot 24 = 192$

The cross products are *unequal*, so the proportion is *not* true.

Replace 6 with x.

$$\frac{x}{8} = \frac{24}{30}$$
$$30 \cdot x = 8 \cdot 24$$
$$x = \frac{8 \cdot \overset{4}{\cancel{24}}}{\underset{5}{\cancel{30}}} = \frac{32}{5} = 6.4, \text{ so}$$
$$\frac{6.4}{8} = \frac{24}{30} \text{ is a true proportion.}$$

Replace 8 with x.

$$\frac{6}{x} = \frac{24}{30}$$
$$24 \cdot x = 6 \cdot 30$$
$$x = \frac{6 \cdot \overset{5}{\cancel{30}}}{\underset{4}{\cancel{24}}} = \frac{30}{4} = 7.5, \text{ so}$$
$$\frac{6}{7.5} = \frac{24}{30} \text{ is a true proportion.}$$

Replace 24 with x.

$$\frac{6}{8} = \frac{x}{30}$$
$$8 \cdot x = 6 \cdot 30$$
$$x = \frac{\overset{3}{\cancel{6}} \cdot \overset{15}{\cancel{30}}}{\underset{\underset{2}{\cancel{4}}}{\cancel{8}}} = \frac{45}{2} = 22.5, \text{ so}$$
$$\frac{6}{8} = \frac{22.5}{30} \text{ is a true proportion.}$$

Replace 30 with x.

$$\frac{6}{8} = \frac{24}{x}$$
$$6 \cdot x = 8 \cdot 24$$
$$x = \frac{8 \cdot \overset{4}{\cancel{24}}}{\underset{1}{\cancel{6}}} = 32, \text{ so}$$
$$\frac{6}{8} = \frac{24}{32} \text{ is a true proportion.}$$

Summary Exercises on Ratios, Rates, and Proportions

1. 2400 freshmen to 2000 juniors

$$\frac{2400}{2000} = \frac{2400 \div 400}{2000 \div 400} = \frac{6}{5}$$

3. 1850 seniors and 2150 sophomores to 2000 juniors

$$\frac{1850 + 2150}{2000} = \frac{4000}{2000} = \frac{4000 \div 2000}{2000 \div 2000} = \frac{2}{1}$$

5. **(i)** 2 million violin players to 22 million piano players

$$\frac{2 \text{ million}}{22 \text{ million}} = \frac{2 \div 2}{22 \div 2} = \frac{1}{11}$$

(ii) 2 million violin players to 20 million guitar players

$$\frac{2 \text{ million}}{20 \text{ million}} = \frac{2 \div 2}{20 \div 2} = \frac{1}{10}$$

(iii) 2 million violin players to 6 million organ players

$$\frac{2 \text{ million}}{6 \text{ million}} = \frac{2 \div 2}{6 \div 2} = \frac{1}{3}$$

(iv) 2 million violin players to 4 million clarinet players

$$\frac{2 \text{ million}}{4 \text{ million}} = \frac{2 \div 2}{4 \div 2} = \frac{1}{2}$$

(v) 2 million violin players to 3 million drum players

$$\frac{2 \text{ million}}{3 \text{ million}} = \frac{2}{3}$$

7. $$\frac{100 \text{ points}}{48 \text{ minutes}} = \frac{100 \div 48 \text{ points}}{48 \div 48 \text{ minutes}}$$
$$= 2.08\overline{3} \approx 2.1 \text{ points/minute}$$

$$\frac{48 \text{ minutes}}{100 \text{ points}} = \frac{48 \div 100 \text{ minutes}}{100 \div 100 \text{ points}}$$
$$= 0.48 \approx 0.5 \text{ minute/point}$$

9. $$\frac{\$652.80}{40 \text{ hours}} = \frac{\$652.80 \div 40}{40 \div 40 \text{ hours}} = \$16.32/\text{hour}$$

Her regular hourly pay rate is $16.32/hour.

$$\frac{\$195.84}{8 \text{ hours}} = \frac{\$195.84 \div 8}{8 \div 8 \text{ hours}} = \$24.48/\text{hour}$$

Her overtime hourly pay rate is $24.48/hour.

11.

Size	Cost per Unit
11 ounces	$\frac{\$6.79}{11 \text{ ounces}} \approx \$0.62$
12 ounces	$\frac{\$7.24}{12 \text{ ounces}} = \$0.60 \ (*)$
16 ounces	$\frac{\$10.99}{16 \text{ ounces}} \approx \$0.69$

The best buy is 12 ounces for $7.24, about $0.60 per ounce.

13. $$\frac{28}{21} = \frac{44}{33}$$

$$\frac{28 \div 7}{21 \div 7} = \frac{4}{3} \text{ and } \frac{44 \div 11}{33 \div 11} = \frac{4}{3}$$

Both ratios are equivalent to $\frac{4}{3}$, so the proportion is *true*.

Chapter 6 Ratio, Proportion, and Line/Angle/Triangle Relationships

15. $\dfrac{2\frac{5}{8}}{3\frac{1}{4}} = \dfrac{21}{26}$

    Cross products:

    $2\dfrac{5}{8} \cdot 26 = \dfrac{21}{\cancel{8}_{4}} \cdot \dfrac{\cancel{26}^{13}}{1} = \dfrac{273}{4}$

    $3\dfrac{1}{4} \cdot 21 = \dfrac{13}{4} \cdot \dfrac{21}{1} = \dfrac{273}{4}$

    The cross products are *equal*, so the proportion is *true*.

17. $\dfrac{15}{8} = \dfrac{6}{x}$

    $15 \cdot x = 8 \cdot 6$ *Cross products are equivalent*

    $\dfrac{\cancel{15}^{1} \cdot x}{\cancel{15}_{1}} = \dfrac{8 \cdot \cancel{6}^{2}}{\cancel{15}_{5}} = \dfrac{16}{5} = 3.2$

    Check: $15(3.2) = 48 = 8 \cdot 6$

19. $\dfrac{10}{11} = \dfrac{x}{4}$

    $11 \cdot x = 10 \cdot 4$ *Cross products are equivalent*

    $\dfrac{\cancel{11}^{1} \cdot x}{\cancel{11}_{1}} = \dfrac{40}{11} \approx 3.64$

    Check: $10 \cdot 4 = 40 = 11 \cdot \frac{40}{11}$

21. $\dfrac{2.6}{x} = \dfrac{13}{7.8}$

    $13 \cdot x = 2.6(7.8)$ *Cross products are equivalent*

    $\dfrac{\cancel{13}^{1} \cdot x}{\cancel{13}_{1}} = \dfrac{20.28}{13} = 1.56$

    Check: $2.6(7.8) = 20.28 = 1.56(13)$

23. $\dfrac{\frac{1}{3}}{8} = \dfrac{x}{24}$

    $8 \cdot x = \dfrac{1}{3} \cdot 24$ *Cross products are equivalent*

    $\dfrac{\cancel{8}^{1} \cdot x}{\cancel{8}_{1}} = \dfrac{8}{8} = 1$

    Check: $\frac{1}{3}(24) = 8 = 8 \cdot 1$

6.4 Problem Solving with Proportions

6.4 Margin Exercises

1. (a) $\dfrac{2 \text{ pounds}}{50 \text{ square feet}} = \dfrac{x \text{ pounds}}{225 \text{ square feet}}$

    $50 \cdot x = 2 \cdot 225$ *Cross products are equivalent*

    $\dfrac{50 \cdot x}{50} = \dfrac{450}{50}$ *Divide both sides by 50*

    $x = 9$

    9 pounds of fertilizer are needed for 225 square feet.

    (b) $\dfrac{1 \text{ inch}}{75 \text{ miles}} = \dfrac{4.75 \text{ inches}}{x \text{ miles}}$

    $1 \cdot x = 75(4.75)$ *Cross products are equivalent*

    $x = 356.25 \approx 356 \text{ miles}$

    The lake's actual length is about 356 miles.

    (c) $\dfrac{30 \text{ milliliters}}{100 \text{ pounds}} = \dfrac{x \text{ milliliters}}{34 \text{ pounds}}$

    $100 \cdot x = 30 \cdot 34$ *Cross products are equivalent*

    $\dfrac{100 \cdot x}{100} = \dfrac{1020}{100}$ *Divide both sides by 100*

    $x = 10.2 \approx 10 \text{ milliliters}$

    A 34-pound child should be given about 10 milliliters of cough syrup.

2. (a) $\dfrac{2 \text{ want to lose}}{3 \text{ surveyed}} = \dfrac{x}{150 \text{ total}}$

    $3 \cdot x = 2 \cdot 150$ *Cross products are equivalent*

    $\dfrac{3 \cdot x}{3} = \dfrac{300}{3}$ *Divide both sides by 3*

    $x = 100$

    In a group of 150, 100 people want to lose weight. This is a reasonable answer because it's more than half the people, but less than all the people.

    Incorrect setup

    $\dfrac{3 \text{ surveyed}}{2 \text{ want to lose}} = \dfrac{x}{150 \text{ total}}$

    $2 \cdot x = 3 \cdot 150$

    $\dfrac{2 \cdot x}{2} = \dfrac{450}{2}$

    $x = 225$

    The incorrect setup gives an answer of 225 people, which is unreasonable since there are only 150 people in the group.

6.4 Problem Solving with Proportions

(b) $\dfrac{3 \text{ FA students}}{5 \text{ students}} = \dfrac{x}{4500 \text{ students}}$

$5 \cdot x = 3 \cdot 4500 \qquad$ *Cross products are equivalent*

$\dfrac{5 \cdot x}{5} = \dfrac{13{,}500}{5} \qquad$ *Divide both sides by 5*

$x = 2700$

At Central Community College 2700 students receive financial aid. This is a reasonable answer because it's more than half the people, but less than all the people.

Incorrect setup

$\dfrac{5 \text{ students}}{3 \text{ FA students}} = \dfrac{x}{4500 \text{ students}}$

$3 \cdot x = 5 \cdot 4500$

$\dfrac{3 \cdot x}{3} = \dfrac{22{,}500}{3}$

$x = 7500$

The incorrect setup gives an answer of 7500 students, which is unreasonable since there are only 4500 students at the college.

(c) $\dfrac{9 \text{ sugarless}}{10 \text{ dentists}} = \dfrac{x}{60 \text{ dentists}}$

$10 \cdot x = 9 \cdot 60 \qquad$ *Cross products are equivalent*

$\dfrac{10 \cdot x}{10} = \dfrac{540}{10} \qquad$ *Divide both sides by 10*

$x = 54$

If the advertisement is true, 54 of the 60 dentists in the city would recommend sugarless gum. This is a reasonable answer since it is almost all of the dentists.

Incorrect setup

$\dfrac{10 \text{ dentists}}{9 \text{ sugarless}} = \dfrac{x}{60 \text{ dentists}}$

$9 \cdot x = 10 \cdot 60$

$\dfrac{9 \cdot x}{9} = \dfrac{600}{9}$

$x = 66.\overline{6} \approx 67$

The incorrect setup gives an answer of 67 dentists, which is unreasonable since there are only 60 dentists in the city.

6.4 Section Exercises

1. $\dfrac{5 \text{ hours}}{4 \text{ cartoon strips}} = \dfrac{x \text{ hours}}{18 \text{ cartoon strips}}$

$\dfrac{5}{4} = \dfrac{x}{18}$

$4 \cdot x = 5 \cdot 18$

$4 \cdot x = 90$

$\dfrac{\cancel{4} \cdot x}{\cancel{4}} = \dfrac{90}{4}$

$x = 22.5$

It will take 22.5 hours to sketch 18 cartoon strips.

3. $\dfrac{60 \text{ newspapers}}{\$27} = \dfrac{16 \text{ newspapers}}{x}$

$60 \cdot x = 27 \cdot 16$

$\dfrac{60 \cdot x}{60} = \dfrac{432}{60}$

$x = 7.2$

The cost of 16 newspapers is $7.20.

5. $\dfrac{3 \text{ pounds}}{350 \text{ square feet}} = \dfrac{x}{4900 \text{ square feet}}$

$350 \cdot x = 3 \cdot 4900$

$\dfrac{350 \cdot x}{350} = \dfrac{14{,}700}{350}$

$x = 42$

42 pounds are needed to cover 4900 square feet.

7. $\dfrac{\$672.80}{5 \text{ days}} = \dfrac{x}{3 \text{ days}}$

$5 \cdot x = (672.80)(3)$

$\dfrac{5 \cdot x}{5} = \dfrac{2018.40}{5}$

$x = 403.68$

In 3 days Tom makes $403.68.

9. $\dfrac{6 \text{ ounces}}{7 \text{ servings}} = \dfrac{x}{12 \text{ servings}}$

$7 \cdot x = 6 \cdot 12$

$\dfrac{7 \cdot x}{7} = \dfrac{72}{7}$

$x = 10\dfrac{2}{7} \approx 10$

You need about 10 ounces for 12 servings.

11. $\dfrac{3 \text{ quarts}}{270 \text{ square feet}} = \dfrac{x}{(350+100) \text{ square feet}}$

$\dfrac{3}{270} = \dfrac{x}{450}$

$270 \cdot x = 3 \cdot 450$

$\dfrac{270 \cdot x}{270} = \dfrac{1350}{270}$

$x = 5$

You will need 5 quarts.

13. First find the length.
$$\frac{1 \text{ inch}}{4 \text{ feet}} = \frac{3.5 \text{ inches}}{x}$$
$$1 \cdot x = 4(3.5)$$
$$x = 14$$

The kitchen is 14 feet long.

Then find the width.
$$\frac{1 \text{ inch}}{4 \text{ feet}} = \frac{2.5 \text{ inches}}{x \text{ feet}}$$
$$1 \cdot x = 4(2.5)$$
$$x = 10$$

The kitchen is 10 feet wide.

15. The length of the dining area is the same as the length of the kitchen, which is 14 feet by Exercise 13.

Find the width of the dining area.
$$4.5 - 2.5 \text{ inches} = 2 \text{ inches}$$
$$\frac{1 \text{ inch}}{4 \text{ feet}} = \frac{2 \text{ inches}}{x}$$
$$1 \cdot x = 4 \cdot 2$$
$$x = 8$$

The dining area is 8 feet wide.

17. Set up and solve a proportion with pieces of chicken and number of guests.
$$\frac{40 \text{ pieces}}{25 \text{ guests}} = \frac{x}{60 \text{ guests}}$$
$$25 \cdot x = 40 \cdot 60$$
$$\frac{25 \cdot x}{25} = \frac{40 \cdot 60}{25}$$
$$x = \frac{40 \cdot 60}{25} = 96 \text{ pieces of chicken}$$

For the other food items, the proportions are similar, so we simply replace "40" in the last step with the appropriate value.

$$\frac{14 \cdot 60}{25} = 33.6 \text{ pounds of lasagna}$$

$$\frac{4.5 \cdot 60}{25} = 10.8 \text{ pounds of deli meats}$$

$$\frac{\frac{7}{3} \cdot 60}{25} = \frac{28}{5} = 5\frac{3}{5} \text{ pounds of cheese}$$

$$\frac{3 \text{ dozen} \cdot 60}{25} = 7.2 \text{ dozen (about 86) buns}$$

$$\frac{6 \cdot 60}{25} = 14.4 \text{ pounds of salad}$$

19. $$\frac{7 \text{ refresher}}{10 \text{ entering}} = \frac{x}{2950 \text{ entering}}$$
$$10 \cdot x = 7 \cdot 2950$$
$$\frac{10 \cdot x}{10} = \frac{20{,}650}{10}$$
$$x = 2065$$

2065 students will probably need a refresher course. This is a reasonable answer because it's more than half the students, but not all the students.

Incorrect setup

$$\frac{10 \text{ entering}}{7 \text{ refresher}} = \frac{x}{2950 \text{ entering}}$$
$$7 \cdot x = 10 \cdot 2950$$
$$\frac{7 \cdot x}{7} = \frac{29{,}500}{7}$$
$$x \approx 4214$$

The incorrect setup gives an estimate of 4214 entering students, which is unreasonable since there are only 2950 entering students.

21. $$\frac{1 \text{ chooses vanilla}}{3 \text{ people}} = \frac{x \text{ choose vanilla}}{238 \text{ people}}$$
$$3 \cdot x = 1 \cdot 238$$
$$\frac{3 \cdot x}{3} = \frac{238}{3}$$
$$x \approx 79.3 \approx 79$$

You would expect about 79 people to choose vanilla ice cream. This is a reasonable answer.

Incorrect setup

$$\frac{3 \text{ people}}{1 \text{ chooses vanilla}} = \frac{x \text{ choose vanilla}}{238 \text{ people}}$$
$$1 \cdot x = 3 \cdot 238$$
$$x = 714$$

With an incorrect setup, 714 people choose vanilla ice cream. This is unreasonable because only 238 people attended the ice cream social.

6.4 Problem Solving with Proportions

23.
$$\frac{98}{100} = \frac{x}{113{,}100{,}000}$$
$$100 \cdot x = 98 \cdot 113{,}100{,}000$$
$$\frac{100 \cdot x}{100} = \frac{11{,}083{,}800{,}000}{100}$$
$$x = 110{,}838{,}000$$

110,838,000 U.S. households had one or more TVs in 2005.

Incorrect setup
$$\frac{100}{98} = \frac{x}{113{,}100{,}000}$$
$$98 \cdot x = 100 \cdot 113{,}100{,}000$$
$$\frac{98 \cdot x}{98} = \frac{11{,}310{,}000{,}000}{98}$$
$$x \approx 115{,}408{,}163$$

The incorrect setup gives about 115,408,163 U.S. households with one or more TVs, but there were only 113,100,000 U.S. households.

25.
$$\frac{5 \text{ stocks up}}{6 \text{ stocks down}} = \frac{x \text{ stocks up}}{750 \text{ stocks down}}$$
$$\frac{5}{6} = \frac{x}{750}$$
$$6 \cdot x = 5 \cdot 750$$
$$\frac{6 \cdot x}{6} = \frac{3750}{6}$$
$$x = 625$$

625 stocks went up.

27.
$$\frac{8 \text{ length}}{1 \text{ width}} = \frac{32.5 \text{ meters length}}{x \text{ meters width}}$$
$$8 \cdot x = 1 \cdot 32.5$$
$$\frac{8 \cdot x}{8} = \frac{32.5}{8}$$
$$x = 4.0625 \approx 4.06$$

The wing must be about 4.06 meters wide.

29.
$$\frac{150 \text{ pounds}}{222 \text{ calories}} = \frac{210 \text{ pounds}}{x}$$
$$150 \cdot x = 222 \cdot 210$$
$$\frac{150 \cdot x}{150} = \frac{46{,}620}{150}$$
$$x = 310.8 \approx 311$$

A 210-pound person would burn about 311 calories.

31.
$$\frac{1.05 \text{ meters}}{1.68 \text{ meters}} = \frac{6.58 \text{ meters}}{x}$$
$$1.05 \cdot x = 1.68(6.58)$$
$$\frac{1.05 \cdot x}{1.05} = \frac{11.0544}{1.05}$$
$$x = 10.528 \approx 10.53$$

The height of the tree is about 10.53 meters.

33. You cannot solve this problem using a proportion because the ratio of age to weight is not constant. As Jim's age increases from 25 to 50 years old, his weight may decrease, stay the same, or increase.

35. First find the number of coffee drinkers.
$$\frac{4 \text{ coffee drinkers}}{5 \text{ students}} = \frac{x}{50{,}500 \text{ students}}$$
$$5 \cdot x = 4 \cdot 50{,}500$$
$$\frac{5 \cdot x}{5} = \frac{202{,}000}{5}$$
$$x = 40{,}400$$

The survey showed that 40,400 students drink coffee. Now find the number of coffee drinkers who use cream.
$$\frac{1 \text{ uses cream}}{8 \text{ coffee drinkers}} = \frac{x}{40{,}400 \text{ coffee drinkers}}$$
$$8 \cdot x = 40{,}400$$
$$\frac{8 \cdot x}{8} = \frac{40{,}400}{8}$$
$$x = 5050$$

According to the survey, 5050 students use cream.

37. First find the number of calories in a $\frac{1}{2}$-cup serving of bran cereal.
$$\frac{\frac{1}{3} \text{ cup}}{80 \text{ calories}} = \frac{\frac{1}{2} \text{ cup}}{x}$$
$$\frac{1}{3} \cdot x = 80 \cdot \frac{1}{2}$$
$$\frac{\frac{1}{3} \cdot x}{\frac{1}{3}} = \frac{40}{\frac{1}{3}} = \frac{40}{1} \cdot \frac{3}{1}$$
$$x = 120$$

Then find the number of grams of fiber in a $\frac{1}{2}$-cup serving of bran cereal.
$$\frac{\frac{1}{3} \text{ cup}}{8 \text{ grams of fiber}} = \frac{\frac{1}{2} \text{ cup}}{x}$$
$$\frac{1}{3} \cdot x = 8 \cdot \frac{1}{2}$$
$$\frac{\frac{1}{3} \cdot x}{\frac{1}{3}} = \frac{4}{\frac{1}{3}} = \frac{4}{1} \cdot \frac{3}{1}$$
$$x = 12$$

A $\frac{1}{2}$-cup serving of bran cereal provides 120 calories and 12 grams of fiber.

39. *Use proportions.*

Water: $$\frac{3\frac{1}{2} \text{ cups}}{12 \text{ servings}} = \frac{x}{6 \text{ servings}}$$
$$12 \cdot x = 3\frac{1}{2} \cdot 6$$
$$12 \cdot x = \frac{7}{2} \cdot 6$$
$$\frac{12 \cdot x}{12} = \frac{21}{12}$$
$$x = \frac{7}{4} = 1\frac{3}{4}$$

Margarine: $$\frac{6 \text{ Tbsp}}{12 \text{ servings}} = \frac{x}{6 \text{ servings}}$$
$$12 \cdot x = 6 \cdot 6$$
$$\frac{12 \cdot x}{12} = \frac{36}{12}$$
$$x = 3$$

Milk: $$\frac{1\frac{1}{2} \text{ cups}}{12 \text{ servings}} = \frac{x}{6 \text{ servings}}$$
$$12 \cdot x = 1\frac{1}{2} \cdot 6$$
$$12 \cdot x = \frac{3}{2} \cdot 6$$
$$\frac{12 \cdot x}{12} = \frac{9}{12}$$
$$x = \frac{3}{4}$$

Potato flakes: $$\frac{4 \text{ cups}}{12 \text{ servings}} = \frac{x}{6 \text{ servings}}$$
$$12 \cdot x = 4 \cdot 6$$
$$\frac{12 \cdot x}{12} = \frac{24}{12}$$
$$x = 2$$

Multiply the quantities by $\frac{1}{2}$ (or divide by 2), since 6 servings is $\frac{1}{2}$ of 12 servings.

Water: $$\frac{3\frac{1}{2}}{2} = \frac{\frac{7}{2}}{2} = \frac{7}{2} \div 2 = \frac{7}{2} \cdot \frac{1}{2}$$
$$= \frac{7}{4} = 1\frac{3}{4}$$

Margarine: $$\frac{6}{2} = 3$$

Milk: $$\frac{1\frac{1}{2}}{2} = \frac{\frac{3}{2}}{2} = \frac{3}{2} \div 2$$
$$= \frac{3}{2} \cdot \frac{1}{2} = \frac{3}{4}$$

Potato flakes: $$\frac{4}{2} = 2$$

For 6 servings, use $1\frac{3}{4}$ cups water, 3 Tbsp margarine, $\frac{3}{4}$ cup milk, and 2 cups potato flakes.

40. *Use proportions.*

Water: $$\frac{3\frac{1}{2} \text{ cups}}{12 \text{ servings}} = \frac{x}{18 \text{ servings}}$$
$$12 \cdot x = 3\frac{1}{2} \cdot 18$$
$$12 \cdot x = \frac{7}{2} \cdot 18$$
$$\frac{12 \cdot x}{12} = \frac{63}{12}$$
$$x = 5\frac{1}{4}$$

Margarine: $$\frac{6 \text{ Tbsp}}{12 \text{ servings}} = \frac{x}{18 \text{ servings}}$$
$$12 \cdot x = 6 \cdot 18$$
$$\frac{12 \cdot x}{12} = \frac{108}{12}$$
$$x = 9$$

Milk: $$\frac{1\frac{1}{2} \text{ cups}}{12 \text{ servings}} = \frac{x}{18 \text{ servings}}$$
$$12 \cdot x = 1\frac{1}{2} \cdot 18$$
$$12 \cdot x = \frac{3}{2} \cdot 18$$
$$\frac{12 \cdot x}{12} = \frac{27}{12}$$
$$x = 2\frac{1}{4}$$

Potato flakes: $$\frac{4 \text{ cups}}{12 \text{ servings}} = \frac{x}{18 \text{ servings}}$$
$$12 \cdot x = 4 \cdot 18$$
$$\frac{12 \cdot x}{12} = \frac{72}{12}$$
$$x = 6$$

Multiply the quantities in Exercise 39 by 3, since 18 servings is 3 times 6 servings.

Water: $1\frac{3}{4} \cdot 3 = \frac{7}{4} \cdot \frac{3}{1} = \frac{21}{4} = 5\frac{1}{4}$

Margarine: $3 \cdot 3 = 9$

Milk: $\frac{3}{4} \cdot 3 = \frac{3}{4} \cdot \frac{3}{1} = \frac{9}{4} = 2\frac{1}{4}$

Potato flakes: $2 \cdot 3 = 6$

For 18 servings, use $5\frac{1}{4}$ cups water, 9 Tbsp margarine, $2\frac{1}{4}$ cups milk, and 6 cups potato flakes.

41. Since 3 servings is $\frac{1}{2}$ of 6 servings, multiply each quantity in Exercise 39 by $\frac{1}{2}$.

Water: $1\frac{3}{4} \cdot \frac{1}{2} = \frac{7}{4} \cdot \frac{1}{2} = \frac{7}{8}$

Margarine: $3 \cdot \frac{1}{2} = \frac{3}{1} \cdot \frac{1}{2} = \frac{3}{2} = 1\frac{1}{2}$

Milk: $\frac{3}{4} \cdot \frac{1}{2} = \frac{3}{8}$

Potato flakes: $2 \cdot \frac{1}{2} = \frac{2}{1} \cdot \frac{1}{2} = 1$

For 3 servings, use $\frac{7}{8}$ cup water, $1\frac{1}{2}$ Tbsp margarine, $\frac{3}{8}$ cup milk, and 1 cup potato flakes.

42. Since 9 servings is 3 times 3 servings, multiply each quantity in Exercise 41 by 3.

Water: $\frac{7}{8} \cdot 3 = \frac{7}{8} \cdot \frac{3}{1} = \frac{21}{8} = 2\frac{5}{8}$

Margarine: $\frac{3}{2} \cdot 3 = \frac{3}{2} \cdot \frac{3}{1} = \frac{9}{2} = 4\frac{1}{2}$

Milk: $\frac{3}{8} \cdot 3 = \frac{3}{8} \cdot \frac{3}{1} = \frac{9}{8} = 1\frac{1}{8}$

Potato flakes: $1 \cdot 3 = 3$

For 9 servings, use $2\frac{5}{8}$ cups water, $4\frac{1}{2}$ Tbsp margarine, $1\frac{1}{8}$ cups milk, and 3 cups potato flakes.

6.5 Geometry: Lines and Angles

6.5 Margin Exercises

1. (a) The figure has two endpoints, so it is a *line segment* named $\overline{EF}$ or $\overline{FE}$.

(b) The figure starts at point S and goes on forever in one direction, so it is a *ray* named $\overrightarrow{SR}$.

(c) The figure goes on forever in both directions, so it is a *line* named $\overleftrightarrow{WX}$ or $\overleftrightarrow{XW}$.

(d) The figure has two endpoints, so it is a *line segment* named $\overline{CD}$ or $\overline{DC}$.

2. (a) The lines cross at E, so they are *intersecting lines*.

(b) The lines never intersect (cross), so they appear to be *parallel lines*.

(c) The lines never intersect (cross), so they appear to be *parallel lines*.

3. (a) The angle is named $\angle 3$, $\angle CQD$, or $\angle DQC$.

(b) Darken rays $\overrightarrow{TW}$ and $\overrightarrow{TZ}$ as shown in the figure.

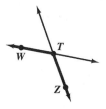

(c) The angle can be named $\angle 1$, $\angle R$, $\angle MRN$, and $\angle NRM$.

4. (a) Since there is a small square at the vertex, the angle measures exactly $90°$, so it is a *right angle*.

(b) The angle measures exactly $180°$, so it is a *straight angle*.

(c) The angle measures between $90°$ and $180°$, so it is *obtuse*.

(d) The angle measures between $0°$ and $90°$, so it is *acute*.

5. The lines in (b) show perpendicular lines, because they intersect at right angles. The lines in (a) show intersecting lines. They cross, but not at right angles.

6. $\angle COD$ and $\angle DOE$ are complementary angles because $70° + 20° = 90°$.
$\angle RST$ and $\angle XPY$ are complementary angles because $45° + 45° = 90°$.

7. (a) The complement of $35°$ is $55°$, because $90° - 35° = 55°$.

(b) The complement of $80°$ is $10°$, because $90° - 80° = 10°$.

8. $\angle CRF$ and $\angle BRF$ are supplementary angles because $127° + 53° = 180°$.
$\angle CRE$ and $\angle ERB$ are supplementary angles because $53° + 127° = 180°$.
$\angle BRF$ and $\angle BRE$ are supplementary angles because $53° + 127° = 180°$.
$\angle CRE$ and $\angle CRF$ are supplementary angles because $53° + 127° = 180°$.

9. (a) The supplement of $175°$ is $5°$, because $180° - 175° = 5°$.

(b) The supplement of $30°$ is $150°$, because $180° - 30° = 150°$.

10. $\angle BOC \cong \angle AOD$ because they each measure $150°$.
$\angle AOB \cong \angle DOC$ because they each measure $30°$.

11. $\angle SPB$ and $\angle MPD$ are vertical angles because they do *not* share a common side and the angles are formed by intersecting lines.
Similarly, $\angle BPD$ and $\angle SPM$ are vertical angles. Vertical angles are congruent (they measure the same number of degrees).

206 Chapter 6 Ratio, Proportion, and Line/Angle/Triangle Relationships

12. **(a)** ∠TOS
    ∠TOS is a right angle, so its measure is 90°.

    (b) ∠QOR
    ∠QOR and ∠SOT are vertical angles, so they are also congruent and have the same measure. The measure of ∠SOT is 90°, so the measure of ∠QOR is 90°.

    (c) ∠VOR
    ∠VOR and ∠SOP are vertical angles, so they are also congruent and have the same measure. The measure of ∠SOP is 38°, so the measure of ∠VOR is 38°.

    (d) ∠POQ
    ∠POQ and ∠SOP are complementary angles, so the sum of their angle measures is 90°. Since the degree measure of ∠SOP is 38°, the measure of ∠POQ equals 90° − 38° = 52°.

    (e) ∠TOV
    ∠TOV and ∠POQ are vertical angles, so they are congruent. The measure of ∠POQ is 52° (from part d), so the measure of ∠TOV is 52°.

    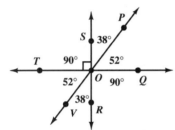

13. **(a)** There are four pairs of corresponding angles:

    ∠1 and ∠5, ∠2 and ∠6,

    ∠7 and ∠3, ∠8 and ∠4

    There are two pairs of alternate interior angles:

    ∠7 and ∠6, ∠8 and ∠5

    (b) There are four pairs of corresponding angles:

    ∠5 and ∠7, ∠6 and ∠8,

    ∠1 and ∠3, ∠2 and ∠4

    There are two pairs of alternate interior angles:

    ∠6 and ∠3, ∠2 and ∠7

14. **(a)** ∠6 ≅ ∠7 ≅ ∠2 ≅ ∠3, so each measures 150°.

    ∠8 measures 180° − 150° = 30°.

    ∠8 ≅ ∠5 ≅ ∠4 ≅ ∠1, so each measures 30°.

    See the figure.

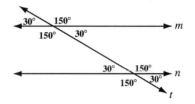

(b) ∠1 ≅ ∠6 ≅ ∠3 ≅ ∠8, so each measures 45°.

∠2 measures 180° − 45° = 135°.

∠2 ≅ ∠5 ≅ ∠4 ≅ ∠7, so each measures 135°.

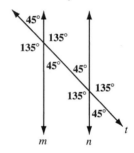

6.5 Section Exercises

1. This is a *line* named $\overleftrightarrow{CD}$ or $\overleftrightarrow{DC}$. A line is a straight row of points that goes on forever in both directions.

3. The figure has two endpoints so it is a *line segment* named $\overline{GF}$ or $\overline{FG}$.

5. This is a *ray* named $\overrightarrow{PQ}$. A ray is a part of a line that has only one endpoint and goes on forever in one direction.

7. The lines are *perpendicular* because they intersect at right angles.

9. These lines appear to be *parallel* lines. Parallel lines are lines in the same plane that never intersect (cross).

11. The lines intersect so they are *not* parallel. At their intersection they *do not* form a right angle so they are not perpendicular. The lines are *intersecting*.

13. The angle can be named ∠AOS or ∠SOA. The middle letter, O, identifies the vertex.

15. The angle can be named ∠CRT or ∠TRC. The middle letter, R, identifies the vertex.

17. The angle can be named ∠AQC or ∠CQA. The middle letter, Q, identifies the vertex.

19. The angle is a *right angle*, as indicated by the small square at the vertex. Right angles measure exactly 90°.

21. The measure of the angle is between 0° and 90°, so it is an *acute angle*.

23. Two rays in a straight line pointing opposite directions measure 180°. An angle that measures 180° is called a *straight angle*.

25. The pairs of complementary angles are: ∠EOD and ∠COD because 75° + 15° = 90°; ∠AOB and ∠BOC because 25° + 65° = 90°.

27. The pairs of supplementary angles are: ∠HNE and ∠ENF because 77° + 103° = 180°; ∠ACB and ∠KOL because 120° + 60° = 180°.

29. The complement of 40° is 50°, because 90° − 40° = 50°.

31. The complement of 86° is 4°, because 90° − 86° = 4°.

33. The supplement of 130° is 50°, because 180° − 130° = 50°.

35. The supplement of 90° is 90°, because 180° − 90° = 90°.

37. ∠SON ≅ ∠TOM because they are vertical angles. ∠TOS ≅ ∠MON because they are vertical angles.

39. Because ∠COE and ∠GOH are vertical angles, they are also congruent. This means they have the same measure. ∠COE measures 63°, so ∠GOH measures 63°.

    The sum of the measures of ∠COE, ∠AOC, and ∠AOH equals 180°. Therefore, ∠AOC measures 180° − (63° + 37°) = 180° − 100° = 80°. Since ∠AOC and ∠GOF are vertical, they are congruent, so ∠GOF measures 80°. Since ∠AOH and ∠EOF are vertical, they are congruent, so ∠EOF measures 37°.

41. "∠UST is 90°" is *true* because $\overleftrightarrow{UQ}$ is perpendicular to $\overleftrightarrow{ST}$.

42. "$\overleftrightarrow{SQ}$ and $\overleftrightarrow{PQ}$ are perpendicular" is *true* because they form a 90° angle, as indicated by the small red square.

43. "The measure of ∠USQ is less than the measure of ∠PQR" is *false*. ∠USQ is a straight angle and so is ∠PQR, therefore each measures 180°. A true statement would be: ∠USQ has the same measure as ∠PQR.

44. "$\overleftrightarrow{ST}$ and $\overleftrightarrow{PR}$ are intersecting" is *false*. $\overleftrightarrow{ST}$ and $\overleftrightarrow{PR}$ are parallel and will never intersect.

45. "$\overleftrightarrow{QU}$ and $\overleftrightarrow{TS}$ are parallel" is *false*. $\overleftrightarrow{QU}$ and $\overleftrightarrow{TS}$ are perpendicular because they intersect at right angles.

46. "∠UST and ∠UQR measure the same number of degrees" is *true* because both angles are formed by perpendicular lines, so they both measure 90°.

47. There are four pairs of corresponding angles:

    ∠1 and ∠8, ∠2 and ∠5,

    ∠3 and ∠6, ∠4 and ∠7

    There are two pairs of alternate interior angles:

    ∠4 and ∠5, ∠3 and ∠8

49. ∠8 ≅ ∠6 ≅ ∠2 ≅ ∠4, so each measures 130°.

    ∠5 measures 180° − 130° = 50°.

    ∠5 ≅ ∠7 ≅ ∠1 ≅ ∠3, so each measures 50°.

51. ∠6 ≅ ∠1 ≅ ∠8 ≅ ∠3, so each measures 47°.

    ∠5 measures 180° − 47° = 133°.

    ∠5 ≅ ∠2 ≅ ∠7 ≅ ∠4, so each measures 133°.

53. ∠6 ≅ ∠8 ≅ ∠4 ≅ ∠2, so each measures 114°.

    ∠7 measures 180° − 114° = 66°.

    ∠7 ≅ ∠5 ≅ ∠3 ≅ ∠1, so each measures 66°.

55. ∠2 and ∠ABC are alternate interior angles, so they have the same measure, 42°. ∠1 and ∠ABC are supplementary angles, so ∠1 = 180° − 42° = 138°. ∠1 and ∠3 are supplements of alternate interior angles, so they have the same measure, 138°.

6.6 Geometry Applications: Congruent and Similar Triangles

6.6 Margin Exercises

1. **(a)** If you picked up △ABC and slid it over on top of △DEF, the two triangles would match.

   The corresponding parts are congruent, so:

∠1 and ∠4	$\overline{AC}$ and $\overline{DF}$
∠2 and ∠5	$\overline{AB}$ and $\overline{DE}$
∠3 and ∠6	$\overline{BC}$ and $\overline{EF}$

   (b) If you rotate △FGH, then slide it on top of △JLK, the two triangles would match.

∠1 and ∠6	$\overline{GF}$ and $\overline{KL}$
∠2 and ∠5	$\overline{FH}$ and $\overline{LJ}$
∠3 and ∠4	$\overline{GH}$ and $\overline{KJ}$

   (c) If you flipped △RST over, then slid it on top of △VWX, the two triangles would match.

∠1 and ∠5	$\overline{RS}$ and $\overline{XV}$
∠2 and ∠4	$\overline{RT}$ and $\overline{XW}$
∠3 and ∠6	$\overline{ST}$ and $\overline{VW}$

2. **(a)** On both triangles, two corresponding angles and the side that connects them measure the same, so the Angle-Side-Angle (ASA) method can be used to prove that the triangles are congruent.

   (b) On both triangles, two corresponding sides and the angle between them measure the same, so the Side-Angle-Side (SAS) method can be used to prove that the triangles are congruent.

   (c) Each pair of corresponding sides has the same length, so the Side-Side-Side (SSS) method can be used to prove that the triangles are congruent.

3. **(a)** Corresponding angles have the same measure.

   The corresponding angles are
   1 and 4, 2 and 5, 3 and 6.
   $\overline{PN}$ and $\overline{ZX}$ are opposite corresponding angles 3 and 6.
   $\overline{PM}$ and $\overline{ZY}$ are opposite corresponding angles 1 and 4.
   $\overline{NM}$ and $\overline{XY}$ are opposite corresponding angles 2 and 5.
   Thus, the corresponding sides are
   $\overline{PN}$ and $\overline{ZX}$, $\overline{PM}$ and $\overline{ZY}$, $\overline{NM}$ and $\overline{XY}$.

   (b) The corresponding angles are
   1 and 6, 2 and 4, 3 and 5.
   The corresponding sides are
   $\overline{AB}$ and $\overline{EF}$, $\overline{BC}$ and $\overline{FG}$, $\overline{AC}$ and $\overline{EG}$.

4. Find the length of $\overline{EF}$.

   $\dfrac{EF}{CB} = \dfrac{5}{15}$ Ratio from Example 3

   Replace EF with x and CB with 33.

   $\dfrac{x}{33} = \dfrac{1}{3}$ Reduce.

   $3 \cdot x = 33 \cdot 1$

   $\dfrac{3 \cdot x}{3} = \dfrac{33}{3}$

   $x = 11$

   $\overline{EF}$ has a length of 11 m.

5. **(a)** Find the length of $\overline{AB}$.
   From Example 4,

   $\dfrac{PR}{AC} = \dfrac{7}{14} = \dfrac{1}{2}$, so $\dfrac{PQ}{AB} = \dfrac{1}{2}$.

   Replace PQ with 3 and AB with y.

   $\dfrac{3}{y} = \dfrac{1}{2}$

   $y \cdot 1 = 3 \cdot 2$

   $y = 6$

   $\overline{AB}$ is 6 ft.
   Find the perimeter.
   Perimeter $= 14$ ft $+ 10$ ft $+ 6$ ft $= 30$ ft

   (b) Set up ratios.

   $\dfrac{PQ}{AB} = \dfrac{10 \text{ m}}{30 \text{ m}} = \dfrac{1}{3}$, so $\dfrac{QR}{BC} = \dfrac{1}{3}$ and $\dfrac{PR}{AC} = \dfrac{1}{3}$.

   Replace QR with x and BC with 18.

   $\dfrac{x}{18} = \dfrac{1}{3}$

   $3 \cdot x = 18 \cdot 1$

   $\dfrac{3 \cdot x}{3} = \dfrac{18}{3}$

   $x = 6$

   $x = \overline{QR}$ is 6 m.
   The perimeter of triangle PQR is

   $10 \text{ m} + 8 \text{ m} + 6 \text{ m} = 24 \text{ m}$.

   Next replace PR with 8 and AC with y.

   $\dfrac{8}{y} = \dfrac{1}{3}$

   $y \cdot 1 = 8 \cdot 3$

   $y = 24$

   $y = \overline{AC}$ is 24 m.
   The perimeter of triangle ABC is

   $30 \text{ m} + 18 \text{ m} + 24 \text{ m} = 72 \text{ m}$.

6. **(a)** Write a proportion.

   $\dfrac{h}{5} = \dfrac{48}{12}$

   $12 \cdot h = 5 \cdot 48$

   $\dfrac{12 \cdot h}{12} = \dfrac{240}{12}$

   $h = 20$

   The flagpole is 20 ft high.

   (b) Write a proportion.

   $\dfrac{h}{7.2} = \dfrac{12.5}{5}$

   $5 \cdot h = 7.2 \cdot 12.5$

   $\dfrac{5 \cdot h}{5} = \dfrac{90}{5}$

   $h = 18$

   The flagpole is 18 m high.

6.6 Section Exercises

1. If you picked up $\triangle ABC$ and slid it over on top of $\triangle DEF$, the two triangles would match.

   The corresponding angles are

   $\angle 1$ and $\angle 4$, $\angle 2$ and $\angle 5$, $\angle 3$ and $\angle 6$.

   The corresponding sides are

   $\overline{AB}$ and $\overline{DE}$, $\overline{BC}$ and $\overline{EF}$, $\overline{AC}$ and $\overline{DF}$.

6.6 Geometry Applications: Congruent and Similar Triangles

3. If you rotate $\triangle TUS$, then slide it on top of $\triangle WXY$, the two triangles would match.

   The corresponding angles are

   $\angle 1$ and $\angle 6$, $\angle 2$ and $\angle 4$, $\angle 3$ and $\angle 5$.

   The corresponding sides are

   $\overline{ST}$ and $\overline{YW}$, $\overline{TU}$ and $\overline{WX}$, $\overline{SU}$ and $\overline{YX}$.

5. If you flipped $\triangle MNL$ over, then slid it on top of $\triangle SRT$, the two triangles would match.

   The corresponding angles are

   $\angle 1$ and $\angle 6$, $\angle 2$ and $\angle 5$, $\angle 3$ and $\angle 4$.

   The corresponding sides are

   $\overline{LM}$ and $\overline{TS}$, $\overline{LN}$ and $\overline{TR}$, $\overline{MN}$ and $\overline{SR}$.

7. On both triangles, two corresponding sides and the angle between them measure the same, so the Side-Angle-Side (SAS) method can be used to prove that the triangles are congruent.

9. Each pair of corresponding sides has the same length, so the Side-Side-Side (SSS) method can be used to prove that the triangles are congruent.

11. On both triangles, two corresponding angles and the side that connects them measure the same, so the Angle-Side-Angle (ASA) method can be used to prove that the triangles are congruent.

13. use SAS: $BC = CE$, $\angle ABC \cong \angle DCE$, $BA = CD$

14. use SSS: $WP = YP$, $ZP = XP$, $WZ = YX$

15. use SAS: $PS = SR$, $m\angle QSP = m\angle QSR = 90°$, $QS = QS$ (common side)

16. use SAS: $LM = OM$, $PM = NM$, $\angle LMP \cong \angle OMN$ (vertical angles)

17. Set up a ratio of corresponding sides.

    $$\frac{6 \text{ cm}}{12 \text{ cm}} = \frac{6}{12} = \frac{1}{2}$$

    Write a proportion to find a.

    $$\frac{a}{12} = \frac{1}{2}$$
    $$2 \cdot a = 12 \cdot 1$$
    $$\frac{2 \cdot a}{2} = \frac{12}{2}$$
    $$a = 6 \text{ cm}$$

    Write a proportion to find b.

    $$\frac{7.5}{b} = \frac{1}{2}$$
    $$1 \cdot b = 2 \cdot 7.5$$
    $$b = 15 \text{ cm}$$

19. Set up a ratio of corresponding sides.

    $$\frac{6 \text{ mm}}{12 \text{ mm}} = \frac{6}{12} = \frac{1}{2}$$

    Write a proportion to find a.

    $$\frac{a}{10} = \frac{1}{2}$$
    $$a \cdot 2 = 10 \cdot 1$$
    $$\frac{a \cdot 2}{2} = \frac{10}{2}$$
    $$a = 5 \text{ mm}$$

    Write a proportion to find b.

    $$\frac{b}{6} = \frac{1}{2}$$
    $$b \cdot 2 = 6 \cdot 1$$
    $$\frac{b \cdot 2}{2} = \frac{6}{2}$$
    $$b = 3 \text{ mm}$$

21. Set up a ratio of corresponding sides.

    $$\frac{18 \text{ in.}}{12 \text{ in.}} = \frac{18}{12} = \frac{3}{2}$$

    Write a proportion to find a.

    $$\frac{a}{16} = \frac{3}{2}$$
    $$a \cdot 2 = 16 \cdot 3$$
    $$\frac{a \cdot 2}{2} = \frac{48}{2}$$
    $$a = 24 \text{ inches}$$

    Write a proportion to find b.

    $$\frac{30}{b} = \frac{3}{2}$$
    $$b \cdot 3 = 30 \cdot 2$$
    $$\frac{b \cdot 3}{3} = \frac{60}{3}$$
    $$b = 20 \text{ inches}$$

23. Write a proportion to find x.

    $$\frac{x}{18.6} = \frac{28}{21} \quad \text{OR} \quad \frac{x}{18.6} = \frac{4}{3}$$
    $$3 \cdot x = 4 \cdot 18.6$$
    $$\frac{3 \cdot x}{3} = \frac{74.4}{3}$$
    $$x = 24.8 \text{ m}$$

    $P = 24.8 \text{ m} + 28 \text{ m} + 20 \text{ m} = 72.8 \text{ m}$

    Write a proportion to find y.

    $$\frac{y}{20} = \frac{21}{28} \quad \text{OR} \quad \frac{y}{20} = \frac{3}{4}$$
    $$4 \cdot y = 20 \cdot 3$$
    $$\frac{4 \cdot y}{4} = \frac{60}{4}$$
    $$y = 15 \text{ m}$$

    $P = 15 \text{ m} + 21 \text{ m} + 18.6 \text{ m} = 54.6 \text{ m}$

25. Since triangles *CDE* and *FGH* are similar, each missing side of triangle *FGH* is 8 cm.

Perimeter of triangle FGH
$= 8 \text{ cm} + 8 \text{ cm} + 8 \text{ cm}$
$= 24 \text{ cm}$

Set up a ratio of corresponding sides to find the height h of triangle *FGH*.

$$\frac{10.4}{12} = \frac{h}{8}$$
$$12 \cdot h = 8 \cdot 10.4$$
$$\frac{12 \cdot h}{12} = \frac{83.2}{12}$$
$$h \approx 6.9 \text{ cm}$$

Area of triangle FGH
$= 0.5 \cdot b \cdot h$
$\approx 0.5 \cdot 8 \text{ cm} \cdot 6.9 \text{ cm}$
$= 27.6 \text{ cm}^2$

27. Write a proportion to find h.

$$\frac{2}{16} = \frac{3}{h}$$
$$2 \cdot h = 3 \cdot 16$$
$$\frac{2 \cdot h}{2} = \frac{48}{2}$$
$$h = 24 \text{ ft}$$

The height of the house is 24 ft.

29. One dictionary definition is "resembling, but not identical." Examples of similar objects are sets of different size pots or measuring cups; small- and large-size cans of beans; child's tennis shoe and adult's tennis shoe.

31. Using the hint, we can write a proportion to find x.

$$\frac{x}{120} = \frac{100}{100 + 140} = \frac{100}{240} \quad \text{OR}$$
$$\frac{x}{120} = \frac{5}{12}$$
$$x \cdot 12 = 120 \cdot 5$$
$$\frac{x \cdot 12}{12} = \frac{600}{12}$$
$$x = 50 \text{ m}$$

33. Write a proportion to find n.

$$\frac{50}{n} = \frac{100}{100 + 120} = \frac{100}{220} \quad \text{OR}$$
$$\frac{50}{n} = \frac{5}{11}$$
$$5 \cdot n = 50 \cdot 11$$
$$\frac{5 \cdot n}{5} = \frac{550}{5}$$
$$n = 110 \text{ m}$$

The length of the lake is 110 m.

Chapter 6 Review Exercises

1. orca whale's length to whale shark's length

$$\frac{30 \text{ ft}}{40 \text{ ft}} = \frac{30}{40} = \frac{30 \div 10}{40 \div 10} = \frac{3}{4}$$

2. blue whale's length to great white shark's length

$$\frac{80 \text{ ft}}{20 \text{ ft}} = \frac{80}{20} = \frac{80 \div 20}{20 \div 20} = \frac{4}{1}$$

3. To get a ratio of $\frac{1}{2}$, we can start with the smallest value in the table and see if there is a value that is two times the smallest. $2 \times 20 = 40$, so the ratio of the *great white shark's* length to the *whale shark's* length is $\frac{20}{40} = \frac{1}{2}$. The length of the orca whale is 30 ft, but $2 \times 30 = 60$ is not in the table. $2 \times 40 = 80$, so the ratio of the *whale shark's* length to the *blue whale's* length is $\frac{40}{80} = \frac{1}{2}$.

4. $2.50 to $1.25

$$\frac{\$2.50}{\$1.25} = \frac{2.50}{1.25} = \frac{2.50 \div 1.25}{1.25 \div 1.25} = \frac{2}{1}$$

5. $0.30 to $0.45

$$\frac{\$0.30}{\$0.45} = \frac{0.30}{0.45} = \frac{0.30 \div 0.15}{0.45 \div 0.15} = \frac{2}{3}$$

6. $1\frac{2}{3}$ cups to $\frac{2}{3}$ cup

$$\frac{1\frac{2}{3} \text{ cups}}{\frac{2}{3} \text{ cup}} = \frac{1\frac{2}{3}}{\frac{2}{3}} = \frac{5}{3} \div \frac{2}{3}$$
$$= \frac{5}{\cancel{3}} \cdot \frac{\cancel{3}}{2} = \frac{5}{2}$$

7. $2\frac{3}{4}$ miles to $16\frac{1}{2}$ miles

$$\frac{2\frac{3}{4} \text{ miles}}{16\frac{1}{2} \text{ miles}} = \frac{2\frac{3}{4}}{16\frac{1}{2}} = \frac{\frac{11}{4}}{\frac{33}{2}} = \frac{11}{4} \div \frac{33}{2}$$
$$= \frac{\cancel{11}}{\cancel{4}} \cdot \frac{\cancel{2}}{\cancel{33}} = \frac{1}{6}$$

8. 5 hours to 100 minutes

5 hours $= 5 \cdot 60$ minutes $= 300$ minutes

$$\frac{300 \text{ minutes}}{100 \text{ minutes}} = \frac{300}{100} = \frac{300 \div 100}{100 \div 100} = \frac{3}{1}$$

9. 9 inches to 2 feet

2 feet $= 24$ inches

$$\frac{9 \text{ inches}}{24 \text{ inches}} = \frac{9}{24} = \frac{9 \div 3}{24 \div 3} = \frac{3}{8}$$

10. 1 ton to 1500 pounds

1 ton $= 2000$ pounds

$$\frac{2000 \text{ pounds}}{1500 \text{ pounds}} = \frac{2000}{1500} = \frac{2000 \div 500}{1500 \div 500} = \frac{4}{3}$$

11. 8 hours to 3 days

    3 days = 3 · 24 hours = 72 hours

    $$\frac{8 \text{ hours}}{72 \text{ hours}} = \frac{8}{72} = \frac{8 \div 8}{72 \div 8} = \frac{1}{9}$$

12. Ramona's $500 to Jake's $350

    $$\frac{\$500}{\$350} = \frac{500 \div 50}{350 \div 50} = \frac{10}{7}$$

    The ratio of Ramona's sales to Jake's sales is $\frac{10}{7}$.

13. $\frac{35 \text{ miles per gallon}}{25 \text{ miles per gallon}} = \frac{35}{25} = \frac{35 \div 5}{25 \div 5} = \frac{7}{5}$

    The ratio of the new car's mileage to the old car's mileage is $\frac{7}{5}$.

14. $\frac{6000 \text{ students}}{7200 \text{ students}} = \frac{6000}{7200} = \frac{6000 \div 1200}{7200 \div 1200} = \frac{5}{6}$

    The ratio of the math students to the English students is $\frac{5}{6}$.

15. $88 for 8 dozen

    $$\frac{\$88 \div 8}{8 \text{ dozen} \div 8} = \frac{\$11}{1 \text{ dozen}}$$

16. 96 children in 40 families

    $$\frac{96 \text{ children} \div 8}{40 \text{ families} \div 8} = \frac{12 \text{ children}}{5 \text{ families}}$$

17. 4 pages in 20 minutes

    (i) $\frac{4 \text{ pages} \div 20}{20 \text{ minutes} \div 20} = \frac{0.2 \text{ page}}{1 \text{ minute}}$
    $= 0.2$ page/minute
    or $\frac{1}{5}$ page/minute

    (ii) $\frac{20 \text{ minutes} \div 4}{4 \text{ pages} \div 4} = \frac{5 \text{ minutes}}{1 \text{ page}}$
    $= 5$ minutes/page

    His rate is $\frac{1}{5}$ page/minute or 5 minutes/page.

18. $24 in 3 hours

    (i) $\frac{\$24 \div 3}{3 \text{ hours} \div 3} = \frac{\$8}{1 \text{ hour}} = \$8/\text{hour}$

    (ii) $\frac{3 \text{ hours} \div 24}{\$24 \div 24} = \frac{0.125 \text{ hour}}{\$1}$
    $= 0.125$ hour/dollar
    or $\frac{1}{8}$ hour/dollar

    Her earnings are $8/hour or $\frac{1}{8}$ hour/dollar.

19. | Size | Cost per Unit |
    |------|---------------|
    | 8 ounces | $\frac{\$4.98}{8 \text{ ounces}} \approx \$0.623$ (∗) |
    | 3 ounces | $\frac{\$2.49}{3 \text{ ounces}} = \$0.83$ |
    | 2 ounces | $\frac{\$1.89}{2 \text{ ounces}} = \$0.945$ |

    The best buy is 8 ounces for $4.98, about $0.623/ounce.

20. 35.2 pounds for $36.96 − $1 (coupon) = $35.96

    $$\frac{\$35.96}{35.2 \text{ pounds}} \approx \$1.022$$

    17.6 pounds for $18.69 − $1 (coupon) = $17.69

    $$\frac{\$17.69}{17.6 \text{ pounds}} \approx \$1.005 \ (*)$$

    3.5 pounds for $4.25

    $$\frac{\$4.25}{3.5 \text{ pounds}} \approx \$1.214$$

    The best buy is 17.6 pounds for $18.69 with the $1 coupon, about $1.005/pound.

21. $\frac{6}{10} = \frac{9}{15}$

    $\frac{6 \div 2}{10 \div 2} = \frac{3}{5}$ and $\frac{9 \div 3}{15 \div 3} = \frac{3}{5}$

    Both ratios are equivalent to $\frac{3}{5}$, so the proportion is *true*.

22. $\frac{6}{48} = \frac{9}{36}$

    Cross products: 6 · 36 = 216; 48 · 9 = 432

    The cross products are *unequal*, so the proportion is *false*.

23. $\frac{47}{10} = \frac{98}{20}$

    Cross products: 47 · 20 = 940; 10 · 98 = 980

    The cross products are *unequal*, so the proportion is *false*.

24. $\frac{1.5}{2.4} = \frac{2}{3.2}$

    Cross products: 1.5(3.2) = 4.8; 2.4 · 2 = 4.8

    The cross products are *equal*, so the proportion is *true*.

25. $\dfrac{3\frac{1}{2}}{2\frac{1}{3}} = \dfrac{6}{4}$

Cross products:

$3\dfrac{1}{2} \cdot 4 = \dfrac{7}{2} \cdot \dfrac{\cancel{4}^{2}}{1} = 14$

$2\dfrac{1}{3} \cdot 6 = \dfrac{7}{\cancel{3}_{1}} \cdot \dfrac{\cancel{6}^{2}}{1} = 14$

The cross products are *equal*, so the proportion is *true*.

26. $\dfrac{4}{42} = \dfrac{150}{x}$ OR $\dfrac{2}{21} = \dfrac{150}{x}$

$2 \cdot x = 21 \cdot 150$ *Cross products are equivalent*

$\dfrac{2 \cdot x}{2} = \dfrac{3150}{2}$

$x = 1575$

Check: $4 \cdot 1575 = 6300 = 42 \cdot 150$

27. $\dfrac{16}{x} = \dfrac{12}{15}$ OR $\dfrac{16}{x} = \dfrac{4}{5}$

$4 \cdot x = 5 \cdot 16$ *Cross products are equivalent*

$\dfrac{4 \cdot x}{4} = \dfrac{80}{4}$

$x = 20$

Check: $16 \cdot 15 = 240 = 20 \cdot 12$

28. $\dfrac{100}{14} = \dfrac{x}{56}$ OR $\dfrac{50}{7} = \dfrac{x}{56}$

$7 \cdot x = 50 \cdot 56$ *Cross products are equivalent*

$\dfrac{7 \cdot x}{7} = \dfrac{2800}{7}$

$x = 400$

Check: $100 \cdot 56 = 5600 = 14 \cdot 400$

29. $\dfrac{5}{8} = \dfrac{x}{20}$

$8 \cdot x = 5 \cdot 20$ *Cross products are equivalent*

$\dfrac{8 \cdot x}{8} = \dfrac{100}{8}$

$x = 12.5$

Check: $5 \cdot 20 = 100 = 8(12.5)$

30. $\dfrac{x}{24} = \dfrac{11}{18}$

$18 \cdot x = 24 \cdot 11$ *Cross products are equivalent*

$\dfrac{18 \cdot x}{18} = \dfrac{264}{18}$

$x = \dfrac{44}{3} \approx 14.67$

Check: $\dfrac{44}{3} \cdot 18 = 264 = 24 \cdot 11$

31. $\dfrac{7}{x} = \dfrac{18}{21}$ OR $\dfrac{7}{x} = \dfrac{6}{7}$

$6 \cdot x = 7 \cdot 7$ *Cross products are equivalent*

$\dfrac{6 \cdot x}{6} = \dfrac{49}{6}$

$x = \dfrac{49}{6} \approx 8.17$

Check: $7 \cdot 21 = 147 = \dfrac{49}{6} \cdot 18$

32. $\dfrac{x}{3.6} = \dfrac{9.8}{0.7}$

$0.7 \cdot x = 9.8(3.6)$ *Cross products are equivalent*

$\dfrac{0.7 \cdot x}{0.7} = \dfrac{35.28}{0.7}$

$x = 50.4$

Check: $50.4(0.7) = 35.28 = 3.6(9.8)$

33. $\dfrac{13.5}{1.7} = \dfrac{4.5}{x}$

$13.5 \cdot x = 1.7(4.5)$ *Cross products are equivalent*

$\dfrac{13.5 \cdot x}{13.5} = \dfrac{7.65}{13.5}$

$x \approx 0.57$

Check: $13.5 \left(\dfrac{7.65}{13.5} \right) = 7.65 = 1.7(4.5)$

34. $\dfrac{0.82}{1.89} = \dfrac{x}{5.7}$

$1.89 \cdot x = 0.82(5.7)$ *Cross products are equivalent*

$\dfrac{1.89 \cdot x}{1.89} = \dfrac{4.674}{1.89}$

$x \approx 2.47$

Check: $0.82(5.7) = 4.674 = 1.89 \left(\dfrac{4.674}{1.89} \right)$

35. $\dfrac{3 \text{ cats}}{5 \text{ dogs}} = \dfrac{x}{45 \text{ dogs}}$

$5 \cdot x = 3 \cdot 45$

$\dfrac{5 \cdot x}{5} = \dfrac{135}{5}$

$x = 27$

There are 27 cats.

36. $\dfrac{8 \text{ hits}}{28 \text{ at bats}} = \dfrac{x}{161 \text{ at bats}}$
$28 \cdot x = 8 \cdot 161$
$\dfrac{28 \cdot x}{28} = \dfrac{1288}{28}$
$x = 46$

She will get 46 hits.

37. $\dfrac{3.5 \text{ pounds}}{\$9.77} = \dfrac{5.6 \text{ pounds}}{x}$
$3.5 \cdot x = 9.77(5.6)$
$\dfrac{3.5 \cdot x}{3.5} = \dfrac{54.712}{3.5}$
$x \approx 15.63$

The cost for 5.6 pounds of ground beef is $15.63 (rounded).

38. $\dfrac{4 \text{ voting students}}{10 \text{ students}} = \dfrac{x}{8247 \text{ students}}$
$10 \cdot x = 4 \cdot 8247$
$\dfrac{10 \cdot x}{10} = \dfrac{32{,}988}{10}$
$x \approx 3299$

They should expect about 3299 students to vote.

39. $\dfrac{1 \text{ inch}}{16 \text{ feet}} = \dfrac{4.25 \text{ inches}}{x}$
$1 \cdot x = 16(4.25)$
$x = 68$

The length of the real boxcar is 68 feet.

40. 2 dozen necklaces = $2 \cdot 12 = 24$ necklaces

$\dfrac{24 \text{ necklaces}}{16\frac{1}{2} \text{ hours}} = \dfrac{40 \text{ necklaces}}{x}$
$24 \cdot x = 16\dfrac{1}{2} \cdot 40 = \dfrac{33}{2} \cdot \dfrac{40}{1}$
$\dfrac{24 \cdot x}{24} = \dfrac{660}{24}$
$x = 27.5 = 27\dfrac{1}{2}$

It will take Marvette $27\frac{1}{2}$ hours or 27.5 hours to make 40 necklaces.

41. $\dfrac{284 \text{ calories}}{25 \text{ minutes}} = \dfrac{x}{45 \text{ minutes}}$
$25 \cdot x = 284 \cdot 45$
$\dfrac{25 \cdot x}{25} = \dfrac{12{,}780}{25}$
$x = 511.2 \approx 511$

A 180-pound person would burn about 511 calories in 45 minutes.

42. $\dfrac{3.5 \text{ milligrams}}{50 \text{ pounds}} = \dfrac{x}{210 \text{ pounds}}$
$50 \cdot x = 3.5 \cdot 210$
$\dfrac{50 \cdot x}{50} = \dfrac{735}{50}$
$x = 14.7$

A patient who weighs 210 pounds should be given 14.7 milligrams of the medicine.

43. The figure has two endpoints so it is a *line segment* named $\overline{AB}$ or $\overline{BA}$.

44. This is a *line* named $\overleftrightarrow{CD}$ or $\overleftrightarrow{DC}$. A line is a straight row of points that goes on forever in both directions.

45. This is a *ray* named $\overrightarrow{OP}$. A ray is a part of a line that has only one endpoint and goes on forever in one direction.

46. These lines appear to be *parallel* lines. Parallel lines are lines in the same plane that never intersect (cross).

47. The lines are *perpendicular* because they intersect at right angles.

48. The lines intersect so they are *not* parallel. At their intersection they *do not* form a right angle so they are not perpendicular. The lines are *intersecting*.

49. The measure of the angle is between 0° and 90°, so it is an *acute angle*.

50. The measure of the angle is between 90° and 180°, so it is an *obtuse angle*.

51. Two rays in a straight line pointing opposite directions measure 180°. An angle that measures 180° is called a *straight angle*.

52. The angle is a *right angle*, as indicated by the small square at the vertex. Right angles measure exactly 90°.

53. (a) The complement of 80° is 10°, because 90° − 80° = 10°.

    (b) The complement of 45° is 45°, because 90° − 45° = 45°.

    (c) The complement of 7° is 83°, because 90° − 7° = 83°.

54. (a) The supplement of 155° is 25°, because 180° − 155° = 25°.

    (b) The supplement of 90° is 90°, because 180° − 90° = 90°.

    (c) The supplement of 33° is 147°, because 180° − 33° = 147°.

55. $\angle 2 \cong \angle 5$, so $\angle 5$ measures $60°$.
    $\angle 6 \cong \angle 3$, so $\angle 6$ and $\angle 3$ measure $90°$.
    $\angle 1$ measures $90° - 60° = 30°$.
    $\angle 1 \cong \angle 4$, so $\angle 4$ measures $30°$.

56. $\angle 8 \cong \angle 3 \cong \angle 6 \cong \angle 1$, so each measures $160°$.
    $\angle 4$ measures $180° - 160° = 20°$.
    $\angle 4 \cong \angle 7 \cong \angle 2 \cong \angle 5$, so each measures $20°$.

57. Each pair of corresponding sides has the same length, so the Side-Side-Side (SSS) method can be used to prove that the triangles are congruent.

58. On both triangles, two corresponding sides and the angle between them measure the same, so the Side-Angle-Side (SAS) method can be used to prove that the triangles are congruent.

59. On both triangles, two corresponding angles and the side that connects them measure the same, so the Angle-Side-Angle (ASA) method can be used to prove that the triangles are congruent.

60. Set up a ratio of corresponding sides.
    $$\frac{40 \text{ ft}}{20 \text{ ft}} = \frac{40}{20} = \frac{2}{1}$$
    Write a proportion to find y.
    $$\frac{y}{15} = \frac{2}{1}$$
    $$y \cdot 1 = 2 \cdot 15$$
    $$y = 30 \text{ ft}$$
    Write a proportion to find x.
    $$\frac{x}{17} = \frac{2}{1}$$
    $$x \cdot 1 = 17 \cdot 2$$
    $$x = 34 \text{ ft}$$
    $P = 34 \text{ ft} + 30 \text{ ft} + 40 \text{ ft}$
    $= 104 \text{ ft}$

61. Set up a ratio of corresponding sides.
    $$\frac{4 \text{ m}}{6 \text{ m}} = \frac{4}{6} = \frac{2}{3}$$
    Write a proportion to find x.
    $$\frac{6}{x} = \frac{2}{3}$$
    $$2 \cdot x = 3 \cdot 6$$
    $$\frac{2 \cdot x}{2} = \frac{18}{2}$$
    $$x = 9 \text{ m}$$
    Write a proportion to find y.
    $$\frac{5}{y} = \frac{2}{3}$$
    $$2 \cdot y = 5 \cdot 3$$
    $$\frac{2 \cdot y}{2} = \frac{15}{2}$$
    $$y = 7.5 \text{ m}$$
    $P = 9 \text{ m} + 7.5 \text{ m} + 6 \text{ m}$
    $= 22.5 \text{ m}$

62. Set up a ratio of corresponding sides.
    $$\frac{16 \text{ mm}}{12 \text{ mm}} = \frac{16}{12} = \frac{4}{3}$$
    Write a proportion to find x.
    $$\frac{x}{9} = \frac{4}{3}$$
    $$3 \cdot x = 9 \cdot 4$$
    $$\frac{3 \cdot x}{3} = \frac{36}{3}$$
    $$x = 12 \text{ mm}$$
    Write a proportion to find y.
    $$\frac{10}{y} = \frac{4}{3}$$
    $$y \cdot 4 = 10 \cdot 3$$
    $$\frac{y \cdot 4}{4} = \frac{30}{4}$$
    $$y = 7.5 \text{ mm}$$
    $P = 12 \text{ mm} + 10 \text{ mm} + 16 \text{ mm}$
    $= 38 \text{ mm}$

63. [6.3] $\quad \dfrac{x}{45} = \dfrac{70}{30} \quad$ OR $\quad \dfrac{x}{45} = \dfrac{7}{3}$
    $$3 \cdot x = 7 \cdot 45 \quad \text{\textit{Cross products}}$$
    $$ \quad \text{\textit{are equivalent}}$$
    $$\frac{3 \cdot x}{3} = \frac{315}{3}$$
    $$x = 105$$
    Check: $105 \cdot 30 = 3150 = 45 \cdot 70$

64. [6.3] $\quad \dfrac{x}{52} = \dfrac{0}{20}$
    Since $\frac{0}{20} = 0$, x *must* equal 0. The denominator 52 could be any number (except 0).

65. [6.3] $\quad \dfrac{64}{10} = \dfrac{x}{20} \quad$ OR $\quad \dfrac{32}{5} = \dfrac{x}{20}$
    $$5 \cdot x = 32 \cdot 20 \quad \text{\textit{Cross products}}$$
    $$ \quad \text{\textit{are equivalent}}$$
    $$\frac{5 \cdot x}{5} = \frac{640}{5}$$
    $$x = 128$$
    Check: $64 \cdot 20 = 1280 = 10 \cdot 128$

66. [6.3] $\dfrac{15}{x} = \dfrac{65}{100}$ OR $\dfrac{15}{x} = \dfrac{13}{20}$

$13 \cdot x = 15 \cdot 20$ *Cross products are equivalent*

$\dfrac{13 \cdot x}{13} = \dfrac{300}{13}$

$x = \dfrac{300}{13} \approx 23.08$

Check: $15 \cdot 100 = 1500 = \dfrac{300}{13} \cdot 65$

67. [6.3] $\dfrac{7.8}{3.9} = \dfrac{13}{x}$ OR $\dfrac{2}{1} = \dfrac{13}{x}$

$2 \cdot x = 1 \cdot 13$

$\dfrac{2 \cdot x}{2} = \dfrac{13}{2}$

$x = 6.5$

Check: $7.8(6.5) = 50.7 = 3.9 \cdot 13$

68. [6.3] $\dfrac{34.1}{x} = \dfrac{0.77}{2.65}$

$0.77 \cdot x = 34.1(2.65)$

$\dfrac{0.77 \cdot x}{0.77} = \dfrac{90.365}{0.77}$

$x \approx 117.36$

Check: $34.1(2.65) = 90.365 = \left(\dfrac{90.365}{0.77}\right)(0.77)$

69. [6.1] 4 dollars to 10 quarters

4 dollars $= 4 \cdot 4 = 16$ quarters

$\dfrac{16 \text{ quarters} \div 2}{10 \text{ quarters} \div 2} = \dfrac{8}{5}$

70. [6.1] $4\tfrac{1}{8}$ inches to 10 inches

$\dfrac{4\tfrac{1}{8} \text{ inches}}{10 \text{ inches}} = \dfrac{\tfrac{33}{8}}{10} = \dfrac{33}{8} \div 10$

$= \dfrac{33}{8} \cdot \dfrac{1}{10} = \dfrac{33}{80}$

71. [6.1] 10 yards to 8 feet

10 yards $= 10 \cdot 3 = 30$ feet

$\dfrac{30 \text{ feet}}{8 \text{ feet}} = \dfrac{30 \div 2}{8 \div 2} = \dfrac{15}{4}$

72. [6.1] $3.60 to $0.90

$\dfrac{\$3.60}{\$0.90} = \dfrac{3.60 \div 0.90}{0.90 \div 0.90} = \dfrac{4}{1}$

73. [6.1] 12 eggs to 15 eggs

$\dfrac{12 \text{ eggs}}{15 \text{ eggs}} = \dfrac{12 \div 3}{15 \div 3} = \dfrac{4}{5}$

74. [6.1] 37 meters to 7 meters

$\dfrac{37 \text{ meters}}{7 \text{ meters}} = \dfrac{37}{7}$

75. [6.1] 3 pints to 4 quarts

4 quarts $= 4 \cdot 2 = 8$ pints

$\dfrac{3 \text{ pints}}{8 \text{ pints}} = \dfrac{3}{8}$

76. [6.1] 15 minutes to 3 hours

3 hours $= 3 \cdot 60 = 180$ minutes

$\dfrac{15 \text{ minutes}}{180 \text{ minutes}} = \dfrac{15 \div 15}{180 \div 15} = \dfrac{1}{12}$

77. [6.1] $4\tfrac{1}{2}$ miles to $1\tfrac{3}{10}$ miles

$\dfrac{4\tfrac{1}{2} \text{ miles}}{1\tfrac{3}{10} \text{ miles}} = \dfrac{4\tfrac{1}{2}}{1\tfrac{3}{10}} = 4\tfrac{1}{2} \div 1\tfrac{3}{10}$

$= \dfrac{9}{2} \div \dfrac{13}{10} = \dfrac{9}{\cancel{2}_1} \cdot \dfrac{\cancel{10}^5}{13}$

$= \dfrac{45}{13}$

78. [6.4]

$\dfrac{7 \text{ buying fans}}{8 \text{ fans}} = \dfrac{x}{28{,}500 \text{ fans}}$

$8 \cdot x = 7 \cdot 28{,}500$

$\dfrac{8 \cdot x}{8} = \dfrac{199{,}500}{8}$

$x \approx 24{,}937.5$

or 24,900

(*rounded to the nearest hundred*)

At today's concert, about 24,900 fans can be expected to buy a beverage.

79. [6.1]

$\dfrac{\$400 \text{ spent on car insurance}}{\$150 \text{ spent on repairs}} = \dfrac{400 \div 50}{150 \div 50} = \dfrac{8}{3}$

The ratio of the amount spent on insurance to the amount spent on repairs is $\tfrac{8}{3}$.

80. [6.2] 25 feet for $0.78

$\dfrac{\$0.78}{25 \text{ feet}} \approx \$0.031$ per foot

75 feet for $1.99 − $0.50 (coupon) = $1.49

$\dfrac{\$1.49}{75 \text{ feet}} \approx \$0.020$ per foot (∗)

100 feet for $2.59 − $0.50 (coupon) = $2.09

$\dfrac{\$2.09}{100 \text{ feet}} \approx \$0.021$ per foot

The best buy is 75 feet for $1.99 with a 50¢ coupon.

81. **[6.4]** First find the length.
$$\frac{0.5 \text{ inch}}{6 \text{ feet}} = \frac{1.75 \text{ inches}}{x}$$
$$0.5 \cdot x = 6(1.75)$$
$$\frac{0.5 \cdot x}{0.5} = \frac{10.5}{0.5}$$
$$x = 21$$

When it is built, the actual length of the patio will be 21 feet.

Then find the width.
$$\frac{0.5 \text{ inch}}{6 \text{ feet}} = \frac{1.25 \text{ inches}}{x}$$
$$0.5 \cdot x = 6(1.25)$$
$$\frac{0.5 \cdot x}{0.5} = \frac{7.5}{0.5}$$
$$x = 15$$

When it is built, the actual width of the patio will be 15 feet.

82. **[6.4]** **(a)**
$$\frac{1000 \text{ milligrams}}{5 \text{ pounds}} = \frac{x}{7 \text{ pounds}}$$
$$5 \cdot x = 1000 \cdot 7$$
$$\frac{5 \cdot x}{5} = \frac{7000}{5}$$
$$x = 1400$$

A 7-pound cat should be given 1400 milligrams.

(b) 8 ounces = $\frac{8}{16}$ pound = 0.5 pound
$$\frac{1000 \text{ milligrams}}{5 \text{ pounds}} = \frac{x}{0.5 \text{ pounds}}$$
$$5 \cdot x = 1000(0.5)$$
$$\frac{5 \cdot x}{5} = \frac{500}{5}$$
$$x = 100$$

An 8-ounce kitten should be given 100 milligrams.

83. **[6.4]**
$$\frac{251 \text{ points}}{169 \text{ minutes}} = \frac{x}{14 \text{ minutes}}$$
$$169 \cdot x = 251 \cdot 14$$
$$\frac{169 \cdot x}{169} = \frac{3514}{169}$$
$$x \approx 20.792899$$

Charles should score 21 points (rounded).

84. **[6.4]**
$$\frac{1\frac{1}{2} \text{ teaspoons}}{24 \text{ pounds}} = \frac{x}{8 \text{ pounds}}$$
$$24 \cdot x = 1\frac{1}{2} \cdot 8 = \frac{3}{2} \cdot \frac{8}{1} = \frac{24}{2} = 12$$
$$\frac{24 \cdot x}{24} = \frac{12}{24}$$
$$x = \frac{1}{2} \text{ or } 0.5$$

The infant should be given $\frac{1}{2}$ or 0.5 teaspoon.

85. **[6.5]** $\overleftrightarrow{WX}$ and $\overleftrightarrow{YZ}$ are parallel lines.

86. **[6.5]** $\overline{QR}$ is a line segment.

87. **[6.5]** $\angle CQD$ is an acute angle.

88. **[6.5]** $\overleftrightarrow{PQ}$ and $\overleftrightarrow{NO}$ are intersecting lines.

89. **[6.5]** $\angle APB$ is a right angle measuring 90°.

90. **[6.5]** $\overrightarrow{AB}$ is a ray.

91. **[6.5]** $\overleftrightarrow{T}$ is a straight angle measuring 180°.

92. **[6.5]** $\angle FEG$ is an obtuse angle.

93. **[6.5]** $\overleftrightarrow{LM}$ and $\overleftrightarrow{JK}$ are perpendicular lines.

94. **[6.5]** **(a)** If your car "did a 360," the car turned around in a complete circle.

(b) If the governor's view on taxes "took a 180° turn," he or she took the opposite view. For example, he or she may have opposed taxes but now supports them.

95. **[6.5]** **(a)** No; obtuse angles are $> 90°$, so their sum would be $> 180°$.

(b) Yes; acute angles are $< 90°$, so their sum could equal 90°.

96. **[6.5]** $\angle 1$ measures $90° - 45° = 45°$.
$\angle 7 \cong \angle 4$, so each measures 55°.
$\angle 3$ measures $90° - 55° = 35°$.
$\angle 3 \cong \angle 6$, so $\angle 6$ measures 35°.
$\angle 5$ measures 90°.

97. **[6.5]** $\angle 5 \cong \angle 2 \cong \angle 7 \cong \angle 4$, so each measures 75°.
$\angle 6$ measures $180° - 75° = 105°$.
$\angle 6 \cong \angle 1 \cong \angle 8 \cong \angle 3$, so each measures 105°.

Chapter 6 Test

1. $15 for 75 minutes

   $$\frac{\$15 \div 15}{75 \text{ minutes} \div 15} = \frac{\$1}{5 \text{ minutes}}$$

2. 3 hours to 40 minutes

   3 hours $= 3 \cdot 60$ minutes $= 180$ minutes

   $$\frac{180 \text{ minutes}}{40 \text{ minutes}} = \frac{180 \div 20}{40 \div 20} = \frac{9}{2}$$

3. $$\frac{1200 \text{ seats}}{320 \text{ seats}} = \frac{1200 \div 80}{320 \div 80} = \frac{15}{4}$$

4. 26 ounces of Brand X for
   $3.89 - \$0.75$ (coupon) $= \$3.14$

   $$\frac{\$3.14}{26 \text{ ounces}} \approx \$0.121 \text{ per ounce}$$

   16 ounces of Brand Y for
   $1.89 - \$0.50$ (coupon) $= \$1.39$

   $$\frac{\$1.39}{16 \text{ ounces}} \approx \$0.087 \text{ per ounce } (*)$$

   14 ounces of Brand Z for $1.29.

   $$\frac{\$1.29}{14 \text{ ounces}} \approx \$0.092 \text{ per ounce}$$

   The best buy is 16 ounces of Brand Y for $1.89 with a 50¢ coupon.

5. $$\frac{5}{9} = \frac{x}{45}$$

   $9 \cdot x = 5 \cdot 45$ *Cross products are equivalent*

   $$\frac{9 \cdot x}{9} = \frac{225}{9}$$

   $x = 25$

   Check: $5 \cdot 45 = 225 = 9 \cdot 25$

6. $$\frac{3}{1} = \frac{8}{x}$$

   $3 \cdot x = 8 \cdot 1$ *Cross products are equivalent*

   $$\frac{3 \cdot x}{3} = \frac{8}{3}$$

   $x \approx 2.67$

   Check: $3\left(\frac{8}{3}\right) = 8 = 1 \cdot 8$

7. $$\frac{x}{20} = \frac{6.5}{0.4}$$

   $0.4 \cdot x = 20(6.5)$ *Cross products are equivalent*

   $$\frac{0.4 \cdot x}{0.4} = \frac{130}{0.4}$$

   $x = 325$

   Check: $325(0.4) = 130 = 20(6.5)$

8. $$\frac{2\frac{1}{3}}{x} = \frac{\frac{8}{9}}{4}$$

   $$\frac{8}{9} \cdot x = 2\frac{1}{3} \cdot 4 = \frac{7}{3} \cdot \frac{4}{1} = \frac{28}{3}$$

   $$\frac{\frac{8}{9} \cdot x}{\frac{8}{9}} = \frac{\frac{28}{3}}{\frac{8}{9}}$$

   $$x = \frac{28}{3} \div \frac{8}{9} = \frac{\overset{7}{\cancel{28}}}{\underset{1}{\cancel{3}}} \cdot \frac{\overset{3}{\cancel{9}}}{\underset{2}{\cancel{8}}} = \frac{21}{2} = 10\frac{1}{2}$$

   Check: $2\frac{1}{3} \cdot 4 = \frac{28}{3} = \frac{21}{2} \cdot \frac{8}{9}$

9. $$\frac{18 \text{ orders}}{30 \text{ minutes}} = \frac{x}{40 \text{ minutes}}$$

   $30 \cdot x = 18 \cdot 40$

   $$\frac{30 \cdot x}{30} = \frac{720}{30}$$

   $x = 24$

   Pedro could enter 24 orders in 40 minutes.

10. $$\frac{2 \text{ left-handed people}}{15 \text{ people}} = \frac{x}{650 \text{ students}}$$

    $15 \cdot x = 2 \cdot 650$

    $$\frac{15 \cdot x}{15} = \frac{1300}{15}$$

    $$x = \frac{260}{3} \approx 87$$

    You could expect about 87 students to be left-handed.

11. $$\frac{8.2 \text{ grams}}{50 \text{ pounds}} = \frac{x}{145 \text{ pounds}}$$

    $50 \cdot x = 8.2(145)$

    $$\frac{50 \cdot x}{50} = \frac{1189}{50}$$

    $x = 23.78 \approx 23.8$

    A 145-pound person should be given 23.8 grams (rounded).

12. $$\frac{1 \text{ inch}}{8 \text{ feet}} = \frac{7.5 \text{ inches}}{x}$$

    $1 \cdot x = 8(7.5)$

    $x = 60$

    The actual height of the building is 60 feet.

13. $\angle LOM$ is an acute angle, so the answer is (e).

14. $\angle YOX$ is a right angle, so the answer is (a). Its measure is 90°.

15. $\overrightarrow{GH}$ is a ray, so the answer is (d).

16. $\overleftrightarrow{W}$ is a straight angle, so the answer is (g). Its measure is 180°.

17. **Parallel lines** are lines in the same plane that never intersect.
    Perpendicular lines intersect to form a right angle.

    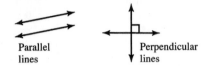

    Parallel lines   Perpendicular lines

18. The complement of an angle measuring 81° is $90° - 81° = 9°$.

19. The supplement of an angle measuring 20° is $180° - 20° = 160°$.

20. $\angle 4 \cong \angle 1$, so each measures 50°.
    $\angle 6 \cong \angle 3$, so each measures 95°.
    $\angle 2$ measures $180° - 50° - 95° = 35°$.
    $\angle 2 \cong \angle 5$, so each measures 35°.

21. $\angle 3 \cong \angle 1 \cong \angle 5 \cong \angle 7$, so each measures 65°.
    $\angle 4$ measures $180° - 65° = 115°$.
    $\angle 4 \cong \angle 2 \cong \angle 6 \cong \angle 8$, so each measures 115°.

22. On both triangles, two corresponding angles and the side that connects them measure the same, so the Angle-Side-Angle (ASA) method can be used to prove that the triangles are congruent.

23. On both triangles, two corresponding sides and the angle between them measure the same, so the Side-Angle-Side (SAS) method can be used to prove that the triangles are congruent.

24. Set up a ratio of corresponding sides.
    $$\frac{10 \text{ cm}}{15 \text{ cm}} = \frac{10}{15} = \frac{2}{3}$$
    Write a proportion to find y.
    $$\frac{y}{18} = \frac{2}{3}$$
    $$y \cdot 3 = 18 \cdot 2$$
    $$\frac{y \cdot 3}{3} = \frac{36}{3}$$
    $$y = 12 \text{ cm}$$
    Write a proportion to find z.
    $$\frac{z}{9} = \frac{2}{3}$$
    $$z \cdot 3 = 9 \cdot 2$$
    $$\frac{z \cdot 3}{3} = \frac{18}{3}$$
    $$z = 6 \text{ cm}$$

25. Set up a ratio of corresponding sides.
    $$\frac{18 \text{ mm}}{15 \text{ mm}} = \frac{18}{15} = \frac{6}{5}$$
    Write a proportion to find x.
    $$\frac{x}{10} = \frac{6}{5}$$
    $$x \cdot 5 = 10 \cdot 6$$
    $$\frac{x \cdot 5}{5} = \frac{60}{5}$$
    $$x = 12 \text{ mm}$$
    Write a proportion to find y.
    $$\frac{16.8}{y} = \frac{6}{5}$$
    $$y \cdot 6 = 16.8 \cdot 5$$
    $$\frac{y \cdot 6}{6} = \frac{84}{6}$$
    $$y = 14 \text{ mm}$$
    The perimeter of the larger triangle:
    $$P = 18 \text{ mm} + 12 \text{ mm} + 16.8 \text{ mm}$$
    $$P = 46.8 \text{ mm}$$
    The perimeter of the smaller triangle:
    $$P = 15 \text{ mm} + 10 \text{ mm} + 14 \text{ mm}$$
    $$P = 39 \text{ mm}$$

Cumulative Review Exercises (Chapters 1–6)

1. **(a)** In words, 77,001,000,805 is seventy-seven billion, one million, eight hundred five.

   (b) In words, 0.02 is two hundredths.

2. **(a)** Three and forty thousandths can be written as 3.040.

   (b) Five hundred million, thirty-seven thousand can be written as 500,037,000.

3. $\dfrac{0}{-16} = 0$

4. $|0| + |-6 - 8|$ Add the opposite.
   $= |0| + |-6 + (-8)|$ Add.
   $= |0| + |-14|$ Absolute values
   $= 0 + 14$ Add.
   $= 14$

Cumulative Review Exercises (Chapters 1–6)   219

5. $4\dfrac{3}{4} - 1\dfrac{5}{6}$ Change to improper fractions.

$= \dfrac{19}{4} - \dfrac{11}{6}$ LCD = 12

$= \dfrac{19 \cdot 3}{4 \cdot 3} - \dfrac{11 \cdot 2}{6 \cdot 2}$

$= \dfrac{57}{12} - \dfrac{22}{12}$

$= \dfrac{57 - 22}{12}$

$= \dfrac{35}{12}$ or $2\dfrac{11}{12}$

6. $\dfrac{h}{5} - \dfrac{3}{10} = \dfrac{2h}{10} - \dfrac{3}{10} = \dfrac{2h - 3}{10}$

7. $100 - 0.0095$

$\overset{99999}{\cancel{1}\cancel{0}\cancel{0}.\cancel{0}\cancel{0}\cancel{0}\cancel{0}10}$
$-\ 0.0095$
$\overline{\ 99.9905}$

8. $\dfrac{-4 + 7}{9 - 3^2} = \dfrac{3}{9 - 9} = \dfrac{3}{0}$, which is undefined.

9. $-6 + 3(0 - 4)$ Parentheses
$= -6 + 3(-4)$ Multiply.
$= -6 + (-12)$ Add.
$= -18$

10. $\dfrac{-8}{\frac{4}{7}} = -\dfrac{8}{1} \div \dfrac{4}{7} = -\dfrac{8}{1} \cdot \dfrac{7}{4} = -\dfrac{2 \cdot \cancel{4} \cdot 7}{1 \cdot \cancel{4}}$

$= -\dfrac{14}{1} = -14$

11. $\dfrac{5n}{6m^3} \div \dfrac{10}{3m^2} = \dfrac{5n}{6m^3} \cdot \dfrac{3m^2}{10}$

$= \dfrac{\cancel{5} \cdot n \cdot \cancel{3} \cdot \cancel{m} \cdot \cancel{m}}{2 \cdot \cancel{3} \cdot \cancel{m} \cdot \cancel{m} \cdot m \cdot 2 \cdot \cancel{5}}$

$= \dfrac{n}{4m}$

12. $(0.06)(-0.007)$ Different signs, negative product

$0.007 \leftarrow$ 3 decimal places
$\underline{\times\ 0.06} \leftarrow$ 2 decimal places
$0.00042 \leftarrow$ 5 decimal places

Answer: -0.00042

13. $\dfrac{x}{14y} \cdot \dfrac{7}{xy} = \dfrac{\cancel{x} \cdot \cancel{7}}{2 \cdot \cancel{7} \cdot y \cdot \cancel{x} \cdot y} = \dfrac{1}{2y^2}$

14. $\dfrac{9}{n} + \dfrac{2}{3} = \dfrac{9 \cdot 3}{n \cdot 3} + \dfrac{2 \cdot n}{3 \cdot n} = \dfrac{27}{3n} + \dfrac{2n}{3n} = \dfrac{27 + 2n}{3n}$

15. $-40 + 8(-5) + 2^4$ Exponent
$= -40 + 8(-5) + 16$ Multiply.
$= -40 + (-40) + 16$ Add.
$= -80 + 16$ Add.
$= -64$

16. $5.8 - (-0.6)^2 \div 0.9$ Exponent
$= 5.8 - 0.36 \div 0.9$ Divide.
$= 5.8 - 0.4$ Subtract.
$= 5.4$

17. $\left(-\dfrac{1}{2}\right)^3 \left(\dfrac{2}{3}\right)^2 = -\dfrac{1}{2}\left(-\dfrac{1}{2}\right)\left(-\dfrac{1}{2}\right)\dfrac{2}{3} \cdot \dfrac{2}{3}$

$= -\dfrac{1 \cdot 1 \cdot 1 \cdot \cancel{2} \cdot \cancel{2}}{2 \cdot \cancel{2} \cdot \cancel{2} \cdot 3 \cdot 3} = -\dfrac{1}{18}$

18.
$7y + 5$	$=$	$-3 - y$	$-y = -1y$
$7y + 5$	$=$	$-3 + (-1y)$	Add $1y$ to both sides.
$\underline{1y}$		$\underline{1y}$	
$8y + 5$	$=$	$-3 + 0$	
$8y + 5$	$=$	-3	Add -5 to both sides.
$\underline{-5}$		$\underline{-5}$	
$8y + 0$	$=$	-8	
$\dfrac{8y}{8}$	$=$	$\dfrac{-8}{8}$	Divide both sides by 8.
y	$=$	-1	

The solution is -1.

19.
$-2 + \dfrac{3}{5}x$	$=$	7	Add 2 to both sides.
$\underline{+2}$		$\underline{+2}$	
$0 + \dfrac{3}{5}x$	$=$	9	Multiply both sides by $\dfrac{5}{3}$.
$\dfrac{5}{3}\left(\dfrac{3}{5}x\right)$	$=$	$\dfrac{5}{3}(9)$	
x	$=$	15	

The solution is 15.

20. $\dfrac{4}{x} = \dfrac{14}{35}$ OR $\dfrac{4}{x} = \dfrac{2}{5}$

$x \cdot 2 = 4 \cdot 5$

$\dfrac{2x}{2} = \dfrac{20}{2}$

$x = 10$

The solution is 10.

21. *Step 1*
    Unknown: starting temperature

    Known: 15° rise, 23° drop, 5° rise, final temperature 71°.

    Step 2
    Let t be the starting temperature.

    Step 3
    $t + 15 + (-23) + 5 = 71$

    Step 4
    $$\begin{aligned} t + (-3) &= 71 \\ +3 & +3 \\ \hline t + 0 &= 74 \\ t &= 74 \end{aligned}$$

    Step 5
    The starting temperature was 74 degrees.

    Step 6
    Check: $74 + 15 - 23 + 5 = 71$

22. *Step 1*
    Unknown: length and width

    Known: perimeter is 100 ft, width is 14 ft less than the length

    Step 2
    Let l represent the length. Then $l - 14$ represents the width.

    Step 3
    $P = 2l + 2w$
    $2(l) + 2(l - 14) = 100$

    Step 4
    $$\begin{aligned} 2(l) + 2(l-14) &= 100 \\ 2l + 2l - 28 &= 100 \\ 4l - 28 &= 100 \\ +28 & +28 \\ \hline 4l + 0 &= 128 \\ \frac{4l}{4} &= \frac{128}{4} \\ l &= 32 \end{aligned}$$

    Step 5
    The length is 32 ft and the width is 14 ft less than the length, or 18 ft.

    Step 6
    Check: $P = 2l + 2w$
    $P = 2 \cdot 32 \text{ ft} + 2 \cdot 18 \text{ ft}$
    $P = 64 \text{ ft} + 36 \text{ ft}$
    $P = 100 \text{ ft}$

23. $P = 2l + 2w$
    $P = 2 \cdot 2.8 \text{ km} + 2 \cdot 0.7 \text{ km}$
    $P = 5.6 \text{ km} + 1.4 \text{ km}$
    $P = 7 \text{ km}$

    $A = l \cdot w$
    $A = 2.8 \text{ km} \cdot 0.7 \text{ km}$
    $A = 1.96 \text{ km}^2$
    $A \approx 2.0 \text{ km}^2$ (rounded to the nearest tenth)

24. $d = 2r$
    $d = 2 \cdot 8.5 \text{ m}$
    $d = 17 \text{ m}$

    The diameter is 17 m.

    $C = \pi \cdot d$
    $C \approx 3.14 \cdot 17 \text{ m}$
    $C \approx 53.38 \text{ m}$

    The circumference is about 53.4 m.

    $A = \pi r^2$
    $A = \pi \cdot r \cdot r$
    $A \approx 3.14 \cdot 8.5 \text{ m} \cdot 8.5 \text{ m}$
    $A \approx 226.865 \text{ m}^2$

    The area is about 226.9 m².

25. $V = \pi r^2 h$
    $V = \pi \cdot r \cdot r \cdot h$
    $V \approx 3.14 \cdot 3 \text{ ft} \cdot 3 \text{ ft} \cdot 12 \text{ ft}$
    $V \approx 339.12 \text{ ft}^3$

    The volume is about 339.1 ft³.

    $SA = 2\pi rh + 2\pi r^2$
    $SA \approx 2 \cdot 3.14 \cdot 3 \text{ ft} \cdot 12 \text{ ft} + 2 \cdot 3.14 \cdot 3 \text{ ft} \cdot 3 \text{ ft}$
    $SA \approx 282.6 \text{ ft}^2$

    The surface area is about 282.6 ft².

26. Find the amount already collected.

    $\dfrac{5}{6} \cdot \dfrac{1500}{1} = 1250$

    $1500 - 1250 = 250$

    They need to collect 250 pounds.

27. $\dfrac{\frac{1}{2} \text{ teaspoon}}{2 \text{ quarts water}} = \dfrac{x}{5 \text{ quarts water}}$

    $2 \cdot x = \dfrac{1}{2} \cdot 5$

    $\dfrac{2 \cdot x}{2} = \dfrac{\frac{5}{2}}{2}$

    $x = \dfrac{5}{2} \div 2 = \dfrac{5}{2} \cdot \dfrac{1}{2} = \dfrac{5}{4} = 1\dfrac{1}{4}$

    Use $1\frac{1}{4}$ teaspoons of plant food.

28. First add the number of times Norma ran in the morning and in the afternoon.

$$4 + 2\frac{1}{2} = 6\frac{1}{2}$$

Then multiply the sum by the distance around Dunning Pond.

$$1\frac{1}{10} \cdot 6\frac{1}{2} = \frac{11}{10} \cdot \frac{13}{2} = \frac{143}{20} = 7\frac{3}{20}$$

Norma ran $7\frac{3}{20}$ miles.

29. Use a calculator to divide the number of miles driven by the number of gallons of gas purchased.

$$\frac{896.5}{49.8} \approx 18.0$$

Rodney's car got 18.0 miles per gallon (rounded) on the vacation.

30. The cost per minute rates (in dollars) from least to greatest are:

0.028, 0.04, 0.043, 0.05, 0.052, 0.106, 0.12

31. $\dfrac{\$10}{\$0.043/\text{minute}} = \dfrac{10}{0.043} \approx 232.6$ minutes

You could make a call to Japan for 233 minutes (rounded) or 232 minutes (truncated).

32. $\dfrac{\$10}{\$0.12/\text{minute}} \approx 83.3$ minutes

You get 83 minutes (rounded) per $10 card, so buy 3 cards (249 minutes) to cover 240 minutes (4 hours).

33. For South Korea:

$$\frac{\$10}{\$0.04/\text{minute}} = 250 \text{ minutes}$$

For Mexico City:

$$\frac{\$10}{\$0.05/\text{minute}} = 200 \text{ minutes}$$

Ratio of minutes to South Korea to minutes to Mexico City:

$$\frac{250 \div 50}{200 \div 50} = \frac{5}{4}$$

The ratio of minutes for a $10 call to South Korea to minutes for a $10 call to Mexico City is $\frac{5}{4}$.

CHAPTER 7 PERCENT

7.1 The Basics of Percent

7.1 Margin Exercises

1. **(a)** $20 per $100 or $\frac{20}{100}$ or 20%

   (b) $5 per $100 or $\frac{5}{100}$ or 5%

   (c) 94 points out of 100 points or $\frac{94}{100}$ or 94%

2. **(a)** $68\% = 68 \div 100 = 0.68$

   (b) $5\% = 5 \div 100 = 0.05$

   (c) $40.6\% = 40.6 \div 100 = 0.406$

   (d) $200\% = 200 \div 100 = 2.00$, or 2

   (e) $350\% = 350 \div 100 = 3.50$, or 3.5

3. **(a)** 90%

   Drop the percent sign and move the decimal point two places to the left.

   $90.\% = 0.90$, or 0.9

   (b) $9\% = 09.\% = 0.09$

   (c) $900\% = 900.\% = 9.00$, or 9

   (d) $9.9\% = 09.9\% = 0.099$

   (e) $0.9\% = 00.9\% = 0.009$

4. **(a)** 0.95

   Move the decimal point two places to the right. (The decimal point is not written with whole number percents.) Attach a percent sign.

   $0.95_\wedge = 95\%$

   (b) $0.16 = 16\%$

   (c) $0.09 = 9\%$

   (d) $0.617 = 61.7\%$

   (e) $0.4 = 40\%$

   (f) $5.34 = 534\%$

   (g) $2.8 = 280\%$

   One zero is attached so the decimal point can be moved two places to the right. Attach a percent sign.

   (h) $4 = 400\%$

   Two zeros are attached so the decimal point can be moved two places to the right. Attach a percent sign.

5. **(a)** $50\% = \dfrac{50}{100} = \dfrac{50 \div 50}{100 \div 50} = \dfrac{1}{2}$

   (b) $19\% = \dfrac{19}{100}$

   (c) $80\% = \dfrac{80}{100} = \dfrac{80 \div 20}{100 \div 20} = \dfrac{4}{5}$

   (d) $6\% = \dfrac{6}{100} = \dfrac{6 \div 2}{100 \div 2} = \dfrac{3}{50}$

   (e) $125\% = \dfrac{125}{100} = \dfrac{125 \div 25}{100 \div 25} = \dfrac{5}{4} = 1\dfrac{1}{4}$

   (f) $300\% = \dfrac{300}{100} = \dfrac{300 \div 100}{100 \div 100} = \dfrac{3}{1} = 3$

6. **(a)** $18.5\% = \dfrac{18.5}{100} = \dfrac{18.5 \cdot 10}{100 \cdot 10} = \dfrac{185}{1000}$
   $= \dfrac{185 \div 5}{1000 \div 5} = \dfrac{37}{200}$

   (b) $87.5\% = \dfrac{87.5}{100} = \dfrac{87.5 \cdot 10}{100 \cdot 10} = \dfrac{875}{1000}$
   $= \dfrac{875 \div 125}{1000 \div 125} = \dfrac{7}{8}$

   (c) $6.5\% = \dfrac{6.5}{100} = \dfrac{6.5 \cdot 10}{100 \cdot 10} = \dfrac{65}{1000}$
   $= \dfrac{65 \div 5}{1000 \div 5} = \dfrac{13}{200}$

   (d) $66\dfrac{2}{3}\% = \dfrac{66\frac{2}{3}}{100} = \dfrac{\frac{200}{3}}{100}$
   $= \dfrac{200}{3} \div 100 = \dfrac{200}{3} \div \dfrac{100}{1}$
   $= \dfrac{200}{3} \cdot \dfrac{1}{100} = \dfrac{2 \cdot \cancel{100} \cdot 1}{3 \cdot \cancel{100}}$
   $= \dfrac{2}{3}$

   (e) $12\dfrac{1}{3}\% = \dfrac{12\frac{1}{3}}{100} = \dfrac{\frac{37}{3}}{100} = \dfrac{37}{3} \div 100$
   $= \dfrac{37}{3} \div \dfrac{100}{1} = \dfrac{37}{3} \cdot \dfrac{1}{100} = \dfrac{37}{300}$

   (f) $62\dfrac{1}{2}\% = \dfrac{62\frac{1}{2}}{100} = \dfrac{\frac{125}{2}}{100} = \dfrac{125}{2} \div 100$
   $= \dfrac{125}{2} \div \dfrac{100}{1} = \dfrac{125}{2} \cdot \dfrac{1}{100}$
   $= \dfrac{5 \cdot 25 \cdot 1}{2 \cdot 4 \cdot 25} = \dfrac{5}{8}$

   OR

   $62\dfrac{1}{2}\% = 62.5\% = \dfrac{62.5}{100} = \dfrac{62.5 \cdot 10}{100 \cdot 10}$
   $= \dfrac{625}{1000} = \dfrac{625 \div 125}{1000 \div 125} = \dfrac{5}{8}$

Chapter 7 Percent

7. **(a)** $\frac{1}{2} = \frac{1}{2} \cdot 100\% = \frac{1}{2} \cdot \frac{100}{1}\% = \frac{1 \cdot \overset{1}{\cancel{2}} \cdot 50}{\underset{1}{\cancel{2}} \cdot 1}\%$

$= \frac{50}{1}\% = 50\%$

(b) $\frac{3}{4} = \frac{3}{4} \cdot 100\% = \frac{3}{4} \cdot \frac{100}{1}\% = \frac{3 \cdot \overset{1}{\cancel{4}} \cdot 25}{\underset{1}{\cancel{4}} \cdot 1}\%$

$= \frac{75}{1}\% = 75\%$

(c) $\frac{1}{10} = \frac{1}{10} \cdot 100\% = \frac{1}{10} \cdot \frac{100}{1}\%$

$= \frac{1 \cdot \overset{1}{\cancel{10}} \cdot 10}{\underset{1}{\cancel{10}} \cdot 1}\% = \frac{10}{1}\% = 10\%$

(d) $\frac{7}{8} = \frac{7}{8} \cdot 100\% = \frac{7}{8} \cdot \frac{100}{1}\% = \frac{7 \cdot \overset{1}{\cancel{4}} \cdot 25}{2 \cdot \underset{1}{\cancel{4}}}\%$

$= \frac{175}{2}\% = 87\frac{1}{2}\%, \quad \text{or} \quad 87.5\%$

(Both are exact answers.)

(e) $\frac{5}{6} = \frac{5}{6} \cdot 100\% = \frac{5}{6} \cdot \frac{100}{1}\% = \frac{5 \cdot \overset{1}{\cancel{2}} \cdot 50}{\underset{1}{\cancel{2}} \cdot 3 \cdot 1}\%$

$= \frac{250}{3}\% = 83\frac{1}{3}\% \text{ (exact)}, \quad \text{or}$

$\approx 83.3\% \text{ (rounded)}$

(f) $\frac{2}{3} = \frac{2}{3} \cdot 100\% = \frac{2}{3} \cdot \frac{100}{1}\% = \frac{200}{3}\%$

$= 66\frac{2}{3}\% \text{ (exact)}, \quad \text{or}$

$\approx 66.7\% \text{ (rounded)}$

8. **(a)** 100% of $4.60 is $4.60.

(b) 100% of 3000 students is 3000 students.

(c) 100% of 7 pages is 7 pages.

(d) 100% of 272 miles is 272 miles.

(e) 100% of $10\frac{1}{2}$ hours is $10\frac{1}{2}$ hours.

9. **(a)** 50% of $4.60 is *half* of $4.60, or $2.30.

(b) 50% of 3000 students is *half* of 3000 students, or 1500 students.

(c) 50% of 7 pages is *half* of 7 pages, or $3\frac{1}{2}$ pages.

(d) 50% of 272 miles is *half* of 272 miles, or 136 miles.

(e) 50% of $10\frac{1}{2}$ hours is *half* of $10\frac{1}{2}$ hours, or $5\frac{1}{4}$ hours.

7.1 Section Exercises

1. $25\% = 25.\% = 0.25$

   Drop the percent sign and move the decimal point two places to the left.

3. $30\% = 30.\% = 0.30, \quad \text{or} \quad 0.3$

   Drop the percent sign and move the decimal point two places to the left.

5. $6\% = 06.\% = 0.06$

   0 is attached so the decimal point can be moved two places to the left.

7. $140\% = 140.\% = 1.40, \quad \text{or} \quad 1.4$

   Drop the percent sign and move the decimal point two places to the left.

9. $7.8\% = 07.8\% = 0.078$

   0 is attached so the decimal point can be moved two places to the left.

11. $100\% = 100.\% = 1.00, \quad \text{or} \quad 1$

13. $0.5\% = 00.5\% = 0.005$

    0 is attached so the decimal point can be moved two places to the left.

15. $0.35\% = 00.35\% = 0.0035$

    0 is attached so the decimal point can be moved two places to the left.

17. $0.5 = 0.50 = 50\%$

    0 is attached so the decimal point can be moved two places to the right. Attach a percent sign. (The decimal point is not written with whole numbers.)

19. $0.62 = 62\%$

    Move the decimal point two places to the right and attach a percent sign.

21. $0.03 = 3\%$

    Move the decimal point two places to the right and attach a percent sign.

23. $0.125 = 12.5\%$

    Move the decimal point two places to the right and attach a percent sign.

25. $0.629 = 62.9\%$

    Move the decimal point two places to the right and attach a percent sign.

27. $2 = 2.00 = 200\%$

    Two zeros are attached so the decimal point can be moved two places to the right. Attach a percent sign.

29. $2.6 = 2.60 = 260\%$

One zero is attached so the decimal point can be moved two places to the right. Attach a percent sign.

31. $0.0312 = 3.12\%$

Move the decimal point two places to the right and attach a percent sign.

33. $20\% = \dfrac{20}{100} = \dfrac{20 \div 20}{100 \div 20} = \dfrac{1}{5}$

35. $50\% = \dfrac{50}{100} = \dfrac{50 \div 50}{100 \div 50} = \dfrac{1}{2}$

37. $55\% = \dfrac{55}{100} = \dfrac{55 \div 5}{100 \div 5} = \dfrac{11}{20}$

39. $37.5\% = \dfrac{37.5}{100} = \dfrac{37.5(10)}{100(10)} = \dfrac{375 \div 125}{1000 \div 125} = \dfrac{3}{8}$

41. $6.25\% = \dfrac{6.25}{100} = \dfrac{6.25(100)}{100(100)} = \dfrac{625 \div 625}{10{,}000 \div 625} = \dfrac{1}{16}$

43. $16\dfrac{2}{3}\% = \dfrac{16\frac{2}{3}}{100} = 16\dfrac{2}{3} \div \dfrac{100}{1} = \dfrac{\overset{1}{\cancel{50}}}{3} \cdot \dfrac{1}{\underset{2}{\cancel{100}}} = \dfrac{1}{6}$

45. $130\% = \dfrac{130}{100} = \dfrac{130 \div 10}{100 \div 10} = \dfrac{13}{10} = 1\dfrac{3}{10}$

47. $250\% = \dfrac{250}{100} = \dfrac{250 \div 50}{100 \div 50} = \dfrac{5}{2} = 2\dfrac{1}{2}$

49. $\dfrac{1}{4} \cdot 100\% = \dfrac{1}{4} \cdot \dfrac{100}{1}\% = \dfrac{1 \cdot \overset{1}{\cancel{4}} \cdot 25}{\underset{1}{\cancel{4}} \cdot 1}\% = \dfrac{25}{1}\% = 25\%$

51. $\dfrac{3}{10} \cdot 100\% = \dfrac{3}{10} \cdot \dfrac{100}{1}\% = \dfrac{3 \cdot \overset{1}{\cancel{10}} \cdot 10}{\underset{1}{\cancel{10}} \cdot 1}\% = \dfrac{30}{1}\%$
$= 30\%$

53. $\dfrac{3}{5} \cdot 100\% = \dfrac{3}{5} \cdot \dfrac{100}{1}\% = \dfrac{3 \cdot \overset{1}{\cancel{5}} \cdot 20}{\underset{1}{\cancel{5}} \cdot 1}\% = \dfrac{60}{1}\%$
$= 60\%$

55. The denominator is already 100.

$\dfrac{37}{100} = 37\%$

57. $\dfrac{3}{8} \cdot 100\% = \dfrac{3}{8} \cdot \dfrac{100}{1}\% = \dfrac{3 \cdot \overset{1}{\cancel{4}} \cdot 25}{2 \cdot \underset{1}{\cancel{4}} \cdot 1}\% = \dfrac{75}{2}\%$
$= 37\dfrac{1}{2}\%, \text{ or } 37.5\%$

59. $\dfrac{1}{20} \cdot 100\% = \dfrac{1}{20} \cdot \dfrac{100}{1}\% = \dfrac{1 \cdot \overset{1}{\cancel{20}} \cdot 5}{\underset{1}{\cancel{20}} \cdot 1}\% = \dfrac{5}{1}\%$
$= 5\%$

61. $\dfrac{5}{9} \cdot 100\% = \dfrac{5}{9} \cdot \dfrac{100}{1}\% = \dfrac{500}{9}\%$
$= 55\dfrac{5}{9}\%, \text{ or } 55.6\% \text{ (rounded)}$

63. $\dfrac{1}{7} \cdot 100\% = \dfrac{1}{7} \cdot \dfrac{100}{1}\% = \dfrac{100}{7}\%$
$= 14\dfrac{2}{7}\%, \text{ or } 14.3\% \text{ (rounded)}$

65. Drop the percent sign and move the decimal point two places to the left.

$8\% \text{ of U.S. homes} = 0.08$

67. Drop the percent sign and move the decimal point two places to the left.

$42\% \text{ of tornadoes} = 0.42$

69. Move the decimal point two places to the right and attach a percent sign.

$0.035 \text{ property tax rate} = 3.5\%$

71. Attach two 0's so that the decimal point can be moved two places to the right. Attach a percent sign.

$2 \text{ times that of the last session} = 200\%$

73. Ninety-five parts of the one hundred parts are shaded.

$\dfrac{95}{100} = 0.95 = 95\%$

Five parts of the one hundred parts are unshaded.

$\dfrac{5}{100} = 0.05 = 5\%$

75. Three of the ten parts are shaded.

$\dfrac{3}{10} = \dfrac{30}{100} = 30\%$

Seven of the ten parts are unshaded.

$\dfrac{7}{10} = \dfrac{70}{100} = 70\%$

77. Three parts of the four parts are shaded.

$\dfrac{3}{4} = \dfrac{3(25)}{4(25)} = \dfrac{75}{100} = 75\%$

One part of the four parts is unshaded.

$\dfrac{1}{4} = \dfrac{1(25)}{4(25)} = \dfrac{25}{100} = 25\%$

79. $\dfrac{1}{100} = 0.01 \quad \leftarrow \text{ Decimal}$
$\phantom{\dfrac{1}{100}} = 1\% \quad \leftarrow \text{ Percent}$

81. $0.2 = 0.20 = \dfrac{20}{100} = \dfrac{1}{5} \quad \leftarrow \text{ Fraction}$
$ = 20\% \quad \leftarrow \text{ Percent}$

83. $30\% = \dfrac{30}{100} = \dfrac{3}{10}$ ← Fraction

$ = 0.30 = 0.3$ ← Decimal

85. $\dfrac{1}{2} = \dfrac{50}{100} = 0.50$ ← Decimal

$\phantom{\dfrac{1}{2}} = 50\%$ ← Percent

87. $90\% = \dfrac{90}{100} = \dfrac{9}{10}$ ← Fraction

$ = 0.90 = 0.9$ ← Decimal

89. $1.5 = \dfrac{150}{100} = \dfrac{3}{2} = 1\dfrac{1}{2}$ ← Fraction

$ = 150\%$ ← Percent

91. $8\% = \dfrac{8}{100} = \dfrac{2}{25}$ ← Fraction

$ = 0.08$ ← Decimal

93. $\dfrac{5 \text{ ft for}}{\text{every } 100 \text{ ft}} = \dfrac{5}{100} = \dfrac{1}{20}$ ← Fraction

$\phantom{\dfrac{5 \text{ ft for}}{\text{every } 100 \text{ ft}}} = 0.05$ ← Decimal

$\phantom{\dfrac{5 \text{ ft for}}{\text{every } 100 \text{ ft}}} = 5\%$ ← Percent

95. There are 8 incisors and 32 total teeth.

$\dfrac{8}{32} = \dfrac{8 \div 8}{32 \div 8} = \dfrac{1}{4}$

$\dfrac{1}{4} = \dfrac{1(25)}{4(25)} = \dfrac{25}{100} = 0.25$

$\phantom{\dfrac{1}{4}} = 25\%$

97. There are 12 molars and 32 total teeth.

$\dfrac{12}{32} = \dfrac{12 \div 4}{32 \div 4} = \dfrac{3}{8}$

$\dfrac{3}{8} = \dfrac{3(125)}{8(125)} = \dfrac{375}{1000} = 0.375$

$\phantom{\dfrac{3}{8}} = 37.5\%$

99. There are 4 canines and 28 total teeth.

$\dfrac{4}{28} = \dfrac{4 \div 4}{28 \div 4} = \dfrac{1}{7}$

$\dfrac{1}{7} = \dfrac{1}{7} \cdot 100\% = \dfrac{100}{7}\% = 14\dfrac{2}{7}\% \approx 14.3\%$

101. (a) The student forgot to move the decimal point in 0.35 two places to the right. So $\dfrac{7}{20} = 35\%$.

(b) The student did the division in the wrong order. Enter $16 \div 25$ to get 0.64 and then move the decimal point two places to the right. So $\dfrac{16}{25} = 0.64 = 64\%$.

103. (a) 100% of $78 is $\underline{\$78}$.

(b) 50% of $78 is half of $78, or $\underline{\$39}$.

105. (a) 100% of 15 inches is $\underline{15 \text{ inches}}$.

(b) 50% of 15 inches is half of 15 inches, or $7\dfrac{1}{2}$ inches.

107. (a) 100% of 2.8 miles is $\underline{2.8 \text{ miles}}$.

(b) 50% of 2.8 miles is half of 2.8 miles, or $\underline{1.4 \text{ miles}}$.

109. 100% of 20 children is 20 children. 20 children are served both meals.

111. 50% of 120 credits is half of 120 credits, or 60 credits.

113. (a) 50% of $285 is half of $285, or $\underline{\$142.50}$.

(b) John will have to pay the remaining 50% of the tuition (100% − 50% = 50%).

(c) John will have to pay the same amount that financial aid will cover, that is, $142.50.

115. (a) 50% of 8200 is 4100, so the number of students who work more than 20 hours per week is *about* 4100.

(b) The percent of students who work 20 hours or less per week is 100% − 50% = 50%.

117. (a) 100% of 35 problems is 35 problems.

(b) He correctly worked all 35 problems, so he missed 0 problems.

119. 50% means 50 parts out of 100 parts. That's half of the number. A shortcut for finding 50% of a number is to divide the number by 2. Examples will vary.

7.2 The Percent Proportion

7.2 Margin Exercises

1. (a) The percent is 15. The number 15 appears with the symbol %.

(b) The word *percent* has no number with it, so the percent is the *unknown* part of the problem.

(c) The percent is $6\dfrac{1}{2}$ because $6\dfrac{1}{2}$ appears with the word *percent*.

(d) The percent is 48. The number 48 appears with the symbol %.

(e) The word *percent* has no number with it, so the percent is the *unknown* part of the problem.

2. (a) The whole is $2000. The number $2000 appears after the word *of*.

(b) The whole is 750 employees. The number 750 appears after the word *of*.

(c) The whole is $590. The number 590 appears after the word *of*.

(d) The whole is *unknown*. The word *of* appears before "what amount."

7.2 The Percent Proportion

(e) The whole is 110 rental cars. The number 110 appears after the word *of*.

3. (a) The part of the $2000 that will be spent on a washing machine is *unknown*.

   (b) The part of the 750 employees is 60 employees.

   (c) The percent is $6\frac{1}{2}$. The whole price is $590. The sales tax would be part of the whole price of $590. The sales tax (part) is *unknown*.

   (d) The part is $30. $30 is part of some unknown dollar value.

   (e) The part is 75 cars. 75 cars is part of the 110 rental cars.

4. (a) Percent: 9%
   Whole: 3250
   Part: unknown (n)

   Step 1 $\quad \dfrac{9}{100} = \dfrac{n}{3250}$
   Step 2 $\quad 100 \cdot n = 9 \cdot 3250$
   Step 3 $\quad \dfrac{100n}{100} = \dfrac{29{,}250}{100}$
   $\qquad\qquad n = 292.5$

   The part is 292.5 miles. 292.5 miles is 9% of 3250 miles.

   (b) Percent: 20%
   Whole: 180
   Part: unknown (n)

   Step 1 $\quad \dfrac{20}{100} = \dfrac{n}{180}$
   Step 2 $\quad 100 \cdot n = 20 \cdot 180$
   Step 3 $\quad \dfrac{100n}{100} = \dfrac{3600}{100}$
   $\qquad\qquad n = 36$

   The part is 36 calories. 36 calories is 20% of 180 calories.

   (c) Percent: 78%
   Whole: $5.50
   Part: unknown (n)

   Step 1 $\quad \dfrac{78}{100} = \dfrac{n}{5.50}$
   Step 2 $\quad 100 \cdot n = 78(5.50)$
   Step 3 $\quad \dfrac{100n}{100} = \dfrac{429}{100}$
   $\qquad\qquad n = 4.29$

   The part is $4.29. $4.29 is 78% of $5.50.

   (d) Percent: $12\frac{1}{2}$% or 12.5%
   Whole: 400
   Part: unknown (n)

   Step 1 $\quad \dfrac{12.5}{100} = \dfrac{n}{400}$
   Step 2 $\quad 100 \cdot n = 12.5(400)$
   Step 3 $\quad \dfrac{100n}{100} = \dfrac{5000}{100}$
   $\qquad\qquad n = 50$

   The part is 50 homes. 50 homes are $12\frac{1}{2}$% of 400.

5. (a) Part: 1200
   Whole: 5000
   Percent: unknown (p)

   Step 1 $\quad \dfrac{p}{100} = \dfrac{1200}{5000}$
   Step 2 $\quad p \cdot 5000 = 100(1200)$
   Step 3 $\quad \dfrac{p \cdot 5000}{5000} = \dfrac{120{,}000}{5000}$
   $\qquad\qquad p = 24$

   The percent is 24%. 1200 is 24% of 5000.

   (b) Part: $0.52
   Whole: $6.50
   Percent: unknown (p)

   Step 1 $\quad \dfrac{p}{100} = \dfrac{0.52}{6.50}$
   Step 2 $\quad 6.50 \cdot p = 100(0.52)$
   Step 3 $\quad \dfrac{6.50p}{6.50} = \dfrac{52}{6.50}$
   $\qquad\qquad p = 8$

   The percent is 8%. $0.52 is 8% of $6.50.

   (c) Part: 20
   Whole: 32
   Percent: unknown (p)

   Step 1 $\quad \dfrac{p}{100} = \dfrac{20}{32}$
   Step 2 $\quad 32 \cdot p = 100(20)$
   Step 3 $\quad \dfrac{32p}{32} = \dfrac{2000}{32}$
   $\qquad\qquad p = 62.5$

   The percent is 62.5% or $62\frac{1}{2}$%. 20 is 62.5% of 32.

6. (a) Part: 37
   Percent: 74%
   Whole: unknown (n)

   Step 1 $\quad \dfrac{74}{100} = \dfrac{37}{n}$
   Step 2 $\quad 74 \cdot n = 100 \cdot 37$
   Step 3 $\quad \dfrac{74n}{74} = \dfrac{3700}{74}$
   $\qquad\qquad n = 50$

   The whole is 50 cars. 37 cars is 74% of 50 cars.

(b) Part: $139.59
Percent: 45%
Whole: unknown (n)

Step 1 $\dfrac{45}{100} = \dfrac{139.59}{n}$
Step 2 $45 \cdot n = 100(139.59)$
Step 3 $\dfrac{45n}{45} = \dfrac{13{,}959}{45}$
$n = 310.20$

The whole is $310.20. $139.59 is 45% of $310.20.

(c) Part: 1.2
Percent: $2\tfrac{1}{2}$% or 2.5%
Whole: unknown (n)

Step 1 $\dfrac{2.5}{100} = \dfrac{1.2}{n}$
Step 2 $2.5 \cdot n = 100(1.2)$
Step 3 $\dfrac{2.5n}{2.5} = \dfrac{120}{2.5}$
$n = 48$

The whole is 48 tons. 1.2 tons is $2\tfrac{1}{2}$% of 48 tons.

7. (a) Percent: 350%
Whole: $6
Part: unknown (n)

Step 1 $\dfrac{350}{100} = \dfrac{n}{6}$
Step 2 $100 \cdot n = 350 \cdot 6$
Step 3 $\dfrac{100n}{100} = \dfrac{2100}{100}$
$n = 21$

The part is $21. 350% of $6 is $21.

(b) Part: 23
Whole: 20 hours (preceded by the word *of*)
Percent: unknown (p)

Step 1 $\dfrac{p}{100} = \dfrac{23}{20}$
Step 2 $20 \cdot p = 100 \cdot 23$
Step 3 $\dfrac{20p}{20} = \dfrac{2300}{20}$
$p = 115$

The percent is 115%.
23 hours is 115% of 20 hours.

(c) Part: $106.47
Whole: $47.32 (preceded by the word *of*)
Percent: unknown (p)

Step 1 $\dfrac{p}{100} = \dfrac{106.47}{47.32}$
Step 2 $47.32 \cdot p = 100(106.47)$
Step 3 $\dfrac{47.32p}{47.32} = \dfrac{10{,}647}{47.32}$
$p = 225$

The percent is 225%. $106.47 is 225% of $47.32.

7.2 Section Exercises

1. (a) The percent is 10%.

(b) The whole is 3000 runners.

(c) The part is unknown (n).

(d) Step 1 $\dfrac{10}{100} = \dfrac{n}{3000}$
Step 2 $100 \cdot n = 10 \cdot 3000$
Step 3 $\dfrac{100n}{100} = \dfrac{30{,}000}{100}$
$n = 300$

10% of 3000 runners is 300 runners.

3. (a) The percent is 4%.

(b) The whole is 120 feet.

(c) The part is unknown (n).

(d) Step 1 $\dfrac{4}{100} = \dfrac{n}{120}$
Step 2 $100 \cdot n = 4 \cdot 120$
Step 3 $\dfrac{100n}{100} = \dfrac{480}{100}$
$n = 4.8$

4% of 120 feet is 4.8 feet.

5. (a) The percent is unknown (p).

(b) The whole is 32 pizzas.

(c) The part is 16 pizzas.

(d) Step 1 $\dfrac{p}{100} = \dfrac{16}{32}$
Step 2 $32 \cdot p = 100 \cdot 16$
Step 3 $\dfrac{32p}{32} = \dfrac{1600}{32}$
$p = 50$

16 pizzas is 50% of 32 pizzas.

7. (a) The percent is unknown (p).

(b) The whole is 200 calories.

(c) The part is 16 calories.

7.2 The Percent Proportion 229

(d) *Step 1* $\dfrac{p}{100} = \dfrac{16}{200}$

Step 2 $200 \cdot p = 100 \cdot 16$

Step 3 $\dfrac{200p}{200} = \dfrac{1600}{200}$

$p = 8$

8% of 200 calories is 16 calories.

9. (a) The percent is 90%.

(b) The whole is unknown (n).

(c) The part is 495 students.

(d) *Step 1* $\dfrac{90}{100} = \dfrac{495}{n}$

Step 2 $90 \cdot n = 100 \cdot 495$

Step 3 $\dfrac{90n}{90} = \dfrac{49{,}500}{90}$

$n = 550$

495 students is 90% of 550 students.

11. (a) The percent is $12\tfrac{1}{2}\% = 12.5\%$.

(b) The whole is unknown (n).

(c) The part is $3.50.

(d) *Step 1* $\dfrac{12.5}{100} = \dfrac{3.50}{n}$

Step 2 $12.5 \cdot n = 100 \cdot 3.50$

Step 3 $\dfrac{12.5n}{12.5} = \dfrac{350}{12.5}$

$n = 28$

$12\tfrac{1}{2}\%$ of $28 is $3.50.

13. Part: unknown (n)
Whole: 7
Percent: 250%

Step 1 $\dfrac{250}{100} = \dfrac{n}{7}$

Step 2 $100 \cdot n = 250 \cdot 7$

Step 3 $\dfrac{100n}{100} = \dfrac{1750}{100}$

$n = 17.5$

250% of 7 hours is 17.5 hours.

15. Part: 32
Whole: 172
Percent: unknown (p)

Step 1 $\dfrac{p}{100} = \dfrac{32}{172}$

Step 2 $172 \cdot p = 100 \cdot 32$

Step 3 $\dfrac{172p}{172} = \dfrac{3200}{172}$

$p \approx 18.6$

$32 is 18.6% (rounded) of $172.

17. Part: 748
Whole: unknown (n)
Percent: 110%

Step 1 $\dfrac{110}{100} = \dfrac{748}{n}$

Step 2 $110 \cdot n = 100 \cdot 748$

Step 3 $\dfrac{110n}{110} = \dfrac{74{,}800}{110}$

$n = 680$

748 books is 110% of 680 books.

19. Part: unknown (n)
Whole: $274
Percent: 14.7%

Step 1 $\dfrac{14.7}{100} = \dfrac{n}{274}$

Step 2 $100 \cdot n = 14.7 \cdot 274$

Step 3 $\dfrac{100n}{100} = \dfrac{4027.8}{100}$

$n \approx 40.28$

14.7% of $274 is $40.28 (rounded).

21. Part: 105
Whole: 54
Percent: unknown (p)

Step 1 $\dfrac{p}{100} = \dfrac{105}{54}$

Step 2 $54 \cdot p = 100 \cdot 105$

Step 3 $\dfrac{54p}{54} = \dfrac{10{,}500}{54}$

$p \approx 194.4$

105 employees is 194.4% (rounded) of 54 employees.

23. Part: $0.33
Whole: unknown (n)
Percent: 4%

Step 1 $\dfrac{4}{100} = \dfrac{0.33}{n}$

Step 2 $4 \cdot n = 100 \cdot 0.33$

Step 3 $\dfrac{4n}{4} = \dfrac{33}{4}$

$n = 8.25$

$0.33 is 4% of $8.25.

25. 150% of $30 cannot be less than $30 because 150% is greater than 100%. The answer must be greater than $30.

25% of $16 cannot be greater than $16 because 25% is less than 100%. The answer must be less than $16.

230 Chapter 7 Percent

27. The correct proportion is:
$$\frac{p}{100} = \frac{14}{8}$$
$$8 \cdot p = 100 \cdot 14$$
$$\frac{8p}{8} = \frac{1400}{8}$$
$$p = 175$$

The answer should be labeled with the % symbol. Correct answer is 175%.

7.3 The Percent Equation

7.3 Margin Exercises

1. **(a)** 110.38 rounds to $100. Divide $100 by 4. The estimate is $25.

(b) 7.6 hours rounds to 8 hours. Divide 8 hours by 4. The estimate is 2 hours.

(c) 34 pounds rounds to 30 pounds. Divide 30 pounds by 4. The estimate is 7.5 pounds. Or, 34 pounds rounds to 32 pounds (a multiple of 4). Divide 32 pounds by 4. The estimate is 8 pounds.

2. **(a)** 10% of $110.38 = $11.038 or $11.04 to the nearest cent.

(b) 10% of 7.6 hours = 0.76 hours

(c) 10% of 34. pounds = 3.4 pounds

3. **(a)** 1% of $110.38 = $1.1038 or $1.10 to the nearest cent.

(b) 1% of 07.6 hours = 0.076 hour

(c) 1% of 34. pounds = 0.34 pound

4. **(a)** 9% of 3250 miles is how many miles?

percent · whole = part
$$(0.09)(3250) = n$$
$$292.5 = n$$

9% of 3250 is 292.5 miles.

(b) 78% of $5.50 is how much?

percent · whole = part
$$(0.78)(5.50) = n$$
$$4.29 = n$$

78% of $5.50 is $4.29.

(c) What is 12.5% of 400 homes?

part = percent · whole
$$n = (0.125)(400)$$
$$n = 50$$

50 homes is 12.5% of 400 homes.

(d) How much is 350% of $6?

part = percent · whole
$$n = (3.5)(6)$$
$$n = 21$$

$21 is 350% of $6.

5. **(a)** 1200 books is what percent of 5000 books?

part = percent · whole
$$1200 = p \cdot 5000$$
$$\frac{1200}{5000} = \frac{5000p}{5000}$$
$$0.24 = p$$

1200 books is 24% of 5000 books.

(b) 23 hours is what percent of 20 hours?

part = percent · whole
$$23 = p \cdot 20$$
$$\frac{23}{20} = \frac{20p}{20}$$
$$1.15 = p$$

23 hours is 115% of 20 hours.

(c) What percent of $6.50 is $0.52?

percent · whole = part
$$p \cdot 6.50 = 0.52$$
$$\frac{6.50p}{6.50} = \frac{0.52}{6.50}$$
$$p = 0.08$$

8% of $6.50 is $0.52.

6. **(a)** 74% of how many cars is 37 cars?

percent · whole = part
$$0.74 \cdot n = 37$$
$$\frac{0.74n}{0.74} = \frac{37}{0.74}$$
$$n = 50$$

74% of 50 cars is 37 cars.

(b) 1.2 tons is $2\frac{1}{2}$% of how many tons?

part = percent · whole
$$1.2 = 0.025 \cdot n$$
$$\frac{1.2}{0.025} = \frac{0.025n}{0.025}$$
$$48 = n$$

1.2 tons is $2\frac{1}{2}$% of 48 tons.

7.3 The Percent Equation

(c) 216 calculators is 160% of how many calculators?

$$\text{part} = \text{percent} \cdot \text{whole}$$
$$216 = 1.60 \cdot n$$
$$\frac{216}{1.60} = \frac{1.60n}{1.60}$$
$$135 = n$$

216 calculators is 160% of 135 calculators.

7.3 Section Exercises

1. 50% of 3000 is the same as half of 3000. Choose 1500 patients.

3. 25% of $60 is the same as dividing $60 by 4.

25% of $60 = $\frac{\$60}{4}$ = $15. Choose $15.

5. 10% of 45 pounds is the same as dividing by 10. Move the decimal one place to the left.

10% of 45. = 4.5 pounds. Choose 4.5 pounds.

7. 200% of $3.50 is the same as multiplying $3.50 by 2.

200% of $3.50 = 2 · $3.50 = $7.00. Choose $7.00.

9. 1% of 5200 students is the same as dividing by 100. Move the decimal two places to the left.

1% of 5200. = 52 students. Choose 52 students.

11. 10% of 8700 cell phones is the same as dividing by 10. Move the decimal one place to the left.

10% of 8700. = 870 cell phones. Choose 870 phones.

13. 25% of 19 hours is the same as dividing 19 by 4.

25% of 19 hours = $\frac{19}{4} = 4\frac{3}{4}$ = 4.75 hours.

Choose 4.75 hours.

15. (a) 10% means $\frac{10}{100}$ or $\frac{1}{10}$. The denominator tells you to divide the whole by 10. The shortcut for dividing by 10 is to move the decimal point one place to the left.

(b) Once you find 10% of a number, multiply the result by 2 for 20% and by 3 for 30%.

17. 35% of 660 is how many programs?

$$0.35 \cdot 660 = n$$
$$231 = n$$

35% of 660 programs is 231 programs.

19. 70 is what percent of 140 truckloads?

$$70 = p \cdot 140$$
$$\frac{70}{140} = \frac{140p}{140}$$
$$0.5 = p \quad (50\%)$$

70 truckloads is 50% of 140 truckloads.

21. 476 circuits is 70% of what number of circuits?

$$476 = 0.70 \cdot n$$
$$\frac{476}{0.70} = \frac{0.70n}{0.70}$$
$$680 = n$$

476 circuits is 70% of 680 circuits.

23. $12\frac{1}{2}\%$ of what number of people is 135 people?

$$0.125 \cdot n = 135$$
$$\frac{0.125n}{0.125} = \frac{135}{0.125}$$
$$n = 1080$$

$12\frac{1}{2}\%$ of 1080 people is 135 people.

25. What is 65% of 1300 species?

$$n = 0.65 \cdot 1300$$
$$n = 845$$

845 species is 65% of 1300 species.

27. 4% of $520 is how much?

$$0.04 \cdot 520 = n$$
$$20.8 = n$$

4% of $520 is $20.80.

29. 38 styles is what percent of 50 styles?

$$38 = p \cdot 50$$
$$\frac{38}{50} = \frac{50p}{50}$$
$$0.76 = p \quad (76\%)$$

38 styles is 76% of 50 styles.

232   Chapter 7   Percent

31. $\underbrace{\text{What percent}}_{p} \cdot \underbrace{\text{of \$264}}_{264} \underbrace{\text{is \$330?}}_{=\ 330}$

$\dfrac{264p}{264} = \dfrac{330}{264}$

$p = 1.25 \quad (125\%)$

125% of \$264 is \$330.

33. $\underbrace{141 \text{ employees}}_{141} \underbrace{\text{is}}_{=} \underbrace{3\%}_{0.03} \underbrace{\text{of}}_{\cdot} \underbrace{\text{what number of employees?}}_{n}$

$\dfrac{141}{0.03} = \dfrac{0.03n}{0.03}$

$4700 = n$

141 employees is 3% of 4700 employees.

35. $\underbrace{32\%}_{0.32} \underbrace{\text{of}}_{\cdot} \underbrace{260 \text{ quarts}}_{260} \underbrace{\text{is}}_{=} \underbrace{\text{how many quarts?}}_{n}$

$83.2 = n$

32% of 260 quarts is 83.2 quarts.

37. $\underbrace{\$1.48}_{1.48} \underbrace{\text{is}}_{=} \underbrace{\text{what percent}}_{p} \underbrace{\text{of}}_{\cdot} \underbrace{\$74?}_{74}$

$\dfrac{1.48}{74} = \dfrac{74p}{74}$

$0.02 = p \quad (2\%)$

\$1.48 is 2% of \$74.

39. $\underbrace{\text{How many tablets}}_{n} \underbrace{\text{is}}_{=} \underbrace{140\%}_{1.40} \underbrace{\text{of}}_{\cdot} \underbrace{500 \text{ tablets?}}_{500}$

$n = 700$

700 tablets is 140% of 500 tablets.

41. $\underbrace{40\%}_{0.40} \underbrace{\text{of}}_{\cdot} \underbrace{\text{what number of salads}}_{n} \underbrace{\text{is}}_{=} \underbrace{130 \text{ salads?}}_{130}$

$\dfrac{0.40n}{0.40} = \dfrac{130}{0.40}$

$n = 325$

40% of 325 salads is 130 salads.

43. $\underbrace{\text{What percent}}_{p} \cdot \underbrace{\text{of 160 liters}}_{160} \underbrace{\text{is}}_{=} \underbrace{2.4 \text{ liters?}}_{2.4}$

$\dfrac{160p}{160} = \dfrac{2.4}{160}$

$p = 0.015 \quad (1.5\%)$

1.5% of 160 liters is 2.4 liters.

45. $\underbrace{225\%}_{2.25} \underbrace{\text{of}}_{\cdot} \underbrace{\text{what number of gallons}}_{n} \underbrace{\text{is}}_{=} \underbrace{11.25 \text{ gallons?}}_{11.25}$

$\dfrac{2.25n}{2.25} = \dfrac{11.25}{2.25}$

$n = 5$

225% of 5 gallons is 11.25 gallons.

47. $\underbrace{\text{What}}_{n} \underbrace{\text{is}}_{=} \underbrace{12.4\%}_{0.124} \underbrace{\text{of}}_{\cdot} \underbrace{8300 \text{ meters?}}_{8300}$

$n = 1029.2$

1029.2 meters is 12.4% of 8300 meters.

49. **(a)** Multiply 0.2 by 100% to change it from a decimal to a percent. So, $0.20 = 20\%$.

(b) The correct equation is $50 = p \cdot 20$.

$\dfrac{50}{20} = \dfrac{20p}{20}$

$2.5 = p$

The solution is 250%.

51. **(a)** $33\dfrac{1}{3}\% = \dfrac{\frac{100}{3}}{3}\% = \dfrac{\frac{100}{3}}{100} = \dfrac{100}{3} \div 100$

$= \dfrac{100}{3} \div \dfrac{100}{1}$

$= \dfrac{100}{3} \cdot \dfrac{1}{100} = \dfrac{1}{3}$

Thus, $33\dfrac{1}{3}\% = \dfrac{1}{3}$.

$\underbrace{33\dfrac{1}{3}\%}_{\frac{1}{3}} \underbrace{\text{of}}_{\cdot} \underbrace{\$162}_{162} \underbrace{\text{is}}_{=} \underbrace{\text{how much?}}_{n}$

$54 = n$

The solution is \$54.

(b) $33\frac{1}{3}\% = \frac{1}{3} \approx 0.333333333$
$(0.333333333)(162) = n$
$54 \approx n$

Depending upon how your calculator rounds numbers, the solution is either $54 or $53.99999995.

(c) There is no difference or the difference is insignificant. The small variation in the solutions is due to truncating or rounding.

52. (a) $66\frac{2}{3}\% = \frac{200}{3}\% = \frac{\frac{200}{3}}{100} = \frac{200}{3} \div 100$
$= \frac{200}{3} \div \frac{100}{1} = \frac{200}{3} \cdot \frac{1}{100} = \frac{2}{3}$

Thus, $66\frac{2}{3}\% = \frac{2}{3}$.

part = percent · whole
$22 = \frac{2}{3} \cdot n$
$\frac{22}{\frac{2}{3}} = \frac{\frac{2}{3}n}{\frac{2}{3}}$
$33 = n$

22 cans is $66\frac{2}{3}\%$ of 33 cans.

(b) From part (a), $66\frac{2}{3}\% = \frac{2}{3} = 0.6666\overline{6}$.

Using the truncated decimal form of $\frac{2}{3}$:

part = percent · whole
$22 = 0.66666666 \cdot n$
$\frac{22}{0.66666666} = \frac{0.66666666n}{0.66666666}$
$33.00000003 = n$

OR

Using the rounded decimal form of $\frac{2}{3}$:

part = percent · whole
$22 = 0.66666667 \cdot n$
$\frac{22}{0.66666667} = \frac{0.66666667n}{0.66666667}$
$32.99999998 = n$

Depending on whether you truncate the decimal form or round it, the solution will be either 33.00000003 or 32.99999998.

(c) There is no difference or the difference is insignificant. The small variation in the solutions is due to truncating or rounding.

Summary Exercises on Percent

1. (a) $\frac{3}{100} = 0.03$ ← Decimal
$= 3\%$ ← Percent

(b) $30\% = \frac{30}{100} = \frac{3}{10}$ ← Fraction
$= 0.30 = 0.3$ ← Decimal

(c) $0.375 = \frac{375}{1000} = \frac{375 \div 125}{1000 \div 125} = \frac{3}{8}$ ← Fraction
$= 37.5\%$ ← Percent

(d) $160\% = \frac{160}{100} = \frac{8}{5} = 1\frac{3}{5}$ ← Fraction
$= 1.6$ ← Decimal

(e) Since 16 doesn't divide evenly into 100 or 1000, we divide 10,000 by 16 to get 625.

$\frac{1}{16} = \frac{625}{10,000} = 0.0625$ ← Decimal
$= 6.25\%$ ← Percent

(f) $5\% = \frac{5}{100} = \frac{1}{20}$ ← Fraction
$= 0.05$ ← Decimal

(g) $2.0 = \frac{200}{100} = \frac{2}{1} = 2$ ← Fraction
$= 200\%$ ← Percent

(h) $\frac{4}{5} = \frac{4(20)}{5(20)} = \frac{80}{100} = 0.8$ ← Decimal
$= 80\%$ ← Percent

(i) $0.072 = \frac{72}{1000} = \frac{72 \div 8}{1000 \div 8} = \frac{9}{125}$ ← Fraction
$= 7.2\%$ ← Percent

3. Part: 9
Whole: 72
Percent: unknown (p)

Step 1 $\quad \frac{p}{100} = \frac{9}{72}$
Step 2 $\quad 72 \cdot p = 100 \cdot 9$
Step 3 $\quad \frac{72p}{72} = \frac{900}{72}$
$p = 12.5$

9 Web sites is 12.5% of 72 Web sites.

5. Part: unknown (n)
Whole: $8.79
Percent: 6%

Step 1 $\quad \frac{6}{100} = \frac{n}{8.79}$
Step 2 $\quad 100 \cdot n = 6 \cdot 8.79$
Step 3 $\quad \frac{100n}{100} = \frac{52.74}{100}$
$n \approx 0.53$

6% of $8.79 is $0.53 (rounded).

234 Chapter 7 Percent

7. Part: unknown (n)
Whole: 168
Percent: $3\frac{1}{2}\%$

Step 1 $\dfrac{3.5}{100} = \dfrac{n}{168}$

Step 2 $100 \cdot n = 3.5 \cdot 168$

Step 3 $\dfrac{100n}{100} = \dfrac{588}{100}$

$n = 5.88$

$3\frac{1}{2}\%$ of 168 pounds is 5.9 pounds (rounded).

9. Part: 40,000
Whole: 80,000
Percent: unknown (p)

Step 1 $\dfrac{p}{100} = \dfrac{40{,}000}{80{,}000}$

Step 2 $80{,}000 \cdot p = 100 \cdot 40{,}000$

Step 3 $\dfrac{80{,}000p}{80{,}000} = \dfrac{4{,}000{,}000}{80{,}000}$

$p = 50$

50% of 80,000 deer is 40,000 deer.

11. Part: unknown (n)
Whole: 35
Percent: 280%

Step 1 $\dfrac{280}{100} = \dfrac{n}{35}$

Step 2 $100 \cdot n = 280 \cdot 35$

Step 3 $\dfrac{100n}{100} = \dfrac{9800}{100}$

$n = 98$

98 golf balls is 280% of 35 golf balls.

13. 9% of what number is 207
  of apartments apartments?

$0.09 \cdot n = 207$

$\dfrac{0.09n}{0.09} = \dfrac{207}{0.09}$

$n = 2300$

9% of 2300 apartments is 207 apartments.

15. $1160 is what percent of $800?

$1160 = p \cdot 800$

$\dfrac{1160}{800} = \dfrac{800p}{800}$

$1.45 = p$ (145%)

$1160 is 145% of $800.

17. What is 300% of 0.007 inch?

$n = 3 \cdot 0.007$

$n = 0.021$

0.021 inch is 300% of 0.007 inch.

19. What percent of 60 yards is 4.8 yards?

$p \cdot 60 = 4.8$

$\dfrac{60p}{60} = \dfrac{4.8}{60}$

$p = 0.08$ (8%)

8% of 60 yards is 4.8 yards.

21. The smallest percentage is $2\frac{1}{2}\%$ for invitations.
$2\frac{1}{2}\%$ of $33,600 = 0.025(\$33{,}600) = \$840$.

23. **(a)** The cost of photograhpy and videography is
11.4% of $\$33{,}600 = 0.114(\$33{,}600) = \$3830.40$.

(b) The amount spent on rings for the bride and groom is
$12\frac{2}{5}\%$ of $\$33{,}600 = 0.124(\$33{,}600) = \$4166.40$.

7.4 Problem Solving with Percent

7.4 Margin Exercises

1. *Step 1*
Unknown: number of students receiving aid
Known: 65% receive aid; 9280 students

Step 2
Let n be the number of students receiving aid.

Step 3
percent $\cdot$ whole = part
$65\% \cdot 9280 = n$

Step 4
$(0.65)(9280) = n$
$6032 = n$

Step 5
6032 students receive financial aid.

Step 6
Check: 10% of 9280. students is 928, which rounds to 900 students. So 60% would be 6 times 900 students = 5400 students, and 70% would be $7 \cdot 900 = 6300$. The solution falls between 5400 and 6300 students, so it is reasonable.

2. *Step 1*
Unknown: number of points earned by Hue
Known: 50 total points; 83% correct

Step 2
Let n be the number of points earned.

Step 3
percent • whole = part
$$83\% \cdot 50 = n$$

Step 4
$$(0.83)(50) = n$$
$$41.5 = n$$

Step 5
Hue earned 41.5 points.

Step 6
Check: 10% of 50. points is 5 points, so 80% would be 8 times 5 points = 40 points. Hue earned a little more than 80%, so 41.5 points is reasonable.

3. *Step 1*
Unknown: percentage of made field goals
Known: 47 made; 80 attempted

Step 2
Let p be the unknown percent.

Step 3
percent • whole = part
$$p \cdot 80 = 47$$

Step 4
$$\frac{80p}{80} = \frac{47}{80}$$
$$p = 0.5875 = 58.75\%$$

Step 5
The Lakers made about 59% of their field goals.

Step 6
Check: The Lakers made a little more than half of their field goals, $\frac{1}{2} = 50\%$, so the solution of 59% is reasonable.

4. *Step 1*
Unknown: percent of the predicted enrollment that the actual enrollment is
Known: 1200 students predicted; 1620 students actual enrollment

Step 2
Let p be the unknown percent.

Step 3
percent • whole = part
$$p \cdot 1200 = 1620$$

Step 4
$$\frac{1200p}{1200} = \frac{1620}{1200}$$
$$p = 1.35 = 135\%$$

Step 5
Enrollment is 135% of the predicted number.

Step 6
Check: More than 1200 students enrolled (more than 100%), so 1620 students must be more than 100% of the predicted number. The solution of 135% is reasonable.

5. **(a)** *Step 1*
Unknown: number of problems on the test
Known: 15 correct; $62\frac{1}{2}\%$ correct

Step 2
Let n be the number of problems on the test.

Step 3
percent • whole = part
$$62\frac{1}{2}\% \cdot n = 15$$

Step 4
$$0.625 \cdot n = 15$$
$$\frac{0.625n}{0.625} = \frac{15}{0.625}$$
$$n = 24$$

Step 5
There were 24 problems on the test.

Step 6
Check: 50% of 24 problems is $24 \div 2 = 12$ problems correct, so it is reasonable that $62\frac{1}{2}\%$ would be 15 problems correct.

(b) *Step 1*
Unknown: total number of calories
Known: 55 calories from fat; 18% of the calories are from fat

Step 2
Let n be the total number of calories

Step 3
percent • whole = part
$$18\% \cdot n = 55$$

Step 4
$$0.18 \cdot n = 55$$
$$\frac{0.18n}{0.18} = \frac{55}{0.18}$$
$$n \approx 306$$

Step 5
There were 306 calories (rounded) in the dinner.

Step 6
Check: 18% is close to 20%. 10% of 306 calories is about 30 calories, so 20% would be $2 \cdot 30 = 60$ calories, which is close to the number given (55 calories).

236   Chapter 7   Percent

6. **(a)** *Step 1*
Unknown: percent of increase
Known: original rent was $650; new rent is $767

Step 2
Let p be the percent increase.

Step 3
Amount of increase = $767 − $650 = $117

$$\underbrace{\text{percent of original}}_{\downarrow \;\;\; \downarrow \;\;\; \downarrow} = \underbrace{\text{amount of}}_{\downarrow}$$
$$\text{value} \qquad \text{increase}$$
$$p \;\; \cdot \;\; \$650 \;\; = \;\; \$117$$

Step 4
$$\frac{650p}{650} = \frac{117}{650}$$
$$p = 0.18 = 18\%$$

Step 5
Duyen's rent increased 18%.

Step 6
Check: 10% increase would be $6\underset{\curvearrowleft}{5}0. = \$65$;
20% increase would be 2 · $65 = $130; so an 18% increase is reasonable.

(b) *Step 1*
Unknown: percent of increase
Known: number of spaces changed from 8 to 20

Step 2
Let p be the percent of increase.

Step 3
Amount of increase = 20 − 8 = 12

$$\underbrace{\text{percent of original}}_{\text{value}} = \underbrace{\text{amount of}}_{\text{increase}}$$
$$p \;\; \cdot \;\; 8 \;\; = \;\; 12$$

Step 4
$$\frac{8p}{8} = \frac{12}{8}$$
$$p = 1.5 = 150\%$$

Step 5
The number of handicapped spaces increased 150%.

Step 6
Check: 150% is $1\frac{1}{2}$, and $1\frac{1}{2} \cdot 8 = 12$ space increase, so it checks.

7. **(a)** *Step 1*
Unknown: percent of decrease
Known: number of students changed from 425 to 200

Step 2
Let p be the percent of decrease.

Step 3
Amount of decrease = 425 − 200 = 225

$$\underbrace{\text{percent of original}}_{\text{value}} = \underbrace{\text{amount of}}_{\text{decrease}}$$
$$p \;\; \cdot \;\; 425 \;\; = \;\; 225$$

Step 4
$$\frac{425p}{425} = \frac{225}{425}$$
$$p \approx 0.53 = 53\%$$

Step 5
Daily attendance decreased 53% (rounded).

Step 6
Check: A 50% decrease would be $425 \div 2 \approx 212$, so a 53% decrease is reasonable.

(b) *Step 1*
Unknown: percent of decrease
Known: number of calories from fat dropped from 70 to 60

Step 2
Let p be the percent of decrease.

Step 3
Amount of decrease = 70 − 60 = 10

$$\underbrace{\text{percent of original}}_{\text{value}} = \underbrace{\text{amount of}}_{\text{decrease}}$$
$$p \;\; \cdot \;\; 70 \;\; = \;\; 10$$

Step 4
$$\frac{70p}{70} = \frac{10}{70}$$
$$p \approx 0.14 = 14\%$$

Step 5
The claim of a 20% decrease is not true; the decrease in calories is about 14%.

Step 6
Check: A 10% decrease would be $7\underset{\curvearrowleft}{0}. = 7$ calories; a 20% decrease would be 2 · 7 = 14 calories, so a 14% decrease is reasonable.

7.4 Section Exercises

1. *Step 1*
Unknown: amount withheld
Known: withhold 18% of $210

Step 2
Let n be the amount withheld.

7.4 Problem Solving with Percent

Step 3
percent · whole = part
$18\% \cdot \$210 = n$

Step 4
$(0.18)(210) = n$
$\$37.80 = n$

Step 5
The amount withheld is $37.80.

Step 6
Check: 10% of $210 is $21, so 20% would be $2 \cdot \$21 = \$42$. $37.80 is slightly less than $42, so it is reasonable.

3. **(a)** *Step 1*
Unknown: percent
Known: $20 withdrawal; $2 fee

Step 2
Let p be the percent.

Step 3
percent · whole = part
$p \cdot 20 = 2$

Step 4
$\dfrac{20p}{20} = \dfrac{2}{20}$
$p = 0.10 = 10\%$

Step 5
The $2 fee is 10% of the $20 withdrawal.

Step 6
Check: 10% of $20. is $2.

Use the same method for parts (b)–(d).

(b) $p = \frac{2}{40} = 0.05 = 5\%$

(c) $p = \frac{2}{100} = 0.02 = 2\%$

(d) $p = \frac{2}{200} = 0.01 = 1\%$

5. **(a)** *Step 1*
Unknown: number of pounds of water
Known: water weight is 61.6% of 165 pounds

Step 2
Let n be the number of pounds of water.

Step 3
percent · whole = part
$61.6\% \cdot 165 = n$

Step 4
$(0.616)(165) = n$
$101.64 = n$

Step 5
101.6 pounds (rounded) of the 165 pounds is water.

Step 6
Check: 50% of 165 pounds is 82.5 pounds, so 101.6 pounds is reasonable.

(b) (minerals) $n = 6.1\%$ of 165
$= (0.061)(165)$
$= 10.065$
$= 10.1$ pounds (rounded)

7. Whole: 335 people
Percent: unknown (p)
Part: 44 female

Percent · Whole = Part
$p \cdot 335 = 44$

$\dfrac{335p}{335} = \dfrac{44}{335}$
$p \approx 0.131$

About 13.1% of the crew is female and 86.9% ($100\% - 13.1\%$) is male.

9. Whole: Total U.S. population is unknown (n)
Part: 42.6 million are 65 years of age or older
Percent: 14.2%

Percent · Whole = Part
$0.142 \cdot n = 42.6$
$\dfrac{0.142n}{0.142} = \dfrac{42.6}{0.142}$
$n = 300$

There were 300 million people in the U.S. at the time of the 2006 census.

11. Whole: $50,000
Percent: unknown (p)
Part: $69,000

Percent · Whole = Part
$p \cdot 50{,}000 = 69{,}000$
$\dfrac{50{,}000p}{50{,}000} = \dfrac{69{,}000}{50{,}000}$
$p = 1.38$ (138%)

The society raised 138% of their goal.

13. Whole: Number of problems on the test is unknown (n)
Part: 34 problems done correctly
Percent: 85%

Percent · Whole = Part
$0.85 \cdot n = 34$
$\dfrac{0.85n}{0.85} = \dfrac{34}{0.85}$
$n = 40$

There were 40 problems on the test.

15. Whole: number of shots tried is unknown (n)
Percent: 47.6%
Part: 638 shots made

$$\text{Percent} \cdot \text{Whole} = \text{Part}$$
$$0.476 \cdot n = 638$$
$$\frac{0.476n}{0.476} = \frac{638}{0.476}$$
$$n \approx 1340$$

He tried 1340 shots (rounded).

17. First calculate her increase in mileage.

$$\begin{array}{ccccc} \text{Percent} & \cdot & \text{Original} & = & \text{Amount of} \\ & & \text{mileage} & & \text{increase} \\ 0.15 & \cdot & 20.6 & = & n \\ & & 3.09 & = & n \end{array}$$

The new tires should increase her mileage by 3.09 miles per gallon.

Her new mileage should be
$20.6 + 3.09 = 23.69 \approx 23.7$ miles per gallon (rounded)

19. Lowest sales month (March) sold 7%.

7% of 350 million cans is how many?

$$0.07 \cdot 350 = n$$
$$24.5 = n$$

24.5 million (24,500,000) cans were sold in March.

21. Highest sales month: January, 15%
Second-highest sales month: February, 11%
Whole: 350 million cans sold each year

Highest sales month sold:
$$0.15 \cdot 350 = 52.5 \text{ million cans}$$

Second-highest month sold:
$$0.11 \cdot 350 = 38.5 \text{ million cans}$$

23. Amount of decrease = $825 - $290 = $535

Let p be the percent of decrease.

$$\begin{array}{ccccc} \text{percent} & \text{of} & \text{original} & = & \text{amount of} \\ & & \text{value} & & \text{decrease} \\ p & \cdot & \$825 & = & \$535 \\ & & \frac{825p}{825} & = & \frac{535}{825} \\ & & p & \approx & 0.648 \quad (64.8\%) \end{array}$$

The percent of decrease was 64.8% (rounded).

25. Original tuition last semester: $1328
New tuition this semester: $1449
Amount of increase = $1449 − $1328 = $121

Let p be the percent of increase.

$$\begin{array}{ccccc} \text{percent} & \text{of} & \text{original} & = & \text{amount of} \\ & & \text{tuition} & & \text{increase} \\ p & \cdot & 1328 & = & 121 \\ & & \frac{1328p}{1328} & = & \frac{121}{1328} \\ & & p & \approx & 0.091 \quad (9.1\%) \end{array}$$

The percent increase is 9.1% (rounded).

27. Original hours = 30 hours
New hours = 18 hours
Amount of decrease = 30 − 18 = 12

Let p be the percent of decrease.

$$\begin{array}{ccccc} \text{percent} & \text{of} & \text{original} & = & \text{amount of} \\ & & \text{amount} & & \text{decrease} \\ p & \cdot & 30 & = & 12 \\ & & \frac{30p}{30} & = & \frac{12}{30} \\ & & p & = & 0.4 \quad (40\%) \end{array}$$

The percent decrease is 40%.

29. Original number = 78
New number = 607
Amount of increase = 607 − 78 = 529

Let p be the percent of increase.

$$\begin{array}{ccccc} \text{percent} & \text{of} & \text{original} & = & \text{amount of} \\ & & \text{number} & & \text{increase} \\ p & \cdot & 78 & = & 529 \\ & & \frac{78p}{78} & = & \frac{529}{78} \\ & & p & \approx & 6.78 \quad (678\%) \end{array}$$

The percent of increase is 678% (rounded).

31. No. 100% is the entire price, so a decrease of 100% would take the price down to 0. Therefore, 100% is the maximum possible decrease in the price of something.

33. George ate more than 65 grams, so the percent must be $> 100\%$. Use $p \cdot 65 = 78$ to get 120%.

34. The team won more than half the games, so the percent must be $> 50\%$. Correct solution is $0.72 = 72\%$.

35. The brain could not weigh 375 pounds, which is more than the person weighs.
$2\frac{1}{2}\% = 2.5\% = 0.025$, so $(0.025)(150) = n$ and $n = 3.75$ pounds.

36. If 80% were absent, then only 20% made it to class. $800 - 640 = 160$ students, or use $n = (0.20)(800) = 160$ students.

7.5 Consumer Applications: Sales Tax, Tips, Discounts, and Simple Interest

7.5 Margin Exercises

1. (a) Percent: tax rate $= 5\frac{1}{2}\%$ or 5.5%
Whole: cost of item $= \$495$
Part: sales tax $=$ unknown (n)

$$\begin{aligned} \text{Percent} \cdot \text{Whole} &= \text{Part} \\ 0.055 \cdot 495 &= n \\ 27.225 &= n \end{aligned}$$

The sales tax is \$27.23 (rounded).

Total cost $= \$495 + \$27.23 = \$522.23$

Check: 1% of \$500. is \$5;
6% is 6 times \$5 $= \$30$

(b) Percent: tax rate $= 7\%$
Whole: cost of item $= \$29.98$
Part: sales tax $=$ unknown (n)

$$\begin{aligned} \text{Percent} \cdot \text{Whole} &= \text{Part} \\ 0.07 \cdot 29.98 &= n \\ 2.0986 &= n \end{aligned}$$

The sales tax is \$2.10 (rounded).

Total cost $= \$29.98 + \$2.10 = \$32.08$

Check: 1% of \$30. is \$0.30;
7% is 7 times \$0.30 $= \$2.10$

(c) Percent: tax rate $= 4\%$
Whole: cost of item $= \$1.19$
Part: sales tax $=$ unknown (n)

$$\begin{aligned} \text{Percent} \cdot \text{Whole} &= \text{Part} \\ 0.04 \cdot 1.19 &= n \\ 0.0476 &= n \end{aligned}$$

The sales tax is \$0.05 (rounded).

Total cost $= \$1.19 + \$0.05 = \$1.24$

Check: 1% of \$1 is \$0.01;
4% is 4 times \$0.01 $= \$0.04$

2. (a) Percent: tax rate $=$ unknown (p)
Whole: cost of item $= \$57$
Part: sales tax $= \$3.42$

$$\begin{aligned} \text{Percent} \cdot \text{Whole} &= \text{Part} \\ p \cdot 57 &= 3.42 \\ \frac{57p}{57} &= \frac{3.42}{57} \\ p &= 0.06 \quad (6\%) \end{aligned}$$

The tax rate is 6%.

Check: 1% of \$60. is \$0.60; 6% would be 6 times \$0.60 or \$3.60 (which is close to \$3.42 given in the problem).

(b) Percent: tax rate $=$ unknown (p)
Whole: cost of item $= \$4$
Part: sales tax $= \$0.18$

$$\begin{aligned} \text{Percent} \cdot \text{Whole} &= \text{Part} \\ p \cdot 4 &= 0.18 \\ \frac{4p}{4} &= \frac{0.18}{4} \\ p &= 0.045 \quad (4.5\%) \end{aligned}$$

The tax rate is 4.5%.

Check: 1% of \$4 is \$0.04; round 4.5% to 5%; 5% of \$4 is 5 times \$0.04, or \$0.20 (which is close to \$0.18 given in the problem).

(c) Percent: tax rate $=$ unknown (p)
Whole: cost of item $= \$998$
Part: sales tax $= \$49.90$

$$\begin{aligned} \text{Percent} \cdot \text{Whole} &= \text{Part} \\ p \cdot 998 &= \$49.90 \\ \frac{998p}{998} &= \frac{49.90}{998} \\ p &= 0.05 \end{aligned}$$

The tax rate is 5%.

Check: 10% of \$1000. is \$100; 5% is half of that, or \$50 (which is close to \$49.90 given in the problem).

3. (a) *Estimate:* Round \$58.37 to \$60. Then 10% of \$60 is \$6. 20% would be 2 times \$6, or \$12.

Exact:
$$\begin{aligned} \text{Percent} \cdot \text{Whole} &= \text{Part} \\ 0.20 \cdot 58.37 &= \text{tip} \\ 11.674 &= \text{tip} \end{aligned}$$

Exact tip amount $= \$11.67$ (rounded).

(b) *Estimate:* Round $11.93 to $12. Then 10% of $12 is $1.20. 5% is half of $1.20, or $0.60. So 15% is $1.20 + $0.60 = $1.80.

Exact: Percent · Whole = Part
$$0.15 \cdot 11.93 = \text{tip}$$
$$1.7895 = \text{tip}$$

Exact tip amount = $1.79 (rounded).

(c) *Estimate:* Round $89.02 to $90. Then 10% of $90 is $9. 5% is half of $9, or $4.50. So 15% is $9 + $4.50 = $13.50.

Exact: Percent · Whole = Part
$$0.15 \cdot 89.02 = \text{tip}$$
$$13.353 = \text{tip}$$

The tip amount is about $13.35, which is close to the estimate of $13.50.

Total cost = $89.02 + $13.35 = $102.37

Amount paid by each friend:

$$\$102.37 \div 4 = \$25.59 \text{ (rounded)}$$

4. **(a)** Percent: rate of discount = 35%
Whole: original price = $950
Part: amount of discount = unknown (n)

Percent · Whole = Part
$$0.35 \cdot 950 = n$$
$$332.50 = n$$

The amount of discount is $332.50.

Sale price = original price − amount of discount
= $950 − $332.50 = $617.50

(b) Percent: rate of discount = 40%
Whole: original price = $68 and $97
Part: amount of discount = unknown (n)

Calculations for the $68 suit:

Percent · Whole = Part
$$0.40 \cdot 68 = n$$
$$27.20 = n$$

The amount of discount is $27.20.

Sale price = $68 − $27.20 = $40.80

Calculations for the $97 suit:

Percent · Whole = Part
$$0.40 \cdot 97 = n$$
$$38.80 = n$$

The amount of discount is $38.80.

Sale price = $97 − $38.80 = $58.20

5. **(a)** $I = p \cdot r \cdot t$ $4\% = 0.04$
$I = (500)(0.04)(1)$
$I = 20$

The interest is $20.

amount due = principal + interest
= $500 + $20
= $520

The total amount due is $520.

(b) $I = p \cdot r \cdot t$ $9\frac{1}{2}\% = 0.095$
$I = (1850)(0.095)(1)$
$I = 175.75$

The interest is $175.75.

amount due = principal + interest
= $1850 + $175.75
= $2025.75

The total amount due is $2025.75.

6. **(a)** $I = p \cdot r \cdot t$ $5\% = 0.05$
$I = (340)(0.05)(3.5)$
$I = 59.50$

The interest is $59.50.

amount due = principal + interest
= $340 + $59.50
= $399.50

The total amount due is $399.50.

(b) $I = p \cdot r \cdot t$ $8\% = 0.08$
$I = (2450)(0.08)(3.25)$
$I = 637$

The interest is $637.

amount due = principal + interest
= $2450 + $637
= $3087

The total amount due is $3087.

(c) $7\frac{1}{2}\% = 7.5\%$ and $2\frac{3}{4}$ yrs = 2.75 yrs

$I = p \cdot r \cdot t$ $7.5\% = 0.075$
$I = (14{,}200)(0.075)(2.75)$
$I = 2928.75$

The interest is $2928.75.

amount due = principal + interest
= $14,200 + $2928.75
= $17,128.75

The total amount due is $17,128.75.

7.5 Consumer Applications: Sales Tax, Tips, Discounts, and Simple Interest

7. (a) $I = p \cdot r \cdot t$
$I = (1600)(0.07)(\frac{4}{12})$ $\frac{4}{12} = \frac{1}{3}$
$I = 112(\frac{1}{3})$
$I = \frac{112}{1} \cdot \frac{1}{3} = \frac{112}{3} = 37\frac{1}{3} \approx 37.33$

The interest is $37.33.

$$\begin{aligned}\text{amount due} &= \text{principal} + \text{interest} \\ &= \$1600 + \$37.33 \\ &= \$1637.33\end{aligned}$$

The total amount due is $1637.33.

(b) $I = p \cdot r \cdot t$ $10\frac{1}{2}\% = 0.105$
$I = (25{,}000)(0.105)(\frac{3}{12})$
$I = (2625)(\frac{1}{4})$
$I = \frac{2625}{1} \cdot \frac{1}{4} = \frac{2625}{4} = 656.25$

The interest is $656.25.

$$\begin{aligned}\text{amount due} &= \text{principal} + \text{interest} \\ &= \$25{,}000 + \$656.25 \\ &= \$25{,}656.25\end{aligned}$$

The total amount due is $25,656.25.

(c) $I = p \cdot r \cdot t$ $12\frac{1}{4}\% = 0.1225$
$= (4350)(0.1225)(\frac{9}{12})$
$= \$399.66$ (rounded interest)

The interest is $399.66.

$$\begin{aligned}\text{amount due} &= \text{principal} + \text{interest} \\ &= \$4350 + \$399.66 \\ &= \$4749.66\end{aligned}$$

The total amount due is $4749.66.

7.5 Section Exercises

1. Cost of Item = $100
Tax Rate = 6% = 0.06
Amount of Tax = 0.06(100) = $6
Total Cost = $100 + $6 = $106

3. Cost of Item = $68
Tax Rate = unknown (p)
Amount of Tax = $2.04
Total Cost = $68 + $2.04 = $70.04

$$\begin{array}{ccccc}\text{Tax} & \cdot & \text{Cost of} & = & \text{Amount} \\ \text{Rate} & & \text{Item} & & \text{of Tax} \\ \downarrow & \downarrow & \downarrow & \downarrow & \downarrow \\ p & \cdot & 68 & = & 2.04 \\ & & \frac{68p}{68} & = & \frac{2.04}{68} \\ & & p & = & 0.03 \quad (3\%)\end{array}$$

The tax rate is 3%.

5. Cost of Item = $365.98
Tax Rate = 8% = 0.08
Amount of Tax = 0.08(365.98) = $29.28 (rounded)
Total Cost = $365.98 + $29.28 = $395.26

7. Cost of Item = $2.10
Tax Rate = $5\frac{1}{2}\% = 0.055$
Amount of Tax = 0.055(2.10) = $0.12 (rounded)
Total Cost = $2.10 + $0.12 = $2.22

9. Cost of Item = $12,600
Tax Rate = unknown (p)
Amount of Tax = $567
Total Cost = $12,600 + $567 = $13,167

$$\begin{array}{ccccc}\text{Tax} & \cdot & \text{Cost of} & = & \text{Amount} \\ \text{Rate} & & \text{Item} & & \text{of Tax} \\ \downarrow & \downarrow & \downarrow & \downarrow & \downarrow \\ p & \cdot & 12{,}600 & = & 567 \\ & & \frac{12{,}600p}{12{,}600} & = & \frac{567}{12{,}600} \\ & & p & = & 0.045 \quad (4.5\%)\end{array}$$

The tax rate is 4.5% or $4\frac{1}{2}\%$.

11. The bill of $32.17 rounds to $30.

Estimate of 15% tip:

10% of $30 is $3. 5% of $30 is half of $3, or $1.50. An estimate is $3 + $1.50 = $4.50.

Exact 15% tip:

$0.15(\$32.17) = \$4.8255 \approx \$4.83$

Estimate of 20% tip:

10% of $30 is $3. 20% is 2 times $3, or $6. An estimate is $6.

Exact 20% tip:

$0.20(\$32.17) = \$6.434 \approx \$6.43$

13. The bill of $78.33 rounds to $80.

Estimate of 15% tip:

10% of $80 is $8. 5% of $80 is half of $8, or $4. An estimate is $8 + $4 = $12.

Exact 15% tip:

$0.15(\$78.33) = \$11.7495 \approx \$11.75$

Estimate of 20% tip:

10% of $80 is $8. 20% is 2 times $8, or $16. An estimate is $16.

Exact 20% tip:

$0.20(\$78.33) = \$15.666 \approx \$15.67$

15. The bill of $9.55 rounds to $10.

Estimate of 15% tip:

10% of $10 is $1. 5% of $10 is half of $1, or $0.50. An estimate is $1 + $0.50 = $1.50.

Exact 15% tip:

$$0.15(\$9.55) = \$1.4325 \approx \$1.43$$

Estimate of 20% tip:

10% of $10 is $1. 20% is 2 times $1, or $2. An estimate is $2.

Exact 20% tip:

$$0.20(\$9.55) = \$1.91$$

17. Original Price = $100
Rate of Discount = 15% = 0.15
Amount of Discount = 0.15(100) = $15
Sale Price = $100 − $15 = $85

19. Original Price = $180
Rate of Discount = unknown (p)
Amount of Discount = $54
Sale Price = $180 − $54 = $126

amount of discount = rate of discount · original price
$$54 = p \cdot 180$$
$$\frac{54}{180} = \frac{180p}{180}$$
$$0.3 = p \quad (30\%)$$

The rate of discount is 30%.

21. Original Price = $17.50
Rate of Discount = 25% = 0.25
Amount of Discount = 0.25(17.50) = $4.375 ≈ $4.38
Sale Price = $17.50 − $4.38 = $13.12

23. Original Price = $37.88
Rate of Discount = 10% = 0.10
Amount of Discount = 0.10($37.88) = $3.788 ≈ $3.79
Sale Price = $37.88 − $3.79 = $34.09

25. $300 at 14% for 1 year

$$I = p \cdot r \cdot t$$
$$= (300)(0.14)(1)$$
$$= 42$$

The interest is $42.

amount due = principal + interest
$$= \$300 + \$42$$
$$= \$342$$

The total amount due is $342.

27. $740 at 6% for 9 months

$$I = p \cdot r \cdot t$$
$$= (740)(0.06)\left(\tfrac{9}{12}\right)$$
$$= 33.30$$

The interest is $33.30.

amount due = principal + interest
$$= \$740 + \$33.30$$
$$= \$773.30$$

The total amount due is $773.30.

29. $1500 at $9\tfrac{1}{2}\%$ for $1\tfrac{1}{2}$ years

$$I = p \cdot r \cdot t$$
$$= (1500)(0.095)(1.5)$$
$$= 213.75$$

The interest is $213.75.

amount due = principal + interest
$$= \$1500 + \$213.75$$
$$= \$1713.75$$

The total amount due is $1713.75.

31. $17,800 at $7\tfrac{3}{4}\%$ for 8 months

$$I = p \cdot r \cdot t$$
$$= (17{,}800)(0.0775)\left(\tfrac{8}{12}\right)$$
$$= 919.67 \text{ (rounded interest)}$$

The interest is $919.67.

amount due = principal + interest
$$= \$17{,}800 + \$919.67$$
$$= \$18{,}719.67$$

The total amount due is $18,719.67.

33. Original Price = $1950
Rate of Discount = 40% = 0.40
Amount of Discount = 0.40(1950) = $780
Sale Price = $1950 − $780 = $1170

The sale price of the ring is $1170.

35. $7500 at $8\tfrac{1}{2}\%$ for 9 months

$$I = p \cdot r \cdot t$$
$$= (7500)(0.085)\left(\tfrac{9}{12}\right)$$
$$= 478.13 \text{ (rounded)}$$

The interest is $478.13.

amount due = principal + interest
$$= \$7500 + \$478.13$$
$$= \$7978.13$$

Rick will owe his mother $7978.13.

7.5 Consumer Applications: Sales Tax, Tips, Discounts, and Simple Interest

37. Cost of Item = $24.99
Tax Rate = $6\frac{1}{2}\% = 0.065$
Amount of Tax = $0.065(24.99) \approx \$1.62$
Total Cost = $\$24.99 + \$1.62 = \$26.61$

The total cost of the headset is $26.61.

39. Cost of Item = $1980
Tax Rate = unknown(p)
Amount of Tax = $99

$$\begin{array}{ccccc} \text{Tax} & \cdot & \text{Cost of} & = & \text{Amount} \\ \text{Rate} & & \text{Item} & & \text{of Tax} \\ \downarrow & \downarrow & \downarrow & \downarrow & \downarrow \\ p & \cdot & 1980 & = & 99 \\ & & \dfrac{1980p}{1980} & = & \dfrac{99}{1980} \\ & & p & = & 0.05 \quad (5\%) \end{array}$$

The sales tax rate is 5%.

41. Original Price = $135
Rate of Discount = 45%
Amount of Discount = $0.45(135) = \$60.75$
Sale Price = $\$135 - \$60.75 = \$74.25$

The sale price of the parka is $74.25.

43.
$$\begin{array}{ccccc} \text{Tip} & \cdot & \text{Original} & = & \text{Tip} \\ \text{rate} & & \text{price} & & \text{amount} \\ \downarrow & \downarrow & \downarrow & \downarrow & \downarrow \\ 0.15 & \cdot & \$43.70 & = & n \\ & & \$6.555 & = & n \end{array}$$

The tip amount is $6.56 (rounded).
Total = $\$43.70 + \$6.56 = \$50.26$
Amount paid by each = $\$50.26 \div 2 = \$25.13$

45. Original Price = $1199.99
Rate of Discount = 18%
Amount of Discount = $0.18(1199.99)$
   = $216.00 (rounded)
Sale Price = $\$1199.99 - \$216.00 = \$983.99$

The discount is $216.00 and the sale price is $983.99.

47. $1900 at $12\frac{1}{4}\%$ for 6 months

$I = p \cdot r \cdot t$
$= (1900)(0.1225)(\frac{6}{12})$
$= 116.375$

The interest is $116.38 (rounded).

amount due = principal + interest
= $1900 + $116.38
= $2016.38

She must pay a total amount of $2016.38.

49.
$$\begin{array}{ccccc} \text{Tip} & \cdot & \text{Original} & = & \text{Tip} \\ \text{rate} & & \text{price} & & \text{amount} \\ \downarrow & \downarrow & \downarrow & \downarrow & \downarrow \\ 0.15 & \cdot & \$17.98 & = & n \\ & & \$2.697 & = & n \end{array}$$

The 15% tip would be about $2.70.
Total = $\$17.98 + \$2.70 = \$20.68$

They would give the delivery person $21 (rounded to the nearest dollar).

51. Computer modem: ($129, 65% off)

Discount amount = $0.65(\$129) = \$83.85$
Sale price = $\$129 - \$83.85 = \$45.15$
Tax amount = $0.06(\$45.15) \approx \$2.71$
Total = $\$45.15 + \$2.71 =$ **$47.86**

Earrings: ($60, 30% off)

Discount amount = $0.30(\$60) = \$18$
Sale price = $\$60 - \$18 = \$42$
Tax amount = $0.06(\$42) = \$2.52$
Total = $\$42 + \$2.52 =$ **$44.52**

Bill = $\$47.86 + \$44.52 =$ **$92.38**

53. Camcorder: ($287.95, 65% off)

Discount amount = $0.65(\$287.95) \approx \$187.17$
Sale price = $\$287.95 - \$187.17 = \$100.78$
Tax amount = $0.06(\$100.78) \approx \$6.05$
Total = $\$100.78 + \$6.05 =$ **$106.83**

Jeans: (2 @ $48, 45% off)

Discount amount = $0.45(\$48) = \$21.60$
Sale price = $\$48 - \$21.60 = \$26.40$
Tax amount: no tax on clothing
Total for 2 pairs = $2(\$26.40) =$ **$52.80**

Ring: ($95, 30% off)

Discount amount = $0.30(\$95) = \$28.50$
Sale price = $\$95 - \$28.50 = \$66.50$
Tax amount = $0.06(\$66.50) = \$3.99$
Total = $\$66.50 + \$3.99 =$ **$70.49**

Bill = $\$106.83 + \$52.80 + \$70.49 =$ **$230.12**

55. (a) Cost of item = $18.50
Discount rate = 6% = 0.06
Amount of discount = $0.06(18.50) = \$1.11$
Cost of discounted book = $\$18.50 - \$1.11$
   = $17.39

Tax rate = 6% = 0.06
Amount of tax = $0.06(17.39) \approx \$1.04$

Final price = $\$17.39 + \$1.04 = \$18.43$

(b) When calculating the discount, the *whole* is $18.50. But when calculating the sales tax, the whole is only $17.39 (the discounted price).

56. **(a)** Cost of item = $398
Discount rate = 7% = 0.07
Amount of discount = 0.07($398) = $27.86
Cost of discounted machine
 = $398 − $27.86 = $370.14

Tax rate = 7% = 0.07
Amount of tax = 0.07($370.14) ≈ $25.91
Final price = $370.14 + $25.91 = $396.05

(b) Tax amount = rate of tax · price of item
$$27.86 = r \cdot 370.14$$
$$\frac{27.86}{370.14} = \frac{370.14r}{370.14}$$
$$0.0753 \approx r \quad (7.53\%)$$

A sales tax of 7.53% would give a tax amount of $27.87 and a final cost of $398.01.

Chapter 7 Review Exercises

1. $25\% = 25 \div 100 = 0.25$
2. $180\% = 180 \div 100 = 1.80$
3. $12.5\% = 12.5 \div 100 = 0.125$
4. $7\% = 7 \div 100 = 0.07$
5. $2.65 = 2.65 \cdot 100\% = 265\%$
6. $0.02 = 0.02 \cdot 100\% = 2\%$
7. $0.3 = 0.3 \cdot 100\% = 30\%$
8. $0.002 = 0.002 \cdot 100\% = 0.2\%$
9. $12\% = \dfrac{12}{100} = \dfrac{12 \div 4}{100 \div 4} = \dfrac{3}{25}$
10. $37.5\% = \dfrac{37.5}{100} = \dfrac{37.5 \cdot 10}{100 \cdot 10} = \dfrac{375}{1000}$
    $= \dfrac{375 \div 125}{1000 \div 125} = \dfrac{3}{8}$
11. $250\% = \dfrac{250}{100} = \dfrac{250 \div 50}{100 \div 50} = \dfrac{5}{2} = 2\dfrac{1}{2}$
12. $5\% = \dfrac{5}{100} = \dfrac{5 \div 5}{100 \div 5} = \dfrac{1}{20}$
13. $\dfrac{3}{4} = \dfrac{3}{4} \cdot 100\% = \dfrac{3}{4} \cdot \dfrac{100}{1}\% = \dfrac{3 \cdot \cancel{4} \cdot 25}{\cancel{4} \cdot 1}\%$
    $= \dfrac{75}{1}\% = 75\%$
14. $\dfrac{5}{8} = \dfrac{5}{8} \cdot 100\% = \dfrac{5}{8} \cdot \dfrac{100}{1}\% = \dfrac{5 \cdot \cancel{4} \cdot 25}{2 \cdot \cancel{4} \cdot 1}\%$
    $= \dfrac{125}{2}\% = 62\dfrac{1}{2}\%, \text{ or } 62.5\%$

15. $3\dfrac{1}{4} = \dfrac{13}{4} = \dfrac{13}{4} \cdot \dfrac{100}{1}\% = \dfrac{13 \cdot \cancel{4} \cdot 25}{\cancel{4} \cdot 1}\%$
    $= \dfrac{325}{1}\% = 325\%$
16. $\dfrac{3}{50} = \dfrac{3}{50} \cdot \dfrac{100}{1}\% = \dfrac{3 \cdot 2 \cdot \cancel{50}}{\cancel{50} \cdot 1}\% = \dfrac{6}{1}\% = 6\%$
17. $\dfrac{1}{8} = 1 \div 8 = 0.125$
18. $0.125 = 0.125 \cdot 100\% = 12.5\%$
19. $0.15 = \dfrac{15}{100} = \dfrac{15 \div 5}{100 \div 5} = \dfrac{3}{20}$
20. $0.15 = 0.15 \cdot 100\% = 15\%$
21. $180\% = \dfrac{180}{100} = \dfrac{18}{10} = \dfrac{9}{5} = 1\dfrac{4}{5}$
22. $180\% = 180 \div 100 = 1.80 = 1.8$
23. 100% of $46 is all of the money, or $46.
24. 50% of $46 is half of the money, or $23.
25. 100% of 9 hours is all of the hours, or 9 hours.
26. 50% of 9 hours is half of the hours, which is $4\dfrac{1}{2}$ or 4.5 hours.
27. Part: 338.8
    Whole: unknown (n)
    Percent: 140%

    Step 1 $\dfrac{140}{100} = \dfrac{338.8}{n}$
    Step 2 $140 \cdot n = 100 \cdot 338.8$
    Step 3 $\dfrac{140n}{140} = \dfrac{33{,}880}{140}$
    $n = 242$

    338.8 meters is 140% of 242 meters.

28. Part: 425
    Whole: unknown (n)
    Percent: 2.5%

    Step 1 $\dfrac{2.5}{100} = \dfrac{425}{n}$
    Step 2 $2.5 \cdot n = 100 \cdot 425$
    Step 3 $\dfrac{2.5n}{2.5} = \dfrac{42{,}500}{2.5}$
    $n = 17{,}000$

    2.5% of 17,000 cases is 425 cases.

29. Part: unknown (n)
Whole: 450
Percent: 6%

Step 1 $\dfrac{6}{100} = \dfrac{n}{450}$

Step 2 $100 \cdot n = 6 \cdot 450$

Step 3 $\dfrac{100n}{100} = \dfrac{2700}{100}$

$n = 27$

6% of 450 cellular phones is 27 cellular phones.

30. Part: unknown (n)
Whole: 1450
Percent: 60%

Step 1 $\dfrac{60}{100} = \dfrac{n}{1450}$

Step 2 $100 \cdot n = 60 \cdot 1450$

Step 3 $\dfrac{100n}{100} = \dfrac{87,000}{100}$

$n = 870$

60% of 1450 reference books is 870 reference books.

31. Part: 36
Whole: 380
Percent: unknown (p)

Step 1 $\dfrac{p}{100} = \dfrac{36}{380}$

Step 2 $380 \cdot p = 100 \cdot 36$

Step 3 $\dfrac{380p}{380} = \dfrac{3600}{380}$

$p \approx 9.5$

36 pairs is 9.5% (rounded) of 380 pairs.

32. Part: 1440
Whole: 640
Percent: unknown (p)

Step 1 $\dfrac{p}{100} = \dfrac{1440}{640}$

Step 2 $640 \cdot p = 100 \cdot 1440$

Step 3 $\dfrac{640p}{640} = \dfrac{144,000}{640}$

$p = 225$

1440 cans is 225% of 640 cans.

33. 11% of $23.60 is how much?
↓   ↓   ↓   ↓   ↓
0.11 · 23.60 = n
     2.596 = n
     $2.60 (rounded) = n

11% of $23.60 is $2.60.

34. What is 125% of 64 days?
↓   ↓   ↓   ↓   ↓
n = 1.25 · 64
n = 80 days

80 days is 125% of 64 days.

35. 1.28 ounces is what percent of 32 ounces?
↓   ↓   ↓   ↓
1.28 = p · 32
1.28 = 32p
$\dfrac{1.28}{32} = \dfrac{32p}{32}$
0.04 = p
4% = p

1.28 ounces is 4% of 32 ounces.

36. $46 is 8% of what number of dollars?
↓   ↓ ↓   ↓
46 = 0.08 · n
46 = 0.08n
$\dfrac{46}{0.08} = \dfrac{0.08n}{0.08}$
$575 = n

$46 is 8% of $575.

37. 8 people is 40% of what number of people?
↓   ↓   ↓   ↓
8 = 0.40 · n
8 = 0.40n
$\dfrac{8}{0.40} = \dfrac{0.40n}{0.40}$
20 people = n

8 people is 40% of 20 people.

38. What percent of 174 ft is 304.5 ft?
↓   ↓   ↓
p · 174 = 304.5
174p = 304.5
$\dfrac{174p}{174} = \dfrac{304.5}{174}$
p = 1.75 = 175%

175% of 174 ft is 304.5 ft.

39. (a) Part: 504 late patients
Whole: total number of patients, unknown (n)
Percent: 16.8%

Percent · Whole = Part
$0.168 \cdot n = 504$
$\dfrac{0.168n}{0.168} = \dfrac{504}{0.168}$
$n = 3000$

There were 3000 patients in January.

(b) Amount of decrease = 504 − 345 = 159

Let p be the percent of decrease.

$$\underbrace{\text{percent of}}_{p} \cdot \underbrace{\text{original value}}_{504} = \underbrace{\text{amount of decrease}}_{159}$$

$$504p = 159$$
$$\frac{504p}{504} = \frac{159}{504}$$
$$p \approx 0.315 \quad (31.5\%)$$

The percent of decrease was 31.5% (rounded).

40. (a) Part: actual amount spent, unknown (n)
Whole: $280 budgeted for food
Percent: 130%

Percent · Whole = Part
$$1.30 \cdot 280 = n$$
$$364 = n$$

$364 was actually spent on food.

(b) Part: amount spent, $112
Whole: amount budgeted, $50
Percent: unknown (p)

Percent · Whole = Part
$$p \cdot 50 = 112$$
$$\frac{50p}{50} = \frac{112}{50}$$
$$p = 2.24 \quad (224\%)$$

She spent 224% of the budgeted amount.

41. (a) Part: 640 trees still living after 1 year
Whole: 800 trees planted
Percent: unknown (p)

Percent · Whole = Part
$$p \cdot 800 = 640$$
$$\frac{800p}{800} = \frac{640}{800}$$
$$p = 0.8 \quad (80\%)$$

80% of the trees were still living.

(b) Amount of increase = 850 − 800 = 50

Let p be the percent of increase.

$$\underbrace{\text{percent of}}_{p} \cdot \underbrace{\text{original value}}_{800} = \underbrace{\text{amount of increase}}_{50}$$

$$800p = 50$$
$$\frac{800p}{800} = \frac{50}{800}$$
$$p = 0.0625 \quad (6.25\%)$$

The percent of increase was 6.3% (rounded).

42. Part: 1000 species in danger of extinction
Whole: 9600 species
Percent: unknown (p)

Percent · Whole = Part
$$p \cdot 9600 = 1000$$
$$\frac{9600p}{9600} = \frac{1000}{9600}$$
$$p \approx 0.104 \quad (10.4\%)$$

10.4% (rounded) of the bird species are in danger of extinction.

43. Part: Sales tax = unknown (n)
Whole: Cost of item = $2.79
Percent: Tax rate = 4%

Percent · Whole = Part
$$0.04 \cdot 2.79 = n$$
$$0.1116 = n$$

Amount of tax = $0.11 (rounded)
Total cost = $2.79 + $0.11 = $2.90

44. Part: Sales tax = $58.50
Whole: Cost of item = $780
Percent: Tax rate = unknown (p)

Percent · Whole = Part
$$p \cdot 780 = 58.50$$
$$\frac{780p}{780} = \frac{58.50}{780}$$
$$p = 0.075 \quad (7.5\%)$$

The tax rate is 7.5% or $7\frac{1}{2}\%$.
Total cost = $780 + $58.50 = $838.50

45. The bill of $42.73 rounds to $40.

Estimate of 15% tip:

10% of $40 is $4. 5% of $40 is half of $4, or $2. An estimate is $4 + $2 = $6.

Exact 15% tip:

$$0.15(\$42.73) = \$6.4095 \approx \$6.41$$

Estimate of 20% tip:

10% of $40 is $4. 20% is 2 times $4, or $8. An estimate is $8.

Exact 20% tip:

$$0.20(\$42.73) = \$8.546 \approx \$8.55$$

46. The bill of $8.05 rounds to $8.

Estimate of 15% tip:

10% of $8 is $0.80. 5% of $8 is half of $0.80, or $0.40. An estimate is $0.80 + $0.40 = $1.20.

Exact 15% tip:

$$0.15(\$8.05) = \$1.2075 \approx \$1.21$$

Estimate of 20% tip:

10% of $8 is $0.80. 20% is 2 times $0.80, or $1.60. An estimate is $1.60.

Exact 20% tip:

$$0.20(\$8.05) = \$1.61$$

47. Original Price = $37.50
Rate of Discount = 10% = 0.10
Amount of Discount = 0.10(37.50) = $3.75
Sale Price = $37.50 − $3.75 = $33.75

48. Original Price = $252
Rate of Discount = unknown (p)
Amount of Discount = $63
Sale Price = $252 − $63 = $189

amount of discount = rate of discount · original price
$$63 = p \cdot 252$$
$$\frac{63}{252} = \frac{252p}{252}$$
$$0.25 = p \qquad (25\%)$$

The rate of discount is 25%.

49. $350 at $6\frac{1}{2}\%$ for 3 years

$$\begin{aligned} I &= p \cdot r \cdot t \\ &= (350)(0.065)(3) \\ &= 68.25 \end{aligned}$$

The interest is $68.25.

$$\begin{aligned} \text{amount due} &= \text{principal} + \text{interest} \\ &= \$350 + \$68.25 \\ &= \$418.25 \end{aligned}$$

The total amount due is $418.25.

50. $1530 at 16% for 9 months

$$\begin{aligned} I &= p \cdot r \cdot t \\ &= (1530)(0.16)\left(\tfrac{9}{12}\right) \\ &= 183.60 \end{aligned}$$

The interest is $183.60.

$$\begin{aligned} \text{amount due} &= \text{principal} + \text{interest} \\ &= \$1530 + \$183.60 \\ &= \$1713.60 \end{aligned}$$

The total amount due is $1713.60.

51. [7.1] $\frac{1}{3} = \frac{1}{3} \cdot 100\% = \frac{100}{3}\% = 33\frac{1}{3}\%$ (exact)

$33\frac{1}{3}\% \approx 0.333 = 33.3\%$ (rounded)

52. [7.1] The largest percent is 68%, which represents cards, so cards is the most popular game.

$$68\% = 0.68 = \frac{68}{100} = \frac{68 \div 4}{100 \div 4} = \frac{17}{25}$$

53. [7.3] $\frac{1}{3} \cdot 830 \approx 276.7 \approx 277$

277 adults (rounded) play electronic/computer games.

54. [7.3] Most popular: cards, 68%

$0.68(277) = 188.36 \approx 188$ adults (rounded)

Least popular: sci-fi/simulation, 37%

$0.37(277) = 102.49 \approx 102$ adults (rounded)

55. [7.4] Part: 2599 dogs placed
Whole: 5371 dogs received
Percent: unknown (p)

Percent · Whole = Part
$$p \cdot 5371 = 2599$$
$$\frac{5371p}{5371} = \frac{2599}{5371}$$
$$p \approx 0.484 \quad (48.4\%)$$

48.4% (rounded) of the dogs were placed in new homes.

56. [7.4] Part: 5371 received so far this year
Whole: number expected, unknown (n)
Percent: 75%

Percent · Whole = Part
$$0.75 \cdot n = 5371$$
$$\frac{0.75n}{0.75} = \frac{5371}{0.75}$$
$$n \approx 7161$$

They expect to receive 7161 dogs (rounded).

57. [7.4] Part: 2346 placed in new homes
Whole: 6447 received
Percent: unknown (p)

Percent · Whole = Part
$$p \cdot 6447 = 2346$$
$$\frac{6447p}{6447} = \frac{2346}{6447}$$
$$p \approx 0.364 \quad (36.4\%)$$

About 36.4% of the cats were placed in new homes.

58. [7.4] During the first 9 months:

Last year = 2300 cats received
This year = 6447 cats received

Amount of increase = 6447 − 2300
= 4147 cats

$$\begin{array}{ccc} \text{Percent} \cdot \text{Original} & = & \text{Amount of} \\ \text{increase} \quad \text{amount} & & \text{increase} \\ p \quad \cdot \quad 2300 & = & 4147 \\ \dfrac{2300p}{2300} & = & \dfrac{4147}{2300} \\ p & \approx & 1.803 \quad (180.3\%) \end{array}$$

The percent increase is 180.3% (rounded).

59. [7.4] Total number of animals received
= 5371 + 6447 + 2223 = 14,041
Total number of animals placed
= 2599 + 2346 + 406 = 5351

Percent · Whole = Part
$$p \cdot 14{,}041 = 5351$$
$$\dfrac{14{,}041p}{14{,}041} = \dfrac{5351}{14{,}041}$$
$$p \approx 0.381 \quad (38.1\%)$$

About 38.1% of the animals received were placed in new homes.

60. [7.4] 40% of all animals received:

$$0.40(14{,}041) \approx 5616$$

Since they placed 5351 animals,
5616 − 5351 = 265 (rounded) animals need to be placed in a home in order to reach their goal.

Chapter 7 Test

1. $75\% = 75 \div 100 = 0.75$

2. $0.6 = 0.60 = 60\%$

3. $1.8 = 1.80 = 180\%$

4. $0.075 = 7.5\%$, or $7\frac{1}{2}\%$

5. $300\% = 300 \div 100 = 3.00$, or 3

6. $2\% = 2 \div 100 = 0.02$

7. $62.5\% = \dfrac{62.5}{100} = \dfrac{62.5 \cdot 10}{100 \cdot 10} = \dfrac{625}{1000}$
$= \dfrac{625 \div 125}{1000 \div 125} = \dfrac{5}{8}$

8. $240\% = \dfrac{240}{100} = \dfrac{24}{10} = \dfrac{12}{5} = 2\dfrac{2}{5}$

9. $\dfrac{1}{20} = \dfrac{1}{20} \cdot \dfrac{100}{1}\% = \dfrac{1 \cdot \cancel{20} \cdot 5}{\cancel{20} \cdot 1}\% = \dfrac{5}{1}\% = 5\%$

10. $\dfrac{7}{8} = \dfrac{7}{8} \cdot \dfrac{100}{1}\% = \dfrac{7 \cdot \cancel{4} \cdot 25}{2 \cdot \cancel{4}}\% = \dfrac{175}{2}\%$
$= 87\dfrac{1}{2}\%, \text{ or } 87.5\%$

11. $1\dfrac{3}{4} = \dfrac{7}{4} = \dfrac{7}{4} \cdot \dfrac{100}{1}\% = \dfrac{7 \cdot \cancel{4} \cdot 25}{\cancel{4} \cdot 1}\% = \dfrac{175}{1}\%$
$= 175\%$

12. 16 laptops is 5% of what number of laptops?

$$\begin{array}{ccc} 16 & = & 0.05 \cdot n \\ \dfrac{16}{0.05} & = & \dfrac{0.05n}{0.05} \\ 320 & = & n \end{array}$$

16 laptops is 5% of 320 laptops.

13. $192 is what percent of $48?

$$\begin{array}{ccc} 192 & = & p \cdot 48 \\ \dfrac{192}{48} & = & \dfrac{48p}{48} \\ 4 & = & p \quad (400\%) \end{array}$$

$192 is 400% of $48.

14. Part: $14,625
Whole: Amount needed = unknown (n)
Percent: 75%

Percent · Whole = Part
$$0.75 \cdot n = 14{,}625$$
$$\dfrac{0.75n}{0.75} = \dfrac{14{,}625}{0.75}$$
$$n = 19{,}500$$

$19,500 is needed for a down payment.

15. Part: Sales tax = unknown (n)
Whole: Cost of item = $7950
Percent: Tax rate = $6\dfrac{1}{2}\% = 6.5\% = 0.065$

Percent · Whole = Part
$$0.065 \cdot 7950 = n$$
$$516.75 = n$$

The sales tax is $516.75.

Total cost of the car = $7950 + $516.75
= $8466.75

16. Last semester = 1440 students
Current semester = 1925 students
Amount of increase
= 1925 − 1440 = 485 students

$$\begin{aligned}\text{Percent} \cdot \text{Original} &= \text{Amount of increase} \\ p \cdot 1440 &= 485 \\ \frac{1440p}{1440} &= \frac{485}{1440} \\ p &\approx 0.34 \ (34\%)\end{aligned}$$

The percent of increase is 34% (rounded).

17. To find 50% of a number, divide the number by 2.
To find 25% of a number, divide the number by 4.
Examples will vary.

18. Round $31.94 to $30. Then 10% of $30 is $3
and 5% of $30 is half of $3 or $1.50, so a 15% tip estimate is $3 + $1.50 = $4.50. A 20% tip estimate is 2($3) = $6.

19. $$\begin{aligned}\text{Percent} \cdot \text{Whole} &= \text{Part} \\ 0.15 \cdot \$31.94 &= \text{Tip amount} \\ \$4.791 &= \text{Tip amount}\end{aligned}$$

The tip amount would be $4.79 (rounded).

$$\begin{aligned}\text{Total expense} &= \$31.94 + \$4.79 \\ &= \$36.73\end{aligned}$$

If the total expense is shared by 3 friends, each person will pay $36.73 ÷ 3 ≈ $12.24.

20. $$\begin{aligned}\text{Discount rate} \cdot \text{Original price} &= \text{Discount amount} \\ 0.08 \cdot \$48 &= \text{Discount amount} \\ \$3.84 &= \text{Discount amount}\end{aligned}$$

$$\begin{aligned}\text{Sale price} &= \text{Original price} - \text{Discount amount} \\ &= \$48 - \$3.84 \\ &= \$44.16\end{aligned}$$

21. $$\begin{aligned}\text{Discount rate} \cdot \text{Original price} &= \text{Discount amount} \\ 0.18 \cdot \$229.95 &= \text{Discount amount} \\ \$41.391 &= \text{Discount amount}\end{aligned}$$

Rounded to the nearest cent, the discount amount is $41.39.

$$\begin{aligned}\text{Sale price} &= \$229.95 - \$41.39 \\ &= \$188.56\end{aligned}$$

22. $$\begin{aligned}\text{Discount rate} \cdot \text{Original price} &= \text{Discount amount} \\ 0.30 \cdot \$1089 &= \text{Discount amount} \\ \$326.70 &= \text{Discount amount}\end{aligned}$$

Discounted price = $1089 − $326.70 = $762.30

$$\begin{aligned}\text{Tax rate} \cdot \text{Price} &= \text{Sales tax} \\ 0.07 \cdot \$762.30 &= \text{Sales tax} \\ \$53.361 &= \text{Sales tax} \\ \$53.36 \ (\text{rounded}) &= \text{Sales tax}\end{aligned}$$

$$\begin{aligned}\text{Total bill} &= \$762.30 + \$53.36 \\ &= \$815.66\end{aligned}$$

Since the payments will be spread out over 6 months, Jamal will need to pay
$815.66 ÷ 6 ≈ $135.94 per month.

23. $5000 at $8\frac{1}{4}\%$ for 4 years

$$\begin{aligned}I &= p \cdot r \cdot t \quad\quad 8\tfrac{1}{4}\% = 0.0825 \\ &= (5000)(0.0825)(4) \\ &= 1650\end{aligned}$$

The simple interest due on the loan is $1650.

$$\begin{aligned}\text{Total amount due} &= \text{principal} + \text{interest} \\ &= \$5000 + \$1650 \\ &= \$6650\end{aligned}$$

24. $860 at 12% for 6 months

$$\begin{aligned}I &= p \cdot r \cdot t \\ &= (860)(0.12)\left(\tfrac{6}{12}\right) \\ &= 51.60\end{aligned}$$

The simple interest due on the loan is $51.60.

$$\begin{aligned}\text{Total amount due} &= \text{principal} + \text{interest} \\ &= \$860 + \$51.60 \\ &= \$911.60\end{aligned}$$

Cumulative Review Exercises (Chapters 1–7)

1. (a) In words, 90.105 is

    ninety and one hundred five thousandths.

    (b) In words, 125,000,670 is

    one hundred twenty-five million, six hundred seventy.

2. (a) Written in digits, thirty billion, five million is

    30,005,000,000.

    (b) Written in digits, seventy-eight ten-thousandths is

    0.0078.

250 Chapter 7 Percent

3. mean
$$= \frac{\text{sum of all values}}{\text{number of values}}$$
$$= \frac{(\$810 + \$880 + \$750 + \$885 + \$1225 + \$795 + \$840 + \$785)}{8}$$
$$= \frac{\$6970}{8} = \$871.25$$

Median: arrange the data from smallest to largest

$750, $785, $795, $810, $840, $880, $885, $1225

Since there is an even number of values, the median is the average of the middle values.

$$\text{median} = \frac{\$810 + \$840}{2} = \$825$$

4. Brand A: $\dfrac{\$2.89}{17 \text{ ounces}} = \$0.170$ per ounce (∗)

   Brand B: $\dfrac{\$3.59}{21 \text{ ounces}} \approx \$0.171$ per ounce

   Brand C: $\dfrac{\$2.79}{15 \text{ ounces}} = \$0.186$ per ounce

   The best buy is Brand A since it is the least expensive per ounce.

5. $50 - 1.099$

   $\overset{\phantom{\cancel{10}}999}{\overset{4\,\cancel{10}\,\cancel{10}\,\cancel{10}\,10}{\cancel{5}\,\cancel{0}\,.\,\cancel{0}\,\cancel{0}\,\cancel{0}}}$
   $-1\,.\,0\,9\,9$
   $\overline{4\,8\,.\,9\,0\,1}$

6. $(-3)^2 + 2^3$
   $= (-3)(-3) + 2 \cdot 2 \cdot 2$
   $= 9 + 8$
   $= 17$

7. $\dfrac{3b}{10a} \cdot \dfrac{15ab}{4} = \dfrac{3 \cdot b \cdot 3 \cdot \overset{1}{\cancel{5}} \cdot \overset{1}{\cancel{a}} \cdot b}{2 \cdot \underset{1}{\cancel{5}} \cdot \underset{1}{\cancel{a}} \cdot 4} = \dfrac{9b^2}{8}$

8. $3\dfrac{3}{10} - 2\dfrac{4}{5}$
   $= \dfrac{33}{10} - \dfrac{14}{5}$
   $= \dfrac{33}{10} - \dfrac{28}{10}$
   $= \dfrac{33 - 28}{10} = \dfrac{5}{10} = \dfrac{1}{2}$

9. $0 + 2(-6 + 1)$
   $= 0 + 2(-5)$
   $= 0 + (-10)$
   $= -10$

10. $\dfrac{4}{5} + \dfrac{x}{4} = \dfrac{4 \cdot 4}{5 \cdot 4} + \dfrac{x \cdot 5}{4 \cdot 5} = \dfrac{16}{20} + \dfrac{5x}{20} = \dfrac{16 + 5x}{20}$

11. $(-0.5)(0.002)$

    $\begin{array}{r} 0.002 \\ \times\, 0.5 \\ \hline 0.0010 \end{array}$ ← 3 decimal places
    ← 1 decimal place
    ← 4 decimal places

    The signs are opposite, so the product is negative.
    $(-0.5)(0.002) = -0.001$

12. $\dfrac{-16 + 2^4}{-3 - 2} = \dfrac{-16 + 16}{-3 + (-2)} = \dfrac{0}{-5} = 0$

13. $\dfrac{7}{8} - \dfrac{2}{m} = \dfrac{7 \cdot m}{8 \cdot m} - \dfrac{2 \cdot 8}{m \cdot 8}$
    $= \dfrac{7m}{8m} - \dfrac{16}{8m} = \dfrac{7m - 16}{8m}$

14. $\dfrac{4.8}{-0.16}$ Different signs, quotient is negative.

    $0.16\overline{)4.\,8\,0} \rightarrow 16\overline{)4\,8\,0}$
    $\underline{4\,8}$
    0

    Answer: -30

15. $\dfrac{8}{9} \div 2n = \dfrac{8}{9} \div \dfrac{2n}{1} = \dfrac{8}{9} \cdot \dfrac{1}{2n}$
    $= \dfrac{\overset{1}{\cancel{2}} \cdot 4 \cdot 1}{9 \cdot \underset{1}{\cancel{2}} \cdot n} = \dfrac{4}{9n}$

16. $1\dfrac{5}{6} + 1\dfrac{2}{3} = \dfrac{11}{6} + \dfrac{5}{3} = \dfrac{11}{6} + \dfrac{10}{6}$
    $= \dfrac{11 + 10}{6} = \dfrac{21}{6}$
    $= \dfrac{\overset{1}{\cancel{3}} \cdot 7}{\underset{1}{\cancel{3}} \cdot 2} = \dfrac{7}{2} = 3\dfrac{1}{2}$

17. $-6w - 5$ Replace w with -4.
    $= -6(-4) - 5$
    $= 24 - 5$
    $= 19$

18. $5x + 3y$ Replace x with -2 and y with 3.
    $= 5(-2) + 3(3)$
    $= -10 + 9$
    $= -1$

19. $x^3 y$
    $= x \cdot x \cdot x \cdot y$ Replace x with -2 and y with 3.
    $= (-2)(-2)(-2)(3)$
    $= (4)(-2)(3)$
    $= (-8)(3)$
    $= -24$

20. $-5w^2x$

$= -5 \cdot w \cdot w \cdot x$ Replace x with -2 and w with -4.

$= (-5)(-4)(-4)(-2)$
$= (20)(-4)(-2)$
$= (-80)(-2)$
$= 160$

21.
$$12 - h = -3h$$
$$12 + (-1h) = -3h$$
$$12 + (-1h) + 1h = -3h + 1h$$
$$12 = -2h$$
$$\frac{12}{-2} = \frac{-2h}{-2}$$
$$-6 = h$$

The solution is -6.

22.
$$5a - 0.6 = 10.4$$
$$5a - 0.6 + 0.6 = 10.4 + 0.6$$
$$5a = 11$$
$$\frac{5a}{5} = \frac{11}{5}$$
$$a = 2.2$$

The solution is 2.2.

23.
$2(x+3) = -4x + 18 + 2x$ Distribute.
$2x + 6 = -4x + 18 + 2x$ Add like terms.
$2x + 6 = -2x + 18$
$2x + 6 + 2x = -2x + 18 + 2x$
$4x + 6 = 18$
$4x + 6 - 6 = 18 - 6$
$4x = 12$
$\frac{4x}{4} = \frac{12}{4}$
$x = 3$

The solution is 3.

24. Let n represent the number.

$$n + 31 = 3n + 1$$
$$n + 31 - n = 3n + 1 - n$$
$$31 = 2n + 1$$
$$31 - 1 = 2n + 1 - 1$$
$$\frac{30}{2} = \frac{2n}{2}$$
$$15 = n$$

The solution is 15.

25. *Step 1*
Unknown: How much Susanna made and how much Neoka made
Known: $1620 total, Neoka earned twice as much

Step 2
Let p be Susanna's pay. Then $2p$ would be Neoka's pay.

Step 3
Susanna's pay + Neoka's pay = Total pay

Step 4
$$p + 2p = 1620$$
$$3p = 1620$$
$$\frac{3p}{3} = \frac{1620}{3}$$
$$p = 540$$

Step 5
Susanna made $540 and Neoka made $1080.

Step 6
Check: $1080 is twice $540 and $1080 + $540 = $1620.

26. $P = 10 \text{ yd} + 10 \text{ yd} + 10 \text{ yd} + 10 \text{ yd} + 10 \text{ yd}$
$P = 50$ yards

The perimeter is 50 yards.

A = area of square + area of triangle
$A = s \cdot s + \frac{1}{2} bh$
$A = 10 \text{ yd} \cdot 10 \text{ yd} + \frac{1}{2} \cdot 10 \text{ yd} \cdot 8.7 \text{ yd}$
$A = 100 \text{ yd}^2 + 43.5 \text{ yd}^2$
$A = 143.5 \text{ yd}^2$

The area is 143.5 yd^2.

27. $C = 2\pi r$
$C \approx 2 \cdot 3.14 \cdot 5$ ft
$C = 31.4$ ft

The circumference is about 31.4 ft.

$A = \pi r^2$
$A = \pi \cdot r \cdot r$
$A \approx 3.14 \cdot 5 \text{ ft} \cdot 5 \text{ ft}$
$A = 78.5 \text{ ft}^2$

The area is about 78.5 ft^2.

28. hypotenuse $= \sqrt{(\text{leg})^2 + (\text{leg})^2}$
$y = \sqrt{(15)^2 + (19)^2}$
$y = \sqrt{225 + 361}$
$y = \sqrt{586}$
$y \approx 24.2$ mm

$A = \frac{1}{2} bh$
$A = \frac{1}{2}(19 \text{ mm})(15 \text{ mm})$
$A = 142.5 \text{ mm}^2$

29. $\frac{3}{8}$ of 5600 students

$$\frac{3}{8} \cdot 5600 = \frac{3}{\cancel{8}} \cdot \frac{\cancel{5600}^{700}}{1} = 2100$$

The survey showed that 2100 students work 20 hours or more per week.

30. The length of the parking space is 18 ft and the width is 9 ft + 5 ft = 14 ft.

$P = 2l + 2w$
$P = 2 \cdot 18 \text{ ft} + 2 \cdot 14 \text{ ft}$
$P = 36 \text{ ft} + 28 \text{ ft}$
$P = 64 \text{ ft}$

$A = l \cdot w$
$A = 18 \text{ ft} \cdot 14 \text{ ft}$
$A = 252 \text{ ft}^2$

31. Add the amounts needed for the recipes.

$$\begin{aligned} 2\tfrac{1}{4} &= 2\tfrac{1}{4} \\ 1\tfrac{1}{2} &= 1\tfrac{2}{4} \\ +\tfrac{3}{4} &= \tfrac{3}{4} \\ \hline & 3\tfrac{6}{4} = 4\tfrac{2}{4} = 4\tfrac{1}{2} \text{ cups} \end{aligned}$$

Find the total amount of brown sugar that is available.

$$\begin{aligned} 2\tfrac{1}{3} \\ +2\tfrac{1}{3} \\ \hline 4\tfrac{2}{3} \text{ cups} \end{aligned}$$

Subtract the amount to be used from the total amount available.

$$\begin{aligned} 4\tfrac{2}{3} &= 4\tfrac{4}{6} \\ -4\tfrac{1}{2} &= 4\tfrac{3}{6} \\ \hline & \tfrac{1}{6} \end{aligned}$$

The Jackson family will have $\frac{1}{6}$ cup more than the amount needed.

32. Discount = Discount · Original
amount   rate   price
       = 0.30 · $189
       = $56.70

Sale = Original − Discount
price   price   amount
    = $189 − $56.70
    = $132.30

Tracy will get a discount of $56.70 and pay $132.30 for the jacket.

33. Let x represent the actual distance from Springfield to Bloomington. Set up a proportion to solve.

$$\frac{1 \text{ cm}}{12 \text{ km}} = \frac{7.8 \text{ cm}}{x \text{ km}} \quad \text{Units must agree.}$$
$$1 \cdot x = 12 \cdot 7.8 \quad \text{Cross products}$$
$$x = 93.6$$

The actual distance is 93.6 km.

34. Use the percent proportion.
part is 31; whole is 35; percent is unknown

$$\frac{31}{35} = \frac{p}{100}$$
$$35 \cdot p = 31 \cdot 100$$
$$\frac{35 \cdot p}{35} = \frac{3100}{35}$$
$$p \approx 88.57 \quad (88.57\%)$$

The percent of the problems that were correct is 88.6% (rounded).

35. Divide the shipping charge by the number of prints to determine how much "per print" the shipping charge adds to the $0.12 cost per print.

$$\frac{\$1.97}{18 \text{ prints}} \approx \$0.11/\text{print}$$

Each print cost $0.12 + $0.11 = $0.23 (rounded).

CHAPTER 8 MEASUREMENT

8.1 Problem Solving with U.S. Customary Measurements

8.1 Margin Exercises

1. (a) 1 c = <u>8</u> fl oz

    (b) <u>4</u> qt = 1 gal

    (c) 1 wk = <u>7</u> days

    (d) <u>3</u> ft = 1 yd

    (e) 1 ft = <u>12</u> in.

    (f) <u>16</u> oz = 1 lb

    (g) 1 ton = <u>2000</u> lb

    (h) <u>60</u> min = 1 hr

    (i) 1 pt = <u>2</u> c

    (j) <u>24</u> hr = 1 day

    (k) 1 min = <u>60</u> sec

    (l) 1 qt = <u>2</u> pt

    (m) <u>5280</u> ft = 1 mi

2. (a) $5\frac{1}{2}$ ft to inches

    You are converting from a *larger* unit to a *smaller* unit, so multiply.

    Because 1 ft = 12 in., multiply by 12.

    $$5\frac{1}{2} \text{ ft} = 5\frac{1}{2} \cdot 12 = \frac{11}{2} \cdot \frac{\cancel{12}^6}{1} = \frac{66}{1} = 66 \text{ in.}$$

    (b) 64 oz to pounds

    You are converting from a *smaller* unit to a *larger* unit, so divide.

    Because 16 oz = 1 lb, divide by 16.

    $$64 \text{ oz} = \frac{64}{16} = 4 \text{ lb}$$

    (c) 6 yd to feet

    You are converting from a *larger* unit to a *smaller* unit, so multiply.

    Because 1 yd = 3 ft, multiply by 3.

    $$6 \text{ yd} = 6 \cdot 3 = 18 \text{ ft}$$

    (d) 2 tons to pounds

    You are converting from a *larger* unit to a *smaller* unit, so multiply.

    Because 1 ton = 2000 lb, multiply by 2000.

    $$2 \text{ tons} = 2 \cdot 2000 = 4000 \text{ lb}$$

    (e) 35 pt to quarts

    You are converting from a *smaller* unit to a *larger* unit, so divide.

    Because 1 qt = 2 pt, divide by 2.

    $$35 \text{ pt} = \frac{35}{2} = 17\frac{1}{2} \text{ qt}$$

    (f) 20 min to hours

    You are converting from a *smaller* unit to a *larger* unit, so divide.

    Because 60 min = 1 hr, divide by 60.

    $$20 \text{ min} = \frac{20}{60} = \frac{20 \div 20}{60 \div 20} = \frac{1}{3} \text{ hr}$$

    (g) 4 wk to days

    You are converting from a *larger* unit to a *smaller* unit, so multiply.

    Because 1 wk = 7 days, multiply by 7.

    $$4 \text{ wk} = 4 \cdot 7 = 28 \text{ days}$$

3. (a) 36 in. to feet

    unit fraction $\Big\}$ $\frac{1 \text{ ft}}{12 \text{ in.}}$

    $$36 \text{ in.} = \frac{\cancel{36}^3 \text{ in.}}{1} \cdot \frac{1 \text{ ft}}{\cancel{12}_1 \text{ in.}} = \frac{3 \cdot 1 \text{ ft}}{1} = 3 \text{ ft}$$

    (b) 14 ft to inches

    unit fraction $\Big\}$ $\frac{12 \text{ in.}}{1 \text{ ft}}$

    $$14 \text{ ft} = \frac{14 \cancel{\text{ ft}}}{1} \cdot \frac{12 \text{ in.}}{1 \cancel{\text{ ft}}} = \frac{14 \cdot 12 \text{ in.}}{1} = 168 \text{ in.}$$

    (c) 60 in. to feet

    unit fraction $\Big\}$ $\frac{1 \text{ ft}}{12 \text{ in.}}$

    $$60 \text{ in.} = \frac{\cancel{60}^5 \text{ in.}}{1} \cdot \frac{1 \text{ ft}}{\cancel{12}_1 \text{ in.}} = \frac{5 \cdot 1 \text{ ft}}{1} = 5 \text{ ft}$$

    (d) 4 yd to feet

    unit fraction $\Big\}$ $\frac{3 \text{ ft}}{1 \text{ yd}}$

    $$4 \text{ yd} = \frac{4 \cancel{\text{ yd}}}{1} \cdot \frac{3 \text{ ft}}{1 \cancel{\text{ yd}}} = \frac{4 \cdot 3 \text{ ft}}{1} = 12 \text{ ft}$$

254 Chapter 8 Measurement

(e) 39 ft to yards

unit fraction: $\dfrac{1 \text{ yd}}{3 \text{ ft}}$

$39 \text{ ft} = \dfrac{\overset{13}{\cancel{39}} \cancel{\text{ft}}}{1} \cdot \dfrac{1 \text{ yd}}{\underset{1}{\cancel{3}} \cancel{\text{ft}}} = \dfrac{13 \cdot 1 \text{ yd}}{1} = 13 \text{ yd}$

(f) 2 mi to feet

unit fraction: $\dfrac{5280 \text{ ft}}{1 \text{ mi}}$

$2 \text{ mi} = \dfrac{2 \cancel{\text{mi}}}{1} \cdot \dfrac{5280 \text{ ft}}{1 \cancel{\text{mi}}} = \dfrac{2 \cdot 5280 \text{ ft}}{1} = 10{,}560 \text{ ft}$

4. **(a)** 16 qt to gallons

unit fraction: $\dfrac{1 \text{ gallon}}{4 \text{ quarts}}$

$16 \text{ qt} = \dfrac{\overset{4}{\cancel{16}} \cancel{\text{qt}}}{1} \cdot \dfrac{1 \text{ gal}}{\underset{1}{\cancel{4}} \cancel{\text{qt}}} = \dfrac{4 \cdot 1 \text{ gal}}{1} = 4 \text{ gal}$

(b) 3 c to pints

unit fraction: $\dfrac{1 \text{ pt}}{2 \text{ c}}$

$3 \text{ c} = \dfrac{3 \cancel{\text{c}}}{1} \cdot \dfrac{1 \text{ pt}}{2 \cancel{\text{c}}} = \dfrac{3 \cdot 1 \text{ pt}}{2} = 1\dfrac{1}{2} \text{ pt or } 1.5 \text{ pt}$

(c) $3\tfrac{1}{2}$ tons to pounds

unit fraction: $\dfrac{2000 \text{ lb}}{1 \text{ ton}}$

$3\dfrac{1}{2} \text{ tons} = \dfrac{3\tfrac{1}{2} \cancel{\text{tons}}}{1} \cdot \dfrac{2000 \text{ lb}}{1 \cancel{\text{tons}}} = \dfrac{7}{\underset{1}{\cancel{2}}} \cdot \dfrac{\overset{1000}{\cancel{2000}}}{1} = 7000 \text{ lb}$

(d) $1\tfrac{3}{4}$ lb to ounces

unit fraction: $\dfrac{16 \text{ oz}}{1 \text{ lb}}$

$1\dfrac{3}{4} \text{ lb} = \dfrac{1\tfrac{3}{4} \cancel{\text{lb}}}{1} \cdot \dfrac{16 \text{ oz}}{1 \cancel{\text{lb}}} = \dfrac{7}{\underset{1}{\cancel{4}}} \cdot \dfrac{\overset{4}{\cancel{16}}}{1} = 28 \text{ oz}$

(e) 4 oz to pounds

unit fraction: $\dfrac{1 \text{ lb}}{16 \text{ oz}}$

$4 \text{ oz} = \dfrac{\overset{1}{\cancel{4}} \cancel{\text{oz}}}{1} \cdot \dfrac{1 \text{ lb}}{\underset{4}{\cancel{16}} \cancel{\text{oz}}} = \dfrac{1}{4} \text{ lb or } 0.25 \text{ lb}$

5. **(a)** 4 tons to ounces

unit fractions: $\dfrac{2000 \text{ lb}}{1 \text{ ton}}, \dfrac{16 \text{ oz}}{1 \text{ lb}}$

$\dfrac{4 \cancel{\text{tons}}}{1} \cdot \dfrac{2000 \cancel{\text{lb}}}{1 \cancel{\text{ton}}} \cdot \dfrac{16 \text{ oz}}{1 \cancel{\text{lb}}} = 4 \cdot 2000 \cdot 16 \text{ oz}$
$= 128{,}000 \text{ oz}$

(b) 3 mi to inches

unit fractions: $\dfrac{5280 \text{ ft}}{1 \text{ mi}}, \dfrac{12 \text{ in.}}{1 \text{ ft}}$

$\dfrac{3 \cancel{\text{mi}}}{1} \cdot \dfrac{5280 \cancel{\text{ft}}}{1 \cancel{\text{mi}}} \cdot \dfrac{12 \text{ in.}}{1 \cancel{\text{ft}}} = 3 \cdot 5280 \cdot 12 \text{ in.}$
$= 190{,}080 \text{ in.}$

(c) 36 pt to gallons

unit fractions: $\dfrac{1 \text{ qt}}{2 \text{ pt}}, \dfrac{1 \text{ gal}}{4 \text{ qt}}$

$\dfrac{36 \cancel{\text{pt}}}{1} \cdot \dfrac{1 \cancel{\text{qt}}}{2 \cancel{\text{pt}}} \cdot \dfrac{1 \text{ gal}}{4 \cancel{\text{qt}}} = \dfrac{\overset{9}{\cancel{36}}}{1} \cdot \dfrac{1}{2} \cdot \dfrac{1}{\underset{1}{\cancel{4}}} \text{ gal} = \dfrac{9}{2} \text{ gal}$
$= 4\dfrac{1}{2} \text{ gal or } 4.5 \text{ gal}$

(d) 2 wk to minutes

unit fractions: $\dfrac{7 \text{ days}}{1 \text{ wk}}, \dfrac{24 \text{ hr}}{1 \text{ day}}, \dfrac{60 \text{ min}}{1 \text{ hr}}$

$\dfrac{2 \cancel{\text{wk}}}{1} \cdot \dfrac{7 \cancel{\text{days}}}{1 \cancel{\text{wk}}} \cdot \dfrac{24 \cancel{\text{hr}}}{1 \cancel{\text{day}}} \cdot \dfrac{60 \text{ min}}{1 \cancel{\text{hr}}} = 2 \cdot 7 \cdot 24 \cdot 60 \text{ min}$
$= 20{,}160 \text{ min}$

6. **(a)** *Step 1*
The problem asks for the price per pound.

Step 2
Convert ounces to pounds. Then divide the cost by the pounds.

Step 3
To estimate, round $3.29 to $3. Then, there are 16 oz in a pound, so 12 oz is about 1 lb. Thus, $3 ÷ 1 lb = $3 per pound is our estimate.

Step 4

$\dfrac{\overset{3}{\cancel{12}} \cancel{\text{oz}}}{1} \cdot \dfrac{1 \text{ lb}}{\underset{4}{\cancel{16}} \cancel{\text{oz}}} = \dfrac{3}{4} \text{ lb} = 0.75 \text{ lb}$

Then divide: $\dfrac{\$3.29}{0.75 \text{ lb}} = 4.38\overline{6} \approx 4.39$

Step 5
The cheese costs $4.39 per pound (to the nearest cent).

8.1 Problem Solving with U.S. Customary Measurements

Step 6
The answer, $4.39, is close to our estimate of $3.

(b) *Step 1*
The problem asks for the number of tons of furnishings they moved.

Step 2
First convert pounds to tons. Then multiply to find the tons for 5 houses.

Step 3
To estimate, round 11,000 lb to 10,000 lb. There are 2000 lb in a ton, so $10,000 \div 2000 = 5$ tons. Multiply by 5 to get 25 tons as our estimate.

Step 4

$$\frac{\overset{11}{\cancel{11{,}000}\ \cancel{\text{lb}}}}{1} \cdot \frac{1 \text{ tons}}{\underset{2}{\cancel{2000}\ \cancel{\text{lb}}}} = \frac{11}{2} \text{ tons} = 5.5 \text{ tons}$$

$$= 5\frac{1}{2} \text{ tons}$$

$5(5.5 \text{ tons}) = 27.5 \text{ tons}$

Step 5
The company moved 27.5 tons or $27\frac{1}{2}$ tons of furnishings.

Step 6
The answer, $27\frac{1}{2}$ tons, is close to our estimate of 25 tons.

8.1 Section Exercises

1. 1 yd = <u>3</u> ft;   <u>12</u> in. = 1 ft

3. <u>8</u> fl oz = 1 c;   1 qt = <u>2</u> pt

5. 1 mi = <u>5280</u> ft;   <u>3</u> ft = 1 yd

7. <u>2000</u> lb = 1 ton;   1 lb = <u>16</u> oz

9. 1 min = <u>60</u> sec;   <u>60</u> min = 1 hr

11. (a) 120 sec to minutes

You are converting from a *smaller* unit to a *larger* unit, so divide. Because 60 sec = 1 min, divide by 60.

$$120 \text{ sec} = \frac{120}{60} = 2 \text{ min}$$

(b) 4 hr to minutes

You are converting from a *larger* unit to a *smaller* unit, so multiply. Because 1 hr = 60 min, multiply by 60.

$$4 \text{ hr} = 4 \cdot 60 = 240 \text{ min}$$

13. (a) 2 qt to gallons

You are converting from a *smaller* unit to a *larger* unit, so divide. Because 4 qt = 1 gal, divide by 4.

$$2 \text{ qt} = \frac{2}{4} = \frac{2 \div 2}{4 \div 2} = \frac{1}{2} \text{ or } 0.5 \text{ gal}$$

(b) $6\frac{1}{2}$ ft to inches

You are converting from a *larger* unit to a *smaller* unit, so multiply. Because 1 ft = 12 inches, multiply by 12.

$$6\frac{1}{2} \text{ ft} = 6\frac{1}{2} \cdot 12 = \frac{13}{\underset{1}{\cancel{2}}} \cdot \frac{\overset{6}{\cancel{12}}}{1} = \frac{78}{1} = 78 \text{ in.}$$

15. 7 to 8 tons to pounds

You are converting from a *larger* unit to a *smaller* unit, so multiply. Because 1 ton = 2000 pounds, multiply by 2000.

$$7 \text{ tons} = 7 \cdot 2000 = 14{,}000 \text{ lb}$$

$$8 \text{ tons} = 8 \cdot 2000 = 16{,}000 \text{ lb}$$

An adult African elephant may weigh 14,000 to 16,000 lb.

17. 9 yd to feet

$$\frac{9 \cancel{\text{ yd}}}{1} \cdot \frac{3 \text{ ft}}{1 \cancel{\text{ yd}}} = 9 \cdot 3 \text{ ft} = 27 \text{ ft}$$

19. 7 lb to ounces

$$\frac{7 \cancel{\text{ lb}}}{1} \cdot \frac{16 \text{ oz}}{1 \cancel{\text{ lb}}} = 7 \cdot 16 \text{ oz} = 112 \text{ oz}$$

21. 5 qt to pints

$$\frac{5 \cancel{\text{ qt}}}{1} \cdot \frac{2 \text{ pt}}{1 \cancel{\text{ qt}}} = 5 \cdot 2 \text{ pt} = 10 \text{ pt}$$

23. 90 min to hours

$$\frac{90 \cancel{\text{ min}}}{1} \cdot \frac{1 \text{ hr}}{60 \cancel{\text{ min}}} = \frac{90}{60} \text{ hr} = 1\frac{1}{2} \text{ or } 1.5 \text{ hr}$$

25. 3 in. to feet

$$\frac{\overset{1}{\cancel{3}} \cancel{\text{ in}}}{1} \cdot \frac{1 \text{ ft}}{\underset{4}{\cancel{12}} \cancel{\text{ in.}}} = \frac{1}{4} \text{ or } 0.25 \text{ ft}$$

27. 24 oz to pounds

$$\frac{24 \cancel{\text{ oz}}}{1} \cdot \frac{1 \text{ lb}}{16 \cancel{\text{ oz}}} = \frac{24}{16} \text{ lb} = 1\frac{1}{2} \text{ or } 1.5 \text{ lb}$$

29. 5 c to pints

$$\frac{5 \cancel{\text{ c}}}{1} \cdot \frac{1 \text{ pt}}{2 \cancel{\text{ c}}} = \frac{5}{2} \text{ pt} = 2\frac{1}{2} \text{ or } 2.5 \text{ pt}$$

31. $\frac{1}{2}$ ft = $\frac{\frac{1}{2}\cancel{\text{ft}}}{1} \cdot \frac{12 \text{ in.}}{1 \cancel{\text{ft}}} = \frac{1}{2} \cdot 12$ in. = 6 in.

The ice will safely support a snowmobile or ATV or a person walking.

33. $2\frac{1}{2}$ tons to lb

$$\frac{2\frac{1}{2} \cancel{\text{tons}}}{1} \cdot \frac{2000 \text{ lb}}{1 \cancel{\text{ton}}} = \frac{5}{\cancel{2}} \cdot \frac{\overset{1000}{\cancel{2000}}}{1} \text{ lb} = 5000 \text{ lb}$$

35. $4\frac{1}{4}$ gal to quarts

$$\frac{4\frac{1}{4} \cancel{\text{gal}}}{1} \cdot \frac{4 \text{ qt}}{1 \cancel{\text{gal}}} = \frac{17}{\cancel{4}} \cdot \frac{\overset{1}{\cancel{4}}}{1} \text{ qt} = 17 \text{ qt}$$

37. $\frac{1}{3}$ ft = $\frac{\frac{1}{3}\cancel{\text{ft}}}{1} \cdot \frac{12 \text{ in.}}{1 \cancel{\text{ft}}} = \frac{1}{\cancel{3}} \cdot \frac{\overset{4}{\cancel{12}}}{1}$ in. = 4 in.

Two-thirds of a foot would be twice as high; that is, 2(4 in.) = 8 in. The cactus could be 4 to 8 in. tall.

39. 6 yd to inches

$$\frac{6 \cancel{\text{yd}}}{1} \cdot \frac{3 \cancel{\text{ft}}}{1 \cancel{\text{yd}}} \cdot \frac{12 \text{ in.}}{1 \cancel{\text{ft}}} = 6 \cdot 3 \cdot 12 \text{ in.} = 216 \text{ in.}$$

41. 112 c to quarts

$$\frac{\overset{28}{\overset{56}{\cancel{112}}} \cancel{\text{c}}}{1} \cdot \frac{1 \cancel{\text{pt}}}{\underset{1}{\cancel{2}} \cancel{\text{c}}} \cdot \frac{1 \text{ qt}}{\underset{1}{\cancel{2}} \cancel{\text{pt}}} = 28 \text{ qt}$$

43. 6 days to seconds

$$\frac{6 \text{ days}}{1} \cdot \frac{24 \text{ hr}}{1 \text{ day}} \cdot \frac{60 \text{ min}}{1 \text{ hr}} \cdot \frac{60 \text{ sec}}{1 \text{ min}}$$
$$= 6 \cdot 24 \cdot 60 \cdot 60 \text{ sec}$$
$$= 518{,}400 \text{ sec}$$

45. $1\frac{1}{2}$ tons to ounces

$$\frac{1\frac{1}{2} \cancel{\text{tons}}}{1} \cdot \frac{2000 \cancel{\text{lb}}}{1 \cancel{\text{ton}}} \cdot \frac{16 \text{ oz}}{1 \cancel{\text{lb}}} = \frac{3}{\cancel{2}} \cdot \frac{\overset{1000}{\cancel{2000}}}{1} \cdot \frac{16}{1} \text{ oz}$$
$$= 48{,}000 \text{ oz}$$

47. **(a)** There is only one relationship that uses 1 to 16: 1 pound = 16 ounces.

(b) 10 to 20 is the same as 1 to 2. There are 2 relationships like this: 1 pint = 2 cups and 1 quart = 2 pints.
10 quarts = 20 pints or 10 pints = 20 cups

(c) 120 to 2 is the same as 60 to 1. There are 2 relationships like this: 60 minutes = 1 hour and 60 seconds = 1 minute.
120 minutes = 2 hours or
120 seconds = 2 minutes

(d) 2 to 24 is the same as 1 to 12. Use 1 foot = 12 inches.
2 feet = 24 inches

(e) 6000 to 3 is the same as 2000 to 1. Use 2000 pounds = 1 ton.
6000 pounds = 3 tons

(f) 35 to 5 is the same as 7 to 1. Use 7 days = 1 week.
35 days = 5 weeks

49. $2\frac{3}{4}$ miles to inches

$$\frac{2\frac{3}{4} \cancel{\text{mi}}}{1} \cdot \frac{5280 \cancel{\text{ft}}}{1 \cancel{\text{mi}}} \cdot \frac{12 \text{ in.}}{1 \cancel{\text{ft}}} = \frac{11}{\cancel{4}} \cdot \frac{5280}{1} \cdot \frac{\overset{3}{\cancel{12}}}{1} \text{ in.}$$
$$= 174{,}240 \text{ in.}$$

51. $6\frac{1}{4}$ gal to fluid ounces

$$\frac{6\frac{1}{4} \cancel{\text{gal}}}{1} \cdot \frac{4 \cancel{\text{qt}}}{1 \cancel{\text{gal}}} \cdot \frac{32 \text{ fl oz}}{1 \cancel{\text{qt}}} = \frac{25}{\cancel{4}} \cdot \frac{\overset{1}{\cancel{4}}}{1} \cdot 32 \text{ fl oz}$$
$$= 800 \text{ fl oz}$$

53. 24,000 oz to tons

$$\frac{24{,}000 \cancel{\text{oz}}}{1} \cdot \frac{1 \cancel{\text{lb}}}{16 \cancel{\text{oz}}} \cdot \frac{1 \text{ ton}}{2000 \cancel{\text{lb}}}$$
$$= \frac{\overset{3}{\underset{}{\cancel{24{,}000}}}}{1} \cdot \frac{1}{\underset{4}{\cancel{16}}} \cdot \frac{1}{\underset{1}{\cancel{2000}}} \text{ ton}$$
$$= \tfrac{3}{4} \text{ or } 0.75 \text{ ton}$$

For Exercises 55–62, the six problem-solving steps should be used, but are only shown for Exercise 55.

55. *Step 1*
The problem asks for the price per pound of strawberries.

Step 2
Convert ounces to pounds. Then divide the cost by the pounds.

Step 3
To estimate, round $2.29 to $2. Then, there are 16 oz in a pound, so 20 oz is a little more than 1 lb. Thus, $2 ÷ 1 lb = $2 per pound is our estimate.

8.1 Problem Solving with U.S. Customary Measurements

Step 4

$$\frac{\overset{5}{\cancel{20}}\text{ oz}}{1} \cdot \frac{1\text{ lb}}{\underset{4}{\cancel{16}}\text{ oz}} = \frac{5}{4}\text{ lb} = 1.25\text{ lb}$$

$$\frac{\$2.29}{1.25\text{ lb}} = 1.832 \approx \$1.83/\text{lb}$$

Step 5
The strawberries are $1.83 per pound (to the nearest cent).

Step 6
The answer, $1.83, is close to our estimate of $2.

57. Find the total number of feet needed.

$$24 \cdot 2 = 48\text{ ft}$$

Then convert feet to yards.

$$\frac{\overset{16}{\cancel{48}}\cancel{\text{ft}}}{1} \cdot \frac{1\text{ yd}}{\underset{1}{\cancel{3}}\cancel{\text{ft}}} = 16\text{ yd}$$

Then multiply to find the cost.

$$\frac{16\text{ yd}}{1} \cdot \frac{\$8.75}{\text{yd}} = \$140$$

It will cost $140 to equip all the stations.

59. (a) Convert seconds per foot to seconds per mile.

$$\frac{1\text{ sec}}{5\cancel{\text{ft}}} \cdot \frac{5280\cancel{\text{ft}}}{1\text{ mi}} = \frac{\overset{1056}{\cancel{5280}}}{\underset{1}{\cancel{5}}}\text{ sec/mi} = 1056\text{ sec/mi}$$

It would take the cockroach 1056 seconds to travel 1 mile.

(b) $$\frac{1056\cancel{\text{sec}}}{1\text{ mi}} \cdot \frac{1\text{ min}}{60\cancel{\text{sec}}} = \frac{\overset{88}{\cancel{1056}}}{\underset{5}{\cancel{60}}}\text{ min/mi}$$
$$= 17.6\text{ min/mi}$$

It would take the cockroach 17.6 minutes to travel 1 mile.

61. (a) Find the total number of cups per week. Then convert cups to quarts.

$$\frac{2}{3} \cdot 15 \cdot 5 = \frac{2}{\underset{1}{\cancel{3}}} \cdot \frac{\overset{5}{\cancel{15}}}{1} \cdot \frac{5}{1} = 50\text{ c}$$

$$\frac{\overset{25}{\cancel{50}}\cancel{\text{c}}}{1} \cdot \frac{1\text{ pt}}{\underset{1}{\cancel{2}}\cancel{\text{c}}} \cdot \frac{1\text{ qt}}{2\cancel{\text{pt}}} = \frac{25}{2}\text{ qt} = 12\frac{1}{2}\text{ qt}$$

The center needs $12\frac{1}{2}$ qt of milk per week.

(b) Convert quarts to gallons.

$$\frac{12\frac{1}{2}\cancel{\text{qt}}}{1} \cdot \frac{1\text{ gal}}{4\cancel{\text{qt}}} = \frac{25}{2} \cdot \frac{1}{4}\text{ gal} = 3.125\text{ gal}$$

The center should order 4 containers, because you can't buy part of a container.

63. (a) Convert 4150 feet to miles.

$$\frac{4150\cancel{\text{ft}}}{1} \cdot \frac{1\text{ mi}}{5280\cancel{\text{ft}}} = \frac{4150}{5280}\text{ mi} \approx 0.8\text{ mi}$$

The distance across the crater is about 0.8 mi.

(b) Multiply the distance across the crater, 4150 feet, by 3.14 to get 13,031 ft. Since 3.14 is an approximation, so is 13,031. The distance around the edge of the crater is about 13,031 ft.

(c) Convert the answer in part (b) to miles.

$$\frac{13{,}031\cancel{\text{ft}}}{1} \cdot \frac{1\text{ mi}}{5280\cancel{\text{ft}}} = \frac{13{,}031}{5280}\text{ mi} \approx 2.5\text{ mi}$$

The distance around the edge of the crater, in miles, is about 2.5 mi.

64. (a) Convert 550 feet to yards.

$$\frac{550\cancel{\text{ft}}}{1} \cdot \frac{1\text{ yd}}{3\cancel{\text{ft}}} = \frac{550}{3}\text{ yd} \approx 183\text{ yd}$$

The depth of the crater is about 183 yd.

(b) Convert 550 feet to inches.

$$\frac{550\cancel{\text{ft}}}{1} \cdot \frac{12\text{ in.}}{1\cancel{\text{ft}}} = 550 \cdot 12\text{ in.} = 6600\text{ in.}$$

The depth of the crater, in inches, is about 6600 in.

(c) Convert 550 feet to miles.

$$\frac{550\cancel{\text{ft}}}{1} \cdot \frac{1\text{ mi}}{5280\cancel{\text{ft}}} = \frac{550}{5280}\text{ mi} \approx 0.1\text{ mi}$$

The depth of the crater, in miles, is about 0.1 mi.

(d) Divide the depth of the crater, 550 feet, by 60 to get about 9.2 feet per story, or 9 ft (rounded to the nearest foot).

65. (a) Convert 18 and 30 inches to feet.

$$\frac{\overset{3}{\cancel{18}}\cancel{\text{in.}}}{1} \cdot \frac{1\text{ ft}}{\underset{2}{\cancel{12}}\cancel{\text{in.}}} = \frac{3}{2}\text{ or }1\frac{1}{2}\text{ ft}$$

$$\frac{\overset{5}{\cancel{30}}\cancel{\text{in.}}}{1} \cdot \frac{1\text{ ft}}{\underset{2}{\cancel{12}}\cancel{\text{in.}}} = \frac{5}{2}\text{ or }2\frac{1}{2}\text{ ft}$$

The trees are $1\frac{1}{2}$ to $2\frac{1}{2}$ ft, or 1.5 to 2.5 ft tall.

(b) Convert 700 years to months.

$$\frac{700\text{ yr}}{1} \cdot \frac{12\text{ mo}}{1\text{ yr}} = 700 \cdot 12\text{ mo} = 8400\text{ months}$$

The trees are 8400 months old.

(c) Three feet is thirty-six inches, which is six inches more than 30 inches. Set up and solve a proportion with age (in years) and tree height (in inches).

$$\frac{700 \text{ years}}{30 \text{ inches}} = \frac{x}{6 \text{ inches}}$$
$$30 \cdot x = 700 \cdot 6$$
$$\frac{30 \cdot x}{30} = \frac{700 \cdot 6}{30}$$
$$x = \frac{700 \cdot 6}{30} = 140 \text{ years}$$

It will take 140 more years to grow 6 inches, for a total of 840 years in all.

66. **(a)** Multiply the distance across the smaller crater, 90 miles, by 3.14 to get 282.6, or about 283 mi. Since 3.14 is an approximation, so is 283.

(b) Multiply the distance across the larger crater, 120 miles, by 3.14 to get 376.8, or about 377 mi.

8.2 The Metric System—Length

8.2 Margin Exercises

1. The length of a baseball bat, the height of a doorknob from the floor, and a basketball player's arm length are each about 1 meter in length.

2. **(a)** The woman's height is 168 cm.

(b) The man's waist is 90 cm around.

(c) Louise ran the 100 m dash in the track meet.

(d) A postage stamp is 22 mm wide.

(e) Michael paddled his canoe 2 km down the river.

(f) The pencil lead is 1 mm thick.

(g) A stick of gum is 7 cm long.

(h) The highway speed limit is 90 km per hour.

(i) The classroom was 12 m long.

(j) A penny is about 18 mm across.

3. **(a)** 3.67 m to cm

unit fraction $\}$ $\frac{100 \text{ cm}}{1 \text{ m}}$

$\frac{3.67 \text{ m}}{1} \cdot \frac{100 \text{ cm}}{1 \text{ m}} = (3.67)(100) \text{ cm} = 367 \text{ cm}$

(b) 92 cm to m

unit fraction $\}$ $\frac{1 \text{ m}}{100 \text{ cm}}$

$\frac{92 \text{ cm}}{1} \cdot \frac{1 \text{ m}}{100 \text{ cm}} = \frac{92}{100} \text{ m} = 0.92 \text{ m}$

(c) 432.7 cm to m

unit fraction $\}$ $\frac{1 \text{ m}}{100 \text{ cm}}$

$\frac{432.7 \text{ cm}}{1} \cdot \frac{1 \text{ m}}{100 \text{ cm}} = \frac{432.7}{100} \text{ m} = 4.327 \text{ m}$

(d) 65 mm to cm

unit fraction $\}$ $\frac{1 \text{ cm}}{10 \text{ mm}}$

$\frac{65 \text{ mm}}{1} \cdot \frac{1 \text{ cm}}{10 \text{ mm}} = \frac{65}{10} \text{ cm} = 6.5 \text{ cm}$

(e) 0.9 m to mm

unit fraction $\}$ $\frac{1000 \text{ mm}}{1 \text{ m}}$

$\frac{0.9 \text{ m}}{1} \cdot \frac{1000 \text{ mm}}{1 \text{ m}} = (0.9)(1000) \text{ mm} = 900 \text{ mm}$

(f) 2.5 cm to mm

unit fraction $\}$ $\frac{10 \text{ mm}}{1 \text{ cm}}$

$\frac{2.5 \text{ cm}}{1} \cdot \frac{10 \text{ mm}}{1 \text{ cm}} = (2.5)(10) \text{ mm} = 25 \text{ mm}$

4. **(a)** $(43.5)(10) = 435$
$43.5_\wedge$ gives 435.

(b) $43.5 \div 10 = 4.35$
$4_\wedge 3.5$ gives 4.35.

(c) $(28)(100) = 2800$
$28.00_\wedge$ gives 2800.

(d) $28 \div 100 = 0.28$
$_\wedge 28.$ gives 0.28.

(e) $(0.7)(1000) = 700$
$0.700_\wedge$ gives 700.

(f) $0.7 \div 1000 = 0.0007$
$_\wedge 000.7$ gives 0.0007.

5. **(a)** 12.008 km to m
Count 3 places to the *right* on the conversion line.
$12.008_\wedge$ km = 12,008 m

(b) 561.4 m to km
Count 3 places to the *left* on the conversion line.
$_\wedge 561.4$ m = 0.5614 km

8.2 The Metric System—Length

(c) 20.7 cm to m
Count 2 places to the *left* on the conversion line.
∧20.7 cm = 0.207 m

(d) 20.7 cm to mm
Count 1 place to the *right* on the conversion line.
20.7∧ cm = 207 mm

(e) 4.66 m to cm
Count 2 places to the *right* on the conversion line.
4.66∧ m = 466 cm

(f) 85.6 mm to cm
Count 1 place to the *left* on the conversion line.
8∧5.6 mm = 8.56 cm

6. **(a)** 9 m to mm
Count 3 places to the *right* on the conversion line.
Three zeros are written in as placeholders.
9.000∧ m = 9000 mm

(b) 3 cm to m
Count 2 places to the *left* on the conversion line.
One zero is written in as a placeholder.
∧03. cm = 0.03 m

(c) 14.6 km to m
Count 3 places to the *right* on the conversion line.
Two zeros are written in as placeholders.
14.600∧ km = 14,600 m

(d) 5 mm to cm
Count 1 place to the *left* on the conversion line.
∧5. mm = 0.5 cm

(e) 70 m to km
Count 3 places to the *left* on the conversion line.
One zero is written in as a placeholder.
∧070. m = 0.07 km

(f) 0.8 m to cm
Count 2 places to the *right* on the conversion line.
One zero is written in as a placeholder.
0.80∧ m = 80 cm

8.2 Section Exercises

1. *Kilo* means 1000, so 1 km = 1000 m.

3. *Milli* means $\frac{1}{1000}$ or 0.001, so

1 mm = $\frac{1}{1000}$ or 0.001 m.

5. *Centi* means $\frac{1}{100}$ or 0.01, so 1 cm = $\frac{1}{100}$ or 0.01 m.

7. The width of your hand in centimeters ■
Answers will vary; about 8 to 10 cm.

9. The width of your thumb in millimeters ■
Answers will vary; about 20 to 25 mm.

11. The child was 91 <u>cm</u> tall.

13. Ming-Na swam in the 200 <u>m</u> backstroke race.

15. Adriana drove 400 <u>km</u> on her vacation.

17. An aspirin tablet is 10 <u>mm</u> across.

19. A paper clip is about 3 <u>cm</u> long.

21. Dave's truck is 5 <u>m</u> long.

23. Some possible answers are: track and field events, metric auto parts, and lead refills for mechanical pencils.

25. 7 m to cm

$$\frac{7 \text{ m}}{1} \cdot \frac{100 \text{ cm}}{1 \text{ m}} = 7 \cdot 100 \text{ cm} = 700 \text{ cm}$$

27. 40 mm to m

$$\frac{40 \text{ mm}}{1} \cdot \frac{1 \text{ m}}{1000 \text{ mm}} = \frac{40}{1000} \text{ m} = 0.040 \text{ m or } 0.04 \text{ m}$$

29. 9.4 km to m

$$\frac{9.4 \text{ km}}{1} \cdot \frac{1000 \text{ m}}{1 \text{ km}} = (9.4)(1000) \text{ m} = 9400 \text{ m}$$

31. 509 cm to m

$$\frac{509 \text{ cm}}{1} \cdot \frac{1 \text{ m}}{100 \text{ cm}} = \frac{509}{100} \text{ m} = 5.09 \text{ m}$$

33. 400 mm to cm
Count 1 place to the *left* on the conversion line.
40∧0. mm = 40.0 cm = 40 cm

35. 0.91 m to mm
Count 3 places to the *right* on the conversion line.
One zero is written in as a placeholder.
0.910∧ m = 910 mm

37. 82 cm to m
Count 2 places to the *left* on the conversion line.
∧82. cm = 0.82 m

0.82 m is less than 1 m, so 82 cm is **less than** 1 m. The difference in length is 1 m − 0.82 m = 0.18 m or 100 cm − 82 cm = 18 cm.

260 Chapter 8 Measurement

39. 5 mm, 1 mm

5 mm to centimeters
Count one place to the *left* on the conversion line.
$\wedge$5. mm = 0.5 cm

Similarly, 1 mm = 0.1 cm.

41. 18 m to km
Count 3 places to the *left* on the conversion line.
One zero is written in as a placeholder.
$\wedge$018. m = 0.018 km

The north fork of the Roe River is just under 0.018 km long.

43. 1.64 m to centimeters and millimeters
Count two (three) places to the *right* on the conversion line.
1.64$\wedge$ m = 164 cm;

1.640$\wedge$ m = 1640 mm

The median height for U.S. females who are 20 to 29 years old is about 164 cm, or equivalently, 1640 mm.

45. 5.6 mm to km

$$\frac{5.6 \text{ mm}}{1} \cdot \frac{1 \text{ m}}{1000 \text{ mm}} \cdot \frac{1 \text{ km}}{1000 \text{ m}} = \frac{5.6}{1{,}000{,}000} \text{ km}$$
$$= 0.0000056 \text{ km}$$

8.3 The Metric System—Capacity and Weight (Mass)

8.3 Margin Exercises

1. The liter would be used to measure the amount of water in the bathtub, the amount of gasoline you buy for your car, and the amount of water in a pail.

2. (a) I bought 8 L of soda at the store.

   (b) The nurse gave me 10 mL of cough syrup.

   (c) This is a 100 L garbage can.

   (d) It took 10 L of paint to cover the bedroom walls.

   (e) My car's gas tank holds 50 L.

   (f) I added 15 mL of oil to the pancake mix.

   (g) The can of orange soda holds 350 mL.

   (h) My friend gave me a 30 mL bottle of expensive perfume.

3. (a) 9 L to mL

   $$\frac{9 \text{ L}}{1} \cdot \frac{1000 \text{ mL}}{1 \text{ L}} = 9000 \text{ mL}$$

   (b) 0.75 L to mL
   Count 3 places to the *right* on the (metric capacity) conversion line.
   0.750$\wedge$ L = 750 mL

   (c) 500 mL to L

   $$\frac{500 \text{ mL}}{1} \cdot \frac{1 \text{ L}}{1000 \text{ mL}} = \frac{500}{1000} \text{ L} = 0.5 \text{ L}$$

   (d) 5 mL to L
   Count 3 places to the *left* on the (metric capacity) conversion line.
   $\wedge$005. mL = 0.005 L

   (e) 2.07 L to mL

   $$\frac{2.07 \text{ L}}{1} \cdot \frac{1000 \text{ mL}}{1 \text{ L}} = 2070 \text{ mL}$$

   (f) 3275 mL to L
   Count 3 places to the *left* on the (metric capacity) conversion line.
   3$\wedge$275. mL = 3.275 L

4. A small paperclip, one playing card from a deck of cards, and the check you wrote to the cable company would each weigh about 1 gram.

5. (a) A thumbtack weights 800 mg.

   (b) A teenager weighs 50 kg.

   (c) This large cast-iron frying pan weighs 1 kg.

   (d) Jerry's basketball weighed 600 g.

   (e) Tamlyn takes a 500 mg calcium tablet every morning.

   (f) On his diet, Greg can eat 90 g of meat for lunch.

   (g) One strand of hair weighs 2 mg.

   (h) One banana might weigh 150 g.

6. (a) 10 kg to g

   $$\frac{10 \text{ kg}}{1} \cdot \frac{1000 \text{ g}}{1 \text{ kg}} = 10{,}000 \text{ g}$$

   (b) 45 mg to g

   $$\frac{45 \text{ mg}}{1} \cdot \frac{1 \text{ g}}{1000 \text{ mg}} = \frac{45}{1000} \text{ g} = 0.045 \text{ g}$$

8.3 The Metric System—Capacity and Weight (Mass) 261

(c) 6.3 kg to g
Count 3 places to the *right* on the [metric weight (mass)] conversion line.
6.300∧ kg = 6300 g

(d) 0.077 g to mg
Count 3 places to the *right* on the [metric weight (mass)] conversion line.
0.077∧ g = 77 mg

(e) 5630 g to kg
Count 3 places to the *left* on the [metric weight (mass)] conversion line.
5∧630. g = 5.63 kg

(f) 90 g to kg
$$\frac{90 \text{ g}}{1} \cdot \frac{1 \text{ kg}}{1000 \text{ g}} = \frac{9}{100} \text{ kg} = 0.09 \text{ kg}$$

7. (a) Gail bought a 4 L can of paint.
Use capacity units.

(b) The bag of chips weighed 450 g.
Use weight units.

(c) Give the child 5 mL of cough syrup.
Use capacity units.

(d) The width of the window is 55 cm.
Use length units.

(e) Akbar drives 18 km to work.
Use length units.

(f) The laptop computer weighs 2 kg.
Use weight units.

(g) A credit card is 55 mm wide.
Use length units.

8.3 Section Exercises

1. The glass held 250 mL of water. (Liquids are measured in mL or L.)

3. Dolores can make 10 L of soup in that pot. (Liquids are measured in mL or L.)

5. Our yellow Labrador dog grew up to weigh 40 kg. (Weight is measured in mg, g, or kg.)

7. Lori caught a small sunfish weighing 150 g. (Weight is measured in mg, g, or kg.)

9. Andre donated 500 mL of blood today. (Blood is a liquid, and liquids are measured in mL or L.)

11. The patient received a 250 mg tablet of medication each hour. (Weight is measured in mg, g, or kg.)

13. The gas can for the lawn mower holds 4 L. (Gasoline is a liquid, and liquids are measured in mL or L.)

15. Pam's backpack weighs 5 kg when it is full of books. (Weight is measured in mg, g, or kg.)

17. This is unreasonable (too much) since 4.1 liters would be about 4 quarts.

19. This is unreasonable (too much) since 5 kilograms of Epsom salts in 1 liter of water would be about 11 pounds with approximately a quart of water.

21. This is reasonable because 15 milliliters would be 3 teaspoons.

23. This is reasonable because 350 milligrams would be a little more than 1 tablet, which is about 325 milligrams.

25. Some capacity examples are 2 L bottles of soda and shampoo bottles marked in mL; weight examples are grams of fat listed on food packages and vitamin doses in milligrams.

27. The unit for your answer (g) is in the numerator. The unit being changed (kg) is in denominator, so it will divide out. The unit fraction is $\frac{1000 \text{ g}}{1 \text{ kg}}$.

29. 15 L to mL
Count 3 places to the *right* on the conversion line.
15.000∧ L = 15,000 mL

31. 3000 mL to L
Count 3 places to the *left* on the conversion line.
3∧000. mL = 3 L

33. 925 mL to L
$$\frac{925 \text{ mL}}{1} \cdot \frac{1 \text{ L}}{1000 \text{ mL}} = \frac{925}{1000} \text{ L} = 0.925 \text{ L}$$

35. 8 mL to L
$$\frac{8 \text{ mL}}{1} \cdot \frac{1 \text{ L}}{1000 \text{ mL}} = \frac{8}{1000} \text{ L} = 0.008 \text{ L}$$

37. 4.15 L to mL
Count 3 places to the *right* on the conversion line.
4.150∧ L = 4150 mL

39. 8000 g to kg
Count 3 places to the *left* on the conversion line.
8∧000. g = 8 kg

41. 5.2 kg to g
Count 3 places to the *right* on the conversion line.
5.200∧ kg = 5200 g

43. 0.85 g to mg
Count 3 places to the *right* on the conversion line.
0.850∧ g = 850 mg

45. 30,000 mg to g
Count 3 places to the *left* on the conversion line.
30∧000. mg = 30 g

47. 598 mg to g
Count 3 places to the *left* on the conversion line.
∧598. mg = 0.598 g

49. 60 mL to L
$$\frac{60 \text{ mL}}{1} \cdot \frac{1 \text{ L}}{1000 \text{ mL}} = \frac{60}{1000} \text{ L} = 0.06 \text{ L}$$

51. 3 g to kg
$$\frac{3 \text{ g}}{1} \cdot \frac{1 \text{ kg}}{1000 \text{ g}} = \frac{3}{1000} \text{ kg} = 0.003 \text{ kg}$$

53. 0.99 L to mL
$$\frac{0.99 \text{ L}}{1} \cdot \frac{1000 \text{ mL}}{1 \text{ L}} = 990 \text{ mL}$$

55. The masking tape is 19 mm wide. (Length is measured in mm, cm, m, or km.)

57. Buy a 60 mL jar of acrylic paint for art class. (Paint is a liquid and liquids are measured in mL or L.)

59. My waist measurement is 65 cm. (Length is measured in mm, cm, m, or km.)

61. A single postage stamp weighs 90 mg. (Weight is measured in mg, g, or kg.)

63. Convert 300 mL to L.
Count 3 places to the *left* on the conversion line.
∧300. mL = 0.3 L

Each day 0.3 L of sweat is released.

65. Convert 1.34 kg to g.
Count 3 places to the *right* on the conversion line.
1.340∧ kg = 1340 g

The average weight of a human brain is 1340 g.

67. Convert 900 mL to L.
$$\frac{900 \text{ mL}}{1} \cdot \frac{1 \text{ L}}{1000 \text{ mL}} = \frac{900}{1000} \text{ L} = 0.9 \text{ L}$$

On average, we breathe in and out roughly 0.9 L of air every 10 seconds.

69. Convert 3000 g to kg and 4000 g to kg.
$$\frac{3000 \text{ g}}{1} \cdot \frac{1 \text{ kg}}{1000 \text{ g}} = \frac{3000}{1000} \text{ kg} = 3 \text{ kg}$$
$$\frac{4000 \text{ g}}{1} \cdot \frac{1 \text{ kg}}{1000 \text{ g}} = \frac{4000}{1000} \text{ kg} = 4 \text{ kg}$$

A small adult cat weighs from 3 kg to 4 kg.

71. There are 1000 milligrams in a gram so 1005 mg is *greater* than 1 g. The difference in weight is 1005 mg − 1000 mg = 5 mg or 1.005 g − 1 g = 0.005 g.

73. Since 1 kg is equivalent to 1000 g, divide 1000 g by the weight of 1 nickel.
$$\frac{1000 \text{ g}}{5 \text{ g}} = 200$$

There are 200 nickels in 1 kg of nickels.

75. (a) 1 Mm = 1,000,000 m

(b) 3.5 Mm to m
$$\frac{3.5 \text{ Mm}}{1} \cdot \frac{1,000,000 \text{ m}}{1 \text{ Mm}} = 3,500,000 \text{ m}$$

76. (a) 1 Gm = 1,000,000,000 m

(b) 2500 m to Gm
$$\frac{2500 \text{ m}}{1} \cdot \frac{1 \text{ Gm}}{1,000,000,000 \text{ m}} = 0.0000025 \text{ Gm}$$

77. (a) 1 Tm = 1,000,000,000,000 m

(b) $\frac{1 \text{ Tm}}{1} \cdot \frac{1,000,000,000,000 \text{ m}}{1 \text{ Tm}} \cdot \frac{1 \text{ Gm}}{1,000,000,000 \text{ m}}$
$= \frac{1,000,000,000,000}{1,000,000,000}$ Gm = 1000 Gm
So 1 Tm = 1000 Gm.

$\frac{1 \text{ Tm}}{1} \cdot \frac{1,000,000,000,000 \text{ m}}{1 \text{ Tm}} \cdot \frac{1 \text{ Mm}}{1,000,000 \text{ m}}$
$= \frac{1,000,000,000,000}{1,000,000}$ Mm = 1,000,000 Mm
So 1 Tm = 1,000,000 Mm.

78. 1 MB = 1,000,000 bytes
1 GB = 1,000,000,000 bytes
2^{20} = 1,048,576
2^{30} = 1,073,741,824

Summary Exercises on U.S. Customary and Metric Units

1.
   U.S. Customary Units

Length	Weight	Capacity
inch	ounce	fluid ounce
foot	pound	cup
yard	ton	pint
mile		quart
		gallon

3. (a) 1 <u>ft</u> = 12 in.

   (b) 3 ft = 1 <u>yd</u>

   (c) 1 mi = <u>5280</u> ft

5. (a) 1 cup = <u>8</u> fl oz

   (b) 4 qt = 1 <u>gal</u>

   (c) <u>2</u> pt = 1 qt

7. My water bottle holds 450 <u>mL</u>.

9. The child weighed 23 <u>kg</u>.

11. The red pen is 14 <u>cm</u> long.

13. Merlene made 12 <u>L</u> of fruit punch for her daughter's birthday party.

15. 45 cm to meters

    $$\frac{45 \text{ cm}}{1} \cdot \frac{1 \text{ m}}{100 \text{ cm}} = \frac{45}{100} \text{ m} = 0.45 \text{ m}$$

17. 0.6 L to milliliters

    $$\frac{0.6 \text{ L}}{1} \cdot \frac{1000 \text{ mL}}{1 \text{ L}} = 600 \text{ mL}$$

19. 300 mm to centimeters

    $$\frac{300 \text{ mm}}{1} \cdot \frac{1 \text{ cm}}{10 \text{ mm}} = \frac{300}{10} \text{ cm} = 30 \text{ cm}$$

21. 50 mL to liters
    Count 3 places to the *left* on the conversion line.
    ₐ050. mL = 0.050 or 0.05 L

    Note: You could also use unit fractions.

23. 7.28 kg to grams
    Count 3 places to the *right* on the conversion line.
    7.280ₐ kg = 7280 g

25. 9 g to kilograms

    $$\frac{9 \text{ g}}{1} \cdot \frac{1 \text{ kg}}{1000 \text{ g}} = \frac{9}{1000} \text{ kg} = 0.009 \text{ kg}$$

27. 2.72 m to centimeters and millimeters
    Count two (three) places to the *right* on the conversion line.
    2.72ₐ m = 272 cm; 2.720ₐ m = 2720 mm

29. The tallest person is 2.72 m tall and the second tallest person is 2.68 m tall. The difference is 2.72 − 2.68 = 0.04 m.

    Now convert 0.04 m to cm and mm.
    Count two (three) places to the *right* on the conversion line.
    0.04ₐ m = 4 cm;   0.040ₐ m = 40 mm

31. We are given dollars and ounces, so we need to use a unit fraction involving ounces and pounds to get an answer in dollars and pounds.

    $$\frac{\$3.49}{\cancel{12} \text{ oz}} \cdot \frac{\cancel{16}^{4} \text{ oz}}{1 \text{ lb}} = \frac{\$13.96}{3 \text{ lb}} \approx \$4.65/\text{lb (rounded)}$$

    To the nearest cent, the price per pound was $4.65.

8.4 Problem Solving with Metric Measurement

8.4 Margin Exercises

1. *Step 1*
   The problem asks for the cost of 75 cm of ribbon.

   Step 2
   The price is $0.89 per meter, but the length wanted is cm. Convert cm to m.

   Step 3
   There are 100 cm in a meter, so 75 cm is $\frac{3}{4}$ of a meter. Round the cost of 1 m from $0.89 to $0.88 so that it is divisible by 4.

   An estimate is: $\frac{3}{4} \cdot \$0.88 = \$0.66$

   Step 4
   Convert 75 cm to m: 75 cm = 0.75 m

   Find the cost.

   $$\frac{0.75 \text{ m}}{1} \cdot \frac{\$0.89}{1 \text{ m}} = \$0.6675 \approx \$0.67$$

   Step 5
   The cost for 75 cm of ribbon is $0.67 (rounded).

   Step 6
   The rounded answer, $0.67, is close to our estimate of $0.66.

264 Chapter 8 Measurement

2. *Step 1*
The problem asks for the number of milligrams in each dose.

Step 2
The total amount of medication to be taken is given in grams. Convert grams to milligrams and divide by the number of doses.

Step 3
There are 1000 mg in a gram, so 1000 divided by 3 doses gives us about 300 mg per dose as an estimate.

Step 4
Convert 1.2 grams to milligrams.
1.200∧ g = 1200 mg

Divide by the number of doses.
$$\frac{1200 \text{ mg}}{3 \text{ doses}} = 400 \text{ mg/dose}$$

Step 5
Each dose should be 400 mg.

Step 6
The estimate was obtained using liberal rounding, so the exact answer of 400 mg is reasonably close to the estimated answer of 300 mg.

3. *Step 1*
The problem asks for the combined number of meters in two pieces of fabric.

Step 2
Convert the given lengths to meters and add them.

Step 3
Round 2 m 35 cm to 2 m and 1 m 85 cm to 2 m. Add them to get 4 m for an estimate.

Step 4
Convert centimeters to meters.
2 m 35 cm = 2.35 m
1 m 85 cm = 1.85 m

Add the measurements of the fabric.

one piece	2.35	m
other piece	+1.85	m
	4.20	m

Step 5
Andrea has 4.2 m of fabric.

Step 6
The exact answer, 4.2 m, is close to our estimate of 4 m.

8.4 Section Exercises

1. Convert 850 g to kg.
$$\frac{850 \text{ g}}{1} \cdot \frac{1 \text{ kg}}{1000 \text{ g}} = 0.85 \text{ kg}$$

$$\frac{\$0.98}{1 \text{ kg}} \cdot \frac{0.85 \text{ kg}}{1} = \$0.833 \approx \$0.83$$

She will pay $0.83 (rounded) for the rice.

3. Convert 500 g to kg.
$$\frac{500 \text{ g}}{1} \cdot \frac{1 \text{ kg}}{1000 \text{ g}} = 0.5 \text{ kg}$$

Find the difference.
90 kg − 0.5 kg = 89.5 kg

The difference in the weights of the two dogs is 89.5 kg.

5. Write 5 L in terms of mL.
$$5 \text{ L} = \frac{5 \text{ L}}{1} \cdot \frac{1000 \text{ mL}}{1 \text{ L}} = 5000 \text{ mL}$$

$$\frac{5000 \text{ mL}}{1} \cdot \frac{1 \text{ beat}}{70 \text{ mL}} \approx 71.42857 \text{ beats} \approx 71 \text{ beats}$$

It takes 71 beats (rounded) to pass all the blood through the heart.

7. Multiply to find the total length in mm, then convert mm to cm.
60 mm · 30 = 1800 mm

$$\frac{1800 \text{ mm}}{1} \cdot \frac{1 \text{ cm}}{10 \text{ mm}} = 180 \text{ cm}$$

The total length of the lead is 180 cm.
Divide to find the cost per cm.

$$\frac{\$3.29}{180 \text{ cm}} \approx \$0.0183/\text{cm} \approx \$0.02/\text{cm}$$

The cost of the lead is $0.02/cm (rounded).

9. Convert centimeters to meters, then add the lengths of the two pieces.

2 m 8 cm	=	2.08 m
2 m 95 cm	=	+2.95 m
		5.03 m

The total length of the two boards is 5.03 m.

11. Convert milliliters to liters.
85 mL = 0.085 L
Multiply by the number of students.
(0.085 L)(45) = 3.825 L
Four one-liter bottles should be ordered.

The amount of acid left over is

$$4 \text{ L} - 3.825 \text{ L} = 0.175 \text{ L}.$$

13. Three cups a day for one week is $3 \cdot 7 = 21$ cups in one week.

$$\frac{21 \text{ cups}}{1} \cdot \frac{90 \text{ mg}}{1 \text{ cup}} = 1890 \text{ mg}$$

Convert mg to g.

$$\frac{1890 \text{ mg}}{1} \cdot \frac{1 \text{ g}}{1000 \text{ mg}} = 1.89 \text{ g}$$

She will consume 1.89 grams of caffeine in one week.

15. (a) $\dfrac{10{,}000 \text{ m}}{1 \text{ sec}} \cdot \dfrac{1 \text{ km}}{1000 \text{ m}} = 10$ km in one second

(b) $\dfrac{60 \text{ sec}}{1 \text{ min}} \cdot \dfrac{10 \text{ km}}{1 \text{ sec}} = 600$ km in one minute

(c) $\dfrac{60 \text{ min}}{1 \text{ hr}} \cdot \dfrac{600 \text{ km}}{1 \text{ min}} = 36{,}000$ km in one hour

17. A box of 50 envelopes weighs 255 g. Subtract the packaging weight to find the net weight.

$$\begin{array}{r} 255 \text{ g} \\ -\ 40 \text{ g} \\ \hline 215 \text{ g} \end{array}$$

There are 50 envelopes. Divide to find the weight of one envelope.

$$\frac{215 \text{ g}}{50 \text{ envelopes}} = \frac{4.3 \text{ g}}{\text{envelope}}$$

One envelope weighs 4.3 grams.
Convert grams to milligrams.

$$\frac{4.3 \text{ g}}{1} \cdot \frac{1000 \text{ mg}}{1 \text{ g}} = 4300 \text{ mg}$$

18. Box of 1000 staples
Subtract the weight of the packaging to find the net weight.

$$\begin{array}{r} 350 \text{ g} \\ -\ 20 \text{ g} \\ \hline 330 \text{ g} \end{array}$$

There are 1000 staples, so divide the net weight by 1000 or move the decimal point 3 places to the left to find the weight of one staple.

$330 \text{ g} \div 1000 = 0.33 \text{ g}$

The weight of one staple is 0.33 g or 330 mg.

19. Given that the weight of one sheet of paper is 3000 mg, find the weight of one sheet in grams.

$3000 \text{ mg} = 3{\wedge}000.\text{ g} = 3 \text{ g}$

To find the net weight of the ream of paper, multiply 500 sheets by 3 g per sheet.

$$\frac{500 \text{ sheets}}{1} \cdot \frac{3 \text{ g}}{1 \text{ sheet}} = 1500 \text{ g}$$

The net weight of the paper is 1500 g.
Since the weight of the packaging is 50 g, the total weight is $1500 \text{ g} + 50 \text{ g} = 1550 \text{ g}$.

20. Box of 100 small paper clips
Use the metric conversion line to find the weight of one paper clip in grams.

$500 \text{ mg} = {\wedge}500.\text{ g} = 0.5 \text{ g}$

The net weight is
$(0.5 \text{ g})(100) = 50 \text{ g}$
The total weight is
$50 \text{ g} + 5 \text{ g} = 55 \text{ g}$.

8.5 Metric—U.S. Customary Conversions and Temperature

8.5 Margin Exercises

1. (a) 23 m to yards

$$\frac{23 \text{ m}}{1} \cdot \frac{1.09 \text{ yd}}{1 \text{ m}} \approx 25.1 \text{ yd}$$

(b) 40 cm to inches

$$\frac{40 \text{ cm}}{1} \cdot \frac{0.39 \text{ in.}}{1 \text{ cm}} \approx 15.6 \text{ in.}$$

(c) 5 mi to kilometers

$$\frac{5 \text{ mi}}{1} \cdot \frac{1.61 \text{ km}}{1 \text{ mi}} \approx 8.1 \text{ km}$$

(d) 12 in. to centimeters

$$\frac{12 \text{ in.}}{1} \cdot \frac{2.54 \text{ cm}}{1 \text{ in.}} \approx 30.5 \text{ cm}$$

2. (a) 17 kg to pounds

$$\frac{17 \text{ kg}}{1} \cdot \frac{2.20 \text{ lb}}{1 \text{ kg}} \approx 37.4 \text{ lb}$$

(b) 5 L to quarts

$$\frac{5 \text{ L}}{1} \cdot \frac{1.06 \text{ qt}}{1 \text{ L}} \approx 5.3 \text{ qt}$$

(c) 90 g to ounces

$$\frac{90 \text{ g}}{1} \cdot \frac{0.035 \text{ oz}}{1 \text{ g}} \approx 3.2 \text{ oz}$$

(d) 3.5 gal to liters

$$\frac{3.5 \text{ gal}}{1} \cdot \frac{3.79 \text{ L}}{1 \text{ gal}} \approx 13.3 \text{ L}$$

(e) 145 lb to kilograms

$$\frac{145 \text{ lb}}{1} \cdot \frac{0.45 \text{ kg}}{1 \text{ lb}} \approx 65.3 \text{ kg}$$

(f) 8 oz to grams

$$\frac{8 \text{ oz}}{1} \cdot \frac{28.35 \text{ g}}{1 \text{ oz}} \approx 226.8 \text{ g}$$

3. **(a)** Set the living room thermostat at:
21 °C

(b) The baby has a fever of:
39 °C

(c) Wear a sweater outside because it's:
15 °C

(d) My iced tea is:
5 °C

(e) Time to go swimming! It's:
35 °C

(f) Inside a refrigerator (not the freezer) it's:
3 °C

(g) There is a blizzard outside. It's:
−20 °C

(h) I need hot water to get these clothes clean. It should be:
55 °C

4. Use $C = \dfrac{5(F - 32)}{9}$.

(a) 72 °F

$$C = \frac{5(72 - 32)}{9} = \frac{5 \cdot 40}{9} = \frac{200}{9} \approx 22$$

72 °F ≈ 22 °C

(b) 20 °F

$$C = \frac{5(20 - 32)}{9} = \frac{5 \cdot (-12)}{9} = \frac{-5 \cdot \overset{1}{\cancel{3}} \cdot 4}{\underset{1}{\cancel{3}} \cdot 3}$$

$$= \frac{-20}{3} \approx -7$$

20 °F ≈ −7 °C

(c) 212 °F

$$C = \frac{5(212 - 32)}{9} = \frac{5 \cdot \overset{20}{\cancel{180}}}{\underset{1}{\cancel{9}}} = 100$$

212 °F = 100 °C

(d) 98.6 °F

$$C = \frac{5(98.6 - 32)}{9} = \frac{5 \cdot 66.6}{9} = 37$$

98.6 °F = 37 °C

5. Use $F = \dfrac{9 \cdot C}{5} + 32$.

(a) 100 °C

$$F = \frac{9 \cdot \overset{20}{\cancel{100}}}{\underset{1}{\cancel{5}}} + 32 = 180 + 32 = 212$$

100 °C = 212 °F

(b) −25 °C

$$F = \frac{-9 \cdot \overset{5}{\cancel{25}}}{\underset{1}{\cancel{5}}} + 32 = -45 + 32 = -13$$

−25 °C = −13 °F

(c) 32 °C

$$F = \frac{9 \cdot 32}{5} + 32 = 57.6 + 32 = 89.6$$

32 °C ≈ 90 °F

(d) −18 °C

$$F = \frac{-9 \cdot 18}{5} + 32 = -32.4 + 32 = -0.4$$

−18 °C ≈ 0 °F (rounded)

8.5 Section Exercises

1. 20 m to yards

$$\frac{20 \text{ m}}{1} \cdot \frac{1.09 \text{ yd}}{1 \text{ m}} \approx 21.8 \text{ yd}$$

3. 80 m to feet

$$\frac{80 \text{ m}}{1} \cdot \frac{3.28 \text{ ft}}{1 \text{ m}} \approx 262.4 \text{ ft}$$

5. 16 ft to meters

$$\frac{16 \text{ ft}}{1} \cdot \frac{0.30 \text{ m}}{1 \text{ ft}} \approx 4.8 \text{ m}$$

7. 150 g to ounces

$$\frac{150 \text{ g}}{1} \cdot \frac{0.035 \text{ oz}}{1 \text{ g}} \approx 5.3 \text{ oz}$$

9. 248 lb to kilograms

$$\frac{248 \text{ lb}}{1} \cdot \frac{0.45 \text{ kg}}{1 \text{ lb}} \approx 111.6 \text{ kg}$$

8.5 Metric—U.S. Customary Conversions and Temperature

11. 28.6 L to quarts

$$\frac{28.6 \text{ L}}{1} \cdot \frac{1.06 \text{ qt}}{1 \text{ L}} \approx 30.3 \text{ qt}$$

13. **(a)** Convert 5 g to ounces.

$$\frac{5 \text{ g}}{1} \cdot \frac{0.035 \text{ oz}}{1 \text{ g}} \approx 0.2 \text{ oz}$$

About 0.2 oz of gold was used.

(b) The extra weight probably did not slow him down.

15. Convert 8.4 gal to liters.

$$\frac{8.4 \text{ gal}}{1} \cdot \frac{3.79 \text{ L}}{1 \text{ gal}} \approx 31.8 \text{ L}$$

The dishwasher uses about 31.8 L.

17. Convert 0.5 in. to centimeters.

$$\frac{0.5 \text{ in.}}{1} \cdot \frac{2.54 \text{ cm}}{1 \text{ in.}} \approx 1.3$$

The dwarf gobie is about 1.3 cm long.

19. Convert 3.5 kg to lb.

$$\frac{3.5 \text{ kg}}{1} \cdot \frac{2.20 \text{ lb}}{1 \text{ kg}} \approx 7.7 \text{ lb}$$

Since the minimum weight is 3.5 kg ≈ 7.7 lb, the 8-lb baby is heavy enough.

Convert 53 cm to in.

$$\frac{53 \text{ cm}}{1} \cdot \frac{0.39 \text{ in.}}{1 \text{ cm}} \approx 20.7 \text{ in.}$$

Since the minimum length is 53 cm ≈ 20.7 in., the 19.5-inch baby is not long enough to be in the carrier.

21. A snowy day
−8 °C is the most reasonable temperature. 12 °C and 28 °C are temperatures well above freezing.

23. A high fever
40 °C is the most reasonable temperature because normal body temperature is about 37 °C.

25. Oven temperature
150 °C is the most reasonable temperature. 50 °C is the temperature of hot bath water, which is clearly not hot enough for an oven temperature.

27. 60 °F

$$C = \frac{5(60 - 32)}{9} = \frac{5 \cdot 28}{9} = \frac{140}{9} \approx 15.6 \approx 16$$

60 °F ≈ 16 °C

29. −4 °F

$$C = \frac{5(-4 - 32)}{9} = \frac{5(-36)}{9} = \frac{-5 \cdot 4 \cdot \cancel{9}}{\cancel{9}} = -20$$

−4 °F = −20 °C

31. 8 °C

$$F = \frac{9 \cdot 8}{5} + 32 = \frac{72}{5} + 32 = 14.4 + 32 = 46.4$$

8 °C ≈ 46 °F

33. −5 °C

$$F = \frac{9(-5)}{5} + 32 = \frac{-9 \cdot \cancel{5}}{\cancel{5}} + 32 = -9 + 32 = 23$$

−5 °C = 23 °F

35. 136 °F

$$C = \frac{5(136 - 32)}{9} = \frac{5 \cdot 104}{9} = \frac{520}{9} \approx 57.8$$

136 °F ≈ 58 °C

−129 °F

$$C = \frac{5(-129 - 32)}{9} = \frac{5(-161)}{9} \approx -89$$

−129 °F ≈ −89 °C

37. 50 °F

$$C = \frac{5(50 - 32)}{9} = \frac{5(18)}{9} = \frac{5 \cdot 2 \cdot \cancel{9}}{\cancel{9}} = 10$$

50 °F = 10 °C

105 °F

$$C = \frac{5(105 - 32)}{9} = \frac{5(73)}{9} \approx 41$$

105 °F ≈ 41 °C

39. **(a)** Since the comfort range of the boots is from 24 °C to 4 °C, you would wear these boots in pleasant weather—above freezing, but not hot.

(b) Change 24 °C to Fahrenheit.

$$F = \frac{9 \cdot C}{5} + 32 = \frac{9 \cdot 24}{5} + 32 = \frac{216}{5} + 32$$
$$= 43.2 + 32 = 75.2$$

Thus, 24 °C ≈ 75 °F.

Change 4 °C to Fahrenheit.

$$F = \frac{9 \cdot C}{5} + 32 = \frac{9 \cdot 4}{5} + 32 = \frac{36}{5} + 32$$
$$= 7.2 + 32 = 39.2$$

Thus, 4 °C ≈ 39 °F.

The boots are designed for Fahrenheit temperatures of about 75 °F to about 39 °F.

(c) The range of metric temperatures in January would depend on where you live. In Minnesota, it's 0°C to −40°C, and in California it's 24°C to 0°C.

41. Length of model plane

$$\frac{6 \,\text{ft}}{1} \cdot \frac{0.30 \text{ m}}{1 \,\text{ft}} \approx 6(0.30 \text{ m}) = 1.8 \text{ m}$$

42. Weight of plane

$$\frac{11 \,\text{lb}}{1} \cdot \frac{0.45 \text{ kg}}{1 \,\text{lb}} \approx 11(0.45 \text{ kg}) = 4.95 \text{ kg}$$
$$\approx 5.0 \text{ kg}$$

43. Length of flight path

$$\frac{1888.3 \,\text{mi}}{1} \cdot \frac{1.61 \text{ km}}{1 \,\text{mi}} \approx 1888.3(1.61 \text{ km})$$
$$\approx 3040.2 \text{ km}$$

44. Time of flight = 38 hr 23 min

45. Cruising altitude

$$\frac{1000 \,\text{ft}}{1} \cdot \frac{0.30 \text{ m}}{1 \,\text{ft}} \approx 1000(0.30 \text{ m}) = 300 \text{ m}$$

46. Fuel at the start

$$\frac{1 \text{ gal}}{1} \cdot \frac{3.79 \text{ L}}{1 \text{ gal}} \cdot \frac{1000 \text{ mL}}{1 \text{ L}} \approx (3.79)(1000 \text{ mL})$$
$$= 3790 \text{ mL}$$

Since the plane carried "less than a gallon of fuel," there would be fewer than 3790 mL.

47. Fuel left after landing

$$\frac{2 \text{ fl oz}}{1} \cdot \frac{1 \text{ qt}}{32 \text{ fl oz}} \cdot \frac{0.95 \text{ L}}{1 \text{ qt}} \cdot \frac{1000 \text{ mL}}{1 \text{ L}}$$
$$= \frac{950}{16} = 59.375 \approx 59.4 \text{ mL}$$

48.
$$\frac{1 \text{ gal}}{1} \cdot \frac{4 \text{ qt}}{1 \text{ gal}} \cdot \frac{32 \text{ fl oz}}{1 \text{ qt}} = 128 \text{ fl oz}$$
$$\frac{2 \text{ fl oz}}{128 \text{ fl oz}} = \frac{1}{64} = 0.015625 \approx 1.6\%$$

Chapter 8 Review Exercises

1. 1 lb = <u>16</u> oz

2. <u>3</u> ft = 1 yd

3. 1 T = <u>2000</u> lb

4. <u>4</u> qt = 1 gal

5. 1 hr = <u>60</u> min

6. 1 c = <u>8</u> fl oz

7. <u>60</u> sec = 1 min

8. <u>5280</u> ft = 1 mi

9. <u>12</u> in. = 1 ft

10. 4 ft to inches

$$\frac{4 \,\text{ft}}{1} \cdot \frac{12 \text{ in.}}{1 \,\text{ft}} = 4 \cdot 12 \text{ in.} = 48 \text{ in.}$$

11. 6000 lb to tons

$$\frac{\overset{3}{\cancel{6000}} \,\text{lb}}{1} \cdot \frac{1 \text{ T}}{\underset{1}{\cancel{2000}} \,\text{lb}} = 3 \text{ T}$$

12. 64 oz to pounds

$$\frac{\overset{4}{\cancel{64}} \,\text{oz}}{1} \cdot \frac{1 \text{ lb}}{\underset{1}{\cancel{16}} \,\text{oz}} = 4 \text{ lb}$$

13. 18 hr to days

$$\frac{\overset{3}{\cancel{18}} \,\text{hr}}{1} \cdot \frac{1 \text{ day}}{\underset{4}{\cancel{24}} \,\text{hr}} = \frac{3}{4} \text{ or } 0.75 \text{ day}$$

14. 150 min to hours

$$\frac{\overset{5}{\cancel{150}} \,\text{min}}{1} \cdot \frac{1 \text{ hr}}{\underset{2}{\cancel{60}} \,\text{min}} = \frac{5}{2} \text{ hr} = 2\frac{1}{2} \text{ or } 2.5 \text{ hr}$$

15. $1\frac{3}{4}$ lb to ounces

$$\frac{1\frac{3}{4} \,\text{lb}}{1} \cdot \frac{16 \text{ oz}}{1 \,\text{lb}} = \frac{7}{\underset{1}{\cancel{4}}} \cdot \frac{\overset{4}{\cancel{16}}}{1} \text{ oz} = 28 \text{ oz}$$

16. $6\frac{1}{2}$ ft to inches

$$\frac{6\frac{1}{2} \,\text{ft}}{1} \cdot \frac{12 \text{ in.}}{1 \,\text{ft}} = \frac{13}{\underset{1}{\cancel{2}}} \cdot \frac{\overset{6}{\cancel{12}}}{1} \text{ in.} = 78 \text{ in.}$$

17. 7 gal to cups

$$\frac{7 \text{ gal}}{1} \cdot \frac{4 \text{ qt}}{1 \text{ gal}} \cdot \frac{2 \text{ pt}}{1 \text{ qt}} \cdot \frac{2 \text{ c}}{1 \text{ pt}} = 7 \cdot 4 \cdot 2 \cdot 2 \text{ c} = 112 \text{ c}$$

18. 4 days to seconds

$$\frac{4 \text{ days}}{1} \cdot \frac{24 \text{ hr}}{1 \text{ day}} \cdot \frac{60 \text{ min}}{1 \text{ hr}} \cdot \frac{60 \text{ sec}}{1 \text{ min}}$$
$$= 4 \cdot 24 \cdot 60 \cdot 60 \text{ sec}$$
$$= 345{,}600 \text{ sec}$$

19. **(a)** 12,460 ft to yards

$$\frac{12{,}460 \text{ ft}}{1} \cdot \frac{1 \text{ yd}}{3 \text{ ft}} = \frac{12{,}460}{3} \text{ yd}$$
$$= 4153\tfrac{1}{3} \text{ yd (exact)}$$
$$\text{or } 4153.3 \text{ yd (rounded)}.$$

The average depth of the world's oceans is $4153\tfrac{1}{3}$ yd (exact) or 4153.3 yd (rounded).

(b) 12,460 ft to miles

$$\frac{12{,}460 \text{ ft}}{1} \cdot \frac{1 \text{ mi}}{5280 \text{ ft}} \approx 2.36 \text{ mi} \approx 2.4 \text{ mi}$$

The average depth of the world's oceans is 2.4 mi (rounded).

20. *Step 1*
The problem asks for the amount of money the company made.

Step 2
Convert pounds to tons. Then multiply to find the total amount.

Step 3
To estimate, round 123,260 lb to 100,000 lb. Then, there are 2000 lb in a ton, so 100,000 lb is 50 T. So 50 · $40 = $2000 is our estimate.

Step 4
$$\frac{123{,}260 \text{ lb}}{1} \cdot \frac{1 \text{ T}}{2000 \text{ lb}} = 61.63 \text{ T}$$
61.63 · $40 = $2465.20

Step 5
The company made $2465.20.

Step 6
The answer, $2465.20, is close to our estimate of $2000.

21. My thumb is 20 <u>mm</u> wide.

22. Her waist measurement is 66 <u>cm</u>.

23. The two towns are 40 <u>km</u> apart.

24. A basketball court is 30 <u>m</u> long.

25. The height of the picnic bench is 45 <u>cm</u>.

26. The eraser on the end of my pencil is 5 <u>mm</u> long.

27. 5 m to cm
Count 2 places to the *right* on the metric conversion line.
5.00˰ m = 500 cm

28. 8.5 km to m

$$\frac{8.5 \text{ km}}{1} \cdot \frac{1000 \text{ m}}{1 \text{ km}} = 8500 \text{ m}$$

29. 85 mm to cm
Count 1 place to the *left* on the metric conversion line.
8˰5. mm = 8.5 cm

30. 370 cm to m

$$\frac{370 \text{ cm}}{1} \cdot \frac{1 \text{ m}}{100 \text{ cm}} = 3.7 \text{ m}$$

31. 70 m to km
Count 3 places to the *left* on the metric conversion line.
˰070. m = 0.07 km

32. 0.93 m to mm

$$\frac{0.93 \text{ m}}{1} \cdot \frac{1000 \text{ mm}}{1 \text{ m}} = 930 \text{ mm}$$

33. The eye dropper holds 1 <u>mL</u>.

34. I can heat 3 <u>L</u> of water in this saucepan.

35. Loretta's hammer weighed 650 <u>g</u>.

36. Yongshu's suitcase weighed 20 <u>kg</u> when it was packed.

37. My fish tank holds 80 <u>L</u> of water.

38. I'll buy the 500 <u>mL</u> bottle of mouthwash.

39. Mara took a 200 <u>mg</u> antibiotic pill.

40. This piece of chicken weighs 100 <u>g</u>.

41. 5000 mL to L
Count 3 places to the *left* on the metric conversion line.
5˰000. mL = 5 L

42. 8 L to mL
Count 3 places to the *right* on the metric conversion line.
8.000˰ L = 8000 mL

43. 4.58 g to mg

$$\frac{4.58 \text{ g}}{1} \cdot \frac{1000 \text{ mg}}{1 \text{ g}} = 4580 \text{ mg}$$

44. 0.7 kg to g

$$\frac{0.7 \text{ kg}}{1} \cdot \frac{1000 \text{ g}}{1 \text{ kg}} = 700 \text{ g}$$

45. 6 mg to g
Count 3 places to the *left* on the metric conversion line.
$\wedge$006. mg = 0.006 g

46. 35 mL to L
Count 3 places to the *left* on the metric conversion line.
$\wedge$035. mL = 0.035 L

47. Convert milliliters to liters.

$$\frac{180 \text{ mL}}{1} \cdot \frac{1 \text{ L}}{1000 \text{ mL}} = \frac{180}{1000} \text{ L} = 0.18 \text{ L}$$

Multiply 0.18 L by the number of servings.
(0.18 L)(175) = 31.5 L

For 175 servings, 31.5 L of punch are needed.

48. Convert kilograms to grams
10 kg = 10,000 g
Divide by the number of people.

$$\frac{10{,}000 \text{ g}}{28 \text{ people}} \approx 357$$

Jason is allowing 357 g (rounded) of turkey for each person.

49. Convert grams to kilograms.
Using the metric conversion line,
4 kg 750 g = 4.75 kg.
Subtract the weight loss from his original weight.

```
  92.00  kg
-  4.75  kg
─────────
  87.25  kg
```

Yerald weighs 87.25 kg.

50. Convert to kilograms.
Using the metric conversion line,
950 g = 0.95 kg.
Multiply by the price per kilogram.

```
    $1.49   ← 2 decimal places
  × 0.95    ← 2 decimal places
  ───────
      745
    1 341
  ───────
   $1.4155  ← 4 decimal places
```

Young-Mi paid $1.42 (rounded) for the onions.

51. 6 m to yards

$$\frac{6 \text{ m}}{1} \cdot \frac{1.09 \text{ yd}}{1 \text{ m}} \approx 6.5 \text{ yd}$$

52. 30 cm to inches

$$\frac{30 \text{ cm}}{1} \cdot \frac{0.39 \text{ in.}}{1 \text{ cm}} \approx 11.7 \text{ in.}$$

53. 108 km to miles

$$\frac{108 \text{ km}}{1} \cdot \frac{0.62 \text{ mi}}{1 \text{ km}} \approx 67.0 \text{ mi}$$

54. 800 mi to kilometers

$$\frac{800 \text{ mi}}{1} \cdot \frac{1.61 \text{ km}}{1 \text{ mi}} \approx 1288 \text{ km}$$

55. 23 qt to liters

$$\frac{23 \text{ qt}}{1} \cdot \frac{0.95 \text{ L}}{1 \text{ qt}} \approx 21.9 \text{ L}$$

56. 41.5 L to quarts

$$\frac{41.5 \text{ L}}{1} \cdot \frac{1.06 \text{ qt}}{1 \text{ L}} \approx 44.0 \text{ qt}$$

57. Water freezes at 0 °C.

58. Water boils at 100 °C.

59. Normal body temperature is about 37 °C.

60. Comfortable room temperature is about 20 °C.

61. 77 °F

$$C = \frac{5(77-32)}{9} = \frac{5(\overset{5}{\cancel{45}})}{\underset{1}{\cancel{9}}} = 25$$

77 °F = 25 °C

62. 5 °F

$$C = \frac{5(5-32)}{9} = \frac{5(-27)}{9} = \frac{-5 \cdot 3 \cdot \overset{1}{\cancel{9}}}{\underset{1}{\cancel{9}}} = -15$$

5 °F = −15 °C

63. −2 °C

$$F = \frac{9(-2)}{5} + 32 = \frac{-18}{5} + 32 = -3.6 + 32$$
$$= 28.4$$

−2 °C ≈ 28 °F

64. 49 °C

$$F = \frac{9 \cdot 49}{5} + 32 = 88.2 + 32 = 120.2 \approx 120$$

49 °C ≈ 120 °F

65. [8.3] I added 1 <u>L</u> of oil to my car.

66. [8.3] The box of books weighed 15 <u>kg</u>.

67. [8.2] Larry's shoe is 30 <u>cm</u> long.

68. [8.3] Jan used 15 <u>mL</u> of shampoo on her hair.

69. [8.2] My fingernail is 10 <u>mm</u> wide.

70. [8.2] I walked 2 <u>km</u> to school.

71. [8.3] The tiny bird weighed 15 <u>g</u>.

72. [8.2] The new library building is 18 <u>m</u> wide.

73. [8.3] The cookie recipe uses 250 <u>mL</u> of milk.

74. [8.3] Renee's pet mouse weighs 30 <u>g</u>.

75. [8.3] One postage stamp weighs 90 <u>mg</u>.

76. [8.3] I bought 30 <u>L</u> of gas for my car.

77. [8.2] 10.5 cm to millimeters
Count 1 place to the *right* on the metric conversion line.
10.5∧ cm = 105 mm

78. [8.1] 45 min to hours
$$\frac{\cancel{45}^{3} \text{ min}}{1} \cdot \frac{1 \text{ hr}}{\cancel{60}_{4} \text{ min}} = \frac{3}{4} \text{ or } 0.75 \text{ hr}$$

79. [8.1] 90 in. to feet
$$\frac{\cancel{90}^{15} \text{ in.}}{1} \cdot \frac{1 \text{ ft}}{\cancel{12}_{2} \text{ in.}} = \frac{15}{2} \text{ ft} = 7\frac{1}{2} \text{ or } 7.5 \text{ ft}$$

80. [8.2] 1.3 m to centimeters
Count 2 places to the *right* on the metric conversion line.
1.30∧ m = 130 cm

81. [8.5] 25 °C to Fahrenheit
$$F = \frac{9 \cdot \cancel{25}^{5}}{\cancel{5}_{1}} + 32 = 45 + 32 = 77$$
25 °C = 77 °F

82. [8.1] $3\frac{1}{2}$ gal to quarts
$$\frac{3\frac{1}{2} \text{ gal}}{1} \cdot \frac{4 \text{ qt}}{1 \text{ gal}} = \frac{7}{\cancel{2}_{1}} \cdot \frac{\cancel{4}^{2}}{1} \text{ qt} = 14 \text{ qt}$$

83. [8.3] 700 mg to grams
Count 3 places to the *left* on the metric conversion line.
∧700. mg = 0.7 g

84. [8.3] 0.81 L to milliliters
Count 3 places to the *right* on the metric conversion line.
0.810∧ L = 810 mL

85. [8.1] 5 lb to ounces
$$\frac{5 \text{ lb}}{1} \cdot \frac{16 \text{ oz}}{1 \text{ lb}} = 5 \cdot 16 \text{ oz} = 80 \text{ oz}$$

86. [8.3] 60 kg to grams
$$\frac{60 \text{ kg}}{1} \cdot \frac{1000 \text{ g}}{1 \text{ kg}} = 60{,}000 \text{ g}$$

87. [8.3] 1.8 L to milliliters
Count 3 places to the *right* on the metric conversion line.
1.800∧ L = 1800 mL

88. [8.5] 86 °F to Celsius
$$C = \frac{5(86 - 32)}{9} = \frac{5(\cancel{54})^{6}}{\cancel{9}_{1}} = 30$$
86 °F = 30 °C

89. [8.2] 0.36 m to centimeters
Count 2 places to the *right* on the metric conversion line.
0.36∧ m = 36 cm

90. [8.3] 55 mL to liters
Count 3 places to the *left* on the metric conversion line.
∧055. mL = 0.055 L

91. [8.4] Convert centimeters to meters.
2 m 4 cm = 2 + 0.04 m =   2.04 m
78 cm   = − 0.78 m
                          ———
                          1.26 m

The board is 1.26 m long.

92. [8.1] 3000 lb to tons
$$\frac{3000 \text{ lb}}{1} \cdot \frac{1 \text{ T}}{2000 \text{ lb}} = \frac{3}{2} \text{ or } 1.5 \text{ T}$$

Multiply by 12 days.
(1.5 T)(12) = 18 T

18 T of cookies are sold in all.

93. [8.5] Convert ounces to grams.

$$\frac{4 \text{ oz}}{1} \cdot \frac{28.35 \text{ g}}{1 \text{ oz}} \approx 113 \text{ g}$$

350 °F to Celsius

$$C = \frac{5(350-32)}{9} = \frac{5(318)}{9} = \frac{1590}{9} \approx 177 \text{ °C}$$

94. [8.5] Convert kilograms to pounds.

$$\frac{80.9 \text{ kg}}{1} \cdot \frac{2.20 \text{ lb}}{1 \text{ kg}} \approx 178.0 \text{ lb}$$

Convert meters to feet.

$$\frac{1.83 \text{ m}}{1} \cdot \frac{3.28 \text{ ft}}{1 \text{ m}} \approx 6.0 \text{ ft}$$

Jalo weighs about 178.0 lb and is about 6.0 ft.

95. [8.5] 44 ft to meters

$$\frac{44 \text{ ft}}{1} \cdot \frac{0.30 \text{ m}}{1 \text{ ft}} \approx 13.2 \text{ m}$$

96. [8.5] 6.7 m to feet

$$\frac{6.7 \text{ m}}{1} \cdot \frac{3.28 \text{ ft}}{1 \text{ m}} \approx 22.0 \text{ ft}$$

97. [8.5] 4 yd to meters

$$\frac{4 \text{ yd}}{1} \cdot \frac{0.91 \text{ m}}{1 \text{ yd}} \approx 3.6 \text{ m}$$

98. [8.5] 102 cm to inches

$$\frac{102 \text{ cm}}{1} \cdot \frac{0.39 \text{ in.}}{1 \text{ cm}} \approx 39.8 \text{ in.}$$

99. [8.5] 220,000 kg to pounds

$$\frac{220,000 \text{ kg}}{1} \cdot \frac{2.20 \text{ lb}}{1 \text{ kg}} \approx 484,000 \text{ lb}$$

100. [8.5] 5 gal to liters

$$\frac{5 \text{ gal}}{1} \cdot \frac{3.79 \text{ L}}{1 \text{ gal}} = 18.95 \approx 19.0 \text{ L}$$

6 gal to liters

$$\frac{6 \text{ gal}}{1} \cdot \frac{3.79 \text{ L}}{1 \text{ gal}} \approx 22.7 \text{ L}$$

101. [8.1] (a) From Exercise 99:

$$\frac{484,000 \text{ lb}}{1} \cdot \frac{1 \text{ T}}{2000 \text{ lb}} = 242 \text{ T}$$

(b) From Exercise 97:

$$\frac{4 \text{ yd}}{1} \cdot \frac{3 \text{ ft}}{1 \text{ yd}} \cdot \frac{12 \text{ in.}}{1 \text{ ft}} = 144 \text{ in.}$$

102. [8.2] (a) From Exercise 98:

$$\frac{102 \text{ cm}}{1} \cdot \frac{1 \text{ m}}{100 \text{ cm}} = 1.02 \text{ m}$$

(b) From Exercise 96:

$$\frac{6.7 \text{ m}}{1} \cdot \frac{100 \text{ cm}}{1 \text{ m}} = 670 \text{ cm}$$

Chapter 8 Test

1. 9 gal to quarts

$$\frac{9 \text{ gal}}{1} \cdot \frac{4 \text{ qt}}{1 \text{ gal}} = 36 \text{ qt}$$

2. 45 ft to yards

$$\frac{45 \text{ ft}}{1} \cdot \frac{1 \text{ yd}}{3 \text{ ft}} = \frac{45}{3} \text{ yd} = 15 \text{ yd}$$

3. 135 min to hours

$$\frac{\overset{27}{\cancel{135} \text{ min}}}{1} \cdot \frac{1 \text{ hr}}{\underset{12}{\cancel{60} \text{ min}}} = \frac{27}{12} \text{ hr} = 2.25 \text{ hr or } 2\frac{1}{4} \text{ hr}$$

4. 9 in. to feet

$$\frac{\overset{3}{\cancel{9} \text{ in.}}}{1} \cdot \frac{1 \text{ ft}}{\underset{4}{\cancel{12} \text{ in.}}} = \frac{3}{4} \text{ ft or 0.75 ft}$$

5. $3\frac{1}{2}$ lb to ounces

$$\frac{3\frac{1}{2} \text{ lb}}{1} \cdot \frac{16 \text{ oz}}{1 \text{ lb}} = \frac{7}{\underset{1}{\cancel{2}}} \cdot \frac{\overset{8}{\cancel{16}}}{1} \text{ oz} = 56 \text{ oz}$$

6. 5 days to minutes

$$\frac{5 \text{ days}}{1} \cdot \frac{24 \text{ hr}}{1 \text{ day}} \cdot \frac{60 \text{ min}}{1 \text{ hr}} = \frac{5 \cdot 24 \cdot 60}{1} \text{ min} = 7200 \text{ min}$$

7. My husband weighs 75 <u>kg</u>.

8. I hiked 5 <u>km</u> this morning.

9. She bought 125 <u>mL</u> of cough syrup.

10. This apple weighs 180 <u>g</u>.

11. This page is 21 <u>cm</u> wide.

12. My watch band is 10 <u>mm</u> wide.

13. I bought 10 <u>L</u> of soda for the picnic.

14. The bracelet is 16 <u>cm</u> long.

15. 250 cm to meters
Count 2 places to the *left* on the metric conversion line.
2∧50. cm = 2.5 m

16. 4.6 km to meters

$$\frac{4.6 \text{ km}}{1} \cdot \frac{1000 \text{ m}}{1 \text{ km}} = 4600 \text{ m}$$

17. 5 mm to centimeters
Count 1 place to the *left* on the metric conversion line.
∧5. mm = 0.5 cm

18. 325 mg to grams

$$\frac{325 \text{ mg}}{1} \cdot \frac{1 \text{ g}}{1000 \text{ mg}} = \frac{325}{1000} \text{ g} = 0.325 \text{ g}$$

19. 16 L to milliliters
Count 3 places to the *right* on the metric conversion line.
16.000∧ L = 16,000 mL

20. 0.4 kg to grams

$$\frac{0.4 \text{ kg}}{1} \cdot \frac{1000 \text{ g}}{1 \text{ kg}} = 400 \text{ g}$$

21. 10.55 m to centimeters
Count 2 places to the *right* on the metric conversion line.
10.55∧ m = 1055 cm

22. 95 mL to liters
Count 3 places to the *left* on the metric conversion line.
∧095. mL = 0.095 L

23. Convert 460 in. to feet.

$$\frac{460 \text{ in}}{1} \cdot \frac{1 \text{ ft}}{12 \text{ in}} = \frac{460}{12} \text{ ft} = \frac{115}{3} \text{ ft}$$

Divide by 12 months

$$\frac{\frac{115}{3} \text{ ft}}{12 \text{ months}} = \frac{115}{3} \cdot \frac{1}{12} \approx 3.2 \text{ ft/month}$$

It rains about 3.2 ft per month.

24. **(a)** Convert 2.9 g to mg.

$$\frac{2.9 \text{ g}}{1} \cdot \frac{1000 \text{ mg}}{1 \text{ g}} = 2900 \text{ mg}$$

Find the difference.
2900 mg − 590 mg = 2310 mg
It has 2310 mg more sodium.

(b) 2900 mg is more than 2400 mg
2900 mg − 2400 mg = 500 mg

$$\frac{500 \text{ mg}}{1} \cdot \frac{1 \text{ g}}{1000 \text{ mg}} = 0.5 \text{ g}$$

The "Super Melt" has 500 mg or 0.5 g more sodium than the 2400 mg recommended daily amount.

25. The water is almost boiling.
On the Celsius scale, water boils at 100°, so 95 °C would be almost boiling.

26. The tomato plants may freeze tonight. Water freezes at 0 °C, so tomatoes would also likely freeze at 0 °C, so choose 0 °C.

27. 6 ft to meters

$$\frac{6 \text{ ft}}{1} \cdot \frac{0.30 \text{ m}}{1 \text{ ft}} \approx 1.8 \text{ m}$$

28. 125 lb to kilograms

$$\frac{125 \text{ lb}}{1} \cdot \frac{0.45 \text{ kg}}{1 \text{ lb}} \approx 56.3 \text{ kg}$$

29. 50 L to gallons

$$\frac{50 \text{ L}}{1} \cdot \frac{0.26 \text{ gal}}{1 \text{ L}} \approx 13 \text{ gal}$$

30. 8.1 km to miles

$$\frac{8.1 \text{ km}}{1} \cdot \frac{0.62 \text{ mi}}{1 \text{ km}} \approx 5.0 \text{ mi}$$

31. 74 °F to Celsius

$$C = \frac{5(74 - 32)}{9} = \frac{5(\overset{14}{\cancel{42}})}{\cancel{9}_3} = \frac{70}{3} \approx 23 \text{ °C}$$

74 °F ≈ 23 °C

32. −12 °C to Fahrenheit

$$F = \frac{9(-12)}{5} + 32 = \frac{-108}{5} + 32$$
$$= -21.6 + 32 = 10.4 \approx 10°$$

−12 °C ≈ 10 °F

274 Chapter 8 Measurement

33. Convert 120 cm to meters.

    Count 2 places to the *left* on the metric conversion line.
    1⌒20. cm = 1.2 m

    Multiply to find the number of meters for 5 pillows.
    $5(1.2 \text{ m}) = 6 \text{ m}$

    Convert 6 m to yards.
    $$\frac{6 \text{ m}}{1} \cdot \frac{1.09 \text{ yd}}{1 \text{ m}} \approx 6.54 \text{ yd}$$

    Multiply by $3.98 to find the cost.
    $$\frac{6.54 \text{ yd}}{1} \cdot \frac{\$3.98}{\text{yd}} \approx \$26.03$$

    It will cost about $26.03.

34. Possible answers: Use same system as rest of the world; easier system for children to learn; less use of fractional numbers; compete internationally.

Cumulative Review Exercises (Chapters 1–8)

1. $\dfrac{7}{6x^2} \cdot \dfrac{9x}{14y} = \dfrac{7 \cdot \overset{1}{\cancel{3}} \cdot 3 \cdot \overset{1}{\cancel{x}}}{2 \cdot \underset{1}{\cancel{3}} \cdot \underset{1}{\cancel{x}} \cdot x \cdot 2 \cdot \underset{1}{\cancel{7}} \cdot y} = \dfrac{3}{4xy}$

2. $\dfrac{3}{c} - \dfrac{5}{6} = \dfrac{3 \cdot 6}{c \cdot 6} - \dfrac{5 \cdot c}{6 \cdot c} = \dfrac{18}{6c} - \dfrac{5c}{6c} = \dfrac{18 - 5c}{6c}$

3. $\dfrac{-3(-4)}{27 - 3^3} = \dfrac{12}{27 - 27} = \dfrac{12}{0}$, which is undefined.

4. $3\dfrac{7}{12} - 4 = \dfrac{43}{12} - \dfrac{4}{1} = \dfrac{43}{12} - \dfrac{48}{12}$
   $= \dfrac{43 - 48}{12} = \dfrac{-5}{12} = -\dfrac{5}{12}$

5. $\dfrac{9y^2}{8x} \div \dfrac{y}{6x^2} = \dfrac{9y^2}{8x} \cdot \dfrac{6x^2}{y}$
   $= \dfrac{3 \cdot 3 \cdot \overset{1}{\cancel{y}} \cdot y \cdot \overset{1}{\cancel{2}} \cdot 3 \cdot \overset{1}{\cancel{x}} \cdot x}{\underset{1}{\cancel{2}} \cdot 4 \cdot \underset{1}{\cancel{x}} \cdot \underset{1}{\cancel{y}}}$
   $= \dfrac{27xy}{4}$

6. $\dfrac{2}{3} + \dfrac{n}{m} = \dfrac{2 \cdot m}{3 \cdot m} + \dfrac{n \cdot 3}{m \cdot 3} = \dfrac{2m}{3m} + \dfrac{3n}{3m}$
   $= \dfrac{2m + 3n}{3m}$

7. $1\dfrac{1}{6} + 1\dfrac{2}{3} = \dfrac{7}{6} + \dfrac{5}{3} = \dfrac{7}{6} + \dfrac{10}{6} = \dfrac{7 + 10}{6}$
   $= \dfrac{17}{6}$ or $2\dfrac{5}{6}$

8. $\dfrac{\frac{14}{15}}{-6} = \dfrac{14}{15} \div (-6) = \dfrac{14}{15} \div \left(-\dfrac{6}{1}\right)$
   $= \dfrac{14}{15} \cdot \left(-\dfrac{1}{6}\right)$
   $= -\dfrac{\overset{1}{\cancel{2}} \cdot 7 \cdot 1}{15 \cdot \underset{1}{\cancel{2}} \cdot 3}$
   $= -\dfrac{7}{45}$

9. $\dfrac{12 \div (2 - 5) + 12(-1)}{2^3 - (-4)^2}$

    Numerator: $12 \div (2 - 5) + 12(-1)$
    $= 12 \div (2 + (-5)) + 12(-1)$
    $= 12 \div (-3) + 12(-1)$
    $= -4 + (-12)$
    $= -16$

    Denominator: $2^3 - (-4)^2$
    $= 8 - 16$
    $= 8 + (-16)$
    $= -8$

    Last step is division: $\dfrac{-16}{-8} = 2$

10. $(-0.8)^2 \div (0.8 - 1)$
    $= (-0.8) \cdot (-0.8) \div (0.8 + (-1))$
    $= (-0.8) \cdot (-0.8) \div (-0.2)$
    $= 0.64 \div (-0.2)$
    $= -3.2$

    Last division step:

    ```
           3 . 2
    0.2⌒⟌0.6⌒4
           6
           0 4
             4
             0
    ```

11. $\left(-\dfrac{1}{3}\right)^2 - \dfrac{1}{4}\left(\dfrac{4}{9}\right)$
    $= \left(-\dfrac{1}{3}\right)\left(-\dfrac{1}{3}\right) - \dfrac{1 \cdot \overset{1}{\cancel{4}}}{\underset{1}{\cancel{4}} \cdot 9}$
    $= \dfrac{1}{9} - \dfrac{1}{9}$
    $= 0$

12.
$$-5n = n - 12$$
$$-5n = 1n - 12$$
$$\underline{-1n \quad\quad -1n}$$
$$-6n = 0 - 12$$
$$\frac{-6n}{-6} = \frac{-12}{-6}$$
$$n = 2$$

The solution is 2.

13.
$$\frac{1.5}{45} = \frac{x}{12}$$
$$45 \cdot x = 1.5 \cdot 12$$
$$\frac{45x}{45} = \frac{18}{45}$$
$$x = 0.4$$

The solution is 0.4.

14.
$$2 = \frac{1}{4}w - 3$$
$$\underline{+3 \quad\quad +3}$$
$$5 = \frac{1}{4}w + 0$$
$$\frac{4}{1}(5) = \frac{4}{1}\left(\frac{1}{4}w\right)$$
$$20 = w$$

The solution is 20.

15.
$$4y - 3 = 7y + 12$$
$$\underline{-4y \quad\quad\quad -4y}$$
$$0 - 3 = 3y + 12$$
$$-3 = 3y + 12$$
$$\underline{-12 \quad\quad -12}$$
$$-15 = 3y + 0$$
$$\frac{-15}{3} = \frac{3y}{3}$$
$$-5 = y$$

The solution is −5.

16. Let n represent the unknown number.
$$2n - 8 = n + 7$$
$$2n - 8 = 1n + 7$$
$$\underline{-1n \quad\quad -1n}$$
$$n - 8 = 0 + 7$$
$$n - 8 = 7$$
$$\underline{+8 \quad\quad +8}$$
$$n + 0 = 15$$
$$n = 15$$

The number is 15.

17. *Step 1*
Unknown: length of each piece
Known: total length is 90 ft; one piece is 6 ft shorter

Step 2
Let p be the length of the longer piece. Then $p - 6$ is the length of the shorter piece.

Step 3
$p + (p - 6) = 90$

Step 4
$$2p - 6 = 90$$
$$\underline{+6 \quad\quad +6}$$
$$\frac{2p}{2} = \frac{96}{2}$$
$$p = 48 \text{ ft}$$

Step 5
The long piece is 48 ft, and the short piece is $48 - 6 = 42$ ft.

Step 6
Check: 42 ft is 6 feet shorter than 48 ft and 42 ft + 48 ft = 90 ft.

18.
$$P = 4s \qquad A = s^2$$
$$P = 4 \cdot 2\frac{1}{4} \text{ ft} \qquad A = s \cdot s$$
$$P = \frac{4}{1} \cdot \frac{9}{4} \text{ ft} \qquad A = 2\frac{1}{4} \text{ ft} \cdot 2\frac{1}{4} \text{ ft}$$
$$P = \frac{\cancel{4}}{1} \cdot \frac{9}{\cancel{4}} \text{ ft} \qquad A = \frac{9}{4} \text{ ft} \cdot \frac{9}{4} \text{ ft}$$
$$P = 9 \text{ ft} \qquad A = \frac{81}{16} \text{ ft}^2 = 5\frac{1}{16} \text{ ft}^2$$
$$\qquad\qquad A \approx 5.1 \text{ ft}^2$$

19.
$$C = \pi \cdot d$$
$$C \approx 3.14 \cdot 9 \text{ mm}$$
$$C = 28.26 \text{ mm}$$
$$C \approx 28.3 \text{ mm}$$
$$r = \frac{9 \text{ mm}}{2} = 4.5 \text{ mm}$$
$$A = \pi r^2$$
$$A = \pi \cdot r \cdot r$$
$$A \approx 3.14 \cdot 4.5 \text{ mm} \cdot 4.5 \text{ mm}$$
$$A = 63.585 \text{ mm}^2$$
$$A \approx 63.6 \text{ mm}^2$$

20. The triangle is a right triangle.
$$\text{Leg} = \sqrt{(\text{hypotenuse})^2 - (\text{leg})^2}$$
$$y = \sqrt{(20)^2 - (16)^2}$$
$$y = \sqrt{400 - 256}$$
$$y = \sqrt{144}$$
$$y = 12 \text{ yd}$$

$$A = \tfrac{1}{2}bh$$
$$A = \tfrac{1}{2}(16 \text{ yd})(12 \text{ yd})$$
$$A = 96 \text{ yd}^2$$

276 Chapter 8 Measurement

21. $V = l \cdot w \cdot h$
$V = 6 \text{ cm} \cdot 4 \text{ cm} \cdot 4 \text{ cm}$
$V = 96 \text{ cm}^3$

$SA = 2lw + 2lh + 2wh$
$SA = 2 \cdot 6 \text{ cm} \cdot 4 \text{ cm} + 2 \cdot 6 \text{ cm} \cdot 4 \text{ cm}$
$\qquad + 2 \cdot 4 \text{ cm} \cdot 4 \text{ cm}$
$SA = 48 \text{ cm}^2 + 48 \text{ cm}^2 + 32 \text{ cm}^2$
$SA = 128 \text{ cm}^2$

22. $4\frac{1}{2}$ ft to inches

$\dfrac{9 \text{ ft}}{2} \cdot \dfrac{12 \text{ in.}}{1 \text{ ft}} = \dfrac{9 \cancel{\text{ft}}}{\cancel{2}_{1}} \cdot \dfrac{\overset{6}{\cancel{12}} \text{ in.}}{1 \cancel{\text{ft}}} = 54 \text{ in.}$

23. 72 hours to days

$\dfrac{72 \text{ hr}}{1} \cdot \dfrac{1 \text{ day}}{24 \text{ hr}} = \dfrac{\overset{3}{\cancel{72}} \cancel{\text{hr}}}{1} \cdot \dfrac{1 \text{ day}}{\cancel{24} \cancel{\text{hr}}_{1}} = 3 \text{ days}$

24. 3.7 kg to grams

From kg to g is 3 places to the right.

3.7 kg = 3.700 kg = 3700 g

25. 60 cm to meters

From cm to m is 2 places to the left.

60 cm = 060. cm = 0.60 m or 0.6 m

26. Part: 35 credits earned
Whole: 60 credits needed
Percent: unknown (p)

Percent $\cdot$ Whole = Part
$p \cdot 60 = 35$
$\dfrac{60p}{60} = \dfrac{35}{60}$
$p = 0.58\overline{3}$

She has 58% (rounded) of the necessary credits.

27. $15\frac{1}{2}$ ounces of Brand T for
$\$2.99 - \$0.30 \text{ (coupon)} = \$2.69$

$\dfrac{\$2.69}{15\frac{1}{2} \text{ ounces}} \approx \$0.174 \text{ per ounce } (*)$

14 ounces of Brand F for $2.49

$\dfrac{\$2.49}{14 \text{ ounces}} \approx \$0.178 \text{ per ounce}$

18 ounces of Brand H for
$\$3.89 - \$0.40 \text{ (coupon)} = \$3.49$

$\dfrac{\$3.49}{18 \text{ ounces}} \approx \$0.194 \text{ per ounce}$

The best buy is Brand T at 15.5 ounces for $2.99 with a $0.30 coupon.

28. **(a)** Set up a proportion.

$\dfrac{1 \text{ cm}}{12 \text{ km}} = \dfrac{7.8 \text{ cm}}{x}$
$x \cdot 1 = (7.8)(12)$
$x = 93.6$

The distance is 93.6 kilometers.

(b) $\dfrac{93.6 \cancel{\text{km}}}{1} \cdot \dfrac{0.62 \text{ mi}}{1 \cancel{\text{km}}} \approx 58.0 \text{ mi}$

The distance is 58.0 mi (rounded).

29. Find the total grams, then convert grams to kilograms.
$450 \text{ g} + 48 \cdot 115 \text{ g} = 450 \text{ g} + 5520 \text{ g} = 5970 \text{ g}$

$\dfrac{5970 \cancel{\text{g}}}{1} \cdot \dfrac{1 \text{ kg}}{1000 \cancel{\text{g}}} = 5.97 \text{ kg}$

The carton would weigh 5.97 kg.

30. First add the amount of canvas material that was used.

$\begin{aligned} 1\tfrac{2}{3} &= 1\tfrac{8}{12} \\ +\,1\tfrac{3}{4} &= 1\tfrac{9}{12} \\ \hline &\,2\tfrac{17}{12} = 3\tfrac{5}{12} \text{ yd} \end{aligned}$

Subtract $3\frac{5}{12}$ from the amount of canvas material that Steven bought.

$\begin{aligned} 4\tfrac{1}{2} &= 4\tfrac{6}{12} \\ -\,3\tfrac{5}{12} &= 3\tfrac{5}{12} \\ \hline &\,1\tfrac{1}{12} \text{ yd} \end{aligned}$

There will be $1\frac{1}{12}$ yd left.

31. Convert 650 g to kg.
$_\wedge 650. \text{ g} = 0.650 \text{ kg}$

Multiply to find the cost.
$(0.650)(\$14.98) \approx \$9.74$

Mark paid $9.74 (rounded).

32. amount of discount = rate of discount $\cdot$ cost of item
$\qquad\qquad\qquad = (0.10)(\$189.94)$
$\qquad\qquad\qquad \approx \$18.99$

The amount of discount is $18.99.
The sale price is $\$189.94 - \$18.99 = \$170.95$.

33. Set up a proportion.

$$\frac{6 \text{ rows}}{5 \text{ cm}} = \frac{x}{100 \text{ cm}}$$
$$5 \cdot x = 6 \cdot 100$$
$$\frac{5 \cdot x}{5} = \frac{600}{5}$$
$$x = 120$$

Akuba will knit 120 rows.

34. $3500 at $7\frac{1}{2}$% for 6 months

$$I = p \cdot r \cdot t \qquad 7.5\% = 0.075$$
$$= (3500)(0.075)\left(\frac{6}{12}\right)$$
$$= 131.25$$

The interest is $131.25.

amount due = principal + interest
$$= \$3500 + \$131.25$$
$$= \$3631.25$$

The total amount due is $3631.25.

35. How many $17 blankets can be purchased from a $400 donation? Use division.

```
      2 3
  17 ) 4 0 0
      3 4
      ─────
        6 0
        5 1
        ───
          9
```

23 blankets can be purchased. $9 will be left over.

36. $\dfrac{\$6}{40 \text{ applications}} = \$0.15/\text{application}$

The cost per application of the spray-on bandage is $0.15.

37. decrease = 8 − 5 = 3 days

$$\frac{3 \text{ days}}{8 \text{ days}} = \frac{x}{100}$$
$$8 \cdot x = 3 \cdot 100$$
$$\frac{8 \cdot x}{8} = \frac{300}{8}$$
$$x = 37.5$$

The length of a hospital stay has decreased by 37.5%.

CHAPTER 9 GRAPHS

9.1 Problem Solving with Tables and Pictographs

9.1 Margin Exercises

1. **(a)** Look across the row labeled Continental to see that 74% of its flights were on time.

   (b) Look down the column headed On-Time Performance.

   To find the best performance, look for the highest percent. The highest percent is 80%. Then look to the left to find the name of the airline: Southwest.

   (c) Look down the column headed Luggage Handling.

   To find the best record, look for the lowest number. The lowest number is 4.1. Then look to the left to find the airline, which is Airtran.

   (d) Look down the column headed Luggage Handling and find all the numbers less than or equal to 5. Then look to the left to find the airlines, which are Airtran and Northwest.

   (e) Add the values and divide by 8.

   $$\frac{4.1 + 7.3 + 5.3 + 7.6 + 5.0 + 5.9 + 5.8 + 8.5}{8}$$
   $$= \frac{49.5}{8} \approx 6.2$$

   The average number of luggage problems was 6.2 (rounded) per 1000 passengers.

2. **(a)** The table shows that the price per mile in Chicago is $1.80, so the cost for 6.5 miles is $6.5(\$1.80) = \$11.70$. Then add the flag drop charge: $\$11.70 + \$2.25 = \$13.95$.

   The table shows that the price per mile in New York is $2.00, so the cost for 6.5 miles is $6.5(\$2.00) = \$13.00$. Then add the flag drop charge: $\$13.00 + \$2.50 = \$15.50$.

   The difference in the fare is $\$15.50 - \$13.95 = \$1.55$.

   The New York fare is $1.55 higher.

   (b) The table shows that the price per mile in Miami is $2.40, so the cost for 12 miles is $12(\$2.40) = \$28.80$. Then, add the flag drop charge of $2.50. Finally, figure out the cost of the wait time.

   $$\frac{\$24}{60 \text{ min}} = \frac{\$x}{10 \text{ min}}$$
   $$60 \cdot x = 24 \cdot 10$$
   $$\frac{60x}{60} = \frac{240}{60}$$
   $$x = 4.00$$

   Total fare $= \$28.80 + \$2.50 + \$4.00 = \$35.30$.

   (c) Price per mile $= \$2.00$
   Price for 4.5 miles $= 4.5(\$2.00) = \$9.00$
   Figure out the cost of the wait time.

   $$\frac{\$22.50}{60 \text{ min}} = \frac{\$x}{30 \text{ min}}$$
   $$60 \cdot x = 22.50 \cdot 30$$
   $$\frac{60x}{60} = \frac{675}{60}$$
   $$x = 11.25$$

   Now add the flag drop charge of $1.60.

   Total fare $= \$9.00 + \$11.25 + \$1.60 = \$21.85$

   Use the percent equation to find the tip.

   $$\text{percent} \cdot \text{whole} = \text{part}$$
   $$(0.15)(\$21.85) = n$$
   $$\$3.28 = n$$

   Rounded to the nearest $0.25, the tip is $3.25. The total cost is $\$21.85 + \$3.25 = \$25.10$.

3. **(a)** The population of Chicago is represented by 5 whole symbols:

   $5 \cdot 2$ million $= 10$ million people (or 10,000,000)

   (b) The population of Atlanta is represented by 2 whole symbols ($2 \cdot 2$ million $= 4$ million) plus half of a symbol

   ($\frac{1}{2}$ of 2 million is 1 million) for a total of 5 million people (or 5,000,000).

   (c) Los Angeles has $6\frac{1}{2}$ symbols and New York has $9\frac{1}{2}$ symbols, so New York has 3 more symbols than Los Angeles. This is $3 \cdot 2$ million $= 6$ million people, so New York's population is 6 million or 6,000,000 greater than Los Angeles' population.

   (d) Dallas has 3 symbols and Atlanta has $2\frac{1}{2}$ symbols, so Dallas has $\frac{1}{2}$ more symbol than Atlanta. This is $\frac{1}{2} \cdot 2$ million $= 1$ million people, so Dallas' population is 1 million or 1,000,000 greater than Atlanta's population.

9.1 Section Exercises

1. **(a)** Look in the points column for Wilt Chamberlain. He scored 31,419 points.

   (b) Look down the points column and find the number(s) greater than 31,419, then read across to the player's name.

   Since 32,292 is greater than 31,419, Michael Jordan scored more points than Wilt Chamberlain.

3. **(a)** Look down the games column and find the largest number, which is 1072. Then read across to find the player's name.

   Michael Jordan has been in the greatest number of games.

   (b) The smallest number in the games column is 747.

   Allen Iverson has been in the fewest number of games.

5. Look down the points column and find the largest and smallest number. Then find the difference.

   $$32{,}292 - 20{,}824 = 11{,}468$$

   The difference is 11,468 points.

7. Round answers to the nearest tenth to match other numbers in column.

   Jerry West: $\dfrac{25{,}192}{932} \approx 27.0$

   Bob Pettit: $\dfrac{20{,}880}{792} \approx 26.4$

   Shaquille O'Neal: $\dfrac{25{,}454}{981} \approx 25.9$

9. The asterisks next to O'Neal's and Iverson's names means that they are still actively playing in the NBA.

11. **(a)** Look down the 140-pounds column and across the aerobic dance activity row.

    The person will burn 255 calories.

    (b) Look down the 140-pounds column for the largest number. Then read across to the activity. Moderate jogging burns the most calories.

13. **(a)** Look down the 110-pounds column for numbers greater than or equal to 200. Then read across to the activities. Moderate jogging, aerobic dance, and racquetball are activities that burn at least 200 calories.

    (b) Use the same method as in part (a), except use the 170-pound column. Moderate jogging, moderate bicycling, aerobic dance, racquetball, and tennis are activities that burn at least 200 calories.

15. 15 minutes is $\frac{1}{2}$ of 30 minutes.
    60 minutes is 2 times 30 minutes.
    Look down the 140-pound column.

    $$\tfrac{1}{2} \cdot 180 + 2 \cdot 140 = 90 + 280 = 370$$

    The person will burn 370 calories.

For Exercises 17 and 19, other proportions are possible.

17. **(a)**
    $$\dfrac{322 \text{ calories}}{110 \text{ pounds}} = \dfrac{x \text{ calories}}{125 \text{ pounds}}$$
    $$110 \cdot x = 322 \cdot 125$$
    $$\dfrac{110x}{110} = \dfrac{40{,}250}{110}$$
    $$x \approx 366$$

    The person would burn approximately 366 calories.

    (b)
    $$\dfrac{210 \text{ calories}}{110 \text{ pounds}} = \dfrac{x \text{ calories}}{125 \text{ pounds}}$$
    $$110 \cdot x = 210 \cdot 125$$
    $$\dfrac{110x}{110} = \dfrac{26{,}250}{110}$$
    $$x \approx 239$$

    The person would burn approximately 239 calories.

19. Aerobic dance: $\dfrac{255 \text{ calories}}{140 \text{ pounds}} = \dfrac{x \text{ calories}}{158 \text{ pounds}}$
    $$140 \cdot x = 255 \cdot 158$$
    $$\dfrac{140x}{140} = \dfrac{40{,}290}{140}$$
    $$x \approx 288$$

    15 min is $\frac{1}{2}$ of 30 min: $\frac{1}{2} \cdot 288 \approx 144$

    Walking: $\dfrac{140 \text{ calories}}{140 \text{ pounds}} = \dfrac{x \text{ calories}}{158 \text{ pounds}}$
    $$140 \cdot x = 140 \cdot 158$$
    $$x = 158$$

    20 min is $\frac{2}{3}$ of 30 min: $\frac{2}{3} \cdot 158 \approx 105$

    The difference is $144 - 105 = 39$ calories.

For Exercises 21–28, each symbol represents 10 million passenger arrivals and departures.

21. **(a)** Chicago is represented by 8 symbols.

    $8(10 \text{ million}) = 80$ million or 80,000,000

    (b) San Francisco is represented by $3\frac{1}{2}$ symbols (or 3.5).

    $3.5(10 \text{ million}) = 35$ million or 35,000,000

23. Atlanta: $8.5(10 \text{ million}) = 85$ million

    Chicago: $8(10 \text{ million}) = 80$ million

    The total number is
    85 million + 80 million = 165 million or 165,000,000.

25. Chicago: $8(10 \text{ million}) = 80$ million

    Atlanta: $8.5(10 \text{ million}) = 85$ million

    The difference is $85 - 80 = 5$ million or 5,000,000.

27. There are $8.5 + 8 + 6 + 4.5 + 3.5 = 30.5$ symbols.

    $30.5(10 \text{ million}) = 305$ million

    The total number is 305 million or 305,000,000.

29. Answers will vary. One possibility: choose Southwest because it has the best on-time performance.

30. Answers will vary. Possibilities include planning more time between each flight, or doing some or all of your business via conference calls or e-mail.

31. Answers will vary. One possibility: choose Airtran because it has the fewest luggage problems.

32. Answers will vary. Possibilities include buying heavy-duty luggage or shipping the golf clubs via a delivery service.

33. Answers will vary. Possibilities include a lot of bad weather, maintenance problems, new computer system. There's nothing they can do about the weather. They could add maintenance staff to fix the problems and add technology staff to fix the computer bugs.

34. Answers will vary. Possibilities include availability of nonstop flights, convenience of departure times, type and size of aircraft, availability of low-cost fares.

9.2 Reading and Constructing Circle Graphs

9.2 Margin Exercises

1. **(a)** The circle graph shows that the greatest number of hours is spent sleeping.

   (b) 6 hours working
         -4 hours studying
         2 hours

   Two more hours are spent working than studying.

   (c) 4 hours studying
         6 hours working
         $+3$ hours attending class
         13 hours

   Thirteen hours are spent studying, working, and attending classes.

2. **(a)** $\dfrac{2 \text{ hours (driving)}}{24 \text{ hours (whole day)}} = \dfrac{2 \text{ hours}}{24 \text{ hours}} = \dfrac{\cancel{2}}{\cancel{2} \cdot 12} = \dfrac{1}{12}$

   (b) $\dfrac{4 \text{ hours (studying)}}{24 \text{ hours (whole day)}} = \dfrac{4 \text{ hours}}{24 \text{ hours}} = \dfrac{\cancel{4}}{\cancel{4} \cdot 6} = \dfrac{1}{6}$

   (c) $\dfrac{7 \text{ hours (sleeping)} + 2 \text{ hours (other)}}{24 \text{ hours (whole day)}}$

   $= \dfrac{9 \text{ hours}}{24 \text{ hours}} = \dfrac{\cancel{3} \cdot 3}{\cancel{3} \cdot 8} = \dfrac{3}{8}$

3. **(a)** $\dfrac{4 \text{ hours (studying)}}{6 \text{ hours (working)}} = \dfrac{4 \text{ hours}}{6 \text{ hours}} = \dfrac{\cancel{2} \cdot 2}{\cancel{2} \cdot 3} = \dfrac{2}{3}$

   (b) $\dfrac{6 \text{ hours (working)}}{7 \text{ hours (sleeping)}} = \dfrac{6 \text{ hours}}{7 \text{ hours}} = \dfrac{6}{7}$

   (c) $\dfrac{4 \text{ hours (studying)}}{2 \text{ hours (driving)}} = \dfrac{4 \text{ hours}}{2 \text{ hours}} = \dfrac{\cancel{2} \cdot 2}{\cancel{2} \cdot 1} = \dfrac{2}{1}$

4. **(a)** The percent for Kraft is 37%.

   percent · whole = part
   $37\% \cdot \$2 \text{ billion} = n$
   $(0.37)(\$2 \text{ billion}) = n$
   $\$0.74 \text{ billion} = n$

   The sales for Kraft are $740,000,000.

   (b) The percent for Van De Kamps is 4%.

   $(0.04)(\$2 \text{ billion}) = n$
   $\$0.08 \text{ billion} = n$

   The sales for Van De Kamps is $80,000,000.

(c) The percent for Tony's Pizza Service is 30%.

$$(0.30)(\$2 \text{ billion}) = n$$
$$\$0.6 \text{ billion} = n$$

The sales for Tony's Pizza Service is $600,000,000.

(d) The percent for Pillsbury Corp. is 9%.

$$(0.09)(\$2 \text{ billion}) = n$$
$$\$0.18 \text{ billion} = n$$

The sales for Pillsbury Corp. is $180,000,000.

5. (a) Ages 10–11:

$$(360°)(25\%) = (360°)(0.25) = 90°$$

(b) Ages 12–13:

$$(360°)(25\%) = (360°)(0.25) = 90°$$

(c) Ages 14–15:

$$(360°)(15\%) = (360°)(0.15) = 54°$$

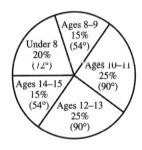

9.2 Section Exercises

1. (a) The total budget for adding the family room is $18,000 + $15,000 + $3000 + $1200 + $2400 + $7200 + $1200 = $48,000

   (b) The greatest single expense in adding the family room is carpentry at $18,000.

3. (a) $\dfrac{\text{amount for carpentry}}{\text{total remodeling budget}}$

   $= \dfrac{\$18,000}{\$48,000} = \dfrac{18,000 \div 6000}{48,000 \div 6000} = \dfrac{3}{8}$

   (b) $\dfrac{\text{amount for materials}}{\text{amount for electrical}}$

   $= \dfrac{\$15,000}{\$3000} = \dfrac{15,000 \div 3000}{3000 \div 3000} = \dfrac{5}{1}$

5. (a) $\dfrac{\text{floor covering} + \text{painting} + \text{window cov.}}{\text{total remodeling budget}}$

   $= \dfrac{\$1200 + \$2400 + \$1200}{\$48,000}$

   $= \dfrac{\$4800}{\$48,000} = \dfrac{4800 \div 4800}{48,000 \div 4800} = \dfrac{1}{10}$

(b) $\dfrac{\text{actual total}}{\text{budget total}}$

$= \dfrac{\$57,600}{\$48,000} = \dfrac{57,600 \div 9600}{48,000 \div 9600} = \dfrac{6}{5}$

7. (a) The least number of people and the smallest sector represent the reason "Don't know."

   (b) Find the second to the smallest sector.

   "Quicker" was the reason given by the second-fewest number of people.

9. Total people $= 1740 + 1200 + 1140 + 180 + 1020 + 720 = 6000$

   "Quicker" to total:

   $$\dfrac{720}{6000} = \dfrac{720 \div 240}{6000 \div 240} = \dfrac{3}{25}$$

11. "Less work/No clean up" to "Atmosphere":

    $$\dfrac{1020}{1200} = \dfrac{1020 \div 60}{1200 \div 60} = \dfrac{17}{20}$$

13. "Wanted food they couldn't cook at home" to "Don't know":

    $$\dfrac{1740}{180} = \dfrac{1740 \div 60}{180 \div 60} = \dfrac{29}{3}$$

15. The percent for "Mingles with the guests" is 35%.

    $$\text{percent} \cdot \text{whole} = \text{part}$$
    $$(0.35)(400) = n$$
    $$140 = n$$

    140 people said their cat mingles with the guests.

17. The response "Raids the buffet table" was given least often at 4%.

    $$(0.04)(400) = n$$
    $$16 = n$$

    16 people said their cat raids the buffet table.

19. Watches from a safe perch: $0.19(400) = 76$
    Runs and hides: $0.35(400) = 140$

    Difference: $140 - 76 = 64$

    64 fewer people said "watches from a safe perch" than said "runs and hides."

21. To determine how many people prefer onions for their hot dog topping, use the percent equation.

    $$\text{percent} \cdot \text{whole} = \text{part}$$
    $$(0.05)(3200) = n$$
    $$160 = n$$

    160 people favored onions.

23. The most popular topping was mustard (30%).

$$(0.30)(3200) = n$$
$$960 = n$$

960 people favored mustard.

25. 12% of 3200 = chili
$$(0.12)(3200) = n$$
$$384 = n$$

10% of 3200 = relish
$$(0.10)(3200) = n$$
$$320 = n$$

There were $384 - 320 = 64$ more people who chose chili than relish.

27. First, find the percent of the total that is represented by each item. Next, multiply the percent by 360° to find the size of each sector. Finally, use a protractor to draw each sector.

29. (a) 25% of total is rent.

Degrees of a circle = 25% of 360°
$$= (0.25)(360)$$
$$= 90°$$

(b) Percent for food $= \dfrac{72°}{360°}$
$$= 0.20 = 20\%$$

(c) Percent for clothing $= \dfrac{\$1092}{\$10{,}920}$
$$= 0.10 = 10\%$$

Degrees of a circle = 10% of 360°
$$= (0.10)(360°)$$
$$= 36°$$

(d) Since the dollar amount, $1092, is the same as part (c), the percent of total is 10% and the number of degrees in the circle is 36°.

(e) Percent for tuition and fees $= \dfrac{\$1638}{\$10{,}920}$
$$= 0.15 = 15\%$$

Degrees of a circle = 15% of 360°
$$= (0.15)(360°)$$
$$= 54°$$

(f) Percent for savings $= \dfrac{\$546}{\$10{,}920}$
$$= 0.05 = 5\%$$

Degrees of a circle = 5% of 360°
$$= (0.05)(360°)$$
$$= 18°$$

(g) Since the dollar amount, $1638, is the same as part (e), the percent of total is 15% and the number of degrees in the circle is 54°.

(h)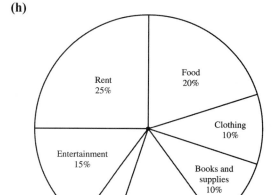

31. (a) Total sales = $12,500 + $40,000
+ $60,000 + $50,000 + $37,500 = $200,000

(b) Adventure classes = $12,500

percent of total $= \dfrac{12{,}500}{200{,}000} = 0.0625 = 6.25\%$

Grocery/provision sales = $40,000

percent of total $= \dfrac{40{,}000}{200{,}000} = 0.2 = 20\%$

Equipment rentals = $60,000

percent of total $= \dfrac{60{,}000}{200{,}000} = 0.3 = 30\%$

Rafting tours = $50,000

percent of total $= \dfrac{50{,}000}{200{,}000} = 0.25 = 25\%$

Equipment sales = $37,500

percent of total $= \dfrac{37{,}500}{200{,}000} = 0.1875 = 18.75\%$

(c) **Adventure classes**:

number of degrees $= (0.0625)(360°) = 22.5°$

Grocery/provision sales:

number of degrees $= (0.2)(360°) = 72°$

Equipment rentals:

number of degrees $= (0.3)(360°) = 108°$

Rafting tours:

number of degrees $= (0.25)(360°) = 90°$

Equipment sales:

number of degrees $= (0.1875)(360°) = 67.5°$

(d)

[Pie chart: Equipment rentals 30%, Rafting tours 25%, Equipment sales 18.75%, Grocery/provision sales 20%, Adventure classes 6.25%]

9.3 Bar Graphs and Line Graphs

9.3 Margin Exercises

1. **(a)** The bar for 2004 rises to 25, showing that the number of college graduates was
$$25 \times 1000 = 25,000.$$
   (b) 2005: $20 \times 1000 = 20,000$ graduates
   (c) 2007: $30 \times 1000 = 30,000$ graduates
   (d) 2008: $35 \times 1000 = 35,000$ graduates

2. **(a)** 1st quarter, 2007:
$$4 \times 1000 = 4000 \text{ installations}$$
   1st quarter, 2008:
$$3 \times 1000 = 3000 \text{ installations}$$
   (b) 3rd quarter, 2007:
$$7 \times 1000 = 7000 \text{ installations}$$
   3rd quarter, 2008:
$$8 \times 1000 = 8000 \text{ installations}$$
   (c) 4th quarter, 2007:
$$5 \times 1000 = 5000 \text{ installations}$$
   4th quarter, 2008:
$$4 \times 1000 = 4000 \text{ installations}$$
   (d) The tallest bar rose to 8, showing 8000 installations for the 3rd quarter of 2008.

3. **(a)** June
   The dot is at 5.5 on the vertical axis.
$$5.5 \times 10,000 = 55,000$$
   (b) May: $3 \times 10,000 = 30,000$
   April: $4 \times 10,000 = 40,000$
   There were $40,000 - 30,000 = 10,000$ fewer trout stocked in May than in April.

(c) July: $6 \times 10,000 = 60,000$
August: $2.5 \times 10,000 = 25,000$

Total (in thousands) =
$$40 + 30 + 55 + 60 + 25 = 210$$

There were 210,000 trout stocked during the five months.

(d) The highest point on the graph corresponds to July. From part (c), 60,000 trout were stocked in July.

4. **(a)** The number of minivans sold:
   2003: $30 \times 1000 = 30,000$
   2005: $40 \times 1000 = 40,000$
   2006: $20 \times 1000 = 20,000$
   2007: $15 \times 1000 = 15,000$

   (b) The number of SUVs sold:
   2003: $10 \times 1000 = 10,000$
   2004: $20 \times 1000 = 20,000$
   2005: $30 \times 1000 = 30,000$
   2006: $50 \times 1000 = 50,000$

   (c) The number of SUVs sold was greater than the number of minivans sold when the red line was above the blue line. The first full year in which this happened was 2006.

   (d) From part (b), the amount of increase in SUV sales from 2005 to 2006 is
   $50,000 - 30,000 = 20,000.$

   percent **of** 2005 sales = amount of increase
$$p \cdot 30,000 = 20,000$$
$$\frac{p \cdot 30,000}{30,000} = \frac{20,000}{30,000}$$
$$p = \frac{2}{3} \approx 0.67 = 67\%$$

   The percent of increase is 67% (rounded).

9.3 Section Exercises

1. Look for the longest bar. The top reason is, "I can shop during off hours." 74% gave this reason.

3. 57% say they find better prices on-line.
$$0.57(600) = 342$$
342 people gave this answer.

5. $\frac{1}{2} = 50\%$

   $\frac{1}{2}$ of the people said, "I can compare products more easily."

   $\frac{3}{4} = 75\% \approx 74\%$

   Nearly $\frac{3}{4}$ of the people said, "I can shop during off hours."

7. May had the greatest number of unemployed workers. The total was 10,000 unemployed workers.

9. Unemployed workers in February of 2008 = 7000
   Unemployed workers in February of 2007 = 5500

   $$7000 - 5500 = 1500$$

   There were 1500 more workers unemployed in February of 2008.

11. The number of unemployed workers increased from 5500 in February of 2007 to 8000 in April of 2008. The increase was $8000 - 5500 = 2500$ workers.

   $$\begin{array}{ccc}
   \text{percent of} & \text{February 2007} & = & \text{amount} \\
    & \text{unemployed workers} & & \text{of increase} \\
   p \cdot & 5500 & = & 2500 \\
   & \dfrac{p \cdot 5500}{5500} & = & \dfrac{2500}{5500} \\
   & p & = & \dfrac{5}{11} \approx 0.45 = 45\%
   \end{array}$$

   The percent of increase was 45% (rounded).

13. 150,000 gallons of supreme unleaded gasoline were sold in 2004.

15. The greatest difference occurred in 2004. The difference was $400,000 - 150,000 = 250,000$ gallons.

17. 700,000 gallons of supreme unleaded gasoline were sold in 2008. 150,000 gallons of supreme unleaded gasoline were sold in 2004.

    $$700,000 - 150,000 = 550,000$$

    There was an increase of 550,000 gallons of supreme unleaded gasoline sales.

   $$\begin{array}{ccc}
   \text{percent of} & \text{supreme} & = & \text{amount} \\
    & \text{unleaded gasoline} & & \text{of increase} \\
    & \text{sales in 2004} & & \\
   p \cdot & 150,000 & = & 550,000 \\
   & \dfrac{p \cdot 150,000}{150,000} & = & \dfrac{550,000}{150,000} \\
   & p & = & \dfrac{11}{3} \approx 3.67 = 367\%
   \end{array}$$

   The percent of increase was 367% (rounded).

19. The number of PCs shipped in 1990 was 24.1 million or 24,100,000.

21. The increase in the number of PCs shipped in 2005 from the number shipped in 1985 was $197.4 - 11.8 = 185.6$ million or 185,600,000.

23. The amount of increase in shipments from 1995 to 2000 was $144.6 - 62.3 = 82.3$ million or 82,300,000 PCs.

   $$\begin{array}{ccc}
   \text{percent of} & 1995 & = & \text{amount} \\
    & \text{shipments} & & \text{of increase} \\
   p \cdot & 62.3 & = & 82.3 \\
   & \dfrac{p \cdot 62.3}{62.3} & = & \dfrac{82.3}{62.3} \\
   & p & \approx & 1.32 = 132\%
   \end{array}$$

   The percent of increase was 132% (rounded).

25. (a) Chain Store A sold $3000 \cdot 1000 = 3,000,000$ DVDs in 2004.

    (b) Chain Store B sold $1500 \cdot 1000 = 1,500,000$ DVDs in 2004.

27. (a) Chain Store A sold $2500 \cdot 1000 = 2,500,000$ DVDs in 2007.

    (b) Chain Store A sold $3000 \cdot 1000 = 3,000,000$ DVDs in 2008.

29. Answers will vary. Possibilities include: Both stores had decreased sales from 2004 to 2005 and increased sales from 2006 to 2008; Store B had lower sales than Store A in 2004-2005 but higher sales than Store A in 2006-2008.

31. On the blue line graph (Sales), the lowest point corresponds to the year 2006 and the amount $25,000.

33. **For 2005:** Profit = $10,000, Sales = $35,000

    $$\text{Percent the profit is of sales} = \dfrac{\$10,000}{\$35,000}$$
    $$\approx 0.29 = 29\%$$

    For 2006: Profit = $5,000, Sales = $25,000

    $$\text{Percent the profit is of sales} = \dfrac{\$5,000}{\$25,000}$$
    $$= 0.20 = 20\%$$

    For 2007: Profit = $5,000, Sales = $30,000

    $$\text{Percent the profit is of sales} = \dfrac{\$5,000}{\$30,000}$$
    $$\approx 0.17 = 17\%$$

    For 2008: Profit = $15,000, Sales = $40,000

    $$\text{Percent the profit is of sales} = \dfrac{\$15,000}{\$40,000}$$
    $$= 0.375 \approx 38\%$$

35. Answers will vary. Possibilities include: The decrease in sales may have resulted from poor service or greater competition; the increase in sales may have been a result of more advertising or better service.

37. Shipments have increased at a rapid rate since 1985.

38. Answers will vary. Some possibilities are: lower prices; more uses and applications for students, home use, and businesses; improved technology.

39. See Exercise 23 for more detail.

   (a) 1985 to 1990:

   $11.8p = 24.1 - 11.8$

   $p = \dfrac{12.3}{11.8} \approx 1.04 =$ **104%**

   (b) 1990 to 1995:

   $24.1p = 62.3 - 24.1$

   $p = \dfrac{38.2}{24.1} \approx 1.59 =$ **159%**

   (c) 1995 to 2000:

   $62.3p = 144.6 - 62.3$

   $p = \dfrac{82.3}{62.3} \approx 1.32 =$ **132%**

   (d) 2000 to 2005:

   $144.6p = 197.4 - 144.6$

   $p = \dfrac{52.8}{144.6} \approx 0.37 =$ **37%**

   (e) 2005 to 2010:

   $197.4p = 289.1 - 197.4$

   $p = \dfrac{91.7}{197.4} \approx 0.46 =$ **46%**

40. Since 2000, the percent of increase for each 5-year period has been much lower than in earlier periods.

41. Answers will vary. Some possibilities are: more people will already own a computer and not want to buy another; some new invention will replace computers.

42. **(a)** The sales for Chain Store A were 3,000,000 in 2004 and in 2008. Since these values are the same, there is no percent of increase or decrease (0%).

   (b) Chain Store B:

   $\begin{aligned} \text{percent of 2004 sales} \cdot 1{,}500{,}000 &= \text{amount of increase} \\ p \cdot 1{,}500{,}000 &= 4{,}000{,}000 - 1{,}500{,}000 \\ \dfrac{p \cdot 1{,}500{,}000}{1{,}500{,}000} &= \dfrac{2{,}500{,}000}{1{,}500{,}000} \\ p &= \dfrac{5}{3} \approx 1.67 = 167\% \end{aligned}$

   From 2004 to 2008, the percent of increase for Chain Store B was 167% (rounded).

43. **(a)** Answers will vary; perhaps 3,500,000 DVDs in 2009.

   (b) Answers will vary; perhaps 4,500,000 DVDs in 2009.

   (c) Answers will vary; one possibility is predicting a continuing increase based on the increases from 2006 to 2008.

44. **(a)** Most people will probably pick Store B because of its greater sales and more consistent upward trend.

   (b) Answers will vary. Some possibilities include: age and physical condition of the store, annual expenses, sales of other products, annual profit.

9.4 The Rectangular Coordinate System

9.4 Margin Exercises

1. **(a)** To plot the point $(1, 4)$ on the grid, start at 0. Move *to the right* along the horizontal axis until you reach 1. Then move *up* 4 units so that you are aligned with 4 on the vertical axis. Make a dot. This is the plot, or graph, of the point $(1, 4)$.

   (b) $(5, 2)$: right 5, up 2

   (c) $(4, 1)$: right 4, up 1

   (d) $(3, 3)$: right 3, up 3

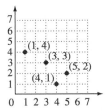

2. **(a)** To plot the point $(5, -3)$ on the coordinate system, start at $(0, 0)$. Move *to the right* 5 units. Now move *down* 3 units. Make a dot.

   (b) $(-5, 3)$: left 5, up 3

   (c) $(0, 3)$: no horizontal movement, up 3

   (d) $(-4, -4)$: left 4, down 4

   (e) $(-2, 0)$: left 2, no vertical movement

9.4 The Rectangular Coordinate System

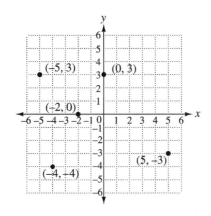

3.

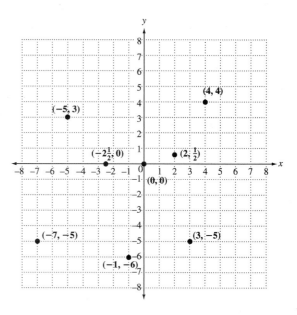

3. A is $(-4, -1)$.
    B is $(3, -3)$.
    C is $(5, 0)$.
    D is $(0, -2)$.
    E is approximately $(-1\frac{1}{2}, 5)$.

4. **(a)** The pattern for all points in Quadrant IV is $(+, -)$.

    Examples: $(3, -10)$, $(1, -4)$, $(12, -21)$

    (b) For $(-2, -6)$, the pattern is $(-, -)$, so the point is in **quadrant III**.

    The point corresponding to $(0, 5)$ is on the y-axis, so it isn't in any quadrant.

    For $(-3, 1)$, the pattern is $(-, +)$, so the point is in **quadrant II**.

    For $(4, -1)$, the pattern is $(+, -)$, so the point is in **quadrant IV**.

5. A is $(3, 4)$; B is $(5, -5)$; C is $(-4, -2)$;
    D is approximately $(4, \frac{1}{2})$; E is $(0, -7)$;
    F is $(-5, 5)$; G is $(-2, 0)$; H is $(0, 0)$.

7. $(-3, -7)$ is in Quadrant III.
    $(0, 4)$ is on the y-axis, so it is not in a quadrant.
    $(10, -16)$ is in Quadrant IV.
    $(-9, 5)$ is in Quadrant II.

9. **(a)** Any *positive* number, because points in Quadrant II have the pattern $(-, +)$.

    (b) Any *negative* number, because points in Quadrant IV have the pattern $(+, -)$.

    (c) 0, because points not in a quadrant have the form $(0, \pm)$ or $(\pm, 0)$.

    (d) Any *negative* number, because points in Quadrant III have the pattern $(-, -)$.

    (e) Any *positive* number, because points in Quadrant I have the pattern $(+, +)$.

11. Starting at the origin, move left or right along the x-axis to the number a; then move up if b is positive or move down if b is negative.

9.4 Section Exercises

1.

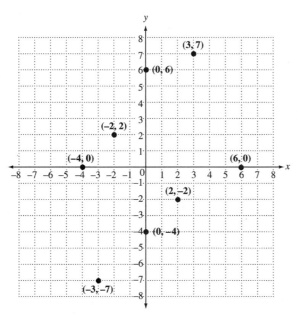

9.5 Introduction to Graphing Linear Equations

9.5 Margin Exercises

1. $x + y = 5$

x	y	Check that $x+y=5$	Ordered Pair (x,y)
0	5	$0 + 5 = 5$	$(0, 5)$
1	4	$1 + 4 = 5$	$(1, 4)$
2	3	$2 + 3 = 5$	$(2, 3)$

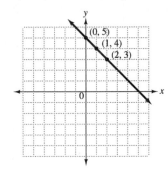

Some other possible solutions are:
$(-1, 6); (3, 2); (4, 1); (5, 0); (6, -1)$

2. $y = 2x$

x	$2 \cdot x = y$	Ordered Pair (x,y)
0	$2 \cdot 0 = 0$	$(0, 0)$
1	$2 \cdot 1 = 2$	$(1, 2)$
2	$2 \cdot 2 = 4$	$(2, 4)$

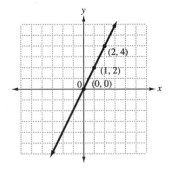

Some other possible solutions are:
$(3, 6); (-1, -2); (-2, -4); (-3, -6)$

3. $y = -\dfrac{1}{2}x$

x	$-\frac{1}{2} \cdot x = y$	Ordered Pair (x,y)
2	$-\frac{1}{2} \cdot 2 = -\frac{2}{2} = -1$	$(2, -1)$
4	$-\frac{1}{2} \cdot 4 = -\frac{4}{2} = -2$	$(4, -2)$
6	$-\frac{1}{2} \cdot 6 = -\frac{6}{2} = -3$	$(6, -3)$

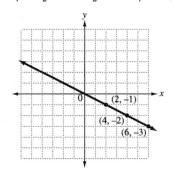

Some other possible solutions are:
$(0, 0); (-2, 1); (-4, 2); (-6, 3);$
$(1, -\frac{1}{2}); (3, -1\frac{1}{2}); (5, -2\frac{1}{2})$

4. $y = x - 5$

x	$x - 5 = y$	Ordered Pair (x,y)
1	$1 - 5 = -4$	$(1, -4)$
2	$2 - 5 = -3$	$(2, -3)$
3	$3 - 5 = -2$	$(3, -2)$

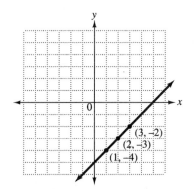

Some other possible solutions are:
$(-1, -6); (0, -5); (4\frac{1}{2}, -\frac{1}{2}); (5, 0); (5\frac{1}{2}, \frac{1}{2})$

5. **(a)** The graph of $y = 2x$ has a *positive* slope. As the value of x increases, the value of y *increases*.

(b) The graph of $y = -\frac{1}{2}x$ has a *negative* slope. As the value of x increases, the value of y *decreases*.

9.5 Section Exercises

1. $x + y = 4$

x	y	Ordered Pair (x, y)
0	4	$(0, 4)$
1	3	$(1, 3)$
2	2	$(2, 2)$

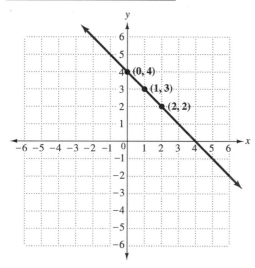

Two other possible solutions are $(3, 1)$ and $(4, 0)$. All points on the line are solutions.

3. $x + y = -1$

x	y	Ordered Pair (x, y)
0	-1	$(0, -1)$
1	-2	$(1, -2)$
2	-3	$(2, -3)$

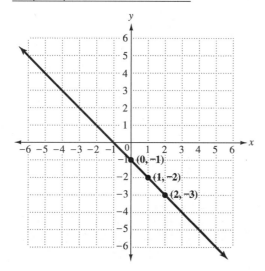

Two other possible solutions are $(-1, 0)$ and $(-2, 1)$. All points on the line are solutions.

5. The line in Exercise 1, $x + y = 4$, crosses the y-axis at $\underline{4}$ or $(0, 4)$.

The line in Exercise 3, $x + y = -1$, crosses the y-axis at $\underline{-1}$ or $(0, -1)$.

Based on these examples:

The line $x + y = -6$ will cross the y-axis at $\underline{-6}$ or $(0, -6)$.

The line $x + y = 99$ will cross the y-axis at $\underline{99}$ or $(0, 99)$.

7. $y = x - 2$

x	$x - 2 = y$	Ordered Pair (x, y)
1	$1 - 2 = -1$	$(1, -1)$
2	$2 - 2 = 0$	$(2, 0)$
3	$3 - 2 = 1$	$(3, 1)$

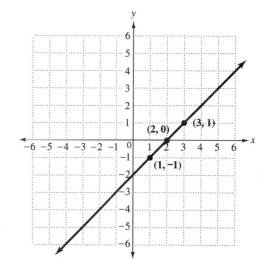

290   Chapter 9   Graphs

9. $y = x + 2$

x	$x + 2 = y$	(x, y)
0	$0 + 2 = 2$	$(0, 2)$
-1	$-1 + 2 = 1$	$(-1, 1)$
-2	$-2 + 2 = 0$	$(-2, 0)$

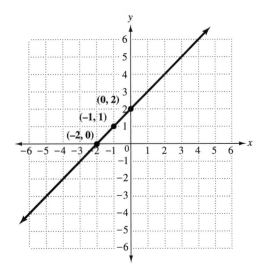

11. $y = -3x$

x	$-3 \cdot x = y$	(x, y)
0	$-3(0) = 0$	$(0, 0)$
1	$-3(1) = -3$	$(1, -3)$
2	$-3(2) = -6$	$(2, -6)$

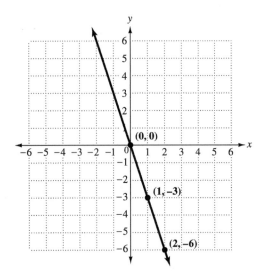

13. The lines in Exercises 7 and 9 have a positive slope. The lines in Exercises 1, 3, and 11 have a negative slope.

15. $y = \frac{1}{3}x$

x	$\frac{1}{3} \cdot x = y$	(x, y)
0	$\frac{1}{3}(0) = 0$	$(0, 0)$
3	$\frac{1}{3}(3) = 1$	$(3, 1)$
6	$\frac{1}{3}(6) = 2$	$(6, 2)$

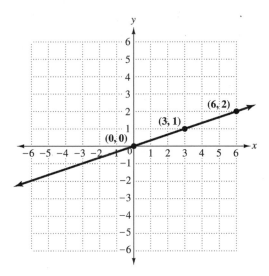

17. $y = x$

x	$y = x$	(x, y)
-1	-1	$(-1, -1)$
-2	-2	$(-2, -2)$
-3	-3	$(-3, -3)$

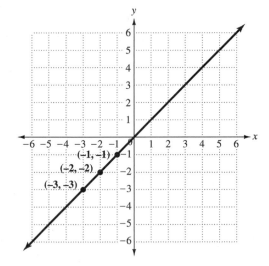

9.5 Introduction to Graphing Linear Equations

19. $y = -2x + 3$

x	$-2 \cdot x + 3 = y$	(x, y)
0	$-2(0) + 3 = 3$	$(0, 3)$
1	$-2(1) + 3 = 1$	$(1, 1)$
2	$-2(2) + 3 = -1$	$(2, -1)$

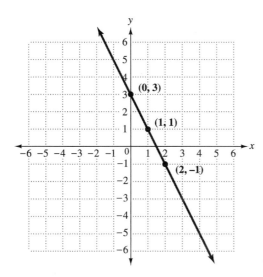

21. $x + y = -3$

One possible table:

x	y	(x, y)
0	-3	$(0, -3)$
1	-4	$(1, -4)$
2	-5	$(2, -5)$

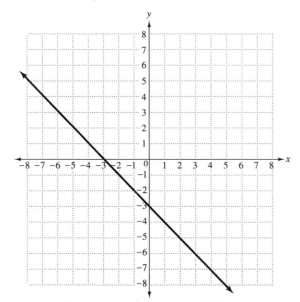

The line has a negative slope (falls).

23. $y = \frac{1}{4}x$

One possible table:

x	$\frac{1}{4} \cdot x = y$	(x, y)
0	$\frac{1}{4}(0) = 0$	$(0, 0)$
4	$\frac{1}{4}(4) = 1$	$(4, 1)$
8	$\frac{1}{4}(8) = 2$	$(8, 2)$

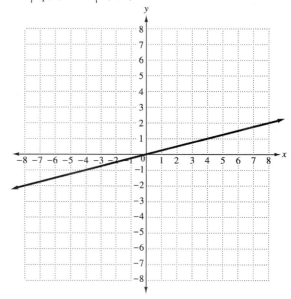

The line has a positive slope (rises).

25. $y = x - 5$

One possible table:

x	$x - 5 = y$	(x, y)
2	$2 - 5 = -3$	$(2, -3)$
3	$3 - 5 = -2$	$(3, -2)$
4	$4 - 5 = -1$	$(4, -1)$

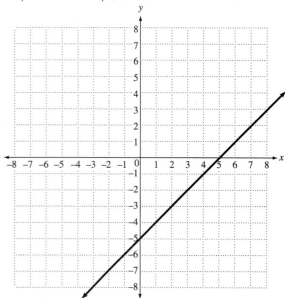

The line has a positive slope (rises).

27. $y = -3x + 1$

One possible table:

x	$-3 \cdot x + 1 = y$	(x, y)
0	$-3(0) + 1 = 1$	$(0, 1)$
1	$-3(1) + 1 = -2$	$(1, -2)$
2	$-3(2) + 1 = -5$	$(2, -5)$

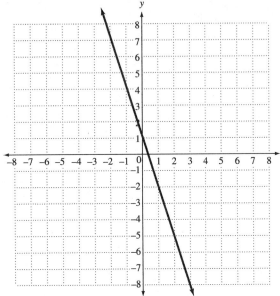

The line has a negative slope (falls).

Chapter 9 Review Exercises

1. **(a)** Look down the column for male teams to find the smallest number (19). Then look across to find the sport. Gymnastics had the fewest men's teams.

   (b) Look down the column for female teams and find the second highest number (982). Then look across to find the sport. Volleyball had the second greatest number of women's teams.

2. **(a)** Look down the column for female athletes for the number closest to 13,000 (12,901). Then look across to find the sport. Cross country had about 13,000 female athletes.

   (b) Look down the column for male athletes for the number closest to 1000 (1161). Then look across to find the sport. Volleyball had about 1000 male athletes.

3. **(a)** From the table, basketball had 16,291 male athletes and cross country had 11,638 male athletes.

   $$16{,}291 - 11{,}638 = 4653$$

   4653 more men participated in basketball than cross country.

   (b) From the table, gymnastics had 1402 female athletes and basketball had 14,686 female athletes.

   $$14{,}686 - 1402 = 13{,}284$$

   13,284 fewer women participated in gymnastics than basketball.

4. The average squad size for men's volleyball is

   $$\frac{1161}{79} \approx 14.7.$$

   The average squad size for women's volleyball is

   $$\frac{13{,}634}{982} \approx 13.9.$$

For Exercises 5–8, each symbol represents 10 inches of snowfall.

5. **(a)** Juneau is represented by 10 symbols.

   $10(10 \text{ inches}) = 100$ inches

   (b) Washington D.C. is represented by $1\frac{1}{2}$ or 1.5 symbols.

   $1.5(10 \text{ inches}) = 15$ inches

6. **(a)** Minneapolis is represented by 5 symbols.

   $5(10 \text{ inches}) = 50$ inches

   (b) Cleveland is represented by $5\frac{1}{2}$ or 5.5 symbols.

   $5.5(10 \text{ inches}) = 55$ inches

7. **(a)** Buffalo: $9(10 \text{ inches}) = 90$ inches
   Cleveland: $5.5(10 \text{ inches}) = 55$ inches

   The difference in average yearly snowfall is
   90 inches $-$ 55 inches $= 35$ inches.

   (b) Memphis: $\frac{1}{2}(10 \text{ inches}) = 5$ inches
   Minneapolis: $5(10 \text{ inches}) = 50$ inches

   The difference in average yearly snowfall is
   50 inches $-$ 5 inches $= 45$ inches.

8. greatest amount: Juneau, 100 inches
   least amount: Memphis, 5 inches

   The difference in average yearly snowfall between Juneau and Memphis is

   100 inches $-$ 5 inches $= 95$ inches.

9. **(a)** Lodging is the largest sector in the circle graph: $560

   (b) Food is the second largest sector in the circle graph: $400

   (c) Total cost $= \$560 + \$400 + \$300 + \$280 + \$160$
   $= \$1700$

10. $\dfrac{\text{lodging}}{\text{food}} = \dfrac{\$560}{\$400} = \dfrac{560 \div 80}{400 \div 80} = \dfrac{7}{5}$

11. $\dfrac{\text{gasoline}}{\text{total cost}} = \dfrac{\$300}{\$1700} = \dfrac{300}{1700} = \dfrac{3}{17}$

12. $\dfrac{\text{sightseeing}}{\text{total cost}} = \dfrac{\$280}{\$1700} = \dfrac{280 \div 20}{1700 \div 20} = \dfrac{14}{85}$

13. $\dfrac{\text{gasoline}}{\text{other}} = \dfrac{\$300}{\$160} = \dfrac{300 \div 20}{160 \div 20} = \dfrac{15}{8}$

14. The most popular project is painting and wallpapering since it has the longest bar. The percent is 63%.

15. The project selected the least is construction work at 33%.

16. 43% of the homeowners in the survey selected carpentry projects.

    percent · whole = part
    $(0.43)(341) = n$
    $n = 146.63 \approx 147$

    About 147 homeowners selected carpentry.

17. 54% of the homeowners in the survey selected landscaping or gardening projects.

    percent · whole = part
    $(0.54)(341) = n$
    $n = 184.14 \approx 184$

    About 184 homeowners selected landscaping or gardening projects.

18. (a) $\tfrac{1}{2} = 0.5 = 50\%$

    51% (about $\tfrac{1}{2}$) of the homeowners selected interior decorating.

    (b) $\tfrac{1}{3} = 0.\overline{3} = 33.\overline{3}\%$

    33% (about $\tfrac{1}{3}$) of the homeowners selected construction work and 34% (about $\tfrac{1}{3}$) of the homeowners selected window treatments.

19. Answers will vary. Possibilities include: painting and wallpapering are easier to do, take less time, or cost less than construction work.

20. In 2008, the greatest amount of water in the lake occurred in March when there were 8,000,000 acre-feet of water.

21. In 2007, the least amount of water in the lake occurred in June when there were 2,000,000 acre-feet of water.

22. In June of 2008, there were 5,000,000 acre-feet of water in the lake.

23. In January of 2007, there were 6,000,000 acre-feet of water in the lake.

24. March 2008: 8,000,000 acre-feet
    June 2008: 5,000,000 acre-feet

    This is a $8{,}000{,}000 - 5{,}000{,}000 = 3{,}000{,}000$ acre-feet decrease.

    percent of March 2008 amount = amount of decrease

    $p \cdot 8{,}000{,}000 = 3{,}000{,}000$
    $\dfrac{p \cdot 8{,}000{,}000}{8{,}000{,}000} = \dfrac{3{,}000{,}000}{8{,}000{,}000}$
    $p = \dfrac{3}{8} = 0.375 = 37.5\%$

    In 2008, the percent of decrease from March to June was 37.5%.

25. April 2007: 5,000,000 acre-feet
    June 2007: 2,000,000 acre-feet

    This is a $5{,}000{,}000 - 2{,}000{,}000 = 3{,}000{,}000$ acre-feet decrease.

    percent of April 2007 amount = amount of decrease

    $p \cdot 5{,}000{,}000 = 3{,}000{,}000$
    $\dfrac{p \cdot 5{,}000{,}000}{5{,}000{,}000} = \dfrac{3{,}000{,}000}{5{,}000{,}000}$
    $p = \dfrac{3}{5} = 0.6 = 60\%$

    In 2007, the percent of decrease from April to June was 60%.

26. In 2005, Center A sold $50,000,000 worth of floor covering.

27. In 2007, Center A sold $20,000,000 worth of floor covering.

28. In 2006, Center B sold $20,000,000 worth of floor covering.

29. In 2008, Center B sold $40,000,000 worth of floor covering.

30. Sales decreased for two years and then moved up slightly. Answers will vary. Perhaps there is less new construction, remodeling, and home improvement in the area near Center A. Also, better product selection and service may have reversed the decline in sales.

31. Sales are increasing. Answers will vary. New construction may have increased in the area near Center B, or greater advertising may attract more attention.

32. Degrees for Plumbing and electrical changes
    = 10% of 360°
    = $(0.10)(360°) = 36°$

294 Chapter 9 Graphs

33. Percent for Work stations (total = $22,400)

$$= \frac{\$7840}{\$22,400} = 0.35 = 35\%$$

Degrees = (0.35)(360°)
= 126°

34. Percent for Small appliances

$$= \frac{\$4480}{\$22,400} = 0.20 = 20\%$$

Degrees = (0.20)(360°)
= 72°

35. Percent for Interior decoration

$$= \frac{\$5600}{\$22,400} = 0.25 = 25\%$$

Degrees = (0.25)(360°)
= 90°

36. The unknowns are dollar amount and percent of total.

$$\text{Percent} = \frac{36°}{360°} = 0.10 \text{ or } 10\%$$

Dollar amount = (0.10)($22,400) = $2240

37.

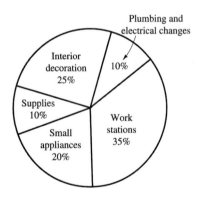

38.

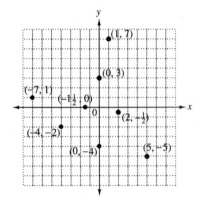

39. A is $(0, 6)$; B is approximately $(-2, 2\frac{1}{2})$; C is $(0, 0)$; D is $(-6, -6)$; E is $(4, 3)$; F is approximately $(3\frac{1}{2}, 0)$; G is $(2, -4)$.

40. $x + y = -2$

x	y	(x, y)
0	-2	$(0, -2)$
1	-3	$(1, -3)$
2	-4	$(2, -4)$

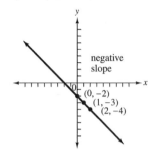

The graph of $x + y = -2$ has a *negative* slope.
Two other solutions of $x + y = -2$ are $(-2, 0)$ and $(-1, -1)$.
All points on the line are solutions.

41. $y = x + 3$

x	$x + 3 = y$	(x, y)
0	$0 + 3 = 3$	$(0, 3)$
1	$1 + 3 = 4$	$(1, 4)$
2	$2 + 3 = 5$	$(2, 5)$

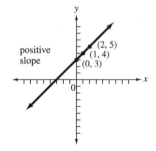

The graph of $y = x + 3$ has a *positive* slope.
Two other solutions for $y = x + 3$ are $(-1, 2)$ and $(-2, 1)$.
All points on the line are solutions.

42. $y = -4x$

x	$-4 \cdot x = y$	(x, y)
-1	$-4(-1) = 4$	$(-1, 4)$
0	$-4(0) = 0$	$(0, 0)$
1	$-4(1) = -4$	$(1, -4)$

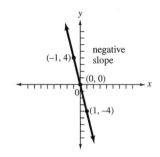

The graph of $y = -4x$ has a *negative* slope. Two other solutions of $y = -4x$ are $\left(-\frac{1}{2}, 2\right)$ and $\left(\frac{1}{2}, -2\right)$.
All points on the line are solutions.

Chapter 9 Test

1. **(a)** Look down the calcium column and find the largest number (371). Then look across to find the food. Sardines have the greatest amount of calcium.

   (b) Look down the calcium column and find the smallest number (23). Then look across to find the food. Cream cheese has the least amount of calcium.

2. Set up a proportion.
$$\frac{95 \text{ calories}}{1 \text{ oz}} = \frac{x \text{ calories}}{1.75 \text{ oz}}$$
$$1 \cdot x = 95 \cdot 1.75$$
$$x \approx 166$$

   There are approximately 166 calories in 1.75 ounces of Swiss cheese.

3. Set up a proportion.
$$\frac{345 \text{ mg}}{8 \text{ oz}} = \frac{x \text{ mg}}{6 \text{ oz}}$$
$$8 \cdot x = 345 \cdot 6$$
$$\frac{8x}{8} = \frac{2070}{8}$$
$$x \approx 259$$

   There are approximately 259 mg of calcium in 6 oz of fruit-flavored yogurt.

   For Exercises 4–6, each symbol represents 10 species.

4. **(a)** Birds and fish are represented by $7\frac{1}{2}$ symbols, more than any other group. Thus, birds and fish are tied for the greatest number of endangered species.

   (b) $7\frac{1}{2}(10 \text{ species}) = 75$ species, so there are 75 bird and 75 fish species in the top groups.

5. Mammals: $7(10 \text{ species}) = 70$ species
   Reptiles: $1.5(10 \text{ species}) = 15$ species

   There are $70 - 15 = 55$ more species of endangered mammals than endangered reptiles.

6. There is a total of $7 + 7.5 + 1.5 + 7.5 + 3.5 = 27$ symbols shown in the pictograph.

   There are $27(10) = 270$ endangered species shown in the pictograph.

7. Look for the largest sector (or largest percent).

   29% of $2,800,000
   $= 0.29(\$2,800,000) = \$812,000$

   Television has the largest budget of $812,000.

8. Look for the smallest sector (or smallest percent).

   3% of $2,800,000
   $= 0.03(\$2,800,000) = \$84,000$

   Miscellaneous has the smallest budget of $84,000.

9. 11% of $2,800,000
   $= 0.11(\$2,800,000) = \$308,000$

   $308,000 is budgeted for internet advertising.

10. 23% of $2,800,000
    $= 0.23(\$2,800,000) = \$644,000$

    $644,000 is budgeted for newspaper ads.

11. In 2007, expenses exceeded income by
    $$\$21{,}000 - \$17{,}000 = \$4000.$$

12. From 2005 to 2006, expenses increased by
    $$\$18{,}000 - \$13{,}000 = \$5000$$

    percent of 2005 expenses = amount of increase

    $$p \cdot 13{,}000 = 5{,}000$$
    $$\frac{p \cdot 13{,}000}{13{,}000} = \frac{5{,}000}{13{,}000}$$
    $$p = \frac{5}{13} \approx 0.38 = 38\%$$

    The amount of increase in the student's expenses was $5000 and the percent of increase was 38%.

13. In 2007, the student's income declined. Explanations will vary. Some possibilities are: laid off from work, changed jobs, was ill, cut down on hours worked.

14. 5500 students were enrolled at College A in 2005.

    3000 students were enrolled at College B in 2006.

15. College B had a higher enrollment in 2008.

College A = 4500 students

College B = 5500 students

College B had 5500 − 4500 = 1000 more students.

16. Explanations will vary. For example, College B may have added new courses or lowered tuition or added child care.

17. Percent = $\dfrac{\$168{,}000}{\$480{,}000} = 0.35 = 35\%$

35% of 360° = 0.35(360°) = 126°

18. Percent = $\dfrac{\$24{,}000}{\$480{,}000} = 0.05 = 5\%$

5% of 360° = 0.05(360°) = 18°

19. Percent = $\dfrac{\$96{,}000}{\$480{,}000} = 0.20 = 20\%$

20% of 360° = 0.20(360°) = 72°

20. Percent = $\dfrac{\$144{,}000}{\$480{,}000} = 0.30 = 30\%$

30% of 360° = 0.30(360°) = 108°

21. Percent = $\dfrac{36°}{360°} = 0.10 = 10\%$

10% of $480,000 = 0.10($480,000) = $48,000

22.

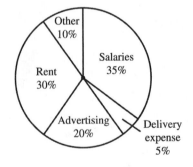

23.–26.

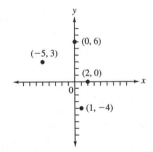

27. Point A is $(0,0)$; no quadrant

28. Point B is $(-5,-4)$; quadrant III

29. Point C is $(3,3)$; quadrant I

30. Point D is $(-2,4)$; quadrant II

31. $y = x - 4$

x	$x-4=y$	(x,y)
0	$0-4=-4$	$(0,-4)$
1	$1-4=-3$	$(1,-3)$
2	$2-4=-2$	$(2,-2)$

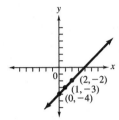

32. **(a)** Answers will vary; all points on the line are solutions. Some possibilities are $(3,-1)$ and $(4,0)$.

(b) For $y = x - 4$, as the value of x increases, the value of y *increases*. Thus, the graph of $y = x - 4$ has a *positive* slope.

Cumulative Review Exercises (Chapters 1–9)

1. $\dfrac{4}{5} + 2\dfrac{1}{3} = \dfrac{4}{5} + \dfrac{7}{3} = \dfrac{12}{15} + \dfrac{35}{15}$

$= \dfrac{12+35}{15} = \dfrac{47}{15}$, or $3\dfrac{2}{15}$

2. $\dfrac{0.8}{-3.2}$ different signs, negative quotient

$3.2\overline{)0.8} \rightarrow 32\overline{)8.00}$
$\phantom{32\overline{)}}\underline{6\,4}$
$\phantom{32\overline{)}}1\,6\,0$
$\phantom{32\overline{)}}\underline{1\,6\,0}$
$\phantom{32\overline{)8.}}0$

Answer: -0.25

3. $5^2 + (-4)^3 = 25 + (-64) = -39$

4. $(0.002)(-0.05)$: different signs, negative product

$0.002 \leftarrow 3$ decimal places
$\times\ 0.05 \leftarrow 2$ decimal places
$\overline{0.00010} \leftarrow 5$ decimal places

Answer: -0.00010, or -0.0001

5. $\dfrac{4a}{9} \cdot \dfrac{6b}{2a^3} = \dfrac{\overset{1}{\cancel{2}}\cdot 2\cdot \overset{1}{\cancel{a}}\cdot 2\cdot \overset{1}{\cancel{3}}\cdot b}{\underset{1}{\cancel{3}}\cdot 3\cdot \underset{1}{\cancel{2}}\cdot \underset{1}{\cancel{a}}\cdot a\cdot a} = \dfrac{4b}{3a^2}$

6. $1\dfrac{1}{4} - 3\dfrac{5}{6} = \dfrac{5}{4} - \dfrac{23}{6} = \dfrac{15}{12} - \dfrac{46}{12} = \dfrac{15-46}{12}$
$= \dfrac{-31}{12} = -\dfrac{31}{12}$, or $-2\dfrac{7}{12}$

7. $-13 + 2.993$

   $|-13| = 13$, $|2.993| = 2.993$

   Since -13 has the larger absolute value, the sum will be negative.

   Subtract:
   $1\overset{\scriptsize 2}{3}.\overset{\scriptsize 9}{\cancel{0}}\overset{\scriptsize 9}{\cancel{0}}\overset{\scriptsize 10}{\cancel{0}}$
   $-2.9\,9\,3$
   $1\,0.0\,0\,7$

   Answer: -10.007

8. $\dfrac{2}{7} - \dfrac{8}{x} = \dfrac{2 \cdot x}{7 \cdot x} - \dfrac{8 \cdot 7}{x \cdot 7} = \dfrac{2x}{7x} - \dfrac{56}{7x} = \dfrac{2x - 56}{7x}$

9. $-3 - 33 = -3 + (-33) = -36$

10. $\dfrac{10m}{3n^2} \div \dfrac{2m^2}{5n} = \dfrac{10m}{3n^2} \cdot \dfrac{5n}{2m^2}$

    $= \dfrac{\overset{1}{\cancel{2}} \cdot 5 \cdot \overset{1}{\cancel{m}} \cdot 5 \cdot \overset{1}{\cancel{n}}}{3 \cdot \underset{1}{\cancel{n}} \cdot \underset{1}{\cancel{2}} \cdot \underset{1}{\cancel{m}} \cdot m}$

    $= \dfrac{25}{3mn}$

11. $\dfrac{-3}{-\tfrac{9}{10}} = -\dfrac{3}{1} \div \left(-\dfrac{9}{10}\right) = -\dfrac{\overset{1}{\cancel{3}}}{1} \cdot \left(-\dfrac{10}{\underset{3}{\cancel{9}}}\right)$

    $= \dfrac{10}{3}$, or $3\dfrac{1}{3}$

12. $\dfrac{3(-7)}{2^4 - 16} = \dfrac{-21}{16 - 16} = \dfrac{-21}{0}$, which is *undefined*.

13. $10 - 2\dfrac{5}{8} = \dfrac{10}{1} - \dfrac{21}{8} = \dfrac{80}{8} - \dfrac{21}{8}$
    $= \dfrac{59}{8}$, or $7\dfrac{3}{8}$

14. $\dfrac{3}{w} + \dfrac{x}{6} = \dfrac{3 \cdot 6}{w \cdot 6} + \dfrac{x \cdot w}{6 \cdot w}$
    $= \dfrac{18}{6w} + \dfrac{wx}{6w} = \dfrac{18 + wx}{6w}$

15. $8 + 4(2 - 5) = 8 + 4(-3) = 8 + (-12) = -4$

16. $\dfrac{7}{8}$ of $960 = \dfrac{7}{\underset{1}{\cancel{8}}} \cdot \dfrac{\overset{120}{\cancel{960}}}{1} = 840$

17. $\dfrac{(-4)^2 + 8(0 - 2)}{8 \div 2(-3 + 5) - 10}$

    Numerator: Denominator:
    $(-4)^2 + 8(0 - 2)$ $8 \div 2(-3 + 5) - 10$
    $= 16 + 8(-2)$ $= 8 \div 2(2) - 10$
    $= 16 + (-16)$ $= 4(2) - 10$
    $= 0$ $= 8 - 10$
                            $= -2$

    The last step is division: $\dfrac{0}{-2} = 0$

18. $0.5 - 0.25(3.2)^2$
    $= 0.5 - 0.25(10.24)$
    $= 0.5 - 2.56$
    $= -2.06$

19. $6\left(-\dfrac{1}{2}\right)^3 + \dfrac{2}{3}\left(\dfrac{3}{5}\right) = \dfrac{6}{1} \cdot \left(-\dfrac{1}{8}\right) + \dfrac{2 \cdot \overset{1}{\cancel{3}}}{\underset{1}{\cancel{3}} \cdot 5}$

    $= -\dfrac{\overset{1}{\cancel{2}} \cdot 3 \cdot 1}{1 \cdot \underset{1}{\cancel{2}} \cdot 4} + \dfrac{2}{5}$

    $= -\dfrac{3}{4} + \dfrac{2}{5}$

    $= -\dfrac{15}{20} + \dfrac{8}{20}$

    $= \dfrac{-15 + 8}{20}$

    $= \dfrac{-7}{20}$, or $-\dfrac{7}{20}$

20. $-12 = 3(y + 2)$
    $-12 = 3y + 6$
    $\underline{-6 -6}$
    $-18 = 3y + 0$
    $\dfrac{-18}{3} = \dfrac{3y}{3}$
    $-6 = y$

    The solution is -6.

21. $6x = 14 - x$
    $6x = 14 - 1x$
    $\underline{+1x +1x}$
    $7x = 14 + 0$
    $\dfrac{7x}{7} = \dfrac{14}{7}$
    $x = 2$

    The solution is 2.

22.
$$-8 = \frac{2}{3}m + 2$$
$$\underline{ -2 \phantom{= \frac{2}{3}m} -2}$$
$$-10 = \frac{2}{3}m + 0$$
$$\frac{3}{2}(-10) = \frac{3}{2}\left(\frac{2}{3}m\right)$$
$$-15 = m$$

The solution is -15.

23.
$$3.4x - 6 = 8 + 1.4x$$
$$\underline{ +6 +6}$$
$$3.4x + 0 = 14 + 1.4x$$
$$3.4x = 14 + 1.4x$$
$$\underline{-1.4x -1.4x}$$
$$2.0x = 14 + 0$$
$$\frac{2x}{2} = \frac{14}{2}$$
$$x = 7$$

The solution is 7.

24.
$$2(h-1) = -3(h+12) - 11$$
$$2h - 2 = -3h - 36 - 11$$
$$2h - 2 = -3h + (-36) + (-11)$$
$$2h - 2 = -3h + (-47)$$
$$\underline{+3h +3h}$$
$$5h - 2 = 0 + (-47)$$
$$5h - 2 = -47$$
$$\underline{+2 +2}$$
$$5h + 0 = -45$$
$$\frac{5h}{5} = \frac{-45}{5}$$
$$h = -9$$

The solution is -9.

25. Let n represent the number.

$\underbrace{\text{If five times a number is subtracted from 12,}}_{12 - 5n}$ $\underbrace{\text{the result is}}_{=}$ $\underbrace{\text{the number.}}_{n}$

$$12 - 5n = n$$
$$\underline{+5n +5n}$$
$$12 + 0 = 6n$$
$$12 = 6n$$
$$\frac{12}{6} = \frac{6n}{6}$$
$$2 = n$$

The number is 2.

26. *Step 1*
Unknown: amount received by each person.
Known: total scholarship is $1800, one person gets $500 more

Step 2
Let m represent the amount of scholarship money received by one person. Then $m + 500$ represents the amount of scholarship money received by the other person.

Step 3
$m + (m + 500) = 1800$

Step 4
$$m + m + 500 = 1800$$
$$2m + 500 = 1800$$
$$\underline{ -500 -500}$$
$$2m = 1300$$
$$\frac{2m}{2} = \frac{1300}{2}$$
$$m = 650$$

Step 5
One person receives $650 and the other receives $650 + $500 = $1150.

Step 6
$1150 is $500 more than $650 and the sum of $650 and $1150 is $1800.

27. $P = 10 \text{ ft} + 8 \text{ ft} + 10 \text{ ft} + 8 \text{ ft} = 36 \text{ ft}$
$A = b \cdot h = 10 \text{ ft} \cdot 7 \text{ ft} = 70 \text{ ft}^2$

28. $C = \pi \cdot d$
$C \approx 3.14 \cdot 6 \text{ ft}$
$C = 18.84 \text{ ft}$
$C \approx 18.8 \text{ ft}$

$A = \pi \cdot r \cdot r$
$A \approx 3.14 \cdot 3 \text{ ft} \cdot 3 \text{ ft}$
$A = 28.26 \text{ ft}^2$
$A \approx 28.3 \text{ ft}^2$

29. The figure is a right triangle with unknown hypotenuse.

$\text{hypotenuse} = \sqrt{(\text{leg})^2 + (\text{leg})^2}$
$x = \sqrt{(7)^2 + (24)^2}$
$x = \sqrt{49 + 576}$
$x = \sqrt{625}$
$x = 25 \text{ mi}$

$P = 24 \text{ mi} + 7 \text{ mi} + 25 \text{ mi}$
$P = 56 \text{ mi}$

$A = \frac{1}{2} \cdot b \cdot h$
$A = 0.5(7 \text{ mi})(24 \text{ mi})$
$A = 84 \text{ mi}^2$

30. If there are 19 nonsmokers out of 25 adults, how many (n) are nonsmokers out of 732 employees?

$$\frac{19}{25} = \frac{n}{732}$$
$$25 \cdot n = 19 \cdot 732$$
$$\frac{25n}{25} = \frac{13{,}908}{25}$$
$$n = 556.32$$

About 556 employees would be nonsmokers.

31. Original price · Discount rate = Discount amount
$$\$129 \cdot 0.15 = \text{Discount amount}$$
$$\$19.35 = \text{Discount amount}$$

Sale price = Original price − Discount amount
$$= \$129 - \$19.35$$
$$= \$109.65$$

Tax amount:

Price · Tax Rate = Tax amount
$$\$109.65 \cdot 0.065 = \text{Tax amount}$$
$$\$7.13 \approx \text{Tax amount}$$

Total cost = $\$109.65 + \$7.13 = \$116.78$

Her total cost for the cell phone was $116.78 (rounded).

32. Part: 167 points
Whole: 180 points
Percent: unknown (p)

percent · whole = part
$$p \cdot 180 = 167$$
$$\frac{180p}{180} = \frac{167}{180}$$
$$p \approx 0.928 = 92.8\%$$

She earned 92.8% (rounded) of the points.

33. Purchased: $2\frac{1}{4}$ pounds

Used: $5\left(\frac{1}{6} \text{ pound}\right) = \frac{5}{1}\left(\frac{1}{6} \text{ pound}\right) = \frac{5}{6}$ pound

Amount left $= 2\frac{1}{4} - \frac{5}{6}$
$$= \frac{9}{4} - \frac{5}{6}$$
$$= \frac{27}{12} - \frac{10}{12}$$
$$= \frac{17}{12}, \text{ or } 1\frac{5}{12} \text{ pounds}$$

There are $1\frac{5}{12}$ pounds of meat left.

34.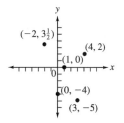

Second quadrant point: $(-2, 3\frac{1}{2})$

Third quadrant points: none

35. $y = x + 6$

x	$x + 6 = y$	(x, y)
0	$0 + 6 = 6$	$(0, 6)$
-1	$-1 + 6 = 5$	$(-1, 5)$
-2	$-2 + 6 = 4$	$(-2, 4)$

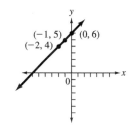

The line has a *positive* slope. All points on the line are solutions, so there are many possibilities such as $(-3, 3)$ and $(-4, 2)$.

CHAPTER 10 EXPONENTS AND POLYNOMIALS

10.1 The Product Rule and Power Rules for Exponents

10.1 Margin Exercises

1. $2 \cdot 2 \cdot 2 \cdot 2 = 2^4$ 2 occurs as a factor 4 times.
    $= 16$

2. (a) $(-2)^5 = (-2)(-2)(-2)(-2)(-2)$
    $= -32$

    Base: -2; exponent: 5

    (b) $-2^5 = -(2 \cdot 2 \cdot 2 \cdot 2 \cdot 2)$
    $= -32$

    Base: 2; exponent: 5

    (c) $-4^2 = -(4 \cdot 4)$
    $= -16$

    Base: 4; exponent: 2

    (d) $(-4)^2 = (-4)(-4)$
    $= 16$

    Base: -4; exponent: 2

3. (a) $8^2 \cdot 8^5 = 8^{2+5}$ Product rule
    $= 8^7$

    (b) $(-7)^5(-7)^3 = (-7)^{5+3}$ Product rule
    $= (-7)^8$

    (c) $y^3 \cdot y \cdot y^3 = y^3 \cdot y^1 \cdot y^3$
    $= y^{3+1+3}$ Product rule
    $= y^7$

    (d) $4^2 \cdot 3^5$: Cannot use the product rule because the bases are different.

    (e) $6^4 + 6^2$: Cannot use the product rule because it is a sum, not a product.

4. (a) $5m^2 \cdot 2m^6 = 5 \cdot 2 \cdot m^2 \cdot m^6 = 10m^{2+6}$
    $= 10m^8$

    (b) $3p^5 \cdot 9p^4 = 3 \cdot 9 \cdot p^5 \cdot p^4 = 27p^{5+4}$
    $= 27p^9$

    (c) $-7p^8 \cdot (3p^8) = -7 \cdot 3 \cdot p^8 \cdot p^8 = -21p^{8+8}$
    $= -21p^{16}$

5. (a) $(5^3)^4 = 5^{3 \cdot 4}$ Power rule (a)
    $= 5^{12}$

    (b) $(6^2)^5 = 6^{2 \cdot 5}$ Power rule (a)
    $= 6^{10}$

    (c) $(x^3)^3 = x^{3 \cdot 3}$ Power rule (a)
    $= x^9$

    (d) $(a^6)^5 = a^{6 \cdot 5}$ Power rule (a)
    $= a^{30}$

6. (a) $5(mn)^3 = 5(m^3 n^3)$ Power rule (b)
    $= 5m^3 n^3$

    (b) $(4ab)^2 = 4^2 a^2 b^2$ Power rule (b)
    $= 16a^2 b^2$

    (c) $(3a^2 b^4)^5 = 3^5 (a^2)^5 (b^4)^5$ Power rule (b)
    $= 3^5 a^{10} b^{20}$ Power rule (a)
    $= 243 a^{10} b^{20}$

    (d) $2(3xy^3)^4 = 2[3^4 x^4 (y^3)^4]$ Power rule (b)
    $= 2(3^4 x^4 y^{12})$ Power rule (a)
    $= 2 \cdot 3^4 x^4 y^{12}$ $3^4 = 81$
    $= 162 x^4 y^{12}$

7. (a) $\left(\dfrac{5}{2}\right)^4 = \dfrac{5^4}{2^4}$ Power rule (c)
    $= \dfrac{625}{16}$

    (b) $\left(\dfrac{p}{q}\right)^2 = \dfrac{p^2}{q^2}$ Power rule (c)

    (c) $\left(\dfrac{r}{t}\right)^3 = \dfrac{r^3}{t^3}$ Power rule (c)

10.1 Section Exercises

1. $xy^2 = x^1 y^2$

    The understood exponent for x is 1.

3. $3^3 = 3 \cdot 3 \cdot 3 = 27$, so the statement $3^3 = 9$ is *false*.

5. $(a^2)^3 = a^{2(3)} = a^6$, so the statement $(a^2)^3 = a^5$ is *false*.

7. $\underbrace{(-2)(-2)(-2)(-2)(-2)}_{\text{5 factors of } -2} = (-2)^5$

9. $\left(\dfrac{1}{2}\right)\left(\dfrac{1}{2}\right)\left(\dfrac{1}{2}\right)\left(\dfrac{1}{2}\right)\left(\dfrac{1}{2}\right)\left(\dfrac{1}{2}\right) = \left(\dfrac{1}{2}\right)^6$

11. $(-8p)(-8p) = (-8p)^2$

13. In $(-3)^4$, -3 is the base.

    $(-3)^4 = (-3)(-3)(-3)(-3) = 81$

    In -3^4, 3 is the base.

    $-3^4 = -(3 \cdot 3 \cdot 3 \cdot 3) = -81$

15. In the exponential expression 3^5, the base is 3 and the exponent is 5.
$$3^5 = 3 \cdot 3 \cdot 3 \cdot 3 \cdot 3 = 243$$

17. In the expression $(-3)^5$, the base is -3 and the exponent is 5.
$$(-3)^5 = (-3)(-3)(-3)(-3)(-3) = -243$$

19. In the expression $(-6x)^4$, the base is $-6x$ and the exponent is 4.

21. In the expression $-6x^4$, -6 is not part of the base. The base is x and the exponent is 4.

23. The product rule does not apply to $5^2 + 5^3$ because the expression is a sum, not a product. The product rule would apply if we had $5^2 \cdot 5^3$.
$$5^2 + 5^3 = 25 + 125 = 150$$

25. $5^2 \cdot 5^6 = 5^{2+6} = 5^8$

27. $4^2 \cdot 4^7 \cdot 4^3 = 4^{2+7+3} = 4^{12}$

29. $(-7)^3(-7)^6 = (-7)^{3+6} = (-7)^9$

31. $t^3 \cdot t \cdot t^{13} = t^{3+1+13} = t^{17}$

33. $(-8r^4)(7r^3) = -8 \cdot 7 \cdot r^4 \cdot r^3$
$= -56r^{4+3}$
$= -56r^7$

35. $(-6p^5)(-7p^5) = (-6)(-7)p^5 \cdot p^5$
$= 42p^{5+5}$
$= 42p^{10}$

37. $5x^4 + 9x^4 = (5+9)x^4 = 14x^4$
$(5x^4)(9x^4) = (5 \cdot 9)x^{4+4} = 45x^8$

39. $-7a^2 + 2a^2 + 10a^2 = (-7 + 2 + 10)a^2 = 5a^2$
$(-7a^2)(2a^2)(10a^2) = (-7 \cdot 2 \cdot 10)a^{2+2+2}$
$= -140a^6$

41. $(4^3)^2 = 4^{3 \cdot 2}$ Power rule (a)
$= 4^6$

43. $(t^4)^5 = t^{4 \cdot 5} = t^{20}$ Power rule (a)

45. $(7r)^3 = 7^3 r^3$ Power rule (b)

47. $(5xy)^5 = 5^5 x^5 y^5$ Power rule (b)

49. $8(qr)^3 = 8q^3 r^3$ Power rule (b)

51. $\left(\dfrac{1}{2}\right)^3 = \dfrac{1^3}{2^3}$ Power rule (c)

53. $\left(\dfrac{a}{b}\right)^3 (b \neq 0) = \dfrac{a^3}{b^3}$ Power rule (c)

55. $\left(\dfrac{9}{5}\right)^8 = \dfrac{9^8}{5^8}$ Power rule (c)

57. $(-2x^2 y)^3 = (-2)^3 (x^2)^3 y^3$
$= (-2)^3 x^6 y^3$

59. $(3a^3 b^2)^2 = 3^2 (a^3)^2 (b^2)^2$
$= 3^2 a^6 b^4$

61. Use the formula for the area of a rectangle, $A = LW$, with $L = 4x^3$ and $W = 3x^2$.
$$A = (4x^3)(3x^2)$$
$$= 4 \cdot 3 \cdot x^3 \cdot x^2$$
$$= 12x^5$$

10.2 Integer Exponents and the Quotient Rule

10.2 Margin Exercises

1. $2^{-3} = \dfrac{1}{8}$ $\left(\dfrac{1}{4} \div 2 = \dfrac{1}{8}\right)$

$2^{-4} = \dfrac{1}{16}$ $\left(\dfrac{1}{8} \div 2 = \dfrac{1}{16}\right)$

2. In every case, the Zero Exponent rule applies.

(a) $28^0 = 1$

(b) $(-16)^0 = 1$

(c) $-7^0 = -(7^0) = -1$

(d) $m^0 = 1$ when $m \neq 0$

3. In every case, the Negative Exponents rule applies.

(a) $4^{-3} = \dfrac{1}{4^3} = \dfrac{1}{64}$

(b) $6^{-2} = \dfrac{1}{6^2} = \dfrac{1}{36}$

(c) $2^{-1} + 5^{-1} = \dfrac{1}{2} + \dfrac{1}{5}$
$= \dfrac{5}{10} + \dfrac{2}{10} = \dfrac{7}{10}$

(d) $m^{-5} = \dfrac{1}{m^5}$ $(m \neq 0)$

4. (a) $\dfrac{5^{11}}{5^8} = 5^{11-8} = 5^3$

(b) $\dfrac{4^7}{4^{10}} = 4^{7-10} = 4^{-3} = \dfrac{1}{4^3}$

(c) $\dfrac{6^{-5}}{6^{-3}} = 6^{-5-(-3)} = 6^{-5+3} = 6^{-2} = \dfrac{1}{6^2}$

(d) $\dfrac{a^4}{a^{-2}} = a^{4-(-2)} = a^{4+2} = a^6$

5. (a) $10^{-7}(10^9) = 10^{-7+9} = 10^2$

(b) $(10^{-3})(10^{-3}) = 10^{-3+(-3)} = 10^{-6} = \dfrac{1}{10^6}$

(c) $x^{-3} \cdot x^{-2} \cdot x^4 = x^{-3+(-2)+4}$
$= x^{-1} = \dfrac{1}{x^1} = \dfrac{1}{x}$

(d) $(a^{-2})(a^{12})(a^{-4}) = a^{-2+12+(-4)} = a^6$

10.2 Section Exercises

1. $(-2)^{-3} = \dfrac{1}{(-2)^3}$ is negative, because $(-2)^3$ is a negative number raised to an odd exponent, which is negative, and the quotient of a positive number and a negative number is a negative number.

3. $-2^4 = -(2^4)$ is negative, because 2^4 is positive.

5. $(-2)^6$ is positive, because the exponent is even.

7. $1 - 5^0 = 1 - 1 = 0$

9. $(-4)^0 = 1$ Definition of zero exponent

11. $-9^0 = -(9^0) = -(1) = -1$

13. $(-2)^0 - 2^0 = 1 - 1 = 0$

15. $\dfrac{0^{10}}{10^0} = \dfrac{0}{1} = 0$

17. $7^0 + 9^0 = 1 + 1 = 2$

19. $b^0 = 1 \ (b \neq 0)$

21. $15x^0 = 15(1) = 15 \ (x \neq 0)$

23. $4^{-3} = \dfrac{1}{4^3}$ Definition of negative exponent
$= \dfrac{1}{64}$

25. $5^{-1} + 3^{-1} = \dfrac{1}{5} + \dfrac{1}{3}$
$= \dfrac{3}{15} + \dfrac{5}{15} = \dfrac{8}{15}$

27. $x^{-4} = \dfrac{1}{x^4} \ (x \neq 0)$

29. $\dfrac{25}{25} = 1$

30. $\dfrac{25}{25} = \dfrac{5^2}{5^2}$

31. $\dfrac{5^2}{5^2} = 5^{2-2} = 5^0$

32. $5^0 = 1$; This supports the definition of an exponent of 0.

33. $\dfrac{6^7}{6^2} = 6^{7-2} = 6^5$

35. $\dfrac{10^4}{10} = \dfrac{10^4}{10^1} = 10^{4-1} = 10^3$

37. $\dfrac{y^2}{y^6} = y^{2-6} = y^{-4} = \dfrac{1}{y^4}$

39. $\dfrac{c^6}{c^5} = c^{6-5} = c^1$ or c

41. $\dfrac{5^3}{5^7} = 5^{3-7} = 5^{-4} = \dfrac{1}{5^4}$

43. $\dfrac{m^7}{m^8} = m^{7-8} = m^{-1} = \dfrac{1}{m^1}$ or $\dfrac{1}{m}$

45. $\dfrac{3^{-4}}{3^{-8}} = 3^{-4-(-8)} = 3^{-4+8} = 3^4$

47. $\dfrac{a^{-2}}{a^{-5}} = a^{-2-(-5)} = a^{-2+5} = a^3$

49. $\dfrac{2^{-10}}{2^{-2}} = 2^{-10-(-2)} = 2^{-10+2} = 2^{-8} = \dfrac{1}{2^8}$

51. $\dfrac{r^{-12}}{r^{-8}} = r^{-12-(-8)} = r^{-12+8} = r^{-4} = \dfrac{1}{r^4}$

53. $\dfrac{10^6}{10^{-4}} = 10^{6-(-4)} = 10^{6+4} = 10^{10}$

55. $\dfrac{10^{-2}}{10^3} = 10^{-2-3} = 10^{-5} = \dfrac{1}{10^5}$

57. $\dfrac{10^3}{10^{-8}} = 10^{3-(-8)} = 10^{3+8} = 10^{11}$

59. $\dfrac{10^{-4}}{10^4} = 10^{-4-4} = 10^{-8} = \dfrac{1}{10^8}$

61. $10^6(10^{-2}) = 10^{6+(-2)} = 10^4$

63. $a^{-4}(a^3) = a^{-4+3} = a^{-1} = \dfrac{1}{a^1}$ or $\dfrac{1}{a}$

65. $(2^{-4})(2^{-4}) = 2^{-4+(-4)} = 2^{-8} = \dfrac{1}{2^8}$

67. $(x^{-10})(x^{-1}) = x^{-10+(-1)} = x^{-11} = \dfrac{1}{x^{11}}$

69. $10^8 \cdot 10^{-2} \cdot 10^{-4} = 10^{8+(-2)+(-4)} = 10^2$

71. $(y^{-3})(y^5)(y^{-4}) = y^{-3+5+(-4)} = y^{-2} = \dfrac{1}{y^2}$

73. $m^{-6} \cdot m^3 \cdot m^{12} = m^{-6+3+12} = m^9$

75. $(10^{-1})(10^{-2})(10^{-3}) = 10^{-1+(-2)+(-3)}$
$= 10^{-6} = \dfrac{1}{10^6}$

Summary Exercises on Exponents

1. $(-3)^2 = (-3)(-3) = 9$

3. $15^0 = 1$

5. $p^0 = 1$

7. $-3^2 = -(3 \cdot 3) = -9$

9. $-2^3 = -(2 \cdot 2 \cdot 2) = -8$

11. $6^{-1} + 2^{-2} = \dfrac{1}{6^1} + \dfrac{1}{2^2}$
$= \dfrac{1}{6} + \dfrac{1}{4}$
$= \dfrac{4}{24} + \dfrac{6}{24}$
$= \dfrac{4+6}{24} = \dfrac{10}{24} = \dfrac{5}{12}$

13. $2^{-5} = \dfrac{1}{2^5} = \dfrac{1}{2 \cdot 2 \cdot 2 \cdot 2 \cdot 2} = \dfrac{1}{32}$

15. $(-2)^3 = (-2)(-2)(-2) = -8$

17. $(6^4)(6^4) = 6^{4+4} = 6^8$

19. $10^2(10^{-10}) = 10^{2+(-10)} = 10^{-8} = \dfrac{1}{10^8}$

21. $(y^{-12})(y^6)(y^6) = y^{-12+6+6} = y^0 = 1$

23. $(-5)^2(-5)^8 = (-5)^{2+8} = (-5)^{10}$

25. $(m^6)^4 = m^{6 \cdot 4} = m^{24}$

27. $(10^{-2})(10^{-2})(10^{-2}) = 10^{-2+(-2)+(-2)}$
$= 10^{-6} = \dfrac{1}{10^6}$

29. $4x^4 \cdot 4x^4 = 4 \cdot 4 x^{4+4} = 16x^8$

31. $(10^4)^2 = 10^{4 \cdot 2} = 10^8$

33. $\left(\dfrac{x}{y}\right)^6 = \dfrac{x^6}{y^6}$

35. $\dfrac{7^{12}}{7^{11}} = 7^{12-11} = 7^1$ or 7

37. $-8a^3(2a) = -8 \cdot 2a^{3+1} = -16a^4$

39. $(5xy)^3 = 5^3 x^3 y^3$

41. $\dfrac{10^{-3}}{10^4} = 10^{-3-4} = 10^{-7} = \dfrac{1}{10^7}$

43. $(b^3)^5 = b^{3 \cdot 5} = b^{15}$

45. $10^{-8}(10^5) = 10^{-8+5} = 10^{-3} = \dfrac{1}{10^3}$

47. $\dfrac{10^5}{10^{-8}} = 10^{5-(-8)} = 10^{5+8} = 10^{13}$

49. $10(ab)^4 = 10(a)^4(b)^4 = 10a^4b^4$

10.3 An Application of Exponents: Scientific Notation

10.3 Margin Exercises

1. (a) $63{,}000 = 6.3 \times 10^4$

   Move the decimal point 4 places. The *original* number was "large" (10 or more), so the exponent is *positive* 4.

   (b) $5{,}870{,}000 = 5.87 \times 10^6$

   Move the decimal point 6 places. The *original* number was "large" (10 or more), so the exponent is *positive* 6.

   (c) $0.0571 = 5.71 \times 10^{-2}$

   Move the decimal point 2 places. The *original* number was "small" (between 0 and 1), so the exponent is *negative* 2.

   (d) $0.000062 = 6.2 \times 10^{-5}$

   Move the decimal point 5 places. The *original* number was "small" (between 0 and 1), so the exponent is *negative* 5.

2. (a) $4.2 \times 10^3 = 4200$

   Since the exponent is *positive* 3, move the decimal point 3 places to the *right*.

   (b) $8.7 \times 10^5 = 870{,}000$

   Since the exponent is *positive* 5, move the decimal point 5 places to the *right*.

   (c) $6.42 \times 10^{-3} = 0.00642$

   Since the exponent is *negative* 3, move the decimal point 3 places to the *left*.

3. (a) $(2.6 \times 10^4)(2 \times 10^{-6})$
   $= (2.6 \times 2)(10^4 \times 10^{-6})$
   $= 5.2 \times 10^{-2} = 0.052$

   (b) $\dfrac{4.8 \times 10^2}{2.4 \times 10^{-3}} = \dfrac{4.8}{2.4} \times \dfrac{10^2}{10^{-3}}$
   $= 2 \times 10^5 = 200{,}000$

4. $\left(\dfrac{4.356 \times 10^4 \text{ ft}^2}{1 \text{ acre}}\right) \cdot \left(\dfrac{1500 \text{ acres}}{1}\right)$
   $= (4.356 \times 10^4)(1.5 \times 10^3) \text{ ft}^2$
   $= (4.356 \times 1.5)(10^4 \times 10^3) \text{ ft}^2$
   $= 6.534 \times 10^7$ square feet,
   or $65{,}340{,}000$ square feet

5. $\dfrac{2.09 \times 10^{-1} \text{ pound}}{950 \text{ spiders}} = \dfrac{2.09 \times 10^{-1}}{9.5 \times 10^2}$
   $= \dfrac{2.09}{9.5} \times \dfrac{10^{-1}}{10^2}$
   $= 0.22 \times 10^{-3}$

   Move decimal point to the *right* one place and change 10^{-3} to 10^{-4}:

   $= 2.2 \times 10^{-4}$ pound per spider,
   or 0.00022 pound per spider

10.3 Section Exercises

1. $0.0553 = 5.53 \times 10^{-2}$

   Move the decimal point 2 places. The *original* number was "small" (between 0 and 1), so the exponent is *negative* 2.

   $317.83 = 3.1783 \times 10^2$

   Move the decimal point 2 places. The *original* number was "large" (10 or more), so the exponent is *positive* 2.

3. $831{,}200{,}000 = 8.312 \times 10^8$

   Move the decimal point 8 places. The *original* number was "large" (10 or more), so the exponent is *positive* 8.

   $319{,}000{,}000 = 3.19 \times 10^8$

   Move the decimal point 8 places. The *original* number was "large" (10 or more), so the exponent is *positive* 8.

5. 4.56×10^3 is written in scientific notation because 4.56 is between 1 and 10, and 10^3 is a power of 10.

7. $5{,}600{,}000$ is not written in scientific notation. It can be written in scientific notation as 5.6×10^6.

9. 0.004 is not written in scientific notation because $|0.004| = 0.004$ is not between 1 and 10. It can be written in scientific notation as 4×10^{-3}.

11. 0.8×10^2 is not written in scientific notation because $|0.8| = 0.8$ is not greater than or equal to 1 and less than 10. It can be written in scientific notation as 8×10^1.

13. To write a number in scientific notation $(a \times 10^n)$ move the decimal point to the right of the first nonzero digit. The absolute value of the exponent, n, is the number of places moved. If the original number is "large," then n is positive. If the original number is "small," then n is negative.

15. $5.876{,}000{,}000 = 5.876 \times 10^9$

    Move the decimal point 9 places. The *original* number was "large" (10 or more), so the exponent is *positive* 9.

17. $8.2{,}350 = 8.235 \times 10^4$

    Move the decimal point 4 places. The *original* number was "large" (10 or more), so the exponent is *positive* 4. (Note that the final zero need not be written.)

19. $0.000007 = 7 \times 10^{-6}$

    Move the decimal point 6 places. The *original* number was "small" (between 0 and 1), so the exponent is *negative* 6.

21. $0.00203 = 2.03 \times 10^{-3}$

    Move the decimal point 3 places. The *original* number was "small" (between 0 and 1), so the exponent is *negative* 3.

23. $7.5 \times 10^5 = 750{,}000$

    Since the exponent is *positive* 5, move the decimal point 5 places to the *right*.

25. $5.677 \times 10^{12} = 5{,}677{,}000{,}000{,}000$

    Since the exponent is *positive* 12, move the decimal point 12 places to the *right*.

27. $6.21 \times 10^0 = 6.21$

    Because the exponent is 0, the decimal point should not be moved. We know this result is correct because $10^0 = 1$.

29. $7.8 \times 10^{-4} = 0.00078$

    Since the exponent is *negative* 4, move the decimal point 4 places to the *left*.

31. $5.134 \times 10^{-9} = 0.000\,000\,005\,134$

    Since the exponent is *negative* 9, move the decimal point 9 places to the *left*.

33. $(2 \times 10^8)(3 \times 10^3)$

    $= (2 \times 3)(10^8 \times 10^3)$ *Commutative and associative properties*

    $= 6 \times 10^{11}$ *Product rule for exponents*

    $= 600{,}000{,}000{,}000$

35. $(5 \times 10^4)(3 \times 10^2)$
    $= (5 \times 3)(10^4 \times 10^2)$
    $= 15 \times 10^6$
    $= 1.5 \times 10^7$
    $= 15{,}000{,}000$

306 Chapter 10 Exponents and Polynomials

37. $(3.15 \times 10^{-4})(2.04 \times 10^8)$
    $= (3.15 \times 2.04)(10^{-4} \times 10^8)$
    $= 6.426 \times 10^4$ *Scientific notation*
    $= 64{,}260$ *Without exponents*

39. $\dfrac{9 \times 10^{-5}}{3 \times 10^{-1}} = \dfrac{9}{3} \times \dfrac{10^{-5}}{10^{-1}}$
    $= 3 \times 10^{-5-(-1)}$
    $= 3 \times 10^{-4}$
    $= 0.0003$

41. $\dfrac{8 \times 10^3}{2 \times 10^2} = \dfrac{8}{2} \times \dfrac{10^3}{10^2}$
    $= 4 \times 10^1$
    $= 40$

43. $\dfrac{(2.6 \times 10^{-3})(7.0 \times 10^{-1})}{(2 \times 10^2)(3.5 \times 10^{-3})}$
    $= \dfrac{(2.6 \times 7.0)(10^{-3} \times 10^{-1})}{(2 \times 3.5)(10^2 \times 10^{-3})}$
    $= \dfrac{18.2 \times 10^{-4}}{7 \times 10^{-1}}$
    $= \dfrac{18.2}{7} \times \dfrac{10^{-4}}{10^{-1}}$
    $= 2.6 \times 10^{-3}$
    $= 0.0026$

45. $3{,}700{,}000 = 3.7 \times 10^6$
    $\dfrac{1.322 \times 10^9 \text{ people}}{3.7 \times 10^6 \text{ mi}^2} = \dfrac{1.322}{3.7} \times \dfrac{10^9}{10^6}$
    $\approx 0.357 \times 10^3$
    $= 3.57 \times 10^2$ (rounded)
    $= 357$ people per square mile

47. **(a)** 1 million $= 1 \times 10^6$
    $(1 \times 10^{-10} \text{ m})(1 \times 10^6) = (1 \times 1)(10^{-10} \times 10^6)$
    $= 1 \times 10^{-4}$
    $= 0.0001$ m

    (b) 1 billion $= 1 \times 10^9$
    $(1 \times 10^{-10} \text{ m})(1 \times 10^9) = (1 \times 1)(10^{-10} \times 10^9)$
    $= 1 \times 10^{-1}$
    $= 0.1$ m

49. $\dfrac{2.1 \times 10^{11}}{2.96 \times 10^8} = \dfrac{2.1}{2.96} \times \dfrac{10^{11}}{10^8}$
    $\approx 0.709 \times 10^3$
    $= 7.09 \times 10^2$
    or about 709 items

10.4 Adding and Subtracting Polynomials

10.4 Margin Exercises

1. **(a)** $5x^4 - 7x^4$
    $= (5-7)x^4$ Distributive property
    $= -2x^4$

    (b) $9pq + 3pq - 2pq$
    $= (9+3-2)pq$ Distributive property
    $= 10pq$

    (c) $r^2 + 3r + 5r^2$
    $= (1r^2 + 5r^2) + 3r$ Commutative and Associative properties
    $= (1+5)r^2 + 3r$ Distributive property
    $= 6r^2 + 3r$

    (d) $8t + 6w$ has unlike terms that cannot be added.

2. **(a)** $3m^5 + 5m^2 - 2m + 1$ is a polynomial written in descending powers, since $m = m^1$ and $1 = m^0$. So **A** and **B** apply.

    (b) $2p^4 + p^6$ is a polynomial, but is not written in descending powers. So **A** applies.

    (c) $\dfrac{1}{x} + 2x^2 + 3$ is not a polynomial, because $\dfrac{1}{x}$ has a variable in the denominator. So **C** applies.

    (d) $x - 3$ is a polynomial written in descending powers, since $x = x^1$ and $3 = 3x^0$. So **A** and **B** apply.

3. **(a)** $3x^2 + 2x - 4$ cannot be simplified, because there are no like terms. There are three terms, and the largest exponent on the variable x is 2. So the polynomial has degree 2 and is a trinomial.

    (b) $x^3 + 4x^3 = 1x^3 + 4x^3$ simplifies to $(1+4)x^3 = 5x^3$. This polynomial has one term, and the exponent on the variable x is 3. So the polynomial has degree 3 and is a monomial.

    (c) $x^8 - x^7 + 2x^8 = 1x^8 + 2x^8 - x^7$ simplifies to $(1+2)x^8 - x^7 = 3x^8 - x^7$. This polynomial has two terms, and the largest exponent on the variable x is 8. So the polynomial has degree 8 and is a binomial.

4. **(a)** $2y^3 + 8y - 6$ Replace y with -1.
    $= 2(-1)^3 + 8(-1) - 6$
    $= 2(-1) + 8(-1) - 6$
    $= -2 - 8 - 6$
    $= -16$

(b) $2y^3 + 8y - 6$ Replace y with 4.
$$= 2(4)^3 + 8(4) - 6$$
$$= 2(64) + 8(4) - 6$$
$$= 128 + 32 - 6$$
$$= 154$$

5. (a) $\quad 4x^3 - 3x^2 + 2x$
$$\underline{\quad 6x^3 + 2x^2 - 3x}$$
$$10x^3 - x^2 - x$$

(b) $\quad x^2 - 2x + 5$
$$\underline{\quad 4x^2 - 2}$$
$$5x^2 - 2x + 3$$

6. (a) Combine like terms.
$$(2x^4 - 6x^2 + 7) + (-3x^4 + 5x^2 + 2)$$
$$= (2x^4 - 3x^4) + (-6x^2 + 5x^2) + (7 + 2)$$
$$= -x^4 - x^2 + 9$$

(b) Combine like terms.
$$(3x^2 + 4x + 2) + (6x^3 - 5x - 7)$$
$$= 6x^3 + 3x^2 + (4x - 5x) + (2 - 7)$$
$$= 6x^3 + 3x^2 - x - 5$$

7. (a) Change subtraction to addition.
$$(14y^3 - 6y^2 + 2y - 5) - (2y^3 - 7y^2 - 4y + 6)$$
$$= (14y^3 - 6y^2 + 2y - 5) + (-2y^3 + 7y^2 + 4y - 6)$$

Combine like terms.
$$= (14y^3 - 2y^3) + (-6y^2 + 7y^2) + (2y + 4y)$$
$$ + (-5 - 6)$$
$$= 12y^3 + y^2 + 6y - 11$$

Check by adding.

$\quad 2y^3 - 7y^2 - 4y + 6 \quad$ *Second polynomial*
$\underline{12y^3 + y^2 + 6y - 11} \quad$ *Answer*
$14y^3 - 6y^2 + 2y - 5 \quad$ *First polynomial*

(b) $2y^2 - 6y + 6$ from $3y^3 - 7y^2 - 11y + 6$

Change subtraction to addition.
$$(3y^3 - 7y^2 - 11y + 6) - (2y^2 - 6y + 6)$$
$$= (3y^3 - 7y^2 - 11y + 6) + (-2y^2 + 6y - 6)$$

Combine like terms.
$$= 3y^3 + (-7y^2 - 2y^2) + (-11y + 6y)$$
$$ + (6 - 6)$$
$$= 3y^3 - 9y^2 - 5y$$

Check by adding.

$\quad\quad\quad 2y^2 - 6y + 6 \quad$ *second poly. in subtraction*
$\underline{3y^3 - 9y^2 - 5y } \quad$ *Answer*
$3y^3 - 7y^2 - 11y + 6 \quad$ *first poly. in subtraction*

10.4 Section Exercises

1. In the term $7x^5$, the coefficient is $\underline{7}$ and the exponent is $\underline{5}$.

3. The degree of the term $-4x^8$ is $\underline{8}$, the exponent.

5. When $x^2 + 10$ is evaluated for $x = 4$, the result is
$$4^2 + 10 = 16 + 10 = \underline{26}.$$

7. The polynomial $6x^4$ has one term. The coefficient of this term is 6.

9. The polynomial t^4 has one term. Since $t^4 = 1 \cdot t^4$, the coefficient of this term is 1.

11. The polynomial $-19r^2 - r$ has two terms. The coefficient of r^2 is -19 and the coefficient of r is -1.

13. $x + 8x^2 - 8x^3 = 1x + 8x^2 - 8x^3$ has 3 terms, and the coefficients are 1, 8, and -8.

15. $-3m^5 + 5m^5 = (-3 + 5)m^5 = 2m^5$

17. $2r^5 + (-3r^5) = [2 + (-3)]r^5$
$$= -1r^5 = -r^5$$

19. The polynomial $0.2m^5 - 0.5m^2$ cannot be simplified. The two terms are unlike because the exponents on the variables are different, so they cannot be combined.

21. $-3x^5 + 2x^5 - 4x^5 = (-3 + 2 - 4)x^5$
$$= -5x^5$$

23. $-4p^7 + 8p^7 + 5p^9 = (-4 + 8)p^7 + 5p^9$
$$= 4p^7 + 5p^9$$

In descending powers of the variable, this polynomial is written $5p^9 + 4p^7$.

25. $-4y^2 + 3y^2 - 2y^2 + y^2$
$$= (-4 + 3 - 2 + 1)y^2$$
$$= -2y^2$$

27. $6x^4 - 9x$

This polynomial has no like terms, so it is already simplified. It is already written in descending powers of the variable x. The highest degree of any nonzero term is 4, so the degree of the polynomial is 4. There are two terms, so this is a *binomial*.

29. $5m^4 - 3m^2 + 6m^5 - 7m^3$

This polynomial is already simplified (no like terms). In descending powers, it is $6m^5 + 5m^4 - 7m^3 - 3m^2$. The degree is 5 (the largest exponent on the variable m). The polynomial has four terms, so it is neither a monomial, nor a binomial, nor a trinomial.

31. $\frac{5}{3}x^4 - \frac{2}{3}x^4 + \frac{1}{3}x^2 - 4$

$= \left(\frac{5}{3} - \frac{2}{3}\right)x^4 + \frac{1}{3}x^2 - 4$

$= x^4 + \frac{1}{3}x^2 - 4$

The resulting polynomial is a *trinomial* of degree 4.

33. $0.8x^4 - 0.3x^4 - 0.5x^4 + 7x$

$= (0.8 - 0.3 - 0.5)x^4 + 7x$

$= 0x^4 + 7x = 7x$

Since $7x$ can be written as $7x^1$, the degree of the polynomial is 1. The simplified polynomial has one term, so it is a *monomial*.

35. (a) $-2x + 3 = -2(2) + 3$ Let $x = 2$

$= -4 + 3$

$= -1$

(b) $-2x + 3 = -2(-1) + 3$ Let $x = -1$

$= 2 + 3$

$= 5$

37. (a) $2x^2 + 5x + 1$

$= 2(2)^2 + 5(2) + 1$ Let $x = 2$

$= 2(4) + 10 + 1$

$= 8 + 10 + 1$

$= 18 + 1$

$= 19$

(b) $2x^2 + 5x + 1$

$= 2(-1)^2 + 5(-1) + 1$ Let $x = -1$

$= 2(1) - 5 + 1$

$= 2 - 5 + 1$

$= -3 + 1$

$= -2$

39. (a) $2x^5 - 4x^4 + 5x^3 - x^2$

$= 2(2)^5 - 4(2)^4 + 5(2)^3 - (2)^2$ Let $x = 2$

$= 2(32) - 4(16) + 5(8) - 4$

$= 64 - 64 + 40 - 4$

$= 36$

(b) $2x^5 - 4x^4 + 5x^3 - x^2$

$= 2(-1)^5 - 4(-1)^4 + 5(-1)^3 - (-1)^2$

Let $x = -1$

$= 2(-1) - 4(1) + 5(-1) - 1$

$= -2 - 4 - 5 - 1$

$= -12$

41. (a) $-4x^5 + x^2$

$= -4(2)^5 + (2)^2$ Let $x = 2$

$= -4(32) + 4$

$= -128 + 4$

$= -124$

(b) $-4x^5 + x^2$

$= -4(-1)^5 + (-1)^2$ Let $x = -1$

$= -4(-1) + 1$

$= 4 + 1$

$= 5$

43. $3m^2 + 5m$ and $2m^2 - 2m$

$3m^2 + 5m$

$\underline{2m^2 - 2m}$

$5m^2 + 3m$

45. $-6x^3 - 4x + 1$ and $5x + 5$

$-6x^3 - 4x + 1$

$\underline{\,5x + 5}$

$-6x^3 + 1x + 6 = -6x^3 + x + 6$

47. $3w^3 - 2w^2 + 8w$ and $2w^2 - 8w + 5$

$3w^3 - 2w^2 + 8w$

$\underline{\,2w^2 - 8w + 5}$

$3w^3 + 0w^2 + 0w + 5 = 3w^3 + 5$

49. $(12x^4 - x^2) - (8x^4 + 3x^2)$

$= (12x^4 - x^2) + (-8x^4 - 3x^2)$

$= (12x^4 - 8x^4) + (-x^2 - 3x^2)$

$= 4x^4 - 4x^2$

51. $(2r^2 + 3r - 12) + (6r^2 + 2r)$

$= (2r^2 + 6r^2) + (3r + 2r) - 12$

$= 8r^2 + 5r - 12$

53. $(8m^2 - 7m) - (3m^2 + 7m - 6)$

$= (8m^2 - 7m) + (-3m^2 - 7m + 6)$

$= (8m^2 - 3m^2) + (-7m - 7m) + 6$

$= 5m^2 - 14m + 6$

55. $(16x^3 - x^2 + 3x) + (-12x^3 + 3x^2 + 2x)$

$= (16x^3 - 12x^3) + (-x^2 + 3x^2) + (3x + 2x)$

$= 4x^3 + 2x^2 + 5x$

57. Subtract $5m^2 - 4$ from $12m^3 - 8m^2 + 6m + 7$.

$(12m^3 - 8m^2 + 6m + 7) - (5m^2 - 4)$

$= (12m^3 - 8m^2 + 6m + 7) + (-5m^2 + 4)$

$= 12m^3 + (-8m^2 - 5m^2) + 6m + (7 + 4)$

$= 12m^3 - 13m^2 + 6m + 11$

59. Subtract $9x^2 - 3x + 7$ from $-2x^2 - 6x + 4$.

$\quad (-2x^2 - 6x + 4) - (9x^2 - 3x + 7)$
$\quad = (-2x^2 - 6x + 4) + (-9x^2 + 3x - 7)$
$\quad = (-2x^2 - 9x^2) + (-6x + 3x) + (4 - 7)$
$\quad = -11x^2 - 3x - 3$

61. Use the formula for the perimeter of a rectangle, $P = 2L + 2W$, with length $L = 4x^2 + 3x + 1$ and width $W = x + 2$.

$\quad P = 2L + 2W$
$\quad = 2(4x^2 + 3x + 1) + 2(x + 2)$
$\quad = 8x^2 + 6x + 2 + 2x + 4$
$\quad = 8x^2 + 8x + 6$

The perimeter of the rectangle is $8x^2 + 8x + 6$.

63. Use the formula for the perimeter of a triangle, $P = a + b + c$, with $a = 3t^2 + 2t + 7$, $b = 5t^2 + 2$, and $c = 6t + 4$.

$\quad P = (3t^2 + 2t + 7) + (5t^2 + 2) + (6t + 4)$
$\quad = (3t^2 + 5t^2) + (2t + 6t) + (7 + 2 + 4)$
$\quad = 8t^2 + 8t + 13$

The perimeter of the triangle is $8t^2 + 8t + 13$.

65. $D = 100t - 13t^2$ $\quad$ *Replace t with 1.*
$\quad D = 100(1) - 13(1)^2$
$\quad D = 100 - 13$
$\quad D = 87$ ft

66. From the previous exercise, $D = 100t - 13t^2$.

t (sec)	D (ft)	(t, D)
0	$100(0) - 13(0)^2 = 0$	$(0, 0)$
0.5	$100(0.5) - 13(0.5)^2$ $= 46.75 \approx 47$	$(0.5, 47)$
1	87 (See Exercise 65)	$(1, 87)$
1.5	$100(1.5) - 13(1.5)^2$ $= 120.75 \approx 121$	$(1.5, 121)$
2	$100(2) - 13(2)^2 = 148$	$(2, 148)$
2.5	$100(2.5) - 13(2.5)^2$ $= 168.75 \approx 169$	$(2.5, 169)$
3	$100(3) - 13(3)^2 = 183$	$(3, 183)$
3.5	$100(3.5) - 13(3.5)^2$ $= 190.75 \approx 191$	$(3.5, 191)$

67.–68.

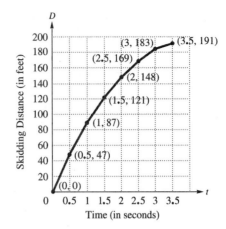

Answers will vary. This graph is a curve instead of the straight lines graphed for linear equations. Also, the rate of change in the distance is not consistent. From 0.5 sec to 1 sec, the distance increases 40 ft, but from 3 sec to 3.5 sec, the distance only increases 8 ft.

10.5 Multiplying Polynomials: An Introduction

10.5 Margin Exercises

1. **(a)** $5m^3(2m + 7)$
$\quad = 5m^3(2m) + 5m^3(7)$ $\quad$ Distributive property
$\quad = 10m^4 + 35m^3$ $\quad$ Multiply monomials.

(b) $2x^4(3x^2 + 2x - 5)$
$\quad = 2x^4(3x^2) + 2x^4(2x) + 2x^4(-5)$
$\quad\quad$ Distributive property
$\quad = 6x^6 + 4x^5 + (-10x^4)$
$\quad\quad$ Multiply monomials.
$\quad = 6x^6 + 4x^5 - 10x^4$

(c) $-4y^2(3y^3 + 2y^2 - 4y + 8)$
$\quad = -4y^2(3y^3) + (-4y^2)(2y^2)$
$\quad\quad + (-4y^2)(-4y) + (-4y^2)(8)$
$\quad\quad$ Distributive property
$\quad = -12y^5 + (-8y^4) + (16y^3) + (-32y^2)$
$\quad\quad$ Multiply monomials.
$\quad = -12y^5 - 8y^4 + 16y^3 - 32y^2$

2. Multiply each term of the second polynomial by each term of the first. Then combine like terms.

(a) $(k + 6)(k + 7)$
$\quad = k(k) + k(7) + 6(k) + 6(7)$
$\quad = k^2 + 7k + 6k + 42$
$\quad = k^2 + 13k + 42$

(b) $(n-2)(4n+3)$
$= n(4n) + n(3) + (-2)(4n) + (-2)(3)$
$= 4n^2 + 3n + (-8n) + (-6)$
$= 4n^2 - 5n - 6$

(c) $(2m+1)(2m^2 + 4m - 3)$
$= 2m(2m^2) + 2m(4m) + 2m(-3)$
$\quad + 1(2m^2) + 1(4m) + 1(-3)$
$= 4m^3 + 8m^2 + (-6m) + 2m^2 + 4m + (-3)$
$= 4m^3 + 10m^2 - 2m - 3$

(d) $(6p^2 - 2p - 4)(3p^2 + 5)$
$= 6p^2(3p^2) + 6p^2(5) + (-2p)(3p^2)$
$\quad + (-2p)(5) + (-4)(3p^2) + (-4)(5)$
$= 18p^4 + 30p^2 + (-6p^3) + (-10p)$
$\quad + (-12p^2) + (-20)$
$= 18p^4 - 6p^3 + 18p^2 - 10p - 20$

3.
$$\begin{array}{r} 2x^3 + 3x^2 + 4x - 5 \\ x + 4 \\ \hline 8x^3 + 12x^2 + 16x - 20 \leftarrow 4(a)* \\ 2x^4 + 3x^3 + 4x^2 - 5x \quad\quad \leftarrow x(a) \\ \hline 2x^4 + 11x^3 + 16x^2 + 11x - 20 \leftarrow \text{Add terms.} \end{array}$$
$*a = 2x^3 + 3x^2 + 4x - 5$

10.5 Section Exercises

1. $5y(8y^2 - 3)$
$= 5y(8y^2) + 5y(-3)$ Distributive property
$= 40y^3 - 15y$ Multiply monomials.

3. $-2m(3m + 2)$
$= -2m(3m) + (-2m)(2)$ Distributive property
$= -6m^2 - 4m$ Multiply monomials.

5. $4x^2(6x^2 - 3x + 2)$
$= 4x^2(6x^2) + 4x^2(-3x) + 4x^2(2)$
$= 24x^4 - 12x^3 + 8x^2$

7. $-3k^3(2k^3 - 3k^2 - k + 1)$
$= -3k^3(2k^3) + (-3k^3)(-3k^2)$
$\quad + (-3k^3)(-k) + (-3k^3)(1)$
$= -6k^6 + 9k^5 + 3k^4 - 3k^3$

9. Multiply each term of the second polynomial by each term of the first. Then combine like terms.

$(n-2)(n+3)$
$= n(n) + n(3) + (-2)(n) + (-2)(3)$
$= n^2 + 3n + (-2n) + (-6)$
$= n^2 + n - 6$

11. Multiply each term of the second polynomial by each term of the first. Then combine like terms.

$(4r+1)(2r-3)$
$= 4r(2r) + 4r(-3) + 1(2r) + 1(-3)$
$= 8r^2 + (-12r) + 2r + (-3)$
$= 8r^2 - 10r - 3$

13. $(6x+1)(2x^2 + 4x + 1)$
$= (6x)(2x^2) + (6x)(4x) + (6x)(1)$
$\quad + (1)(2x^2) + (1)(4x) + (1)(1)$
$= 12x^3 + 24x^2 + 6x + 2x^2 + 4x + 1$
$= 12x^3 + 26x^2 + 10x + 1$

15. $(4m+3)(5m^3 - 4m^2 + m - 5)$
Multiply vertically.

$$\begin{array}{r} 5m^3 - 4m^2 + m - 5 \\ 4m + 3 \\ \hline 15m^3 - 12m^2 + 3m - 15 \\ 20m^4 - 16m^3 + 4m^2 - 20m \\ \hline 20m^4 - m^3 - 8m^2 - 17m - 15 \end{array}$$

17. $(5x^2 + 2x + 1)(x^2 - 3x + 5)$
Multiply vertically.

$$\begin{array}{r} 5x^2 + 2x + 1 \\ x^2 - 3x + 5 \\ \hline 25x^2 + 10x + 5 \\ -15x^3 - 6x^2 - 3x \\ 5x^4 + 2x^3 + x^2 \\ \hline 5x^4 - 13x^3 + 20x^2 + 7x + 5 \end{array}$$

19. $(-x-3)(-x-4)$
Multiply vertically.

$$\begin{array}{r} -x - 3 \\ -x - 4 \\ \hline 4x + 12 \\ x^2 + 3x \\ \hline x^2 + 7x + 12 \end{array}$$

21. $(2x+1)(x^2 + 3x)$
Multiply vertically.

$$\begin{array}{r} x^2 + 3x \\ 2x + 1 \\ \hline x^2 + 3x \\ 2x^3 + 6x^2 \\ \hline 2x^3 + 7x^2 + 3x \end{array}$$

23. $(a+4)(a+5)$
$= a(a) + a(5) + 4(a) + 4(5)$
$= a^2 + 5a + 4a + 20$
$= a^2 + 9a + 20$ **Choice D**

$(a-4)(a+5)$
$= a(a) + a(5) + (-4)(a) + (-4)(5)$
$= a^2 + 5a + (-4a) + (-20)$
$= a^2 + a - 20$ **Choice B**

$(a+4)(a-5)$
$= a(a) + a(-5) + 4(a) + 4(-5)$
$= a^2 + (-5a) + 4a + (-20)$
$= a^2 - a - 20$ **Choice A**

$(a-4)(a-5)$
$= a(a) + a(-5) + (-4)(a) + (-4)(-5)$
$= a^2 + (-5a) + (-4a) + 20$
$= a^2 - 9a + 20$ **Choice C**

25. Area = Width · length
$= 10(3x+6)$
$= 10(3x) + 10(6)$
$= (30x + 60)$ yd^2

26. Area = 600
$30x + 60 = 600$ Subtract 60.
$30x = 540$ Divide by 30.
$\dfrac{30x}{30} = \dfrac{540}{30}$
$x = 18$

27. With $x = 18$, $3x + 6 = 3(18) + 6 = 60$.

The rectangle measures 10 yd by 60 yd.

28. ($3.50 per yd^2)(600 yd^2) = $2100

29. Perimeter = 2(length) + 2(width)
$= 2(60) + 2(10)$
$= 120 + 20$
$= 140$ yd

30. ($9.00 per yd)(140 yd) = $1260

31. $(x+4)(x-4) = x(x) + x(-4) + 4(x) + 4(-4)$
$= x^2 + (-4x) + 4x + (-16)$
$= x^2 - 16$

$(y+3)(y-3) = y(y) + y(-3) + 3(y) + 3(-3)$
$= y^2 + (-3y) + 3y + (-9)$
$= y^2 - 9$

$(r+7)(r-7) = r(r) + r(-7) + 7(r) + 7(-7)$
$= r^2 + (-7r) + 7r + (-49)$
$= r^2 - 49$

Each answer is the difference of the squares of the first terms and the squares of the last terms from the binomials.

Chapter 10 Review Exercises

1. $(-5)^6(-5)^5 = (-5)^{6+5} = (-5)^{11}$

2. $(-8x^4)(9x^3) = (-8)(9)(x^4)(x^3)$
$= -72x^{4+3} = -72x^7$

3. $(2x^2)(5x^3)(x^9) = (2)(5)(x^2)(x^3)(x^9)$
$= 10x^{2+3+9} = 10x^{14}$

4. $(19x)^5 = 19^5 x^5$

5. $5(pt)^4 = 5p^4 t^4$

6. $\left(\dfrac{7}{5}\right)^6 = \dfrac{7^6}{5^6}$

7. $(3x^2 y^3)^3$
$= 3^3 (x^2)^3 (y^3)^3$
$= 3^3 x^{2 \cdot 3} y^{3 \cdot 3}$
$= 3^3 x^6 y^9$

8. The product rule for exponents does not apply because it is the *sum* of 7^2 and 7^4, not their *product*.

9. $5^0 + 8^0 = 1 + 1 = 2$

10. $2^{-5} = \dfrac{1}{2^5} = \dfrac{1}{32}$

11. $10w^0 = 10 \cdot 1 = 10$

12. $4^{-2} + 4^{-1} = \dfrac{1}{4^2} + \dfrac{1}{4^1}$
$= \dfrac{1}{16} + \dfrac{1}{4}$
$= \dfrac{1}{16} + \dfrac{4}{16} = \dfrac{5}{16}$

13. $\dfrac{p^{-8}}{p^4} = p^{-8-4} = p^{-12} = \dfrac{1}{p^{12}}$

14. $\dfrac{r^{-2}}{r^{-6}} = r^{-2-(-6)} = r^{-2+6} = r^4$

15. $\dfrac{5^4}{5^5} = 5^{4-5} = 5^{-1} = \dfrac{1}{5^1}$ or $\dfrac{1}{5}$

16. $10^3(10^{-10}) = 10^{3+(-10)} = 10^{-7} = \dfrac{1}{10^7}$

17. $n^{-2} \cdot n^3 \cdot n^{-4} = n^{-2+3+(-4)} = n^{-3} = \dfrac{1}{n^3}$

18. $(10^4)(10^{-1}) = 10^{4+(-1)} = 10^3$

19. $(2^{-4})(2^{-4}) = 2^{-4+(-4)} = 2^{-8} = \dfrac{1}{2^8}$

20. $\dfrac{x^2}{x^{-4}} = x^{2-(-4)} = x^{2+4} = x^6$

21. $48{,}000{,}000 = 4.8 \times 10^7$

    Move the decimal point 7 places. The *original* number was "large" (10 or more), so the exponent is *positive* 7.

22. $0.0000000824 = 8.24 \times 10^{-8}$

    Move the decimal point 8 places. The *original* number was "small" (between 0 and 1), so the exponent is *negative* 8.

23. $2.4 \times 10^4 = 24{,}000$

    Since the exponent is *positive* 4, move the decimal point 4 places to the *right*.

24. $9.95 \times 10^{-12} = 0.000\,000\,000\,009\,95$

    Since the exponent is *negative* 12, move the decimal point 12 places to the *left*.

25. $(2 \times 10^{-3})(4 \times 10^5)$
    $= (2 \times 4)(10^{-3} \times 10^5)$
    $= 8 \times 10^{-3+5} = 8 \times 10^2$
    $= 800$

26. $\dfrac{12 \times 10^{-8}}{4 \times 10^{-3}} = \dfrac{12}{4} \times \dfrac{10^{-8}}{10^{-3}}$
    $= 3 \times 10^{-8-(-3)}$
    $= 3 \times 10^{-5}$
    $= 0.00003$

27. $\dfrac{(2.5 \times 10^5)(4.8 \times 10^{-4})}{(7.5 \times 10^8)(1.6 \times 10^{-5})}$
    $= \dfrac{(2.5 \times 4.8)(10^5 \times 10^{-4})}{(7.5 \times 1.6)(10^8 \times 10^{-5})}$
    $= \dfrac{12 \times 10^1}{12 \times 10^3}$
    $= \dfrac{12}{12} \times \dfrac{10^1}{10^3}$
    $= 1 \times 10^{-2}$
    $= 0.01$

28. $\dfrac{3.04 \times 10^8 \text{ people}}{3.54 \times 10^6 \text{ mi}^2} = \dfrac{3.04}{3.54} \times \dfrac{10^8}{10^6}$
    $\approx 0.86 \times 10^2$
    $= 8.6 \times 10^1$ (rounded)
    $= 86$ people per square mile

29. $(36 \text{ sec})(3.0 \times 10^5 \text{ km per sec}) = (36 \times 3.0)(10^5)$
    $= 108 \times 10^5$
    $= 1.08 \times 10^7$
    $= 10{,}800{,}000$ km

30. $9m^2 + 11m^2 + 2m^2 = (9 + 11 + 2)m^2$
    $= 22m^2$

    The degree is 2.

    To determine if the polynomial is a monomial, binomial, or trinomial, count the number of terms in the final expression.

    There is one term, so this is a *monomial*.

31. $-7y^5 - 8y^4 - y^5 + y^4 + 9y$
    $= -7y^5 - 1y^5 - 8y^4 + 1y^4 + 9y$
    $= (-7 - 1)y^5 + (-8 + 1)y^4 + 9y$
    $= -8y^5 - 7y^4 + 9y$

    The degree is 5.

    There are three terms, so the polynomial is a *trinomial*.

32. $(-2a^3 + 5a^2) + (-3a^3 - a^2)$

    $\begin{array}{rr} -2a^3 & + 5a^2 \\ -3a^3 & - a^2 \\ \hline -5a^3 & + 4a^2 \end{array}$

33. Subtract $-5y^2 + 2y - 7$ from $6y^2 - 8y + 2$.

    $\begin{array}{rrr} 6y^2 & - 8y & + 2 \\ -5y^2 & + 2y & - 7 \end{array}$

    Change all signs in the second row and then add.

    $\begin{array}{rrr} 6y^2 & - 8y & + 2 \\ 5y^2 & - 2y & + 7 \\ \hline 11y^2 & - 10y & + 9 \end{array}$

34. $(-5y^2 + 3y + 11) + (4y^3 - 7y + 15)$
    $= 4y^3 - 5y^2 + (3y - 7y) + (11 + 15)$
    $= 4y^3 - 5y^2 - 4y + 26$

35. $(12r^4 - 7r^3 + 2r^2) - (5r^4 - 3r^3 + 2r^2 + 1)$
    $= (12r^4 - 7r^3 + 2r^2) + (-5r^4 + 3r^3 - 2r^2 - 1)$
    $= (12r^4 - 5r^4) + (-7r^3 + 3r^3) + (2r^2 - 2r^2) - 1$
    $= 7r^4 - 4r^3 - 1$

36. $5x(2x + 14)$
$= 5x(2x) + 5x(14)$
$= 10x^2 + 70x$

37. $-3p^3(2p^2 - 5p)$
$= -3p^3(2p^2) + (-3p^3)(-5p)$
$= -6p^5 + 15p^4$

38. $(3r - 2)(2r^2 + 4r - 3)$
Multiply vertically.

$$\begin{array}{r} 2r^2 + 4r - 3 \\ 3r - 2 \\ \hline -4r^2 - 8r + 6 \\ 6r^3 + 12r^2 - 9r \\ \hline 6r^3 + 8r^2 - 17r + 6 \end{array}$$

39. $(3k - 6)(2k + 1)$
$= (3k)(2k) + (3k)(1) + (-6)(2k) + (-6)(1)$
$= 6k^2 + 3k - 12k - 6$
$= 6k^2 - 9k - 6$

40. $(6p - 3q)(2p - 7q)$
$= 6p(2p) + 6p(-7q) + (-3q)(2p) + (-3q)(-7q)$
$= 12p^2 + (-42pq) + (-6pq) + (21q^2)$
$= 12p^2 - 48pq + 21q^2$

41.
$$\begin{array}{r} m^2 + m - 9 \\ 2m^2 + 3m - 1 \\ \hline -m^2 - m + 9 \\ 3m^3 + 3m^2 - 27m \\ 2m^4 + 2m^3 - 18m^2 \\ \hline 2m^4 + 5m^3 - 16m^2 - 28m + 9 \end{array}$$

42. [10.2] $19^0 - 3^0 = 1 - 1 = 0$

43. [10.1] $(3p)^4 = 3^4 p^4$

44. [10.2] $7^{-2} = \dfrac{1}{7^2}$

45. [10.5] $-m^5(8m^2 + 10m + 6)$
$= -m^5(8m^2) + (-m^5)(10m) + (-m^5)(6)$
$= -8m^7 + (-10m^6) + (-6m^5)$
$= -8m^7 - 10m^6 - 6m^5$

46. [10.2] $2^{-1} + 4^{-1} = \dfrac{1}{2^1} + \dfrac{1}{4^1}$
$= \dfrac{1}{2} + \dfrac{1}{4}$
$= \dfrac{2}{4} + \dfrac{1}{4} = \dfrac{3}{4}$

47. [10.5] $(a + 2)(a^2 - 4a + 1)$
Multiply vertically.

$$\begin{array}{r} a^2 - 4a + 1 \\ a + 2 \\ \hline 2a^2 - 8a + 2 \\ a^3 - 4a^2 + a \\ \hline a^3 - 2a^2 - 7a + 2 \end{array}$$

48. [10.4] $(5y^3 - 8y^2 + 7) - (-3y^3 + y^2 + 2)$
$= (5y^3 - 8y^2 + 7) + (3y^3 - y^2 - 2)$
$= (5y^3 + 3y^3) + (-8y^2 - y^2) + (7 - 2)$
$= 8y^3 - 9y^2 + 5$

49. [10.5] Perimeter $= 2 \cdot$ length $+ 2 \cdot$ width
$= 2(2x - 3) + 2(x + 2)$
$= 2(2x) + 2(-3) + 2(x) + 2(2)$
$= 4x + (-6) + 2x + 4$
$= 6x - 2$

Area $=$ length $\cdot$ width
$= (2x - 3)(x + 2)$
$= 2x(x) + 2x(2) + (-3)(x) + (-3)(2)$
$= 2x^2 + 4x + (-3x) + (-6)$
$= 2x^2 + x - 6$

Chapter 10 Test

1. $5^{-4} = \dfrac{1}{5^4} = \dfrac{1}{625}$

2. $(-3)^0 + 4^0 = 1 + 1 = 2$

3. $4^{-1} + 3^{-1} = \dfrac{1}{4^1} + \dfrac{1}{3^1} = \dfrac{3}{12} + \dfrac{4}{12} = \dfrac{7}{12}$

4. $6^{-3} \cdot 6^4 = 6^{-3+4} = 6^1$ or 6

5. $12(xy)^3 = 12x^3y^3$

6. $\dfrac{10^5}{10^9} = 10^{5-9} = 10^{-4} = \dfrac{1}{10^4}$

7. $r^{-4} \cdot r^{-4} \cdot r^3 = r^{-4+(-4)+3} = r^{-5} = \dfrac{1}{r^5}$

8. $(7x^2)(-3x^5) = 7(-3)(x^2 \cdot x^5)$
$= -21x^{2+5}$
$= -21x^7$

9. $\dfrac{m^{-5}}{m^{-8}} = m^{-5-(-8)} = m^{-5+8} = m^3$

10. $(3a^4b)^2 = 3^2(a^4)^2 b^2$
$= 3^2 a^{4 \cdot 2} b^2$
$= 3^2 a^8 b^2$ or $9a^8b^2$

11. $3^4 + 3^2 = 81 + 9 = 90$
Note that this is a *sum*, not a *product*, so we cannot use the product rule for exponents.

12. $300{,}000{,}000{,}000 = 3.0 \times 10^{11}$

Move the decimal point 11 places. The *original* number was "large" (10 or more), so the exponent is *positive* 11.

13. $0.00000557 = 5.57 \times 10^{-6}$

Move the decimal point 6 places. The *original* number was "small" (between 0 and 1), so the exponent is *negative* 6.

14. $2.9 \times 10^7 = 29{,}000{,}000$

Since the exponent is *positive* 7, move the decimal point 7 places to the *right*.

15. $6.07 \times 10^{-8} = 0.000\,000\,060\,7$

Since the exponent is *negative* 8, move the decimal point 8 places to the *left*.

16. $(330{,}000)(5.98 \times 10^{24} \text{ kg})$
$= (3.3 \times 10^5)(5.98 \times 10^{24})$
$= (3.3 \times 5.98)(10^5 \times 10^{24})$
$= 19.734 \times 10^{29}$
$\approx 1.97 \times 10^{30}$ kg (rounded)

17. $5x^2 + 8x - 12x^2 = (5-12)x^2 + 8x$
$= -7x^2 + 8x$

The degree is 2 (the largest exponent on the variable x). The polynomial has two terms, so it is a binomial.

18. $13n^3 - n^2 + n^4 + 3n^4 - 9n^2$
$= (1+3)n^4 + 13n^3 + (-1-9)n^2$
$= 4n^4 + 13n^3 - 10n^2$

The degree is 4 (the largest exponent on the variable n). The polynomial has three terms, so it is a trinomial.

19. $(5t^4 - 3t^2 + 3) - (t^4 - t^2 + 3)$
$= (5t^4 - 3t^2 + 3) + (-t^4 + t^2 - 3)$
$= (5t^4 - t^4) + (-3t^2 + t^2) + (3-3)$
$= 4t^4 - 2t^2$

20. Add. $\quad 2y^4 \qquad\;\; - 8y + 8$
$\qquad\qquad \underline{\quad\;\; -3y^2 + 2y - 8\;}$
$\qquad\qquad 2y^4 - 3y^2 - 6y$

21. Subtract $9t^3 + 8t^2 - 6$ from $9t^3 - 4t^2 + 2$.

$(9t^3 - 4t^2 + 2) - (9t^3 + 8t^2 - 6)$
$= (9t^3 - 4t^2 + 2) + (-9t^3 - 8t^2 + 6)$
$= (9t^3 - 9t^3) + (-4t^2 - 8t^2) + (2+6)$
$= -12t^2 + 8$

22. $-4x^3(3x^2 - 5x)$
$= -4x^3(3x^2) + (-4x^3)(-5x)$
$= -12x^5 + 20x^4$

23. $(y+2)(3y-1)$
$= y(3y) + y(-1) + 2(3y) + 2(-1)$
$= 3y^2 + (-y) + 6y + (-2)$
$= 3y^2 + 5y - 2$

24. $(2r - 3)(r^2 + 2r - 5)$
Multiply vertically.

$$\begin{array}{r} r^2 + 2r - 5 \\ 2r - 3 \\ \hline -3r^2 - 6r + 15 \\ 2r^3 + 4r^2 - 10r \\ \hline 2r^3 + r^2 - 16r + 15 \end{array}$$

25. Use the formula for the perimeter of a square, $P = 4s$, with $s = 3x + 9$.

$P = 4(3x + 9)$
$\;\;\;= 4(3x) + 4(9)$
$\;\;\;= 12x + 36$

Use the formula for the area of a square, $A = s^2$, with $s = 3x + 9$.

$A = (3x+9)^2$
$\;\;\;= (3x+9)(3x+9)$
$\;\;\;= (3x)(3x) + (3x)(9) + (9)(3x) + (9)(9)$
$\;\;\;= 9x^2 + 27x + 27x + 81$
$\;\;\;= 9x^2 + 54x + 81$

Cumulative Review Exercises (Chapters 1–10)

1. (a) In words, 10.035 is written as ten and thirty-five thousandths.

(b) In words, 410,000,351,109 is written as four hundred ten billion, three hundred fifty-one thousand, one hundred nine.

2. (a) Using digits, three hundred million, six thousand, eighty is written as 300,006,080.

(b) Using digits, fifty-five ten-thousandths is written as 0.0055.

3. (a) 0.80<u>2</u>9 to the nearest hundredth

2 is less than 5, so drop all digits to the right of 0 in the hundredths place. **0.80**

(b) 340,<u>5</u>19,000 to the nearest million

Since the number in the hundred-thousands place is 5 or greater, round 519,000 up to 1,000,000. **341,000,000**

Cumulative Review Exercises (Chapters 1–10)   315

    (c) 1$\underline{4}$.973 to the nearest tenth

    Since the number in the hundredths place, 7, is 5 or greater, round 0.973 up to 1. **15.0**

4.   mean $= \dfrac{\text{sum of all values}}{\text{number of values}}$

$$= \dfrac{(41 + 65 + 37 + 90 + 41 + 65 + 48 + 41 + 59 + 40)}{10}$$

$$= \dfrac{527}{10} = 52.7$$

    The mean is $52.70.

    Arrange in increasing order:

        37, 40, 41, 41, 41, 48, 59, 65, 65, 90

    The median is the average of the 5th and 6th values.

$$\text{median} = \dfrac{41 + 48}{2} = \dfrac{89}{2} = 44.5$$

    The median is $44.50.

    $41 occurs the greatest number of times (3), so the mode is $41.

5. $-12 + 7.829$

    The addends have unlike signs.

    $|-12| = 12; \; |7.829| = 7.829$

    -12 has the larger absolute value, so the sum will be negative. Subtract the absolute values.

$$\begin{array}{r} 12.000 \\ -\,7.829 \\ \hline 4.171 \end{array}$$

    $-12 + 7.829 = -4.171$

6. $7 + 5(3 - 8) = 7 + 5(-5) = 7 - 25 = -18$

7. $1\dfrac{2}{3} + \dfrac{3}{5} = \dfrac{5}{3} + \dfrac{3}{5} = \dfrac{25}{15} + \dfrac{9}{15} = \dfrac{34}{15} = 2\dfrac{4}{15}$

8. $\dfrac{(-6)^2 + 9(0 - 4)}{10 \div 5(-7 + 5) - 10} = \dfrac{36 + 9(-4)}{10 \div 5(-2) - 10}$

$$= \dfrac{36 - 36}{2(-2) - 10}$$

$$= \dfrac{0}{-4 - 10}$$

$$= \dfrac{0}{-14}$$

$$= 0$$

9. $\dfrac{8x^2}{9} \cdot \dfrac{12w}{2x^3} = \dfrac{\cancel{8}x^2}{\cancel{9}} \cdot \dfrac{\cancel{12}w}{\cancel{2}x^3}$ (reducing 8/2=4, 12/9... showing 4 and 4, with 3 and 1)

$$= \dfrac{16x^2 w}{3x^3}$$

$$= \dfrac{16w}{3x}$$

10. $\dfrac{-0.7}{5.6} = \dfrac{-0.7}{8(0.7)} = -\dfrac{1}{8} = -0.125$

11. $\dfrac{10^5}{10^6} = 10^{5-6} = 10^{-1} = \dfrac{1}{10}$

12. $2\dfrac{1}{4} - 2\dfrac{5}{6} = \dfrac{9}{4} - \dfrac{17}{6} = \dfrac{27}{12} - \dfrac{34}{12} = -\dfrac{7}{12}$

13. $(-6)^2 + (-2)^3 = 36 + (-8) = 28$

14. $(-4y^3)(3y^4) = -4 \cdot 3 y^{3+4} = -12y^7$

15. $\dfrac{5b}{2a^2} \div \dfrac{3b^2}{10a} = \dfrac{5b}{2a^2} \cdot \dfrac{10a}{3b^2}$

$$= \dfrac{25ab}{3a^2 b^2}$$

$$= \dfrac{25}{3ab}$$

16. $\dfrac{r}{8} + \dfrac{6}{t} = \dfrac{r \cdot t}{8 \cdot t} + \dfrac{6 \cdot 8}{t \cdot 8} = \dfrac{rt}{8t} + \dfrac{48}{8t} = \dfrac{rt + 48}{8t}$

17. $\dfrac{-4(6)}{3^3 - 27} = \dfrac{-24}{27 - 27} = \dfrac{-24}{0}$, which is *undefined*.

18. $\dfrac{5}{6}$ of $900 = \dfrac{5}{6} \cdot 900$

$$= \dfrac{5 \cdot 900}{6}$$

$$= \dfrac{5 \cdot 6 \cdot 150}{6}$$

$$= 5 \cdot 150$$

$$= 750$$

19. $(-0.003)(0.04) = (-3 \times 10^{-3})(4 \times 10^{-2})$

$$= (-3 \times 4)(10^{-3 + (-2)})$$

$$= -12 \times 10^{-5}$$

$$= -0.00012$$

20. $(2a^3 b^2)^4 = 2^4 (a^3)^4 (b^2)^4$

$$= 2^4 \cdot a^{3 \cdot 4} b^{2 \cdot 4}$$

$$= 2^4 a^{12} b^8 \text{ or } 16 a^{12} b^8$$

21. $\dfrac{4}{9} - \dfrac{6}{m} = \dfrac{4 \cdot m}{9 \cdot m} - \dfrac{6 \cdot 9}{m \cdot 9} = \dfrac{4m}{9m} - \dfrac{54}{9m} = \dfrac{4m - 54}{9m}$

22. $t^5 \cdot t^{-2} \cdot t^{-4} = t^{5-2-4} = t^{-1} = \dfrac{1}{t^1}$ or $\dfrac{1}{t}$

23. $\dfrac{-\frac{15}{16}}{-6} = -\dfrac{15}{16} \div (-6) = -\dfrac{15}{16}\left(-\dfrac{1}{6}\right)$

$= \dfrac{\cancel{3}\cdot 5}{16\cdot 2\cdot \cancel{3}} = \dfrac{5}{32}$

24. $\left(-\dfrac{1}{2}\right)^4 + \cancel{6}^2\left(\dfrac{2}{\cancel{3}}\right) = \dfrac{1}{16} + 4$

$= \dfrac{1}{16} + \dfrac{64}{16}$

$= \dfrac{65}{16}, \text{ or } 4\dfrac{1}{16}$

25. $\dfrac{n^{-3}}{n^{-4}} = n^{-3-(-4)} = n^{-3+4} = n^1$ or n

26. $-8 - 88 = -8 + (-88) = -96$

27. $8^0 + 2^{-1} = 1 + \dfrac{1}{2} = 1\dfrac{1}{2}$

28. $9 - 7\dfrac{4}{5} = \dfrac{45}{5} - \dfrac{39}{5} = \dfrac{6}{5}$, or $1\dfrac{1}{5}$

29. When $x = 4$ and $y = -1$,

$-3xy^3 = -3(4)(-1)^3$
$= -3(4)(-1)$
$= 12$

30. When $y = -1$ and $w = -2$,

$15y - 6w = 15(-1) - 6(-2)$
$= -15 - (-12)$
$= -3$

31. When $x = 4$ and $y = -1$,

$x^2 - 2xy + 6 = 4^2 - 2(4)(-1) + 6$
$= 16 + 8 + 6$
$= 30$

32. $4 + h = 3h - 6$ Subtract 4.
$h = 3h - 10$ Subtract $3h$.
$-2h = -10$ Divide by -2.
$\dfrac{-2h}{-2} = \dfrac{-10}{-2}$
$h = 5$

The solution is 5.

33. $-1.65 = 0.5x + 2.3$ Subtract 2.3.
$-3.95 = 0.5x$ Divide by 0.5.
$\dfrac{-3.95}{0.5} = \dfrac{0.5x}{0.5}$
$x = -7.9$

The solution is -7.9.

34. $3(a - 5) = -3 + a$ Distributive property
$3a - 15 = -3 + a$ Add 15.
$3a = 12 + a$ Subtract a.
$2a = 12$ Divide by 2.
$\dfrac{2a}{2} = \dfrac{12}{2}$
$a = 6$

The solution is 6.

35. Let n represent the number.

$20 - 3n = 2n$ Add $3n$.
$20 = 5n$ Divide by 5.
$\dfrac{20}{5} = \dfrac{5n}{5}$
$n = 4$

The number is 4.

36. *Step 2*
Let b represent the number of elephant breaths. Then $16b$ represents the number of mouse breaths.

Step 3
$b + 16b = 170$

Step 4
$17b = 170$
$\dfrac{17b}{17} = \dfrac{170}{17}$
$b = 10$

Step 5
An elephant takes 10 breaths. A mouse takes 16 times that, or 160 breaths.

Step 6
Check: 16 times 10 is 160 and the sum of 10 and 160 is 170.

37. $\dfrac{78 \text{ runs}}{234 \text{ innings}} = \dfrac{n \text{ runs}}{9 \text{ innings}}$

$234 \cdot n = 78 \cdot 9$
$\dfrac{234n}{234} = \dfrac{702}{234}$
$n = 3$

At this rate, he will give up 3 runs in a 9-inning game.

38. percent $\cdot$ whole $=$ part
$p \cdot 1.8 = 2.61$
$\dfrac{1.8p}{1.8} = \dfrac{2.61}{1.8}$
$p = 1.45$ (145%)

2.61 inches is 145% of 1.8 inches.

39. Tax = ($64.95)(0.075) ≈ $4.87
The amount of tax was $4.87 (rounded).

Total = $64.95 + $4.87 = $69.82
The total cost of the camera was $69.82.

40. $\dfrac{5.87 \times 10^{12} \text{ miles}}{1 \text{ year}} \cdot \dfrac{1 \text{ year}}{365 \text{ days}}$

$= \dfrac{5.87}{365} \times 10^{12}$

$\approx 0.0161 \times 10^{12}$

$= 1.61 \times 10^{10}$

Light will travel 1.61×10^{10} (rounded) or 16,100,000,000 miles in one day.

41. (a) Size 6 shoes correspond to a foot length of $5\frac{1}{8}$ in.

(b) $5\frac{7}{16} > 5\frac{1}{8} \left[= 5\frac{2}{16} \right]$, so size 6 is too small. She should order size 7.

(c) A size 6 is $5\frac{1}{8}$ in. and a size 5 is $4\frac{13}{16}$ in.

$$5\frac{1}{8} = 4\frac{9}{8} = 4\frac{18}{16}$$
$$-4\frac{13}{16} \phantom{= 4\frac{9}{8}} = 4\frac{13}{16}$$
$$\phantom{-4\frac{13}{16} = 4\frac{9}{8} =} \frac{5}{16} \text{ in.}$$

The difference in length between a size 5 and a size 6 shoe is $\frac{5}{16}$ inch.

42. Perimeter = 4 m + 6 m + 3.9 m + 3.9 m + 6 m
= 23.8 m

Area = area of rectangle + area of triangle
$= l \cdot w + \frac{1}{2} \cdot b \cdot h$
= (6 m)(4 m) + $\frac{1}{2}$(4 m)(3.3 m)
= 24 m² + 6.6 m²
= 30.6 m²

43. Circumference = π · diameter = π(9 cm)
≈ 3.14(9)
≈ 28.3 cm

Area = π · (radius)² = π($\frac{9}{2}$ cm)²
≈ 3.14(4.5)²
≈ 63.6 cm²

WHOLE NUMBERS COMPUTATION: PRETEST

Adding Whole Numbers

1. $\overset{1}{3}68$
   $\underline{+22}$
   390

3. $\overset{111}{85}$
   $\underline{+2968}$
   3053

5. $714 + 3728 + 9 + 683{,}775$

   $\overset{212}{714}$
   $3\,728$
   9
   $\underline{+683{,}775}$
   $688{,}226$

Subtracting Whole Numbers

1. $\overset{312}{\cancel{4}\cancel{2}6}$
   $\underline{-76}$
   350

3. $\overset{9159}{\overset{2\cancel{10}\cancel{3}\cancel{10}12}{\cancel{3}\cancel{0}{,}\cancel{6}\cancel{0}\cancel{2}}}$
   $\underline{-5\,708}$
   $24{,}894$

5. $679{,}420 - 88{,}033$

   $\overset{11}{\overset{5173\cancel{1}10}{\cancel{6}79{,}4\cancel{2}\cancel{0}}}$
   $\underline{-88{,}033}$
   $591{,}387$

Multiplying Whole Numbers

1. $3 \times 3 \times 0 \times 6 = 0$ because 0 times any number is 0.

3. $(520)(3000)$

   520
   $\underline{\times\,3\,000}$
   $1{,}560{,}000$

5. Multiply 359 and 48.

   $\overset{23}{}$
   $\overset{47}{359}$
   $\underline{\times48}$
   $2\,872 \quad \leftarrow \quad 8 \times 359$
   $\underline{14\,36} \quad \leftarrow \quad 4 \times 359$
   $17{,}232$

Dividing Whole Numbers

1. 23
   $3\overline{)69}$
   $\underline{6}$
   9
   $\underline{9}$
   0

3. $\dfrac{25{,}036}{4}$

   $6\,2\,5\,9$
   $4\overline{)2\,5{,}0\,3\,6}$
   $\underline{2\,4}$
   $1\,0$
   $\underline{8}$
   $2\,3$
   $\underline{2\,0}$
   $3\,6$
   $\underline{3\,6}$
   0

5. $3\,4$
   $52\overline{)1\,7\,6\,8}$
   $\underline{1\,5\,6}$
   $2\,0\,8$
   $\underline{2\,0\,8}$
   0

7. $6\,0$
   $38\overline{)2\,3\,0\,0}$
   $\underline{2\,2\,8}$
   $2\,0$
   $\underline{0}$
   $2\,0$

   Answer: 60 **R**20

CHAPTER R WHOLE NUMBERS REVIEW

R.1 Adding Whole Numbers

R.1 Margin Exercises

1. (a) $3 + 4 = 7; 4 + 3 = 7$

   (b) $9 + 9 = 18;$
   No change occurs when the commutative property is used.

   (c) $7 + 8 = 15; 8 + 7 = 15$

   (d) $6 + 9 = 15; 9 + 6 = 15$

2. (a) $\begin{array}{r} 5 \\ 4 \\ 6 \\ 9 \\ +2 \\ \hline 26 \end{array}$ $\begin{array}{l} 5 + 4 = 9 \\ 9 + 6 = 15 \\ 15 + 9 = 24 \\ 24 + 2 = 26 \end{array}$

   (b) $\begin{array}{r} 7 \\ 5 \\ 1 \\ 2 \\ +6 \\ \hline 21 \end{array}$ $\begin{array}{l} 7 + 5 = 12 \\ 12 + 1 = 13 \\ 13 + 2 = 15 \\ 15 + 6 = 21 \end{array}$

   (c) $\begin{array}{r} 9 \\ 2 \\ 1 \\ 3 \\ +4 \\ \hline 19 \end{array}$ $\begin{array}{l} 9 + 2 = 11 \\ 11 + 1 = 12 \\ 12 + 3 = 15 \\ 15 + 4 = 19 \end{array}$

   (d) $\begin{array}{r} 3 \\ 8 \\ 6 \\ 4 \\ +8 \\ \hline 29 \end{array}$ $\begin{array}{l} 3 + 8 = 11 \\ 11 + 6 = 17 \\ 17 + 4 = 21 \\ 21 + 8 = 29 \end{array}$

3. (a) $\begin{array}{r} 25 \\ +73 \\ \hline 98 \end{array}$ (b) $\begin{array}{r} 364 \\ +532 \\ \hline 896 \end{array}$ (c) $\begin{array}{r} 42{,}305 \\ +11{,}563 \\ \hline 53{,}868 \end{array}$

4. (a) $\begin{array}{r} \overset{1}{6}9 \\ +26 \\ \hline 95 \end{array}$ $9 + 6 = 15$

   (b) $\begin{array}{r} \overset{1}{7}6 \\ +18 \\ \hline 94 \end{array}$ $6 + 8 = 14$

   (c) $\begin{array}{r} \overset{1}{5}6 \\ +37 \\ \hline 93 \end{array}$ $6 + 7 = 13$

   (d) $\begin{array}{r} \overset{1}{3}4 \\ +49 \\ \hline 83 \end{array}$ $4 + 9 = 13$

5. (a) $\begin{array}{r} \overset{22}{4}81 \\ 79 \\ 38 \\ +395 \\ \hline 993 \end{array}$

   Add the numbers in the ones column: 23. Write 3, carry 2 to the tens column. Add the numbers in the tens column including the regrouped 2: 29. Write 9, carry 2 to the hundreds column. Add the numbers in the hundreds column including the regrouped 2: 9.

   (b) $\begin{array}{r} \overset{1\,2\,1}{4\,2}71 \\ 372 \\ 8\,976 \\ +162 \\ \hline 13{,}781 \end{array}$

   (c) $\begin{array}{r} \overset{22}{5}7 \\ 4 \\ 392 \\ 804 \\ 51 \\ +27 \\ \hline 1335 \end{array}$

   (d) $\begin{array}{r} \overset{2\,21}{7\,8}21 \\ 435 \\ 72 \\ 305 \\ +1\,693 \\ \hline 10{,}326 \end{array}$

   (e) $\begin{array}{r} \overset{1\;25}{15{,}8}29 \\ 765 \\ 78 \\ 15 \\ 9 \\ 7 \\ +13{,}179 \\ \hline 29{,}882 \end{array}$

6. From Conway to Pine Hills through Clear Lake:

$$\begin{array}{r} 7 \\ 11 \\ 8 \\ +5 \\ \hline 31 \end{array} \begin{array}{l} \text{Conway to Shadow Hills} \\ \text{Shadow Hills to Clear Lake} \\ \text{Clear Lake to Orlando} \\ \text{Orlando to Pine Hills} \end{array}$$

From Conway to Pine Hills through Belle Isle:

$$\begin{array}{r} 6 \\ 3 \\ 6 \\ 9 \\ +5 \\ \hline 29 \end{array} \begin{array}{l} \text{Conway to Belle Isle} \\ \text{Belle Isle to Pine Castle} \\ \text{Pine Castle to Resort Area} \\ \text{Resort Area to Orlando} \\ \text{Orlando to Pine Hills} \end{array}$$

From Conway to Pine Hills through Casselberry:

$$\begin{array}{r} 7 \\ 9 \\ 6 \\ 5 \\ +8 \\ \hline 35 \end{array} \begin{array}{l} \text{Conway to Shadow Hills} \\ \text{Shadow Hills to Bertha} \\ \text{Bertha to Casselberry} \\ \text{Casselberry to Altamonte Springs} \\ \text{Altamonte Springs to Pine Hills} \end{array}$$

The shortest way from Conway to Pine Hills is through Belle Isle.

7. The next shortest route from Orlando to Clear Lake is as follows:

$$\begin{array}{r} 5 \\ 8 \\ 5 \\ 6 \\ 7 \\ +7 \\ \hline 38 \end{array} \begin{array}{l} \text{Orlando to Pine Hills} \\ \text{Pine Hills to Altamonte Springs} \\ \text{Altamonte Springs to Casselberry} \\ \text{Casselberry to Bertha} \\ \text{Bertha to Winter Park} \\ \text{Winter Park to Clear Lake} \\ \text{miles} \end{array}$$

The shortest way from Orlando to Clear Lake with the closed roads is through Casselberry.

8. (a)
$$\begin{array}{r} 59 \\ \hline 32 \uparrow \\ 8 \\ 5 \\ +14 \\ \hline 59 \end{array}$$ Adding up and carrying mentally.

Sum of 59 is correct.

(b)
$$\begin{array}{r} 1609 \\ \hline 872 \uparrow \\ 539 \\ 46 \\ +152 \\ \hline 1609 \end{array}$$ Adding up and carrying mentally.

Sum of 1609 is correct.

(c)
$$\begin{array}{r} 943 \\ \hline 79 \uparrow \\ 218 \\ 7 \\ +639 \\ \hline 953 \end{array} \text{Adding up} \quad \begin{array}{r} \overset{13}{79} \\ 218 \\ 7 \\ +639 \downarrow \\ \hline 943 \end{array} \text{Adding down}$$

The correct answer is 943.

(d)
$$\begin{array}{r} 77{,}563 \\ \hline 21{,}892 \uparrow \\ 11{,}746 \\ +43{,}925 \\ \hline 79{,}563 \end{array} \text{Adding up} \quad \begin{array}{r} \overset{2\ 11}{21{,}892} \\ 11{,}746 \\ +43{,}925 \downarrow \\ \hline 77{,}563 \end{array} \text{Adding down}$$

The correct answer is 77,563.

R.1 Section Exercises

1. (a)
$$\begin{array}{r} 5 \\ 7 \\ 6 \\ +5 \\ \hline 23 \end{array} \begin{array}{l} 5 + 7 = 12 \\ 12 + 6 = 18 \\ 18 + 5 = 23 \end{array}$$

(b)
$$\begin{array}{r} 9 \\ 2 \\ 1 \\ 3 \\ +4 \\ \hline 19 \end{array} \begin{array}{l} 9 + 2 = 11 \\ 11 + 1 = 12 \\ 12 + 3 = 15 \\ 15 + 4 = 19 \end{array}$$

3. (a) $3213 + 5715 = 8928$; $5715 + 3213 = 8928$

(b) $38{,}204 + 21{,}020 = 59{,}224$
$21{,}020 + 38{,}204 = 59{,}224$

5.
$$\begin{array}{r} \overset{1}{67} \\ +83 \\ \hline 150 \end{array}$$

7.
$$\begin{array}{r} \overset{1}{746} \\ +905 \\ \hline 1651 \end{array}$$

9.
$$\begin{array}{r} \overset{11}{798} \\ +206 \\ \hline 1004 \end{array}$$

11.
$$\begin{array}{r} \overset{111}{7968} \\ +1285 \\ \hline 9253 \end{array}$$

13. $\overset{1\;1\;1}{7\,896}$
     $+\;\;3\,728$
     $\overline{11{,}624}$

15. $\overset{\;\;1\;\;\;1}{3\,705}$
     $\;\;\;3\,916$
     $+\,9\,037$
     $\overline{16{,}658}$

17. $\overset{1\;1}{\;\;\;\;32}$
     $+\,4\,977$
     $\overline{\;\;5\,009}$

19. $\overset{\;\;1\;1}{3\,0\,7\,7}$
     $\;\;\;\;\;\;\;8$
     $+\;\;\;421$
     $\overline{\;\;3\,506}$

21. $\overset{\;\;\;2\;2}{9\,0\,5\,6}$
     $\;\;\;\;\;78$
     $\;\;6\,089$
     $+\;\;\;731$
     $\overline{15{,}954}$

23. $\overset{1\;1\;2}{\;\;\;\;18}$
     $\;\;\;708$
     $\;9\,286$
     $+\;\;\;636$
     $\overline{10{,}648}$

25. Add up to check addition.

    $\overline{769}$
    179
    214
    $+\,376$
    $\overline{759}$ incorrect; should be 769

27. Add up to check addition.

    $\overline{5420}$
    $\overline{4713}$
    28
    615
    $+\;64$
    $\overline{5420}$ correct

29. The shortest route between Southtown and Rena is through Thomasville.

21	Southtown to Thomasville
+ 12	Thomasville to Rena
33	miles

31. The shortest route between Thomasville and Murphy is through Rena and Austin.

12	Thomasville to Rena
15	Rena to Austin
+ 11	Austin to Murphy
38	miles

33. $\overset{\;\;2}{\$99}$ auto tune-up
     $\;\;24$ tire rotation
     $+\,29$ oil change
     $\overline{\$152}$ total cost

35. 413 women
     $+\,286$ men
     $\overline{699}$ total people

37. $\overset{\;\;\;\;\;\;11\;11}{36{,}457{,}549}$
     $+\,23{,}507{,}783$
     $\overline{59{,}965{,}332}$

    The total population of the two states is 59,965,332 people.

39. The perimeter is the sum of all of the sides in the figure.

    $\overset{3}{98}$
    49
    98
    $+\,49$
    $\overline{294}$

    294 inches is the perimeter of the figure.

41. The perimeter is the sum of all of the sides in the figure.

    $\overset{11}{286}$
    308
    $+\,114$
    $\overline{708}$

    708 feet is the perimeter of the figure.

R.2 Subtracting Whole Numbers

R.2 Margin Exercises

1. **(a)** $4 + 3 = 7$: $7 - 3 = 4$ or $7 - 4 = 3$

   (b) $6 + 5 = 11$: $11 - 5 = 6$ or $11 - 6 = 5$

   (c) $150 + 220 = 370$:
   $370 - 220 = 150$ or $370 - 150 = 220$

   (d) $623 + 55 = 678$:
   $678 - 55 = 623$ or $678 - 623 = 55$

2. **(a)** $5 - 3 = 2$: $5 = 3 + 2$

(b) $8 - 3 = 5$: $8 = 3 + 5$

(c) $21 - 15 = 6$: $21 = 15 + 6$

(d) $58 - 42 = 16$: $58 = 42 + 16$

3. (a)
$$\begin{array}{r} 56 \\ -31 \\ \hline 25 \end{array} \quad \begin{array}{l} 6-1=5 \\ 5-3=2 \end{array}$$

(b)
$$\begin{array}{r} 38 \\ -14 \\ \hline 24 \end{array} \quad \begin{array}{l} 8-4=4 \\ 3-1=2 \end{array}$$

(c)
$$\begin{array}{r} 378 \\ -235 \\ \hline 143 \end{array}$$

(d)
$$\begin{array}{r} 3927 \\ -2614 \\ \hline 1313 \end{array}$$

(e)
$$\begin{array}{r} 5464 \\ -324 \\ \hline 5140 \end{array}$$

4. (a)
$$\begin{array}{r} 65 \\ \underline{23} \\ 42 \end{array} \text{ Subtraction problem} \qquad \begin{array}{r} 23 \\ +19 \\ \hline 65 \end{array} \text{ Addition problem}$$

42 is correct.

(b)
$$\begin{array}{r} 46 \\ -32 \\ \hline 24 \end{array} \text{ Subtraction problem} \qquad \begin{array}{r} 32 \\ +24 \\ \hline 56 \end{array} \text{ Addition problem}$$

$56 \neq 46$, so 24 is incorrect.

Rework.
$$\begin{array}{r} 46 \\ -32 \\ \hline 14 \end{array} \text{ is correct.}$$

(c)
$$\begin{array}{r} 374 \\ -251 \\ \hline 113 \end{array} \text{ Subtraction problem} \qquad \begin{array}{r} 251 \\ +113 \\ \hline 364 \end{array} \text{ Addition problem}$$

113 is incorrect.

Rework.
$$\begin{array}{r} 374 \\ -251 \\ \hline 123 \end{array} \text{ is correct.}$$

(d)
$$\begin{array}{r} 7531 \\ -4301 \\ \hline 3230 \end{array} \text{ Subtraction problem} \qquad \begin{array}{r} 4301 \\ +3230 \\ \hline 7531 \end{array} \text{ Addition problem}$$

3230 is correct.

5. (a)
$$\begin{array}{r} \overset{5\,17}{\cancel{6}\,\cancel{7}} \\ -3\,8 \\ \hline 2\,9 \end{array}$$

(b)
$$\begin{array}{r} \overset{8\,17}{\cancel{9}\,\cancel{7}} \\ -2\,9 \\ \hline 6\,8 \end{array}$$

(c)
$$\begin{array}{r} \overset{2\,11}{3\,\cancel{1}} \\ -1\,7 \\ \hline 1\,4 \end{array}$$

(d)
$$\begin{array}{r} \overset{5\,13}{8\,\cancel{6}\,\cancel{3}} \\ -4\,7 \\ \hline 8\,1\,6 \end{array}$$

(e)
$$\begin{array}{r} \overset{5\,12}{7\,\cancel{6}\,\cancel{2}} \\ -1\,5\,7 \\ \hline 6\,0\,5 \end{array}$$

6. (a)
$$\begin{array}{r} \overset{2\,15}{3\,\cancel{5}\,4} \\ -8\,2 \\ \hline 2\,7\,2 \end{array}$$

(b)
$$\begin{array}{r} \overset{3\,14\,17}{\cancel{4}\,\cancel{5}\,\cancel{7}} \\ -6\,8 \\ \hline 3\,8\,9 \end{array}$$

(c)
$$\begin{array}{r} \overset{7\,16\,14}{\cancel{8}\,\cancel{7}\,\cancel{4}} \\ -4\,8\,6 \\ \hline 3\,8\,8 \end{array}$$

(d)
$$\begin{array}{r} \overset{0\,13\,12\,17}{\cancel{1}\,\cancel{4}\,\cancel{3}\,\cancel{7}} \\ -9\,8\,8 \\ \hline 4\,4\,9 \end{array}$$

(e)
$$\begin{array}{r} \overset{7\,16\,13}{8\,\cancel{7}\,\cancel{3}\,9} \\ -3\,8\,9\,2 \\ \hline 4\,8\,4\,7 \end{array}$$

7. (a)
$$\begin{array}{r} \overset{2\,10}{3\,\cancel{0}\,8} \\ -2\,8\,5 \\ \hline 2\,3 \end{array}$$

(b)
$$\begin{array}{r} \overset{1\,\overset{9}{\cancel{10}}\,16}{\cancel{2}\,\cancel{0}\,\cancel{6}} \\ -1\,4\,8 \\ \hline 5\,8 \end{array}$$

(c)
$$\begin{array}{r} \overset{4\,10}{\cancel{5}\,\cancel{0}\,7\,3} \\ -1\,6\,3\,2 \\ \hline 3\,4\,4\,1 \end{array}$$

R.2 Subtracting Whole Numbers

8. (a) $\overset{9}{\overset{3\,10\,15}{\cancel{4}\,\cancel{0}\,\cancel{5}}}$
$\underline{-2\,6\,7}$
$1\,3\,8$

(b) $\overset{6\,10}{3\,7\,\cancel{0}}$
$\underline{-1\,6\,3}$
$2\,0\,7$

(c) $\overset{0\,14\,16\,10}{\cancel{1}\,\cancel{5}\,\cancel{7}\,\cancel{0}}$
$\underline{-9\,8\,3}$
$5\,8\,7$

(d) $\overset{9\,9}{\overset{6\,10\,10\,11}{\cancel{7}\,\cancel{0}\,\cancel{0}\,\cancel{1}}}$
$\underline{-5\,1\,9\,3}$
$1\,8\,0\,8$

(e) $\overset{9\,9}{\overset{3\,10\,10\,10}{\cancel{4}\,\cancel{0}\,\cancel{0}\,\cancel{0}}}$
$\underline{-1\,7\,8\,2}$
$2\,2\,1\,8$

9. (a) 425 $\overset{11}{368}$ *Check by*
$\underline{-368}$ *Subtraction* $\underline{+57}$ *addition.*
57 *problem* 425

Match: 57 is correct.

(b) 670 $\overset{1}{439}$ *Check by*
$\underline{-439}$ *Subtraction* $\underline{+241}$ *addition.*
241 *problem* 680

Not a match: 241 is incorrect.

Rework.

$\overset{6\,10}{6\,7\,\cancel{0}}$
$\underline{-4\,3\,9}$ *Subtraction*
$2\,3\,1$ *problem*

Match: 231 is correct.

(c) $\overset{13\,16\,11\,16}{\cancel{14},\cancel{7}\,\cancel{2}\,\cancel{6}}$ $\overset{1\,11}{8\,839}$
$\underline{-8\,8\,3\,9}$ $\underline{+5\,887}$
$5\,8\,8\,7$ $14{,}726$

Match: 5887 is correct.

10. (a) Lopez made 147 deliveries on Friday, but only 126 on Tuesday.

147 *Deliveries on Friday*
$\underline{-126}$ *Deliveries on Tuesday*
21

Lopez made 21 fewer deliveries on Tuesday than she made on Friday.

(b) Lopez made 126 deliveries on Tuesday, but only 119 on Wednesday.

$\overset{116}{1\,2\,\cancel{6}}$ *Deliveries on Tuesday*
$\underline{-1\,1\,9}$ *Deliveries on Wednesday*
7

Lopez made 7 fewer deliveries on Wednesday than she made on Tuesday.

(c) Lopez made 89 deliveries on Thursday and none on Saturday, so she made 89 more deliveries on Thursday than on Saturday.

R.2 Section Exercises

1. 89 $\overset{1}{27}$ *Check by*
$\underline{-27}$ *Given* $\underline{+63}$ *addition.*
63 90

$90 \neq 89$, so 63 is incorrect. Rework.

89
$\underline{-27}$
62 is correct.

3. 382 261 *Check by*
$\underline{-261}$ *Given* $\underline{+131}$ *addition.*
131 392

$392 \neq 382$, so 131 is incorrect. Rework.

382
$\underline{-261}$
121 is correct.

5. $\overset{2\,16}{3\,\cancel{6}}$
$\underline{-2\,8}$
8

7. $\overset{7\,13}{8\,\cancel{3}}$
$\underline{-5\,8}$
$2\,5$

9. $\overset{3\,15}{4\,\cancel{5}}$
$\underline{-2\,9}$
$1\,6$

11. $\overset{6\,11}{7\,\cancel{1}\,9}$
$\underline{-6\,5\,8}$
$6\,1$

13. $\overset{6\,11}{7\,7\,\cancel{1}}$
$\underline{-2\,5\,2}$
$5\,1\,9$

15.
$$\begin{array}{r}\overset{7\ 15 11}{9\,8\,6\,\rlap{/}1}\\-6\,8\,4\\\hline 9\,1\,7\,7\end{array}$$

17.
$$\begin{array}{r}\overset{8\ 17 18}{9\,9\,8\,8}\\-2\,3\,9\,9\\\hline 7\,5\,8\,9\end{array}$$

19.
$$\begin{array}{r}\overset{2\ 17\ 12 12 15}{3\,8\,,3\,3\,5}\\-2\,9\,,4\,7\,6\\\hline 8\,,8\,5\,9\end{array}$$

21.
$$\begin{array}{r}\overset{3\ 10}{4\,\rlap{/}0}\\-3\,7\\\hline 3\end{array}$$

23.
$$\begin{array}{r}\overset{5\ 10}{6\,\rlap{/}0}\\-3\,7\\\hline 2\,3\end{array}$$

25.
$$\begin{array}{r}\overset{9}{5\,\rlap{/}{10}\,11\,10}\\\rlap{/}6\,\rlap{/}0\,\rlap{/}2\,\rlap{/}0\\-4\,0\,7\,8\\\hline 1\,9\,4\,2\end{array}$$

27.
$$\begin{array}{r}\overset{9}{7\,14\,\rlap{/}{10}\,13}\\\rlap{/}8\,\rlap{/}5\,\rlap{/}0\,\rlap{/}3\\-2\,8\,1\,6\\\hline 5\,6\,8\,7\end{array}$$

29.
$$\begin{array}{r}\overset{109}{7\,\rlap{/}{10}\,6\,\rlap{/}{10}\,15}\\\rlap{/}8\,\rlap{/}0\,,\rlap{/}7\,\rlap{/}0\,\rlap{/}5\\-6\,1\,,6\,6\,7\\\hline 1\,9\,,0\,3\,8\end{array}$$

31.
$$\begin{array}{r}\overset{99}{5\,\rlap{/}{10}\,\rlap{/}{10}\,10}\\\rlap{/}6\,\rlap{/}6\,,\rlap{/}0\,\rlap{/}0\,\rlap{/}0\\-4\,4\,4\\\hline 6\,5\,,5\,5\,6\end{array}$$

33.
$$\begin{array}{r}\overset{99}{1\,\rlap{/}{10}\,\rlap{/}{10}\,17\,10}\\\rlap{/}2\,\rlap{/}0\,,\rlap{/}0\,\rlap{/}8\,\rlap{/}0\\-9\,6\\\hline 1\,9\,,9\,8\,4\end{array}$$

35.
$$\begin{array}{r}3070\\-576\quad\text{Subtraction}\\\hline 2596\quad\text{problem}\end{array}\qquad\begin{array}{r}\overset{1 1 1}{576}\quad\text{Check by}\\+2596\quad\text{addition.}\\\hline 3172\end{array}$$

$3172 \neq 3070$, so 2596 is incorrect. Rework.

$$\begin{array}{r}\overset{9}{2\,\rlap{/}{10}\,16\,10}\\\rlap{/}3\,\rlap{/}0\,\rlap{/}7\,\rlap{/}0\\-5\,7\,6\\\hline 2\,4\,9\,4\end{array}\quad\text{is correct.}$$

37.
$$\begin{array}{r}27{,}600\\-807\quad\text{Subtraction}\\\hline 26{,}793\quad\text{problem}\end{array}\qquad\begin{array}{r}\overset{1\ 11}{807}\quad\text{Check by}\\+26{,}793\quad\text{addition.}\\\hline 27{,}600\end{array}$$

Matches: 26,793 is correct.

39.
$$\begin{array}{r}\overset{1\ 15}{2\,\rlap{/}5\,5}\quad\text{calories burned by swimming}\\-1\,8\,5\quad\text{calories burned by hiking}\\\hline 7\,0\quad\text{fewer calories burned}\end{array}$$

70 fewer calories are burned in 30 minutes by hiking than by swimming.

41.
$$\begin{array}{r}\overset{1\ 15}{2\,\rlap{/}5\,4}\quad\text{number of passengers}\\-1\,8\,3\quad\text{passengers departing in Atlanta}\\\hline 7\,1\quad\text{passengers remaining}\end{array}$$

$$\begin{array}{r}\overset{1}{7}1\quad\text{passengers remaining}\\-109\quad\text{passengers got on}\\\hline 180\quad\text{passengers on the plane}\end{array}$$

There were 180 passengers on the plane.

43. **(a)** From the table, the occupation with the highest earnings is computer programmer at $65,510 and the occupation with the lowest earnings is medical secretary at $28,090.

(b)
$$\begin{array}{r}\overset{5 15\ 4 11}{\$6\,\rlap{/}5\,,\rlap{/}5\,\rlap{/}1\,0}\quad\text{comp. programmer earnings}\\-2\,8\,,0\,9\,0\quad\text{medical secretary earnings}\\\hline \$3\,7\,,4\,2\,0\quad\text{difference}\end{array}$$

The earnings of a computer programmer are $37,420 per year more than those of a medical secretary.

45.
$$\begin{array}{r}\overset{7\ 13}{1\,\rlap{/}8\,\rlap{/}3\,1}\quad\text{CN Tower height}\\-1\,4\,5\,1\quad\text{Sears Tower height}\\\hline 3\,8\,0\quad\text{difference}\end{array}$$

There is a difference in height of 380 feet.

R.3 Multiplying Whole Numbers

R.3 Margin Exercises

1. **(a)**
$$\begin{array}{r}3\quad\text{factor}\\\times\,6\quad\text{factor}\\\hline 18\quad\text{product}\end{array}$$

(b)
$$\begin{array}{r}8\quad\text{factor}\\\times\,4\quad\text{factor}\\\hline 32\quad\text{product}\end{array}$$

(c)
$$\begin{array}{r}5\quad\text{factor}\\\times\,7\quad\text{factor}\\\hline 35\quad\text{product}\end{array}$$

(d) $\begin{array}{r} 3 \\ \times\ 9 \\ \hline 27 \end{array}$ *factor*
factor
product

2. (a) $4 \times 7 = 28$; $7 \times 4 = 28$

   (b) $0 \times 9 = 0$; $9 \times 0 = 0$

   (c) $8 \cdot 6 = 48$; $6 \cdot 8 = 48$

   (d) $5 \cdot 5 = 25$; there is no change if the order is switched.

   (e) $(3)(8) = 24$; $(8)(3) = 24$

3. (a) $2 \times 3 \times 4 = (2 \times 3) \times 4 = 6 \times 4 = 24$

   (b) $6 \cdot 1 \cdot 5 = (6 \cdot 1) \cdot 5 = 6 \cdot 5 = 30$

   (c) $(8)(3)(0) = 0$ since zero times any number is 0.

   (d) $3 \times 3 \times 7 = (3 \times 3) \times 7 = 9 \times 7 = 63$

   (e) $4 \cdot 2 \cdot 8 = (4 \cdot 2) \cdot 8 = 8 \cdot 8 = 64$

   (f) $(2)(2)(9) = [(2)(2)](9) = (4)(9) = 36$

4. (a) $\begin{array}{r} \overset{1}{5}2 \\ \times\ 5 \\ \hline 260 \end{array}$

   $5 \cdot 2 = 10$ Write 0, carry 1 ten.
   $5 \cdot 5 = 25$ Add 1 to get 26. Write 26.

   (b) $\begin{array}{r} 79 \\ \times\ 0 \\ \hline 0 \end{array}$ Any number times 0 is 0.

   (c) $\begin{array}{r} \overset{51}{8}62 \\ \times\ 9 \\ \hline 7758 \end{array}$

   $9 \cdot 2 = 18$ Write 8, carry 1 ten.
   $9 \cdot 6 = 54$ Add 1 to get 55. Write 5, carry 5.
   $9 \cdot 8 = 72$ Add 5 to get 77. Write 77.

   (d) $\begin{array}{r} \overset{52}{2}831 \\ \times\ 7 \\ \hline 19{,}817 \end{array}$

   $7 \cdot 1 = 7$ Write 7.
   $7 \cdot 3 = 21$ Write 1, carry 2 tens.
   $7 \cdot 8 = 56$ Add 2 to get 58. Write 8, carry 5.
   $7 \cdot 2 = 14$ Add 5 to get 19. Write 19.

   (e) $\begin{array}{r} \overset{513}{4}714 \\ \times\ 8 \\ \hline 37{,}712 \end{array}$

   $8 \cdot 4 = 32$ Write 2, carry 3 tens.
   $8 \cdot 1 = 8$ Add 3 to get 11. Write 1, carry 1.
   $8 \cdot 7 = 56$ Add 1 to get 57. Write 7, carry 5.
   $8 \cdot 4 = 32$ Add 5 to get 37. Write 37.

5. (a) $45 \times 10 = 450$ *Attach* 0.

   (b) $102 \times 100 = 10{,}200$ *Attach* 00.

   (c) $571 \times 1000 = 571{,}000$ *Attach* 000.

   (d) $3625 \times 100 = 362{,}500$ *Attach* 00.

   (e) $69 \times 1000 = 69{,}000$ *Attach* 000.

6. (a) 14×50

   $\begin{array}{r} 14 \\ \times\ 5 \\ \hline 70 \end{array}$ $14 \times 50 = 700$ *Attach* 0.

   (b) $(68)(400)$

   $\begin{array}{r} 68 \\ \times\ 4 \\ \hline 272 \end{array}$ $68 \times 400 = 27{,}200$ *Attach* 00.

   (c) $\begin{array}{r} 180 \\ \times\ 30 \end{array}$

   $\begin{array}{r} 18 \\ \times\ 3 \\ \hline 54 \end{array}$ $180 \times 30 = 5400$ *Attach* 00.

   (d) $\begin{array}{r} 6100 \\ \times\ 90 \end{array}$

   $\begin{array}{r} 61 \\ \times\ 9 \\ \hline 549 \end{array}$ $6100 \times 90 = 549{,}000$ *Attach* 000.

   (e) $\begin{array}{r} 800 \\ \times\ 200 \end{array}$

   $\begin{array}{r} 8 \\ \times\ 2 \\ \hline 16 \end{array}$ $800 \times 200 = 160{,}000$ *Attach* 0000.

   (f) $(5000)(700)$

   $\begin{array}{r} 5 \\ \times\ 7 \\ \hline 35 \end{array}$ $5000 \times 700 = 3{,}500{,}000$ *Attach* 00000.

(g) $(9)(20{,}000)$

$$\begin{array}{r} 9 \\ \times\ 2 \\ \hline 18 \end{array}\qquad 9 \times 20{,}000 = 180{,}000 \quad \textit{Attach } 0000.$$

7. (a)
$$\begin{array}{r} 38 \\ \times\ 15 \\ \hline 190 \\ 38 \\ \hline 570 \end{array}\quad \begin{array}{l} \leftarrow 5 \times 38 \\ \leftarrow 1 \times 38 \end{array}$$

(b)
$$\begin{array}{r} 31 \\ \times\ 43 \\ \hline 93 \\ 124 \\ \hline 1333 \end{array}\quad \begin{array}{l} \leftarrow 3 \times 31 \\ \leftarrow 4 \times 31 \end{array}$$

(c)
$$\begin{array}{r} 67 \\ \times\ 59 \\ \hline 603 \\ 335 \\ \hline 3953 \end{array}\quad \begin{array}{l} \leftarrow 9 \times 67 \\ \leftarrow 5 \times 67 \end{array}$$

(d)
$$\begin{array}{r} 234 \\ \times\ 73 \\ \hline 702 \\ 1638 \\ \hline 17{,}082 \end{array}\quad \begin{array}{l} \leftarrow 3 \times 234 \\ \leftarrow 7 \times 234 \end{array}$$

(e)
$$\begin{array}{r} 835 \\ \times\ 189 \\ \hline 7515 \\ 6680 \\ 835 \\ \hline 157{,}815 \end{array}\quad \begin{array}{l} \leftarrow 9 \times 835 \\ \leftarrow 8 \times 835 \\ \leftarrow 1 \times 835 \end{array}$$

8. (a)
$$\begin{array}{r} 28 \\ \times\ 60 \\ \hline 1680 \end{array}$$

(b)
$$\begin{array}{r} 817 \\ \times\ 30 \\ \hline 24{,}510 \end{array}$$

(c)
$$\begin{array}{r} 481 \\ \times\ 206 \\ \hline 2886 \\ 9620 \\ \hline 99{,}086 \end{array}\quad 2 \times 481 = 962 \quad \textit{Insert } 0.$$

(d)
$$\begin{array}{r} 3526 \\ \times\ 6002 \\ \hline 7052 \\ 21156\,00 \\ \hline 21{,}163{,}052 \end{array}\quad 6 \times 3526 = 21{,}156 \quad \textit{Insert } 00.$$

9. (a)
$$\begin{array}{r} 36 \\ \times\ 79 \\ \hline 324 \\ 252 \\ \hline 2844 \end{array}\quad \begin{array}{l} \textit{months} \\ \textit{dollars per month} \end{array}$$

The total cost of 36 months of cable TV is $2844.

(b)
$$\begin{array}{r} 1090 \\ \times\ 15 \\ \hline 5450 \\ 1090 \\ \hline 16{,}350 \end{array}\quad \begin{array}{l} \textit{dollars per computer} \\ \textit{computers} \end{array}$$

The total cost of 15 laptop computers is $16,350.

(c)
$$\begin{array}{r} 389 \\ \times\ 60 \\ \hline 23{,}340 \end{array}\quad \begin{array}{l} \textit{dollars per month} \\ \textit{months} \end{array}$$

The total cost of 60 months of car payments is $23,340.

R.3 Section Exercises

1. $3 \times 1 \times 3 = (3 \times 1) \times 3 = 3 \times 3 = 9$

3. $9 \times 1 \times 7 = (9 \times 1) \times 7 = 9 \times 7 = 63$

5. $9 \cdot 5 \cdot 0 = (9 \cdot 5) \cdot 0 = 45 \cdot 0 = 0$
The product of any number and 0 is 0.

7. $(4)(1)(6) = [(4)(1)](6) = (4)(6) = 24$

9. $(2)(3)(6) = [(2)(3)](6) = (6)(6) = 36$

11.
$$\begin{array}{r} \overset{3}{3}5 \\ \times\ 7 \\ \hline 245 \end{array}$$

$7 \cdot 5 = 35$ Write 5, carry 3 tens.
$7 \cdot 3 = 21$ Add 3 to get 24. Write 24.

13.
$$\begin{array}{r} \overset{4}{2}8 \\ \times\ 6 \\ \hline 168 \end{array}$$

$6 \cdot 8 = 48$ Write 8, carry 4 tens.
$6 \cdot 2 = 12$ Add 4 to get 16. Write 16.

R.3 Multiplying Whole Numbers

15. $\overset{141}{3182}$
    $\underline{\times6}$
    $19{,}092$

    $6 \cdot 2 = 12$ Write 2, carry 1 ten.
    $6 \cdot 8 = 48$ Add 1 to get 49. Write 9, carry 4.
    $6 \cdot 1 = 6$ Add 4 to get 10. Write 0, carry 1.
    $6 \cdot 3 = 18$ Add 1 to get 19. Write 19.

17. $\overset{46\,1}{36{,}921}$
    $\underline{\times7}$
    $258{,}447$

    $7 \cdot 1 = 7$ Write 7.
    $7 \cdot 2 = 14$ Write 4, carry 1.
    $7 \cdot 9 = 63$ Add 1 to get 64. Write 4, carry 6.
    $7 \cdot 6 = 42$ Add 6 to get 48. Write 8, carry 4.
    $7 \cdot 3 = 21$ Add 4 to get 25. Write 25.

19.   125   125   125
    $\underline{\times\,100}$ $\underline{\times\,1}$ $\underline{\times\,100}$
                  125   12,500   *Attach* 00.

21.   1485   1485   1485
    $\underline{\times30}$ $\underline{\times3}$ $\underline{\times30}$
              4455   44,550   *Attach* 0.

23.   900   9   900
    $\underline{\times\,300}$ $\underline{\times\,3}$ $\underline{\times\,300}$
              27   270,000   *Attach* 0000.

25.   43,000   43   43,000
    $\underline{\times\,2000}$ $\underline{\times\,2}$ $\underline{\times\,2000}$
              86   86,000,000   *Attach* 000000.

27.   68
    $\underline{\times22}$
      136   ← 2 × 68
      136   ← 2 × 68
    $\overline{1496}$

29.   83
    $\underline{\times45}$
      415   ← 5 × 83
      332   ← 4 × 83
    $\overline{3735}$

31. (32)(475)   475
                 $\underline{\times32}$
                  950   ← 2 × 475
                1425   ← 3 × 475
                $\overline{15{,}200}$

33. (729)(45)   729
                  $\underline{\times45}$
                3645   ← 5 × 729
                2916   ← 4 × 729
                $\overline{32{,}805}$

35.   538
    $\underline{\times342}$
      1076   ← 2 × 538
      2152   ← 4 × 538
      1614   ← 3 × 538
    $\overline{183{,}996}$

37.   8162
    $\underline{\times407}$
       57134   ← 7 × 8162
      326480   ← 40 × 8162
    $\overline{3{,}321{,}934}$

39.   6310
    $\underline{\times\,3008}$
       50480   ← 8 × 6310
    1893000   ← 300 × 6310
    $\overline{18{,}980{,}480}$

41.   18   *inches per day*
    $\underline{\times14}$ *days*
       72
       18
    $\overline{252}$ *inches in two weeks*

       18   *inches per day*
    $\underline{\times30}$ *days*
      540   *inches in 30 days*

43.   48   *flats*
    $\underline{\times12}$ *tomato plants per flat*
       96
       48
    $\overline{576}$ *tomato plants*

    The total number of tomato plants is 576.

45.   45   *miles per gallon*
    $\underline{\times11}$ *gallons*
       45
       45
    $\overline{495}$ *miles*

    The Prius can travel 495 miles on 11 gallons of gas.

330   Chapter R   Whole Numbers Review

47. 2695 Reno to Atlantic Ocean
    -255 Reno to Pacific Ocean
    2440 difference

It is 2440 miles farther from Reno to the Atlantic Ocean than it is from Reno to the Pacific Ocean.

    2695 miles per trip
    $\times6$ trips (3 round trips)
    $16{,}170$ miles

You will earn 16,170 frequent flier miles.

49. $\overset{916}{140\cancel{0}}$ calories per high-fat meal
    -348 calories per low-fat meal
    1058 calories difference

    $\overset{45}{1058}$ more calories per meal
    $\times7$ meals
    7406 more calories

There are 7406 more calories in seven high-fat meals than in seven low-fat meals.

R.4 Dividing Whole Numbers

R.4 Margin Exercises

1. (a) $48 \div 6 = 8$: $6\overline{)48}^{\,8}$ and $\dfrac{48}{6} = 8$

   (b) $24 \div 6 = 4$: $6\overline{)24}^{\,4}$ and $\dfrac{24}{6} = 4$

   (c) $9\overline{)36}^{\,4}$: $36 \div 9 = 4$ and $\dfrac{36}{9} = 4$

   (d) $\dfrac{42}{6} = 7$: $42 \div 6 = 7$ and $6\overline{)42}^{\,7}$

2. (a) $10 \div 2 = 5$

   dividend: 10; divisor: 2; quotient: 5

   (b) $6 = 30 \div 5$

   dividend: 30; divisor: 5; quotient: 6

   (c) $\dfrac{28}{7} = 4$

   dividend: 28; divisor: 7; quotient: 4

   (d) $2\overline{)36}^{\,18}$

   dividend: 36; divisor: 2; quotient: 18

3. (a) $0 \div 9 = 0$

   (b) $\dfrac{0}{36} = 0$

   (c) $57\overline{)0}^{\,0}$

4. (a) $6\overline{)18}^{\,3}$; $6 \cdot 3 = 18$ or $3 \cdot 6 = 18$

   (b) $\dfrac{28}{4} = 7$; $4 \cdot 7 = 28$ or $7 \cdot 4 = 28$

   (c) $48 \div 8 = 6$; $8 \cdot 6 = 48$ or $6 \cdot 8 = 48$

5. (a) $\dfrac{8}{0}$; undefined

   (b) $\dfrac{0}{8} = 0$

   (c) $0\overline{)32}$; undefined

   (d) $32\overline{)0}^{\,0}$

   (e) $100 \div 0$; undefined

   (f) $0 \div 100 = 0$

6. (a) $5 \div 5 = 1$

   (b) $14\overline{)14}^{\,1}$

   (c) $\dfrac{37}{37} = 1$

7. (a) $2\overline{)18}^{\,9}$

   (b) $3\overline{)39}^{\,13}$ $\dfrac{3}{3} = 1$, $\dfrac{9}{3} = 3$

   (c) $4\overline{)88}^{\,22}$ $\dfrac{8}{4} = 2$, $\dfrac{8}{4} = 2$

   (d) $2\overline{)462}^{\,231}$ $\dfrac{4}{2} = 2$, $\dfrac{6}{2} = 3$, $\dfrac{2}{2} = 1$

8. (a) $2\overline{)225}^{\,112\,\text{R1}}$ $\dfrac{2}{2} = 1$, $\dfrac{2}{2} = 1$, $\dfrac{5}{2} = 2\,\text{R1}$

   (b) $3\overline{)275}^{\,91\,\text{R2}}$ $\dfrac{27}{3} = 9$, $\dfrac{5}{3} = 1\,\text{R2}$

   (c) $4\overline{)5^{1}3^{1}8}^{\,1\ 3\ 4\,\text{R2}}$ $\dfrac{5}{4} = 1\,\text{R1}$, $\dfrac{13}{4} = 3\,\text{R1}$,

   $\dfrac{18}{4} = 4\,\text{R2}$

R.4 Dividing Whole Numbers

(d) $\dfrac{819}{5}$

$5\overline{)8^31^19}$ gives 1 6 3 R4; $\dfrac{8}{5} = 1\,\mathbf{R}3,\ \dfrac{31}{5} = 6\,\mathbf{R}1,$

$\dfrac{19}{5} = 3\,\mathbf{R}4$

9. (a) $4\overline{)83\,^37}$ gives 20 9 R1

(b) $7\overline{)74\,^47}$ gives 10 6 R5

(c) $5\overline{)453\,^38}$ gives 90 7 R3

(d) $8\overline{)244\,^40}$ gives 30 5

10. (a) $3\overline{)115}$ gives 38 R1

divisor × *quotient* + *remainder* = *dividend*
$\downarrow\quad\ \ \downarrow\quad\ \ \downarrow\quad\ \ \downarrow$
3 × 38 + 1 = 115
114 + 1 = 115

The answer is correct.

(b) $8\overline{)743}$ gives 92 R2

divisor × *quotient* + *remainder* = *dividend*
$\downarrow\quad\ \ \downarrow\quad\ \ \downarrow\quad\ \ \downarrow$
8 × 92 + 2
736 + 2 = 738
↑
incorrect

$8\overline{)74\,^23}$ gives 9 2 R7

The correct answer is 92 **R**7.

(c) $4\overline{)1312}$ gives 328

divisor × *quotient* + *remainder* = *dividend*
$\downarrow\quad\ \ \downarrow\quad\ \ \downarrow\quad\ \ \downarrow$
4 × 328 + 0 = 1312

The answer is correct.

(d) $5\overline{)2033}$ gives 46 **R**3

divisor × *quotient* + *remainder* = *dividend*
$\downarrow\quad\ \ \downarrow\quad\ \ \downarrow\quad\ \ \downarrow$
5 × 46 + 3
230 + 3 = 233
↑
incorrect

$5\overline{)203\,^33}$ gives 40 6 **R**3

The correct answer is 406 **R**3.

11. (a) 612: ends in 2, divisible by 2

(b) 315: ends in 5, not divisible by 2

(c) 2714: ends in 4, divisible by 2

(d) 36,000: ends in 0, divisible by 2

12. (a) 836: The sum of the digits is $8 + 3 + 6 = 17$. Because 17 is *not* divisible by 3, the number 836 is *not* divisible by 3.

(b) 7545: The sum of the digits is $7 + 5 + 4 + 5 = 21$. Because 21 is divisible by 3, the number 7545 is divisible by 3.

(c) 242,913: The sum of the digits is $2 + 4 + 2 + 9 + 1 + 3 = 21$. Because 21 is divisible by 3, the number 242,913 is divisible by 3.

(d) 102,484: The sum of the digits is $1 + 0 + 2 + 4 + 8 + 4 = 19$. Because 19 is *not* divisible by 3, the number 102,484 is *not* divisible by 3.

13. (a) 160: ends in 0, divisible by 5

(b) 635: ends in 5, divisible by 5

(c) 3381: ends in 1, not divisible by 5

(d) 108,605: ends in 5, divisible by 5

14. (a) 290: ends in 0, divisible by 10

(b) 218: ends in 8, not divisible by 10

(c) 2020: ends in 0, divisible by 10

(d) 11,670: ends in 0, divisible by 10

R.4 Section Exercises

1. $\dfrac{12}{12} = 1;\ \ 12\overline{)12}\ \ \text{or}\ \ 12 \div 12$

3. $24 \div 0$ is undefined; $\dfrac{24}{0}\ \ \text{or}\ \ 0\overline{)24}$

5. $\dfrac{0}{4} = 0;\ \ 4\overline{)0}\ \ \text{or}\ \ 0 \div 4$

332 Chapter R Whole Numbers Review

7. $0 \div 12 = 0$; $\dfrac{0}{12}$ or $12\overline{)0}$

9. $0\overline{)21}$ is undefined; $\dfrac{21}{0}$ or $21 \div 0$

11. $\begin{array}{r}21\\4\overline{)84}\end{array}$

The dividend is 84, the divisor is 4, and the quotient is 21.

Check: $4 \times 21 = 84$

13. $\begin{array}{r}3\ 6\\9\overline{)32\,^54}\end{array}$

The dividend is 324, the divisor is 9, and the quotient is 36.

Check: $9 \times 36 = 324$

15. $\begin{array}{r}1\ 5\ 20\ \mathbf{R5}\\6\overline{)9\,^31\,^125}\end{array}$

Check: $6 \times 1520 + 5 = 9120 + 5 = 9125$

17. $\begin{array}{r}30\ 9\\6\overline{)185\,^54}\end{array}$

Check: $6 \times 309 = 1854$

19. $4024 \div 4$ $\begin{array}{r}100\ 6\\4\overline{)402\,^24}\end{array}$

Check: $4 \times 1006 = 4024$

21. $15{,}019 \div 3$ $\begin{array}{r}5\ 00\ 6\,\mathbf{R1}\\3\overline{)15{,}01\,^19}\end{array}$

Check: $3 \times 5006 + 1 = 15{,}018 + 1 = 15{,}019$

23. $\dfrac{26{,}684}{4}$ $\begin{array}{r}6\ 6\ 71\\4\overline{)26{,}\,^26\,^28\,4}\end{array}$

Check: $4 \times 6671 = 26{,}684$

25. $\dfrac{74{,}751}{6}$ $\begin{array}{r}1\ 2{,}4\ 5\ 8\ \mathbf{R3}\\6\overline{)7\,^14{,}\,^27\,^35\,^51}\end{array}$

Check: $6 \times 12{,}458 + 3 = 74{,}748 + 3 = 74{,}751$

27. $\dfrac{71{,}776}{7}$ $\begin{array}{r}10{,}2\ 5\ 3\ \mathbf{R5}\\7\overline{)71{,}\,^17\,^37\,^26}\end{array}$

Check: $7 \times 10{,}253 + 5 = 71{,}771 + 5 = 71{,}776$

29. $\dfrac{128{,}645}{7}$ $\begin{array}{r}1\ 8{,}3\ 7\ 7\ \mathbf{R6}\\7\overline{)12\,^58{,}\,^26\,^54\,^55}\end{array}$

Check: $7 \times 18{,}377 + 6 = 128{,}639 + 6 = 128{,}645$

31. $\begin{array}{r}67\ \mathbf{R2}\\7\overline{)4692}\end{array}$

Check: $7 \times 67 + 2 = 469 + 2 = 471$ *incorrect*

Rework:

$\begin{array}{r}6\ 70\ \mathbf{R2}\\7\overline{)46\,^492}\end{array}$

Check: $7 \times 670 + 2 = 4690 + 2 = 4692$ *correct*

33. $\begin{array}{r}3\ 568\ \mathbf{R2}\\6\overline{)21{,}409}\end{array}$

Check: $6 \times 3568 + 2 = 21{,}408 + 2 = 21{,}410$ *incorrect*

Rework:

$\begin{array}{r}3\ 5\ 6\ 8\ \mathbf{R1}\\6\overline{)21{,}\,^34\,^40\,^49}\end{array}$

Check: $6 \times 3568 + 1 = 21{,}408 + 1 = 21{,}409$

35. $\begin{array}{r}3\ 003\ \mathbf{R5}\\6\overline{)18{,}023}\end{array}$

Check: $6 \times 3003 + 5 = 18{,}018 + 5 = 18{,}023$ *correct*

37. $\begin{array}{r}11{,}523\ \mathbf{R2}\\6\overline{)69{,}140}\end{array}$

Check: $6 \times 11{,}523 + 2 = 69{,}138 + 2 = 69{,}140$ *correct*

39. $\begin{array}{r}2\ 3\\8\overline{)18\,^24}\end{array}$ $\begin{array}{r}1\ 4\\8\overline{)11\,^32}\end{array}$ $\begin{array}{r}1\ 9\\8\overline{)15\,^72}\end{array}$

Kaci earns \$23/hour. Her workers earn \$14/hour and \$19/hour.

41. $\begin{array}{r}1\ 6{,}600\\6\overline{)9\,^39{,}\,^3600}\end{array}$

Each van costs \$16,600.

43. $\begin{array}{r}3\ 7\ 8\\5\overline{)18\,^39\,^40}\end{array}$ (378 for \$5 tickets)

$\begin{array}{r}2\ 7\ 0\\7\overline{)18\,^490}\end{array}$ (270 for \$7 tickets)

$\begin{array}{r}2\ 1\ 0\\9\overline{)1890}\end{array}$ (210 for \$9 tickets)

45. $300 \times \$5 = \$1500$ (too little for the $1890 budget)

$300 \times \$7 = \$2100$ (enough to cover the $1890 budget)

$$\begin{array}{r} \overset{10}{1}\overset{10}{\cancel{0}}10 \\ \$\cancel{2}\cancel{1}\cancel{0}0 \\ -\,1890 \\ \hline \$210 \end{array}$$

Selling 300 $7 tickets would cover the $1890 budget with $210 extra.

47. **(a)** Circle the numbers that end in 0, 2, 4, 6, or 8; that is, circle 358 and 190.

(b) Find the sum of the digits for each number.

736: $7 + 3 + 6 = 16$
10,404: $1 + 0 + 4 + 0 + 4 = 9$
5603: $5 + 6 + 0 + 3 = 14$
78: $7 + 8 = 15$

Since the sums 16 and 14 *are not* divisible by 3, the numbers 736 and 5603 *are not* divisible by 3. Since the sums 9 and 15 *are* divisible by 3, the numbers 10,404 and 78 *are* divisible by 3 and should be circled.

(c) Circle the numbers that end in 0 or 5; that is, circle 13,740 and 985.

49. 30 ends in 0, so it is divisible by 2, 5, and 10. The sum of its digits, 3, is divisible by 3, so 30 is divisible by 3.

51. 184 ends in 4, so it is divisible by 2, but not divisible by 5 or 10. The sum of its digits, 13, is not divisible by 3, so 184 is not divisible by 3.

53. 445 ends in 5, so it is divisible by 5, but not divisible by 2 or 10. The sum of its digits, 13, is not divisible by 3, so 445 is not divisible by 3.

55. 903 ends in 3, so it is not divisible by 2, 5, or 10. The sum of its digits, 12, is divisible by 3, so 903 is divisible by 3.

57. 5166 ends in 6, so it is divisible by 2, but not divisible by 5 or 10. The sum of its digits, 18, is divisible by 3, so 5166 is divisible by 3.

59. 21,763 ends in 3, so it is not divisible by 2, 5, or 10. The sum of its digits, 19, is not divisible by 3, so 21,763 is not divisible by 3.

R.5 Long Division

R.5 Margin Exercises

1. (a)
$$\begin{array}{r} 72 \\ 64\overline{)4608} \\ 448 \leftarrow 7 \times 64 \\ \hline 128 \\ 128 \leftarrow 2 \times 64 \\ \hline 0 \end{array}$$

(b) Because 32 is closer to 30 than to 40, use 3 as a trial divisor.

$$\begin{array}{r} 56 \\ 32\overline{)1792} \\ 160 \leftarrow 5 \times 32 \\ \hline 192 \\ 192 \leftarrow 6 \times 32 \\ \hline 0 \end{array}$$

(c) Because 51 is closer to 50 than to 60, use 5 as a trial divisor.

$$\begin{array}{r} 45 \\ 51\overline{)2295} \\ 204 \leftarrow 4 \times 51 \\ \hline 255 \\ 255 \leftarrow 5 \times 51 \\ \hline 0 \end{array}$$

(d) Because 83 is closer to 80 than to 90, use 8 as a trial divisor.

$$\begin{array}{r} 77 \\ 83\overline{)6391} \\ 581 \leftarrow 7 \times 83 \\ \hline 581 \\ 581 \leftarrow 7 \times 83 \\ \hline 0 \end{array}$$

2. (a)
$$\begin{array}{r} 42 \\ 56\overline{)2352} \\ 224 \leftarrow 4 \times 56 \\ \hline 112 \\ 112 \leftarrow 2 \times 56 \\ \hline 0 \end{array}$$

(b)
$$\begin{array}{r} 42 \text{R}3 \\ 38\overline{)1599} \\ 152 \leftarrow 4 \times 38 \\ \hline 79 \\ 76 \leftarrow 2 \times 38 \\ \hline 3 \end{array}$$

(c)
$$\begin{array}{r}83\ \textbf{R21}\\65\overline{\smash{)}5416}\\520\quad\leftarrow 8\times 65\\\hline 216\\195\ \leftarrow 3\times 65\\\hline 21\end{array}$$

(d)
$$\begin{array}{r}74\ \textbf{R63}\\89\overline{\smash{)}6649}\\623\quad\leftarrow 7\times 89\\\hline 419\\356\ \leftarrow 4\times 89\\\hline 63\end{array}$$

3. (a)
$$\begin{array}{r}130\ \textbf{R7}\\24\overline{\smash{)}3127}\\24\quad\leftarrow 1\times 24\\\hline 72\\72\quad\leftarrow 3\times 24\\\hline 07\\0\ \leftarrow 0\times 24\\\hline 7\end{array}$$

(b)
$$\begin{array}{r}205\\52\overline{\smash{)}10{,}660}\\104\quad\leftarrow 2\times 52\\\hline 260\\260\ \leftarrow 5\times 52\\\hline 0\end{array}$$

(c)
$$\begin{array}{r}408\ \textbf{R21}\\39\overline{\smash{)}15{,}933}\\156\quad\leftarrow 4\times 39\\\hline 333\\312\ \leftarrow 8\times 39\\\hline 21\end{array}$$

(d)
$$\begin{array}{r}300\ \textbf{R62}\\78\overline{\smash{)}23{,}462}\\234\quad\leftarrow 3\times 78\\\hline 62\end{array}$$

4. (a) $5\underline{0}\div 1\underline{0}=5$
One zero is dropped.

(b) $18\underline{00}\div 1\underline{00}=18$
Two zeros are dropped.

(c) $305{,}\underline{000}\div 1\underline{000}=305$
Three zeros are dropped.

5. (a) $60\overline{\smash{)}7200}$

Drop 1 zero from the divisor and the dividend.

$$\begin{array}{r}120\\6\overline{\smash{)}720}\\6\\\hline 12\\12\\\hline 0\\0\\\hline 0\end{array}$$

The quotient is 120.

(b) $130\overline{\smash{)}131{,}040}$

Drop 1 zero from the divisor and the dividend.

$$\begin{array}{r}1008\\13\overline{\smash{)}13{,}104}\\13\\\hline 104\\104\\\hline 0\end{array}$$

The quotient is 1008.

(c) $2600\overline{\smash{)}195{,}000}$

Drop 2 zeros from the divisor and the dividend.

$$\begin{array}{r}75\\26\overline{\smash{)}1950}\\182\\\hline 130\\130\\\hline 0\end{array}$$

The quotient is 75.

6. (a)
$$\begin{array}{r}43\\18\overline{\smash{)}774}\\72\\\hline 54\\54\\\hline 0\end{array}\qquad\begin{array}{r}43\\\times\ 18\\\hline 344\\43\\\hline 774\end{array}$$

Multiply the quotient and the divisor.

← *correct; the result matches the dividend.*

R.5 Long Division   **335**

(b)
$$
\begin{array}{r}
42\ \mathbf{R}178 \\
426\overline{\smash{)}19{,}170} \\
\underline{17\ 04} \\
1\ 13\ 0 \\
\underline{9\ 5\ 2} \\
1\ 7\ 8
\end{array}
\qquad
\begin{array}{r}
426 \\
\times\ \ 42 \\
\hline
852 \\
17\ 04 \\
\hline
17{,}892 \\
+\ \ \ 178 \\
\hline
18{,}070
\end{array}
$$

The result does not match the dividend. Rework.

$$
\begin{array}{r}
45 \\
426\overline{\smash{)}19{,}170} \\
\underline{17\ 04} \\
2\ 13\ 0 \\
\underline{2\ 13\ 0} \\
0
\end{array}
$$

The quotient is 45.

R.5 Section Exercises

1. Because 24 is closer to 20 than to 30, use 2 as a trial divisor.

$$24\overline{\smash{)}768}$$ with 3 above the 6

3 goes over the 6 because $\frac{76}{24}$ is about 3.

3. Because 18 is closer to 20 than to 10, use 2 as a trial divisor.

$$18\overline{\smash{)}4500}$$ with 2 above the 5

2 goes over the 5 because $\frac{45}{18}$ is about 2.

5. Because 86 is closer to 90 than to 80, use 9 as a trial divisor.

$$86\overline{\smash{)}10{,}327}$$ with 1 above the 3

1 goes over the 3 because $\frac{103}{86}$ is about 1.

7. Because 52 is closer to 50 than to 60, use 5 as a trial divisor.

$$52\overline{\smash{)}38{,}025}$$ with 7 above the 0

7 goes over the 0 because $\frac{380}{52}$ is about 7.

9. Because 77 is closer to 80 than to 70, use 8 as a trial divisor.

$$77\overline{\smash{)}249{,}826}$$ with 3 above the 9

3 goes over the 9 because $\frac{249}{77}$ is about 3.

11. Because 420 is closer to 400 than to 500, use 4 as a trial divisor.

$$420\overline{\smash{)}470{,}800}$$ with 1 above the first 0

1 goes over the first 0 because $\frac{470}{420}$ is about 1.

13.
$$
\begin{array}{r}
64\ \mathbf{R}3 \\
29\overline{\smash{)}1859} \\
\underline{174} \\
119 \\
\underline{116} \\
3
\end{array}
\qquad
\text{Check:}\quad
\begin{array}{r}
64 \\
\times\ 29 \\
\hline
576 \\
128 \\
\hline
1856 \\
+\ \ \ 3 \\
\hline
1859
\end{array}
$$

15.
$$
\begin{array}{r}
236\ \mathbf{R}29 \\
47\overline{\smash{)}11{,}121} \\
\underline{94} \\
1\ 72 \\
\underline{1\ 41} \\
31\ 1 \\
\underline{28\ 2} \\
29
\end{array}
\qquad
\text{Check:}\quad
\begin{array}{r}
236 \\
\times\ \ 47 \\
\hline
1652 \\
944 \\
\hline
11{,}092 \\
+\ \ \ 29 \\
\hline
11{,}121
\end{array}
$$

17.
$$
\begin{array}{r}
2407\ \mathbf{R}1 \\
26\overline{\smash{)}62{,}583} \\
\underline{52} \\
10\ 5 \\
\underline{10\ 4} \\
1\ 83 \\
\underline{1\ 82} \\
1
\end{array}
\qquad
\text{Check:}\quad
\begin{array}{r}
2407 \\
\times\ \ \ 26 \\
\hline
14442 \\
4814 \\
\hline
62{,}582 \\
+\ \ \ \ 1 \\
\hline
62{,}583
\end{array}
$$

19.
$$
\begin{array}{r}
1239\ \mathbf{R}15 \\
63\overline{\smash{)}78{,}072} \\
\underline{63} \\
15\ 0 \\
\underline{12\ 6} \\
2\ 47 \\
\underline{1\ 89} \\
58\ 2 \\
\underline{56\ 7} \\
15
\end{array}
\qquad
\text{Check:}\quad
\begin{array}{r}
1239 \\
\times\ \ \ 63 \\
\hline
3717 \\
7434 \\
\hline
78{,}057 \\
+\ \ \ 15 \\
\hline
78{,}072
\end{array}
$$

21. $150 \overline{)499,760}$ Drop 1 zero.

```
      3 3 3 1  R11*    Check:    3 3 3 1
15 )4 9, 9 7 6                 ×     1 5
   4 5                           1 6 6 5 5
     4 9                         3 3 3 1
     4 5                         4 9, 9 6 5
       4 7                     +      1 1
       4 5                       4 9, 9 7 6
         2 6
         1 5
           1 1
```

*Note: If you get a nonzero remainder when dropping zeros, you must add the same number of zeros to the remainder after you divide. Hence, the answer is 3331 **R110**.

23. $400 \overline{)340,000}$ Drop 2 zeros.

```
       8 5 0         Check:     8 5 0
4 )3 4 0 0                    ×       4
   3 2                          3 4 0 0
     2 0
     2 0
       0
```

25. $56 \overline{)5943}$ = 1 0 6 **R17**   Check: 1 0 6
                                        ×   5 6
                                          6 3 6
                                          5 3 0
                                          5 9 3 6
                                        +   1 7
                                          5 9 5 3 *incorrect*

Rework.

```
       1 0 6  R7     Check:      1 0 6
56 )5 9 4 3                    ×    5 6
   5 6                           6 3 6
     3 4 3                       5 3 0
     3 3 6                       5 9 3 6
           7                   +       7
                                 5 9 4 3
```

The correct answer is 106 **R7**.

27. $600 \overline{)394,800}$ = 6 5 8 **R9**

Check: 6 5 8
         ×   6 0 0
         3 9 4, 8 0 0
         +   9
         3 9 4, 8 0 9 *incorrect*

From the check, we can see that the correct answer is 658.

29. $410 \overline{)25,420}$ = 6 2 **R3**

Check: 4 1 0
         ×   6 2
             8 2 0
         2 4 6 0
         2 5, 4 2 0
         +   3
         2 5, 4 2 3 *incorrect*

From the check, we can see that the correct answer is 62.

31. $72 \overline{)32,465}$ = 4 5 0 **R65**

Check: 4 5 0
         ×   7 2
             9 0 0
         3 1 5 0
         3 2 4 0 0
         +   6 5
         3 2, 4 6 5 *correct*

33.
```
  4                     4
  59  hours per week    38  hours per week
× 50  weeks per year  × 50  weeks per year
2950                   1900
```

$2950 - 1900 = 1050$ more hours were worked per year in 1900 than today.

Alternatively, we see that there are $59 - 38 = 21$ more hours of work per week in 1900 than today, so we get $50 \times 21 = 1050$ more hours per year.

35.
```
  32                        4
  475  dollars per night    69  dollars per night
×   5  number of nights   ×  5  number of nights
2375                       345
```

  2375   The difference in cost between the
− 345   most expensive and least expensive
  2030   rooms for a five-night stay is $2030.

Alternatively, the difference is
$475 − $69 = $406 per night, so for 5 nights, the total savings is $5 \times \$406 = \$2030$.

37. Divide.

```
       3 0 8      Check:    3 0 8
36 )1 1, 0 8 8            ×    3 6
   1 0 8                    1 8 4 8
       2 8                    9 2 4
        0                   1 1, 0 8 8
       2 8 8
       2 8 8
           0
```

Judy's monthly payment is $308.

39.
$$\begin{array}{r} 42 \\ \times\ 8 \\ \hline 336 \\ \times\ 5 \\ \hline 1680 \end{array}\ \begin{array}{l} \textit{circuits per hour} \\ \textit{hours per day} \\ \textit{circuits per day} \\ \textit{days per week} \\ \textit{circuits per week} \end{array}$$

He can assemble 1680 circuits in a 40-hr workweek.

41. First subtract.

$$\begin{array}{r} \overset{6\,1\,5}{\$7\,\cancel{5}\,88} \\ -\ \ 8\,38 \\ \hline \$6\,7\,50 \end{array}\ \begin{array}{l} \textit{money raised} \\ \textit{expenses} \\ \textit{remaining money} \end{array}$$

Then divide.

Number of teams → $18\overline{)6750}$ ← Remaining money

$$\begin{array}{r} 375 \\ 18\overline{)6750} \\ \underline{54} \\ 135 \\ \underline{126} \\ 90 \\ \underline{90} \\ 0 \end{array}$$

Each team received $375.

APPENDIX: INDUCTIVE AND DEDUCTIVE REASONING

Margin Exercises

1. $2, 8, 14, 20, \ldots$

   Find the difference between each pair of successive numbers.

   $$8 - 2 = 6$$
   $$14 - 8 = 6$$
   $$20 - 14 = 6$$

   Note that each number is 6 greater than the previous number. So the next number is $20 + 6 = 26$.

2. $6, 11, 7, 12, 8, 13, \ldots$

   The pattern involves addition and subtraction.

   $$6 + 5 = 11$$
   $$11 - 4 = 7$$
   $$7 + 5 = 12$$
   $$12 - 4 = 8$$
   $$8 + 5 = 13$$

   Note that we add 5, then subtract 4. To obtain the next number, 4 should be subtracted from 13, to get 9.

3. $2, 6, 18, 54, \ldots$

   To see the pattern, use division.

   $$6 \div 2 = 3$$
   $$18 \div 6 = 3$$
   $$54 \div 18 = 3$$

   To obtain the next number, multiply the previous number by 3. So the next number is $(54)(3) = 162$.

4. The next figure is obtained by rotating the previous figure clockwise the same amount of rotation ($\frac{1}{4}$ turn) as the second figure was from the first, and as the third figure was from the second.

5. All cars have four wheels.
   All Fords are cars.
   ∴ All Fords have four wheels.

   The statement "All cars have four wheels" is shown by a large circle that represents all items that have 4 wheels with a small circle inside that represents cars.

   The statement "All Fords are cars" is represented by adding a third circle representing Fords inside the circle representing cars.

   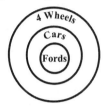

   Since the circle representing Fords is completely inside the circle representing items with 4 wheels, it follows that:

   All Fords have four wheels.

   The conclusion follows from the premises; it is valid.

6. (a) All animals are wild.
   All cats are animals.
   ∴ All cats are wild.

   "All animals are wild" is represented by a large circle representing wild creatures with a smaller circle inside representing animals.

   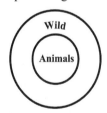

   "All cats are animals" is represented by a small circle representing cats, inside the circle representing animals.

   Since the circle representing cats is completely inside the circle representing wild creatures, it follows that:

   All cats are wild.

   continued

The conclusion follows from the premises; it is valid. (Note that correct deductive reasoning may lead to false conclusions if one of the premises is false, in this case, all animals are wild.)

(b) All students use math.
All adults use math.
∴ All adults are students.

A larger circle is used to represent people who use math. A small circle inside the larger circle represents students who use math.

Another small circle inside the larger circle represents adults. The circles should overlap since some students could be adults.

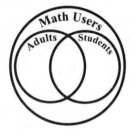

From the diagram, we can see that some adults are not students. Thus, the conclusion does *not* follow from the premises; it is invalid.

7. All 100 students in the class are represented by a large circle.

Students taking history are represented by a small circle inside and students taking math are also represented by a small circle inside. The small circles overlap since some students take both math and history.

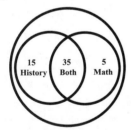

Since 35 students take history and math and 50 take history, $50 - 35 = 15$ students take history, but not math. Since 40 students take math, $40 - 35 = 5$ students take math, but not history. So $15 + 35 + 5 = 55$ students take history, math or both subjects. Therefore, $100 - 55 = 45$ students take neither math nor history.

8. A Chevy, BMW, Cadillac, and Ford are parked side by side.

1. The Ford is on the right side (fact a), so write "Ford" at the right of a line.

   Ford

2. The Chevy is between the Ford and the Cadillac (fact c). So write "Chevy" between Ford and Cadillac.

   Cadillac   Chevy   Ford

3. The BMW is next to the Cadillac (fact b), so the BMW must be on the other side of the Cadillac.

   BMW   Cadillac   Chevy   Ford

Therefore, the BMW is parked on the left end.

Appendix Exercises

1. $2, 9, 16, 23, 30, \ldots$

   Inspect the sequence and note that 7 is added to a term to obtain the next term. So the term immediately following 30 is $30 + 7 = 37$.

3. $0, 10, 8, 18, 16, \ldots$

   Inspect the sequence and note that 10 is added, then 2 is subtracted, then 10 is added, then 2 is subtracted, and so on. So the term immediately following 16 is $16 + 10 = 26$.

5. $1, 2, 4, 8, \ldots$

   Inspect the sequence and note that 2 is multiplied times a term to obtain the next term. So, the term immediately following 8 is $(8)(2) = 16$.

7. $1, 3, 9, 27, 81, \ldots$

   Inspect the sequence and note that 3 is multiplied times a term to obtain the next term. So, the term immediately following 81 is $(81)(3) = 243$.

9. $1, 4, 9, 16, 25, \ldots$

   Inspect the sequence and note that the pattern is add 3, add 5, add 7, etc., or $1^1, 2^2, 3^2$, etc. So, the term immediately following 25 is $6^2 = 36$.

11. The first three figures are unique. The fourth figure is the same as the first figure except that it is reversed, and reversing the position of the second figure gives the fifth figure. So, the next figure will be the reverse of the third figure.

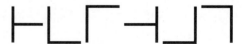